GW00640927

ASSOCIATIVE NEURAL MEMORIES

Associative Neural Memories

Theory and Implementation

Edited by
MOHAMAD H. HASSOUN

New York   Oxford
OXFORD UNIVERSITY PRESS
1993

970461

Oxford University Press

Oxford   New York   Toronto
Delhi   Bombay   Calcutta   Madras   Karachi
Kuala Lumpur   Singapore   Hong Kong   Tokyo
Nairobi   Dar es Salaam   Cape Town
Melbourne   Auckland   Madrid

and associated companies in
Berlin   Ibadan

Copyright © 1993 by Oxford University Press, Inc.

Published by Oxford University Press, Inc.
200 Madison Avenue, New York, New York 10016

Oxford is a registered trademark of Oxford University Press

All rights reserved. No part of this publication may be reproduced,
stored in a retrieval system, or transmitted, in any form or by any means,
electronic, mechanical, photocopying, recording, or otherwise,
without the prior permission of Oxford University Press

A catalogue record for this book is available from the British Library

Library of Congress Cataloging in Publication Data
Associative neural memories : theory and implementation / edited by
Mohamad H. Hassoun.
p. cm.   Includes bibliographical references and index.
ISBN 0–19–507682–6
1. Neural networks (Computer science)   2. Associative storage.
I. Hassoun, Mohamad H.
QA76.87.A88 1993
006.3—dc20   92-30660

9 8 7 6 5 4 3 2 1

Printed in the United States of America
on acid-free paper

*To a safe, just, and
peaceful world*

Preface

This edited volume brings together significant works on associative neural memory theory (architecture, learning, analysis, and design) and hardware implementation (VLSI and opto-electronic) by leading international researchers. The purpose of this book is to integrate recent fundamental and significant research results on associative neural memories into a single volume, and present the material in a clear and organized format that makes it accessible to researchers and students.

Associative neural memories are a class of artificial neural networks (connectionist nets) that have gained substantial attention relative to other neural-net paradigms. Associative memories have been the subject of research since the early 1970s. Recent interest in these memories has been spurred by the seminal work of Hopfield in the early 1980s, who has shown how a simple discrete nonlinear dynamical system can exhibit associative recall of stored binary patterns through collective computing. Since then, a number of important contributions have appeared in conference proceedings and technical journals addressing various issues of associative neural memories, including multiple-layer architectures, recording/storage algorithms, capacity, retrieval dynamics, fault-tolerance, and hardware implementation. Currently, associative neural memories are among the most extensively studied and best-understood neural paradigms. They have been studied as possible models of biological associative phenomena, as models of cognition and categorical perception, as high-dimensional nonlinear dynamical systems, as collective computing nets, as error-correcting nets, and as fault-tolerant content-addressable computer memories.

The book is organized into an introductory chapter and four parts: Biological and Psychological Connections, Artificial Associative Neural Memory Models, Analysis of Memory Dynamics and Capacity, and Implementation. The group of chapters in the first part deals with associative neural models that have close connections to biological and/or psychological aspects of memory. This group consists of three chapters by D. Alkon et al., P. Kanerva, and J. Anderson. The second part consists of three chapters by Y.-F. Wang et al., A. Dembo, and B. Baird and F. Eeckman. These chapters present more complex extensions of the simple associative memory

models covered in the introductory chapter, and study their recall capabilities. The analysis of artificial associative neural memory dynamics, capacity, and error-correction capabilities are addressed in Part III, which comprises the seven chapters by S-I. Amari and H.-F. Yanai, J. Komlós and R. Paturi, F. Waugh et al., S. Hui et al., G. Pancha and S. Venkatesh, S. Yoshizawa et al., and P.-C. Chung and T. Krile. The last part of the book deals with hardware implementation of associative neural memories (some of these memories and/or their associated recording algorithms constitute variations and/or extensions to those discussed in earlier chapters). Here, three chapters by K. A. Boahen and A. G. Andreou, M. Verleysen et al., and T.-D. Chiueh and R. Goodman address electronic VLSI implementations. Two additional chapters, one by F. T. S. Yu and the other by K. Kyuma et al., present optoelectronic implementations.

Chapter 1 is an introduction to basic dynamic associative memory (DAM) architectures and their associated learning/recording algorithms. DAMs are treated as collective nonlinear dynamical systems in which information retrieval is accomplished through an evolution of the system's state in a high-dimensional (binary) state space. This chapter reviews some basic supervised learning/recording algorithms and derives the necessary conditions for perfect storage and retrieval of memories in a simple DAM. The characteristics of high-performance DAMs are outlined and such general issues as stability, capacity, and retrieval dynamics are discussed. Also, references to other chapters of the book are made so as to point out architecture extensions, additional learning algorithms, formal analyses, and other associative neural memory issues that are briefly described in this introductory chapter.

Chapter 2 demonstrates the important contributions that neurobiology can make to the design of artificial neural networks in general and associative learning nets in particular. It describes some results of biochemical and biophysical experiments that elucidate the properties of the *Hermissenda* visual-vestibular network, followed by descriptions of two computer models, each representing a different level of aggregation of the essential features of learning and memory. A biologically-based computer model called Dystal is presented. The model demonstrates efficient associative learning and is applied to problems in face recognition and optical character recognition.

Chapter 3 describes and analyzes a class of associative memories known as "sparse distributed memory" and relates it to associative memory models of the cerebellum, digital random-access memory, and other sparse memory models reported in the literature. The chapter presents a unified formulation of a broad class of two-layer feed-forward associative memory architectures, and advances the concept of "pattern computing" as a new computing model, as contrasted to numeric computing and symbolic computing.

Chapter 4 focuses on the brain-state-in-a-box (BSB) associative memory as a low-order approximation to a broad range of human cognitive operations. The chapter presents the theory of the BSB model and informally describes some of its mathematical properties. Simulations are presented that show how BSB can model psychological response time.

Chapters 5 to 7 present more complex models of artificial associative neural memories as compared to those reviewed in Chapter 1. Chapter 5 addresses bidirectional associative memory (BAM), originally proposed independently by B. Kosko and Okajima et al. in 1987, and proposes several alternative recording schemes for improved recall. In Chapter 6, a class of high-density associative memory models with such desirable properties as high capacity, controllable basins of attraction, and fast convergence speed is proposed and analyzed. In Chapter 7, the "projection-algorithm"-based network and its extensions are proposed for the guaranteed associative memory storage of analog patterns, continuous sequences, and chaotic attractors in the same network, with no spurious attractors. In this chapter, the authors concentrate on mathematical analysis and engineering-oriented applications of the projection-algorithm-based memory. The chapter also attempts to relate the emergent dynamical behavior of interconnected modules of the proposed network to those of cortical computations by taking the view that oscillatory and possibly chaotic network modules form the actual cortical substrate of diverse sensory, motor, and cognitive operations.

The next group of seven chapters, Chapters 8 to 14, deals with the analysis of various aspects of associative memory, such as capacity, convergence dynamics, shaping basins of attraction, effects of nonmonotonic activation functions on retrieval dynamics, and fault tolerance. Chapter 8 formulates and presents a unified approach to the analysis of various architectures of associative memories based on a statistical neurodynamical method that allows for the analysis of associative recall dynamics and storage capacity. The method is applied to cross-correlation, cascaded, cyclic, autocorrelation, and associative sequence-generator memories. Chapter 9 presents a detailed, rigorous mathematical analysis of convergence in the synchronous and asynchronous updated Hopfield associative memory. Theorems are presented, along with their proofs, on the amount of error-correction and the rate of convergence as a function of the number of fundamental memories. The stability and dynamics of analog parallel-updated associative memories are studied in Chapter 10. The operation of these dynamic analog nets as associative memories is explained in terms of phase diagrams, relating memory loading to neuron activation function slopes, for correlation and generalized inverse recording. Chapter 11 examines the stability of the generalized BSB associative memory. It characterizes the stability and location of all fixed points of the BSB model for different weight matrices. In Chapter 12, a family of algorithms is discussed for DAM recording in terms of memory capacity and algorithm complexity. The chapter also emphasizes recording schemes for controlling the basins of attraction of selected memories and/or controlling the basins of attraction of selected memory vector/pattern components. Chapter 13 adopts a piecewise linear nonmonotonic neuron activation function in a dynamic autocorrelation associative memory and investigates the existence and stability of equilibrium states. This chapter also gives theoretical estimates on the capacity of such memories. Chapter 14 analyzes the effects of the faults characteristic of optical and electronic implementation technologies on implemented associative memory retrieval characteristics.

Chapters 15 to 19 cover hardware implementations of associative neural memories. In Chapter 15, some basic building blocks for VLSI implementation of neural circuitry are described. The basic principles of analog VLSI architectures are discussed in connection with precision limitations encountered with such technology. The chapter describes how to overcome some of these limitations by appropriate designs of artificial neurons and adaptation of basic associative learning algorithms. Chapter 16 describes a hybrid analog/digital CMOS chip implementation of a high-capacity exponential correlation associative memory (ECAM), along with simulations and experimental validation. The chapter also analyzes the storage capacity and error-correction characteristics of ECAMs, and the effect of hardware-limited exponentiation dynamic range on capacity. In Chapter 17, a scalable, efficient, and fault-tolerant chip design of a novel bidirectional associative memory architecture, based on subthreshold current-mode MOS transistors, is described and validated through simulations. The design techniques employed in this chapter allow for compact implementation that can potentially lead to associative memory chips with densities approaching that of static random-access memory (RAM). An optical implementation of a dynamic single-layer associative memory based on a liquid-crystal television (LCTV) spatial light modulator (SLM) is described in Chapter 18. Robust associative retrieval is demonstrated in this optical memory for several recording algorithms. This chapter also describes how a high-dimensional set of memories can be handled by employing a space–time sharing architecture. Finally, Chapter 19 considers the implementation of neural network architectures employing 3-D optical neurochips with on-chip analog memory capabilities based on integrated LED and variable-sensitivity photodetector (VSPD) arrays. Experimental results are reported for a two-layer winner-take-all-based associative memory for stamp classification and a two-layer perceptron net employing back error-propagation learning.

In putting this book together, an effort was made to include those researchers who are in many cases the originators of significant ideas on associative neural memories. However, as in any book like this, it would be impossible to include all significant work on this topic. Therefore, my strategy was to invite contributions by leading researchers who are able to relate their ideas to others in the literature, and who, in some cases, present a unifying framework for the study of associative neural memories.

I would like to take this opportunity to thank those who have contributed in various ways to the completion of this book. This project would not have been successful without the enthusiasm and professional cooperation of the contributing authors and their high-quality chapter contributions. I would like to thank the National Science Foundation for support of my work on associative neural memories through a Presidential Young Investigator Award (Grant ECE-9057896). In particular, my thanks go to Dr. Paul Werbos of NSF for his support of my research ideas since 1988. Also, thanks go to Ford Motor Company, Sun Microsystems, Unisys Corporation, and Zenith Data Systems for their valuable support, which has contributed directly or indirectly to the success of this project. Special thanks to Donald C. Jackson of Oxford University Press for his interest and help in publishing this book. I also take this opportunity to thank my wife Amal for her understanding and support, and thank my daughter Lamees for her patience.

Mohamad H. Hassoun
Detroit, June 1992

Contents

Part I

Biological and Psychological Connections

Part II

Artificial Associative Neural Memory Models

Part III

Analysis of Memory Dynamics and Capacity

Part IV

Implementation

List of Contributors

DANIEL L. ALKON
Neural Systems Section
National Institute of Neurological Disorders
   and Stroke
National Institutes of Health
Bethesda, MD 20892, USA

SHUN-ICHI AMARI
Instrumentation Physics
Faculty of Engineering
University of Tokyo
Bunkyo-ku Tokyo, 113 Japan

JAMES A. ANDERSON
Department of Cognitive and Linguistic
   Sciences
Brown University
Providence, RI 02912, USA

ANDREAS G. ANDREOU
Department of Electrical and Computer
   Engineering
Johns Hopkins University
Baltimore, MD 21218, USA

BILL BAIRD
Department of Mathematics
University of California at Berkeley
Berkeley, CA 94720, USA

KIM T. BLACKWELL
Environmental Research Institute of Michigan
1101 Wilson Blvd., Suite 1100
Arlington, VA 22209-2248, USA

KWABENA A. BOAHEN
Computation and Neural Systems
California Institute of Technology
Pasadena, CA 91125, USA

TZI-DAR CHIUEH
Department of Electrical Engineering
National Taiwan University
Taipei, Taiwan 10617

PAU-CHOO CHUNG
Department of Electrical Engineering
National Cheng Kung University
Tainan 70101, Taiwan

JOSÉ B. CRUZ
College of Engineering
Ohio State University
Columbus, OH 43210, USA

AMIR DEMBO
Department of Statistics and
   Department of Mathematics
Stanford University
Stanford, CA 94305, USA

FRANK EECKMAN
Lawrence Livermore National Laboratory
P.O. Box 808 (L-426)
Livermore, CA 94550, USA

RODNEY M. GOODMAN
Department of Electrical Engineering
California Institute of Technology
Pasadena, CA 91125, USA

MOHAMAD H. HASSOUN
Department of Electrical and Computer
   Engineering
Wayne State University
Detroit, MI 48202, USA

STEFEN HUI
Department of Mathematical Sciences
San Diego State University
San Diego, CA 92182-0314, USA

PAUL G. A. JESPERS
Microelectronics Laboratory
Department of Electrical Engineering
Université Catholique de Louvain
1348 Louvain-la-Neuve
Belgium

PENTTI KANERVA
Research Institute for Advanced Computer
  Science
NASA Ames Research Center, T-041-5
Moffett Field, CA 94035, USA

JÁNOS KOMLÓS
Department of Mathematics
Hill Center, Busch Campus
Rutgers University
New Brunswick, NJ 08903, USA

THOMAS F. KRILE
Optical Systems Laboratory
Department of Electrical Engineering
Texas Tech University
Lubbock, TX 79409-3102, USA

KAZUO KYUMA
Mitsubishi Electric Corporation
Central Research Laboratory
8-1-1 Tsukaguchi-Honmachi
Amagasaki-city, 661 Japan

JEAN-DIDIER LEGAT
Microelectronics Laboratory
Department of Electrical Engineering
Université Catholique de Louvain
1348 Louvain-la-Neuve
Belgium

WALTER E. LILLO
The Aerospace Corporation
PO Box 92957
Los Angeles, CA 90009-2957, USA

CHARLES M. MARCUS
Department of Physics
Stanford University
Palo Alto, CA 94305, USA

MASAHIKO MORITA
Institute of Information Sciences and Electronics
University of Tsukuba
Ibaragi 305, Japan

JAMES H. MULLIGAN, Jr.
Department of Electrical and Computer
  Engineering
University of California, Irvine
Irvine, CA 92717, USA

YOSHIKAZU NITTA
Mitsubishi Electric Corporation
Central Research Laboratory
8-1-1 Tsukaguchi-Honmachi
Amagasaki-city, 661 Japan

JUN OHTA
Mitsubishi Electric Corporation
Central Research Laboratory
8-1-1 Tsukaguchi-Honmachi
Amagasaki-city, 661 Japan

GIRISH PANCHA
Oracle Corporation
500 Oracle Parkway
PO Box 659504
Redwood Shores, CA 94065, USA

RAMAMOHAN PATURI
Department of Computer Science and
  Engineering
University of California at San Diego
Mail Code 0114
La Jolla, CA 92093-0114, USA

SANTOSH S. VENKATESH
Department of Electrical Engineering
University of Pennsylvania
Philadelphia, PA 19104-6390, USA

MICHEL VERLEYSEN
Microelectronics Laboratory
Department of Electrical Engineering
Université Catholique de Louvain
1348 Louvain-la-Neuve
Belgium

THOMAS P. VOGL
Environmental Research Institute of
  Michigan
1101 Wilson Blvd., Suite 1100
Arlington, VA 22209-2248, USA

YEOU-FANG WANG
Department of Electrical and Computer
  Engineering
University of California, Irvine
Irvine, CA 92717, USA

FREDERICK R. WAUGH
Division of Applied Sciences and Department
  of Physics
Harvard University
Cambridge, MA 02138, USA

SUSAN A. WERNESS
Environmental Research Institute of Michigan
PO Box 134001
An Arbor, MI 48113-4001, USA

ROBERT M. WESTERVELT
Division of Applied Sciences and Department
  of Physics
Harvard University
Cambridge, MA 02138, USA

HIRO-FUMI YANAI
Department of Information and
  Communication Engineering
Faculty of Engineering
Tamagawa Univeristy
Tamagawa-Gakuen, Tokyo 194, Japan

SHUJI YOSHIZAWA
Department of Mechano-Informatics
Faculty of Engineering
University of Tokyo
Bunkyo-ku, Tokyo 113, Japan

FRANCIS T. S. YU
Department of Electrical Engineering and
  Center for Electro-Optics Research
The Pennsylvania State University
121 Electrical Engineering East
University Park, PA 16802, USA

STANISLAW H. ŻAK
School of Electrical Engineering
Purdue University
West Lafayette, IN 47907, USA

ASSOCIATIVE NEURAL MEMORIES

1

Dynamic Associative Neural Memories[1]

MOHAMAD H. HASSOUN

1.1. INTRODUCTION

This chapter is an introduction to basic dynamic associative neural memory (DAM) architectures and their associated learning algorithms. These memories are treated as nonlinear dynamical systems where information retrieval is performed as an evolution of the system's state in a high-dimensional (binary) state space. The chapter reviews some basic supervised learning/recording algorithms and derives the necessary conditions for perfect storage and retrieval of a given memory set. The characteristics of high-performance DAMs are outlined, and DAM stability, capacity, and retrieval dynamics are analyzed.

Associative learning and retrieval of information in parallel neural-like systems is a powerful processing technique with a wide range of applications ranging from content addressable memories to robust pattern classification and control. Dynamic associative memories are a class of artificial neural networks which utilize a recording/learning algorithm to store information as stable memory states. The retrieval of stored memories is accomplished by first initializing the DAM state with a noisy or partial input pattern (key) and then allowing the memory to perform a collective relaxation search to find the closest associated stored memory.

A simple DAM is characterized by a regular layered architecture of highly distributed and densely interconnected processing layers with feedback. Each processing layer consists of a set of noninteracting nodes; each node receives the same set of data (input pattern or output from a preceding layer), processes this data, and then broadcasts its single output to the next processing layer. The transfer function of a given DAM node can vary in complexity; however, all nodes are assumed to have the same functional form. The most common node transfer function is equivalent to a weighted sum of the input data followed by a nonlinear activation function. The weighted sum processing step represents a local identification of the input data based on a similarity computation (or projection) between the data vector and a locally-stored weight vector. The nodes' weight vectors also describe an interconnection (communication) pattern between the nodes of adjacent layers. The node weights are assumed to be synthesized, during a learning/recording phase, from a given memory set. On the other hand, a node's activation function is usually a monotone-increasing function with saturation (e.g., a tanh or a threshold function) which can be thought of as implementing a "local decision" on the preceding similarity computation. In theory, DAM mapping dynamics can be understood and controlled through the network architecture, the learning/recording algorithm used, and the encoding of stored memories/associations.

Several associative neural memories have been proposed over the last two decades (Amari, 1972; Anderson, 1972; Nakano, 1972; Kohonen, 1972, 1974; Kohonen and Ruohonen, 1973; Hopfield, 1982; Kosko, 1987, 1988; Okajima et al., 1987). These memories can be classified in various ways depending on their retrieval mode (dynamic vs. static and/or synchronous vs. asynchronous), the nature of the stored associations (autoassociative vs. heteroassociative and/or binary vs. continuous), the type of learning algorithm (adaptive vs. nonadaptive), and/or the complexity and capability of the learning algorithm. In this chapter, dynamic synchronous binary-state neural memories are emphasized. These memories have been extensively studied and analyzed by several researchers (Uesaka and Ozeki, 1972; Wigstrom, 1973; Amari, 1977; Hopfield, 1982; Kohonen, 1984; Amit et al., 1985; Abu-Mustafa and Jacques, 1985; Weisbuch and Fogelman, 1985; Montgomery and Kumar, 1986; Stiles and Denq, 1987; Meir and Domany, 1987; McEliece et al., 1987; Amari and Maginu, 1988; Newman, 1988; Komlos and Paturi, 1988; Dembo, 1989; Hassoun

[1] An earlier, shorter version of this chapter appeared in the book *Artificial Neural Networks and Statistical Pattern Recognition*, I. K. Sethi and A. K. Jain, Eds. (North-Holland: Amsterdam, 1991).

and Youssef, 1988, 1989; Hassoun, (1989a); Amari, 1989, 1990; Yanai and Sawada, 1990; see also Chapters 8, 9, and 10 in this book.

This chapter is intended as a review of the fundamental concepts relating to basic DAM architectures, the various learning algorithms and recording strategies, and DAM capacity and performance. Section 1.2 presents the basic architectures, transfer characteristics, and general retrieval dynamics for auto- and heteroassociative DAMs. Section 1.3 summarizes several desirable characteristics of associative memories which serve as DAM performance measures. Several DAM recording/learning algorithms, including correlation, generalized-inverse, and Ho–Kashyap algorithms, are presented and analyzed in Section 1.4. General recording strategies for controlling and enhancing DAM dynamics are discussed in Section 1.5. In Section 1.6, DAM capacity is analyzed and retrieval dynamics are compared for various learning algorithms. The effects of restricted weight accuracy on DAM performance are considered in Section 1.7. The chapter concludes with a summary in Section 1.8.

1.2. DAM ARCHITECTURES AND GENERAL MEMORY DYNAMICS

The simplest associative neural memory architectures exhibiting dynamical behavior are considered and their transfer characteristics are formulated in this section. Potential DAM state-space trajectories are also indicated. This section deals with two basic DAM architectures: autoassociative and heteroassociative. Some important effects of various activation functions and state update strategies on DAM stability and dynamics are also outlined.

1.2.1. Autoassociative DAM

The autoassociative DAM is basically a discrete memory (Nakano, 1972; Amari, 1977; Hopfield, 1982) employing a single layer of perceptrons with hard-clipping activations. The perceptrons are fully interconnected through a feedback path, as shown in Fig. 1.1a, and may operate in a synchronous (parallel) or sequential (random) retrieval mode. Figure 1.1b depicts a block diagram of such an autoassociative DAM. Theoretically, the interconnection weight matrix $\mathbf{W}$ has real valued components w_{ij} connecting the jth perceptron to the ith perceptron. It is to be noted that, due to the hard-clipping nature of the activation function operator F and

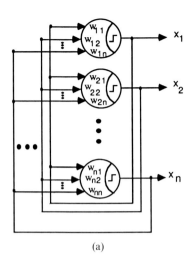

(a)

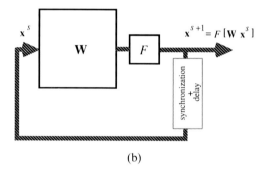

(b)

Fig. 1.1. (a) Interconnection pattern for an n-perceptron autoassociative DAM. (b) A block diagram representation of the DAM in (a).

the presence of feedback, the memory can only store and retrieve binary patterns.

Let the DAM output of Fig. 1 be represented by an n-dimensional binary valued pattern (column vector) $\mathbf{x}^s$ and call it the state of the DAM at discrete times s. Hence, the DAM's state evolves according to the first-order nonlinear difference equation

$$\mathbf{x}^{s+1} = F[\mathbf{W}\mathbf{x}^s] \qquad (1)$$

where F operates component-wise on its n-dimensional vector argument. The operator $F[\mathbf{W}\mathbf{x}(\cdot)]$ is referred to as the state transition operator. The weight matrix and the threshold activations are computed during a learning session in such a way as to store a set of m binary (bipolar or unipolar) patterns $\{\mathbf{x}_1, \mathbf{x}_2, \ldots, \mathbf{x}_k, \ldots, \mathbf{x}_m\}$, with $\mathbf{x}_k = [x_1^k x_2^k \cdots x_i^k \cdots x_n^k]^\mathrm{T}$ where x_i^k is the ith bit of the kth memory pattern, satisfying the condition $\mathbf{x}^{s+1} = \mathbf{x}^s = \mathbf{x}_k$; i.e., synthesized $\mathbf{W}$ and F guarantee that $\mathbf{x}_k$ is a fixed DAM state. The memory pattern $\mathbf{x}_k$ will be referred to as a fundamental memory. All other fixed states that are not fundamental memories will be referred to as spurious memories. In addition to the above two types of dynamics, the DAM can also converge to a limit cycle. It has been shown by Fogelman (1985) that if $\mathbf{W}$ is symmetric, the parallel updated DAM has limit cycle of period 2, at most.

The autoassociative mapping performed by the above DAM may seem trivial. However, the DAM is intended to act as a filter which corrects noisy, distorted, and/or partial input versions, $\mathbf{x}'_k$, of the fundamental memories $\{\mathbf{x}_k\}$. Theoretically, the DAM converges to $\mathbf{x}_k$, when initialized with $\mathbf{x}^0 = \mathbf{x}'_k$. This suggests that a basin of attraction exists around each one of the fundamental memories; i.e., the $\mathbf{x}_k$ memories are attractors of Eq. (1). Under certain conditions, discussed later in this chapter, the above DAM is capable of realizing such basins of attraction. Unfortunately, the complex dynamics of the DAM also give rise to spurious attractor memories, thus degrading performance. These and additional DAM characteristics are considered in Section 1.3.

1.2.2. Heteroassociative DAM

A heteroassociative DAM (Kosko, 1987, 1988; Okajima et al., 1987) may be thought of as an extension of the autoassociative DAM described above. Here, two single-layer feed-forward neural nets are connected in a closed loop as shown in Fig. 1.2. This architecture allows for simultaneous hetero- and autoassociative recollection of stored data. Ideally, a heteroassociative DAM realizes the two mappings $\mathbf{M}$ and $\mathbf{M}^*$ between a set of m binary input patterns $\{\mathbf{x}_1, \mathbf{x}_2, \ldots, \mathbf{x}_k, \ldots, \mathbf{x}_m\}$ and another corresponding set of m output

patterns $\{\mathbf{y}_1, \mathbf{y}_2, \ldots, \mathbf{y}_k, \ldots, \mathbf{y}_m\}$ according to

$$\mathbf{M}: \mathbf{x}_k \Rightarrow \mathbf{y}_k \quad \text{and} \quad \mathbf{M}^*: \mathbf{y}_k \Rightarrow \mathbf{x}_k; k = 1, 2, \ldots, m \tag{2}$$

The above DAM consists of a forward processing path and a feedback processing path. The forward path, considered alone, constitutes a unidirectional (static) heteroassociative memory that is potentially capable of realizing the mapping $\mathbf{M}$ of Eq. (2) by recalling $\mathbf{y}_k$ from $\mathbf{x}_k$ according to

$$\mathbf{y}_k = F[\mathbf{W}_1 \mathbf{x}_k]; \quad k = 1, 2, \ldots, m \tag{3}$$

where $\mathbf{y}_k$ and $\mathbf{x}_k$ are assumed to be binary column vector patterns of dimensions L and n, respectively, and $\mathbf{W}_1$ is an $L \times n$ weight matrix which is assumed to have been computed during a training session. F is the same activation function operator defined earlier.

One of the most appealing features of an associative memory is its ability to tolerate noisy and/or partial inputs; that is, given an input $\mathbf{x}'_k$ that is *similar* to the pattern $\mathbf{x}_k$ of the stored association pair $\{\mathbf{x}_k, \mathbf{y}_k\}$, the memory will respond with the correct association $\mathbf{y}_k$ according to

$$\mathbf{y}_k = F[\mathbf{W}_1 \mathbf{x}'_k] \tag{4}$$

However, the above equation may not hold true when relatively large numbers of associations are stored and/or the test input pattern is *slightly similar* to $\mathbf{x}_k$. This problem can be partially alleviated in the case of autoassociative retrieval ($\mathbf{y}_k = \mathbf{x}_k$) by feeding the output of the unidirectional memory directly into the input and simultaneously removing the original input $\mathbf{x}'_k$. This gives rise to the autoassociative DAM architecture of Fig. 1.1. However, in the heteroassociative case ($\mathbf{x}_k$ has a different size and/or encoding than $\mathbf{y}_k$), direct feedback is not compatible and a natural and simple remedy would

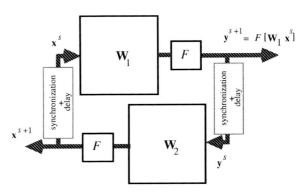

Fig. 1.2. A block diagram representation of a dynamic heteroassociative neural memory

be to feed the output back through the inverse of the **M** mapping, **M***, defined in Eq. (2), or, explicitly, by the equation

$$\mathbf{x}_k = F[\mathbf{W}_2\mathbf{y}_k]; \quad \mathbf{k} = 1, 2, \ldots, m \qquad (5)$$

where $\mathbf{W}_2$ is an $n \times L$ real-valued matrix [Kosko (1987) employs $\mathbf{W}_2 = \mathbf{W}_1^T$]. The resulting DAM is the one shown in Fig. 1.2.

The nonlinear difference equations governing the dynamics of the heteroassociative DAM in Fig. 1.2 are given by

$$\mathbf{y}^{s+1} = F[\mathbf{W}_1 F[\mathbf{W}_2\mathbf{y}^s]] \qquad (6)$$
$$\mathbf{x}^{s+1} = F[\mathbf{W}_2 F[\mathbf{W}_1\mathbf{x}^s]] \qquad (7)$$

where s is the iteration number (a positive integer), $\mathbf{x}^0$ is the initial state, and $\mathbf{y}^0$ is given from $\mathbf{x}^0$ by Eq. (3). The DAM dynamics can also be completely described through Eqs. (5) and (6) or (3) and (7). This DAM has two outputs $\mathbf{x}$ and $\mathbf{y}$. Output $\mathbf{y}$ represents heteroassociative recollections and output $\mathbf{x}$ represents autoassociative recall. The physical interpretation of these two outputs is determined by the application at hand. This class of DAMs is potentially powerful in robust pattern classification and pattern identification applications. Here, the output $\mathbf{y}_k$ may encode a classification of the initial state $\mathbf{x}'$ or it may encode a specific action or decision. On the other hand, the output $\mathbf{x}_k$ gives a reconstruction/correction of the initial DAM state (input pattern); this output may be used as an identification output that verifies the classification/action output by acting as a confidence measure. The $\mathbf{x}_k$ output process may also be viewed as a filtering process. A detailed analysis of the BAM and its extensions is treated in Chapter 5.

1.2.3. Stability and DAM Variations

In the above, the architectures of simple DAMs were described, for both auto- and hetero-

associative retrieval. Variations of such DAMs have been proposed and analyzed in the literature. These variations deal with DAM state update strategies, and assume various types of activation functions. The dynamics and stability of such DAMs are obviously affected by these variations.

In his original neural memory model, Hopfield (1982) employed an asynchronous (random) state update in the autoassociative DAM of Fig. 1.1. Each perceptron is assumed to update its bipolar binary state stochastically, independently of the times of firing of the other $n - 1$ neurons in the network. This asynchrony was introduced in order to model the propagation delays and noise in real neural systems. The discrete dynamics of the ith state of this model is given by

$$\left.\begin{array}{l} x_i^{s+1} = \text{sgn}(\mathbf{W}_i\mathbf{x}^s + I_i) \\[2mm] \text{and} \quad x_j^{s+1} = x_j^s \quad \forall j \neq i \end{array}\right\} \qquad (8)$$

where $\mathbf{W}_i$ is the ith row of $\mathbf{W}$ and I_i is an external bias (not shown in Fig. 1.1), which exists for all times s. The perceptron label i in Eq. (8) is stochastically determined, and thus only one perceptron is allowed to change its activity in the transition from time s to $s + 1$. "sgn" is the sign function shown in Fig. 1.3a. Hopfield, by employing an energy function approach, showed that a sufficient condition for the stability of DAMs with the dynamics of Eq. (8) is to have a symmetric interconnection matrix $\mathbf{W}$ [for stability analysis of a more general class of associative nets, the reader is referred to the work of Cohen and Grossberg (1983)]. This can be seen by noting that the following function

$$E^s = -\frac{1}{2}\sum_{i=1}^{n}\sum_{j=1}^{n} w_{ij}x_i^s x_j^s - \sum_{i=1}^{n} I_i x_i^s$$

is an energy function (expressed at time s) for the system described in Eq. (8). Hopfield showed that

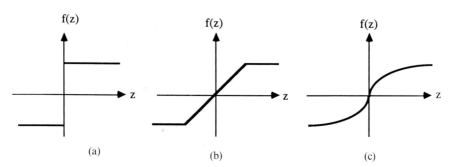

Fig. 1.3. Various types of perceptron activation functions employed in DAMs: (a) hard clipping ($f(x) = \text{sgn}(x)$); (b) saturation piecewise linear; and (c) sigmoidal activations (e.g., $f(x) = \tanh(x)$).

the asynchronous dynamics of (8) evolves in such a way as to locally minimize E as long as $\mathbf{W}$ is symmetric. This can be seen by assuming that the pth perceptron state flips at time $s + 1$, thus moving the system into energy E^{s+1}, and evaluating the change in the system's energy ΔE as

$$\Delta E = E^{s+1} - E^s$$

$$= \frac{1}{2}\left[\sum_{i=1}^{n} w_{ip}x_i^s x_p^s + \sum_{j=1}^{n} w_{pj}x_p^s x_j^s\right]$$

$$- \frac{1}{2}\left[\sum_{i \neq p}^{n} w_{ip}x_i^s x_p^{s+1} + \sum_{j \neq p}^{n} w_{pj}x_p^{s+1}x_j^s\right.$$

$$\left. + w_{pp}x_p^{s+1}x_p^{s+1}\right] + I_p x_p^s - I_p x_p^{s+1}$$

A sufficient condition on ΔE to be negative or zero is that $\mathbf{W}$ is symmetric. In this latter case, the change in energy is expressed as

$$\Delta E = -\left[\sum_{i}^{n} w_{pi}x_i^s + I_p\right]\Delta x_p;$$

where

$$\Delta x_p = x_p^{s+1} - x_p^s$$

which is always negative or zero, since it can easily be shown, employing Eq. (8), that the sign of the term in brackets is the same as that of $\Delta \mathbf{x}_p$. Now, since E is bounded, the dynamics in Eq. (8) may only converge to an attractor state. (For extensions of the above ideas on stability, the reader is referred to Chapter 12).

Fogelman (1985) shows that when the sharp threshold activation function (Fig. 1.3a) is replaced by a saturation piece-wise linear function, shown in Fig. 1.3b, the resulting randomly and sequentially updated DAM is also stable if $\mathbf{W}$ is symmetric with a nonnegative diagonal. Golden (1986) and Greenberg (1988) extend the above stability results to the "brain-state-in-a-box" (BSB) [Anderson (1972); see also Chapter 4] DAM described by the dynamics

$$\mathbf{x}^{s+1} = F[\rho \mathbf{W}\mathbf{x}^s + \mathbf{x}^s] \qquad (9)$$

where ρ is a positive constant and F is the activation function operator shown in Fig. 1.3b. It was shown by Greenberg that if $\mathbf{W}$ is symmetric and diagonal-dominant (w_{jj} is larger than the sum of the absolute values of all off-diagonal elements w_{ij}, for $j = 1, 2, \ldots, n$) then the BSB DAM is stable and the only stable points are the corners of the n-curve. (Hui, Lillo, and Żak study the stability and dynamics of a yet more general form of Eq. (9) in Chapter 11.)

In his continuous DAM model, Hopfield (1984) assumes an analog electronic amplifier-like implementation of a perceptron which results in the following deterministic DAM retrieval dynamics:

$$\rho \frac{d\mathbf{v}(t)}{dt} = \mathbf{W}\mathbf{x}(t) - \sigma\mathbf{v}(t) + \mathbf{I}; \quad \text{with } \mathbf{v} = F^{-1}(\mathbf{x})$$
$$(10)$$

where ρ and σ are positive constants and the activation operator F takes the form of a sigmoid function (e.g., $\tanh(\beta v)$) as depicted in Fig. 3c. Here, F^{-1} is the inverse of F; that is, the ith component of $\mathbf{v}$ is given by $v_i = f^{-1}(x_i)$. Hopfield shows that if $\mathbf{W}$ is symmetric with zero diagonal, then the continuous autoassociative DAM is stable. Furthermore, if the amplifier gains (slopes of the activation functions) are very large, then the only stable states of the continuous DAM have a simple correspondence with the stable states of the stochastic Hopfield DAM described in Eq. (8). Marcus and Westervelt (1989) have investigated a synchronous discrete-time variation of the continuous Hopfield model having the same form as Eq. (1). It was shown that if $\mathbf{W}$ is symmetric and if the activation functions are single-valued, monotonically increasing, and rise less rapidly than linearly for large arguments, then all attractors are either fixed points or period-2 oscillations. Furthermore, if the system obeys the condition $\beta < |1/\lambda_{\min}|$, where $\beta > 0$ is the maximum slope of the activation function and $\lambda_{\min}$ is the most negative eigenvalue of $\mathbf{W}$, then all period-2 oscillations are eliminated and convergence to stable attractors is guaranteed (the reader is referred to Chapter 10 for additional interesting results on the dynamics of this latter DAM).

1.3. CHARACTERISTICS OF A HIGH-PERFORMANCE DAM

A DAM is expected to exhibit a number of characteristics, such as noise and/or distortion tolerance, high capacity of stored associations, and well-behaved dynamics. In general, after recording a DAM, a number of fundamental memories are learned which are expected to behave as stable attractive states of the system. However, and in addition to the learned fundamental memories, spurious and/or oscillatory attractors can exist which negatively affect the performance of a DAM. A spurious attractor is a stable memory which is not part of the learned memories. These spurious memories are not

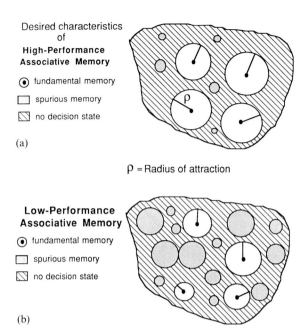

Desired characteristics
of
**High-Performance
Associative Memory**

⊙ fundamental memory

☐ spurious memory

▨ no decision state

(a)

ρ = Radius of attraction

**Low-Performance
Associative Memory**

⊙ fundamental memory

☐ spurious memory

▨ no decision state

(b)

Fig. 1.4. A conceptual diagram comparing (a) high-performance and (b) low-performance autoassociative DAMs.

desirable, yet they exist in all of the DAMs discussed above. Therefore, if one cannot train a DAM to exhibit no spurious states, then it is highly desirable to have them exist in a region of state-space that is far from the regions of fundamental memories (Chapter 7 presents a projection algorithm which allows for associative memories exhibiting no spurious states). Oscillations may be controlled by employing some of the DAM variations discussed in Section 1.2.3.

A set of performance characteristics must be met in order for a DAM to be efficient for associative processing applications. Depending on the encoding of the training memory associations, two classes of DAM mappings are distinguished: linearly separable and nonlinearly separable. Due to its single-layer architecture, the autoassociative DAM of Fig. 1.1 can only realize linearly separable mappings; i.e., it can only store linearly separable memory associations. This is also true for the heteroassociative DAM of Fig. 1.2, since the stability of a given association pair hinges on the ability of the single-layer forward and backward subnets to realize, perfectly, the m $\mathbf{x}_k$-to-$\mathbf{y}_k$ and $\mathbf{y}_k$-to-$\mathbf{x}_k$ associations, respectively. On the other hand, multiple-layer architectures (Rumelhart et al., 1986; Psaltis and Park, 1986; Hassoun, 1988) are needed to store a set of nonlinearly separable associations (the reader may also refer to the sparse associative

memories in Chapter 3). Multiple-layer DAMs are more difficult to analyze than single-layer ones and are not considered here. In the rest of this chapter, training associations are assumed to be linearly separable.

The following is a summary of some of the important characteristics of a DAM: (1) Tolerance to noisy, partial, and distorted inputs. This implies a high probability of convergence to fundamental memories. (2) High capacity of stored memories. (3) Existence of relatively few spurious memories and few or no orbits, and a low convergence to such states. (4) Convergence within a small number of retrieval cycles. (5) Provision for a *no decision* (ground) state. DAM inputs with relatively low signal-to-noise ratios must have a high probability of convergence to this state. (6) Autoassociative and/or heteroassociative processing capabilities. Depending on the nature of the application, association of identical or distinct data vectors may be required. Some of these desirable dynamics are depicted in Fig. 1.4a for an autoassociative DAM. On the other hand, Fig. 1.4b depicts the state-space of a low-performance DAM.

These characteristics can be used to compare different DAM architectures and/or recording (or learning) algorithms. It is to be noted that, for a given memory association and architecture, all of the above characteristics are dependent on the recording algorithm used.

1.4. ASSOCIATIVE LEARNING IN A DAM

In theory, there exist an infinite number of inter-connection matrices and thresholds that realize the mapping of a fixed set of associations in a DAM. However, different solutions may lead to different DAM dynamics that affect storage capacity, convergence rates to fundamental memories, number and location of spurious attractors, number and location of orbits, robustness, and other characteristics. For high-performance DAMs, the best solution is the one that gives rise to the desirable characteristics discussed in Section 1.3. Such a solution is very difficult to achieve, since it requires an optimization process that involves many constraints and parameters and which may be too complicated and computationally expensive to implement [for a general theoretical discussion of associative memory design in the context of nonlinear dynamical systems, see Cohen (1992)]. Another alternative is to synthesize interconnection weights that will guarantee a perfect recording of only the $\{\mathbf{x}_k\}$ memories ($\{\mathbf{x}_k, \mathbf{y}_k\}$ associations in the heteroassociative case) and hope that such a solution will also give rise to acceptable DAM performance. In fact, most of the existing DAM recording/learning techniques proposed in the literature are based on this latter approach (the "normal form projection DAM" of Chapter 7 is an example of an exception).

The training phase is responsible for synthesizing the interconnection matrix $\mathbf{W}$ from a training set of associations of the form $\{\mathbf{x}_k, \mathbf{y}_k\}$, for $k = 1, 2, \ldots, m$, for the heteroassociative case (notice that autoassociative training can be arrived at as a special case of heteroassociative training by setting $\mathbf{y}_k = \mathbf{x}_k$). Here, $\mathbf{x}_k$ and $\mathbf{y}^k$ belong to the n- and L-dimensional binary spaces, respectively. Therefore, the objective here is to solve the set of equations

$$\mathbf{y}_k = \mathbf{W}\mathbf{x}_k; \quad k = 1, 2, \ldots, m \qquad (11)$$

or in matrix form,

$$\mathbf{Y} = \mathbf{W}\mathbf{X} \qquad (12)$$

where $\mathbf{Y} = [\mathbf{y}_1 \mathbf{y}_2 \cdots \mathbf{y}_k \cdots \mathbf{y}_m]$ and $\mathbf{X} = [\mathbf{x}_1 \mathbf{x}_2 \cdots \mathbf{x}_k \cdots \mathbf{x}_m]$. The assumption of binary valued associations and the presence of a clipping nonlinearity F operating on $\mathbf{W}\mathbf{X}$ relaxes some of the constraints imposed by Eq. (12); that is, it is sufficient to solve the equation

$$\mathbf{Z} = \mathbf{W}\mathbf{X} \quad \text{and} \quad \mathbf{Y} = F[\mathbf{Z}] \qquad (13)$$

Next, several DAM recording techniques will be derived and analyzed.

1.4.1. Correlation Recording

One of the earliest associative memory recording techniques is the correlation technique (Anderson, 1972; Nakano, 1972; Kohonen, 1972), which was originally proposed for the synthesis of a static linear associative memory (LAM). This is a simple recording technique for generating $\mathbf{W}$ according to

$$\mathbf{W} = \mathbf{Y}\mathbf{X}^{\mathrm{T}} \qquad (14)$$

where "T" is the transpose operator. This is a *direct* method for computing the correlation weight matrix $\mathbf{W}$ which assumes that all the associations are present simultaneously during recording. A more practical method for computing the correlation matrix is to use the following equivalent form of Eq. (14):

$$\mathbf{W} = \sum_{k=1}^{m} \mathbf{y}_k \mathbf{x}_k^{\mathrm{T}} \qquad (15)$$

where the term inside the summation is the outer product of the vectors $\mathbf{y}_k$ and $\mathbf{x}_k$. This in turn allows us to derive the *adaptive* correlation recording technique according to

$$\mathbf{W}^{\mathrm{new}} = \mathbf{W}^c + \mathbf{y}_k \mathbf{x}_k^{\mathrm{T}} \quad \text{for } k = 1, 2, \ldots, m \qquad (16)$$

where $\mathbf{W}^c$ is the current weight matrix (initialized as the zero matrix). This makes it very convenient if, at some time after the initial recording phase is complete, one wants to add a new memory or delete an already-recorded memory. The correlation recording of autoassociations is identical to the above but with $\mathbf{y}_k = \mathbf{x}_k$.

Let us now investigate the requirements on the $\{\mathbf{x}_k, \mathbf{y}_k\}$ associations which will guarantee the successful retrieval of all recorded memories $\mathbf{y}_k$ from perfect key inputs $\mathbf{x}_k$, in a correlation-recorded LAM. Employing Eqs. (11) and (15) and assuming that the key input $\mathbf{x}_h$ is one of the $\mathbf{x}_k$ vectors, we get an expression for the retrieved pattern as

$$\tilde{\mathbf{y}}_h = \left[\sum_{k=1}^{m} \mathbf{y}_k \mathbf{x}_k^{\mathrm{T}} \right] \mathbf{y}_h = \mathbf{y}_h \mathbf{x}_h^{\mathrm{T}} \mathbf{x}_h + \sum_{k \neq h}^{m} (\mathbf{y}_k \mathbf{x}_k^{\mathrm{T}}) \mathbf{x}_h$$

$$= \mathbf{y}_h \|\mathbf{x}_h\|^2 + \sum_{k \neq h}^{m} (\mathbf{y}_k \mathbf{x}_k^{\mathrm{T}} \mathbf{x}_h) \qquad (17)$$

The second term on the right-hand side of Eq. (17) represents the "cross-talk" between the input key $\mathbf{x}_h$ and the remaining $(m - 1)$ $\mathbf{x}_k$ patterns. This term can be reduced to zero if the $\mathbf{x}_k$ vectors are orthogonal. The first term is proportional to the desired memory $\mathbf{y}_h$, with a proportionality constant equal to the square of the norm of the key vector $\mathbf{x}_h$. Hence, a sufficient condition for the retrieved memory to be the desired perfect recollection is to have *orthonormal* key vectors $\mathbf{x}_k$, independent of the encoding of the $\mathbf{y}_k$ (note how the $\mathbf{y}_k$ affects the cross-talk if the $\mathbf{x}_k$ are not orthogonal). However, recalling the nonlinear nature of the DAM reflected in Eq. (13), perfect recall of the binary $\mathbf{y}_k$ vectors is, in general, possible even when the key vectors are only pseudo-orthogonal. This can be seen in the following analysis.

Assuming the associative memory retrieval equation (13) with the activation function shown in Fig. 1.3a, and that we store bipolar patterns, we can employ a normalized version of correlation recording given by

$$\mathbf{W} = \frac{1}{n} \mathbf{Y} \mathbf{X}^{\mathrm{T}} \quad (18)$$

which automatically normalizes the $\mathbf{x}_k$ vectors (note that the square of the norm of an n-dimensional bipolar binary vector is n). Now, if we present one of the recorded key patterns, $\mathbf{x}_h$, as input, we arrive at the following expression for the retrieved memory pattern:

$$\tilde{\mathbf{y}}_h = F\left[\frac{1}{n} \mathbf{x}_h^{\mathrm{T}} \mathbf{x}_h \mathbf{y}_h + \frac{1}{n} \sum_{k \neq h}^{m} (\mathbf{y}_k \mathbf{x}_k^{\mathrm{T}}) \mathbf{x}_h \right]$$

$$= F\left[\mathbf{y}_h + \frac{1}{n} \sum_{k \neq h}^{m} \mathbf{y}_k (\mathbf{x}_k^{\mathrm{T}} \mathbf{x}_h) \right] = F[\mathbf{y}_h + \Delta_h] \quad (19)$$

and for the ith bit, we get

$$\tilde{y}_i^h = \mathrm{sgn}\left[y_i^h + \frac{1}{n} \sum_{j=1}^{n} \sum_{k \neq h}^{m} y_i^k x_j^k x_j^h \right]$$

$$= \mathrm{sgn}[y_i^h + \Delta_i^h] \quad (20)$$

from which it can be seen that the condition for perfect recall is given by the requirement

$$\text{and} \quad \left. \begin{array}{l} \Delta_i^h > -1 \text{ for } y_i^h = +1 \\ \Delta_i^h < +1 \text{ for } y_i^h = -1 \end{array} \right\} \quad (21)$$

which is a less restrictive requirement than the orthonormality of the $\mathbf{x}_k$ in a LAM. Personnaz et al. (1986) have shown, for the autoassociative case, that any state lying within a Hamming distance of $n/2m$ from a fundamental memory will converge to this fundamental memory in one step (direct convergence). Which implies that the basin of attraction of a fundamental memory falls sharply as m approaches $n/2$.

Amari (1977, 1990) has analyzed the correlation-recorded DAM extensively. Some of his results for correlation-recorded static auto-associative memory, with uniformly distributed random bipolar memories and sgn activation

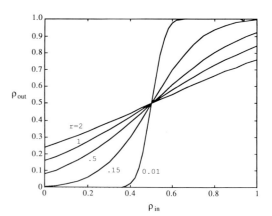

Fig. 1.5. Single-pass error correction in a correlation-recorded DAM as a function of pattern ratio $r = m/n$.

functions, are interpreted in the plot of Fig. 1.5, which depicts the single-pass error correction curves predicted for a parallel updated DAM, in the limit of large n. Here, ρ_{in} and ρ_{out} are the probabilities of input and output bit error, respectively, and $r = m/n$ is the pattern ratio. Note how the ability of the DAM to retrieve fundamental memories from noisy inputs is reduced as r approaches the value of 0.15 (this limit has also been reported by Hopfield and others and will be derived in Section 1.6. For lower loading levels (smaller r values), the error correction capabilities of the memory improves, and for $m \ll n$ the memory can tolerate up to 50 percent error in the input patterns. Note that for large n and $m \ll n$, the random memory set $\{\mathbf{x}_k\}$ becomes orthogonal with a probability approaching 1; hence these results agree with the results of Personnaz et al. reported above.

1.4.2. Generalized-Inverse Recording

The correlation recording technique is restrictive in many applications, due to the requirement that the $\mathbf{x}_k$ be orthonormal (for LAM) or pseudo-orthogonal (for Hopfield-like memory). This technique does not make optimal use of the DAM interconnection weights. A more optimal recording technique is the generalized-inverse recording (orthogonal recording) technique proposed by Kohonen and Ruohonen (1973). The generalized-inverse technique was originally proposed for the synthesis of the $\mathbf{W}$ matrix of the optimal linear associative memory (OLAM) (Kohonen, 1984), employing perceptrons with linear activations and no feedback. Starting from Eq. (12) and multiplying both sides of the equation by $\mathbf{X}^T$ one gets

$$\mathbf{YX}^T = \mathbf{WXX}^T \tag{22}$$

The motivation behind the choice of the multiplier $\mathbf{X}^T$ is that the $\mathbf{W}$ which satisfies Eq. (22) is the mean-square-error (MSE) solution (here we assume that the training set is non-linearly independent or that it has more associations than the number of components of $\mathbf{x}_k$; i.e., $m > n$) that minimizes the objective function $J(\mathbf{W})$ given by

$$\begin{aligned} J(\mathbf{W}) &= \|\mathbf{Y} - \mathbf{WX}\|_E^2 \\ &= \text{tr}[(\mathbf{Y} - \mathbf{WX})^T(\mathbf{Y} - \mathbf{WX})] \\ &= \sum_{k=1}^{m}\sum_{i=1}^{L}(\varepsilon_i^k)^2 \end{aligned} \tag{23}$$

where $\|\cdot\|_E$ is the Euclidean norm, tr is the trace operator, and ε_i^k is the error between the ith component of the desired vector $\mathbf{y}_k$ and the estimated one. Going back to Eq. (22) and multiplying both sides of the equation by the inverse of $\mathbf{XX}^T$, the following solution for $\mathbf{W}$ is achieved:

$$\mathbf{W} = \mathbf{YX}^T(\mathbf{XX}^T)^{-1} = \mathbf{YX}^+ \tag{24}$$

where $\mathbf{X}^+$ is the pseudo- or generalized-inverse of $\mathbf{X}$. This solution is only valid if the inverse of $\mathbf{XX}^T$ exists, which requires that $\mathbf{XX}^T$ be of rank n.

Next, let us investigate the retrieval characteristics of the OLAM, presented by an input $\mathbf{x}_h$, by substituting Eq. (24) in (11) and arriving at

$$\tilde{\mathbf{y}}_h = \mathbf{Wx}_h = \mathbf{YX}^T[(\mathbf{XX}^T)^{-1}\mathbf{x}_h] \tag{25}$$

which shows that the OLAM can be viewed as a correlation-recorded associative memory with a preprocessing stage attached, as shown in Fig. 1.6. The preprocessing stage performs a projection, defined by the term inside brackets in Eq. (25), of $\mathbf{x}_h$ onto the space spanned by the $\mathbf{x}_k$ vectors. This projection maps the hth training vector $\mathbf{x}_h$ into a new vector $\mathbf{x}_h'$ which is less correlated with the recorded $\mathbf{x}_k$ (k different than h) memories than $\mathbf{x}_h$, thus outputting a vector $\mathbf{y}$ which is closest to $\mathbf{y}_h$, in the MSE sense.

The ability of the OLAM to store all memory patterns m and to retrieve them from noisy versions is directly related to the rank r_m of the m memory patterns. It can be shown (Personnaz et al., 1986) that if $r_m = n$ (with $m > n$), the weight

matrix in (25) is the identity matrix and all possible states of the memory are fixed points; the memory is degenerate. If $r_m < n$, then associative retrieval of fundamental memories is possible and the basin of attraction of these memories will fall sharply to zero as r_m/n approaches 0.5 (here, $m - r_m$ of the m memory patterns to be stored are linear combinations of r_m linearly independent memory patterns).

It was assumed in the above discussion that the training set is a general linearly dependent set. It is shown next that the m memories must be linearly independent if all memories are to be fixed points of the associative memory. Therefore, it is important to consider the problem of recording an underdetermined ($m < n$) set of associations. When the memory set is linearly independent (implying that m is less than or equal to n), Eq. (12) has multiple exact solutions $\mathbf{W}^*$. Here, the minimum-norm solution $\mathbf{W} = \min\|\mathbf{W}^*\|$ is selected as the solution leading to the most robust associative retrieval (Kohonen 1984). Assuming that the $\mathbf{x}_k$ are linearly independent, a direct computation of $\mathbf{W}$ is given by

$$\mathbf{W} = \mathbf{Y}(\mathbf{X}^T\mathbf{X})^{-1}\mathbf{X}^T = \mathbf{YX}^+ \tag{26}$$

For an arbitrary $\mathbf{Y}$, if the $\mathbf{x}_k$ are orthonormal, then $\mathbf{X}^T\mathbf{X} = \mathbf{I}$ and Eq. (26) reduces to the correlation recording algorithm discussed above. Again, for this underdetermined case, associative retrieval can be thought of as that of a correlation-recorded memory with preprocessed inputs. To see this, the identity $\mathbf{X}^+ = \mathbf{X}^T(\mathbf{XX}^T)^+$ is used in Eq. (26), which upon substitution in Eq. (11) gives

$$\tilde{\mathbf{y}}_h = \mathbf{Wx}_h = \mathbf{YX}^T[(\mathbf{XX}^T)^+\mathbf{x}_h] \tag{27}$$

where the term inside the brackets is the n-dimensional preprocessed input vector. Since $\mathbf{X}^T(\mathbf{XX}^T)^+\mathbf{X} = \mathbf{I}$, it can be concluded that the operator $(\mathbf{XX}^T)^+$ maps the hth training vector $\mathbf{x}_h$ which is linearly independent from the remaining $m - 1$ training vectors $\mathbf{x}_k$ into a vector $\mathbf{x}_h'$ which is orthogonal to the $\mathbf{x}_k$ vectors and has an inner product of unity with the $\mathbf{x}_h$ vector. On the other hand, if one inputs a noisy version of one of the training key vectors, say $\mathbf{x}_h$, then the preprocessed output $\mathbf{x}_h'$ will be rotated more in the direction of $\mathbf{x}_h$ and at the same time made more orthogonal to the remaining training vectors.

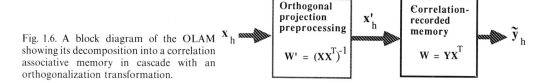

Fig. 1.6. A block diagram of the OLAM showing its decomposition into a correlation associative memory in cascade with an orthogonalization transformation.

When the dimensions m and/or n are large, the direct method for solving for the generalized-inverse in Eqs. (24) and (26) becomes impractical from a computational point of view. Furthermore, in many practical applications, the nature of the memory vectors is such that the matrix $\mathbf{XX}^T$ (or $\mathbf{X}^T\mathbf{X}$ for the underdetermined case) may be ill-conditioned, leading to numerical instabilities when computing the inverse. Therefore, it is desirable to replace the direct computation of the generalized-inverse with a more practical-to-compute stable method. This can be achieved by using an iterative method (Kohonen, 1984) based on Greville's theorem (Albert, 1972) which leads to the exact weight matrix after m iterations through the memory set. This method is convenient since new memories can be learned (and old memories can be deleted) in a single update step without involving other earlier-learned memories.

An alternative adaptive learning method is to employ gradient descent on $J(\mathbf{W})$ in Eq. (23) and iteratively solving Eq. (12). Here, the weight matrix is incremented (starting from a zero-valued $\mathbf{W}$ matrix) according to the equation

$$\Delta\mathbf{W} = \mathbf{W}^{\text{new}} - \mathbf{W}^c = -\tfrac{1}{2}\rho\, \frac{\partial J(\mathbf{W})}{\partial\mathbf{W}}\bigg|_{\mathbf{W}^c}$$
$$= \rho[\mathbf{Y} - \mathbf{W}^c\mathbf{X}]\mathbf{X}^T \qquad (28)$$

where the value of ρ for practical problems should be in the range $0 < \rho \ll 1$. We will refer to this algorithm as the *batch-mode adaptive* generalized-inverse training technique. Convergence can be speeded by initializing $\mathbf{W}^c$ as the correlation matrix $\mathbf{YX}^T$, since it can be shown (Kohonen, 1984) that this correlation term is the lowest-order term in a von Neumann expansion (Rao and Mitra, 1971) of the matrix $\mathbf{YX}^+$ given by

$$\mathbf{YX}^+ = \alpha\mathbf{Y}\sum_{N=0}^{\infty}[\mathbf{I} - \alpha\mathbf{X}^T\mathbf{X}]^N\mathbf{X}^T \qquad (29)$$

where $0 < \alpha < 2/\lambda$ and λ is the largest eigenvalue of $\mathbf{X}^T\mathbf{X}$.

Equation (28) may also be modified further into *continuous-* or *local-mode*, which allows for adaptive updating of the interconnection weights every time a new association $\{\mathbf{x}_k, \mathbf{y}_k\}$ is presented. The following are two versions of this type of *continuous* adaptive generalized-inverse training:

$$\mathbf{W}^{\text{new}} = \mathbf{W}^c + \rho[\mathbf{y}_k - \mathbf{W}^c\mathbf{x}_k]\mathbf{x}_k^T \qquad (30)$$
and
$$\mathbf{w}_i^{\text{new}} = \mathbf{w}_i^c + \rho[y_i^k - \mathbf{w}_i^c\mathbf{x}_k]\mathbf{x}_k^T \qquad (31)$$

where $\mathbf{w}_i$ is the ith row of matrix $\mathbf{W}$ representing the weight vector of the ith perceptron ($i = 1, 2, \ldots, L$) and y_i^k is the ith bit of the kth

association vector $\mathbf{y}_k$. In these equations, k is incremented after each iteration and the whole training set is cycled through multiple times until convergence is achieved. Equation (31) is used to synthesize the L memory perceptrons, separately, and is known in the literature as the μ-LMS or Widrow–Hoff learning rule (Widrow and Hoff, 1960; Widrow and Lehr, 1990). (In fact, the LMS rule differs slightly from Eq. (31) in that it employs an additional perceptron bias bit, which results in an extra weight that could be used in adjusting the threshold of the hard-clipping nonlinearity in a DAM.) The choice of ρ is critical in determining stability and convergence time of the LMS algorithm. Choosing a large ρ speeds up convergence, but may lead to instability. Horowitz and Senne (1981) proved that the choice $0 < \rho < 1/[6\,\text{tr}(\mathbf{X}^T\mathbf{X})]$ guarantees the stability of Eq. (31), for $\mathbf{x}_k$ patterns generated by a zero-mean Gaussian process independent over time.

A recording technique closely related to the underdetermined generalized-inverse technique of Eq. (26) for the recording of autoassociative DAMs is the *spectral technique* (Personnaz et al., 1985; Venkatesh and Psaltis, 1985, 1989; also, see Chapter 12). Here, the weight matrix for auto-associative recording is defined as follows:

$$\mathbf{W} = \mathbf{XD}(\mathbf{X}^T\mathbf{X})^{-1}\mathbf{X}^T = \mathbf{XDX}^+ \qquad (32)$$

where $\mathbf{D} = \text{diag}[\lambda_1, \lambda_2, \ldots, \lambda_k, \ldots, \lambda_m]$ is an $m \times m$ diagonal matrix of positive eigenvalues ($\lambda_k > 0$). Note that $\mathbf{W}$ is well defined if the inverse of $\mathbf{X}^T\mathbf{X}$ exists; i.e., if the $\mathbf{x}_k$ vectors are linearly independent. Furthermore, $\mathbf{W}$ is symmetric if $\lambda_k = \lambda$ for $k = 1, 2, \ldots, m$. Multiplying Eq. (32) by $\mathbf{X}$ gives

$$\mathbf{WX} = \mathbf{XD} \quad \text{or} \quad \mathbf{Wx}_k = \lambda_k\mathbf{x}_k, \quad k = 1, 2, \ldots, m \qquad (33)$$

Assuming arbitrary positive λ_k, the above equation is a weighted minimum-norm solution where a fundamental memory (eigenvector) $\mathbf{x}_k$ having a large eigenvalue λ_k tends to have an enlarged basin of attraction compared to other memories with corresponding smaller eigenvalues; that is, more attention is shifted towards $\mathbf{x}_k$.

Equation (33) suggests that the autoassociative storage of the memory patterns $\mathbf{x}_k$ as LAM fixed points can be achieved by a weight matrix $\mathbf{W}$ having eigenvectors $\mathbf{x}_k$ with corresponding eigenvalues $\lambda_k = 1$. If we further assume a DAM architecture as in Fig. 1.1 with linear activation functions, then upon initializing with a random state, the DAM will tend to evolve into an infinite norm state vector that is parallel to the most dominant eigenvector (having the largest eigen-value) of the weight matrix $\mathbf{W}$. If all the eigenvalues are equal and positive, the DAM will evolve its vector state in the direction of its eigenvector having the smallest angle with the initial DAM state. The use of saturating non-linear activation functions, as shown in Fig. 1.3, potentially helps clamp DAM states into stored memories (refer to Chapter 4 for a discussion of BSB DAM dynamics in terms of the eigenvalues and eigenvectors of $\mathbf{W}$).

1.4.3. Ho–Kashyap Recording

Higher DAM performance can be accomplished if the perceptron activation functions are taken into account during the recording phase. An optimal recording technique employing the above feature has been proposed by Hassoun and Youssef (1988, 1989) and Hassoun (1989a) for autoassociative and heteroassociative DAMs, respectively. This technique is based on the Ho–Kashyap algorithm (Ho and Kashyap, 1965) for the optimal MSE solution of a set of linear inequalities. One major difference between the Ho–Kashyap recording rule and the earlier recording techniques is that the weight vector and the activation function threshold are inde-pendently optimized for each neuron. (It should be noted here that the Ho–Kashyap technique normally results in an asymmetric $\mathbf{W}$.)

According to the Ho–Kashyap algorithm, the ith row of the weight matrix $\mathbf{W}$ and its corresponding threshold T_i are formulated as the weight vector $\mathbf{w}_i = [w_{i0}\, w_{i1}\, w_{i2} \cdots w_{in}]^T$, where $T_i = -w_{i0}$. Then, and upon the presentation of the kth association pair, the ith perceptron can be trained to classify a given training set $\{\mathbf{x}_k, \mathbf{y}_k\}$ correctly by computing the $(n + 1)$-dimensional weight vector $\mathbf{w}_i$ satisfying the following set of m inequalities:

$$\mathbf{x}_k^T \mathbf{w}_i \begin{cases} > 0 & \text{if } y_i^k = 1 \\ < 0 & \text{if } y_i^k = 0 \ (\text{or} -1) \end{cases}$$
$$\text{for } k = 1, 2, \ldots, m \quad (34)$$

where the vector $\mathbf{x}_k$ is derived from the original $\mathbf{x}_k$ by augmenting it with a bias of "1" as its x_0^k component. Next, if we define a set of m new vectors $\mathbf{z}_k$ according to

$$\mathbf{z}_k = \begin{cases} +\mathbf{x}_k & \text{if } y_i^k = 1 \\ -\mathbf{x}_k & \text{if } y_i^k = 0 \ (\text{or} -1) \end{cases}$$
$$\text{for } k = 1, 2, \ldots, m \quad (35)$$

and let

$$\mathbf{Z} = [\mathbf{z}_1\, \mathbf{z}_2 \cdots \mathbf{z}_m] \quad (36)$$

then Eq. (34) may be rewritten (the subscript i is dropped in order to simplify notation) as

$$\mathbf{Z}^T \mathbf{w} > 0 \quad (37)$$

Now, if we define an m-dimensional positive-valued margin vector $\mathbf{b}$ ($\mathbf{b} > 0$) and use it in Eq. (36), we arrive at the following equivalent form of Eq. (34):

$$\mathbf{Z}^T \mathbf{w} = \mathbf{b} > 0 \quad (38)$$

Thus the training of the perceptron is now equivalent to solving Eq. (38) for $\mathbf{w}$, subject to the constraint $\mathbf{b} > 0$. Ho and Kashyap have proposed an iterative algorithm for solving Eq. (38). In the Ho–Kashyap algorithm, the components of the margin vector are first initialized to small positive values and the Moore–Penrose pseudoinverse is used to gen-erate an MSE solution for $\mathbf{w}$ (based on the initial guess for $\mathbf{b}$) which minimizes the objective function $J(\mathbf{w}, \mathbf{b}) = \|\mathbf{Z}^T \mathbf{w} - \mathbf{b}\|^2$

$$\mathbf{w} = (\mathbf{Z}^T)^+ \mathbf{b} \quad (39)$$

Next, a new estimate for the margin vector is computed by performing the constrained ($\mathbf{b} > 0$) descent

$$\mathbf{b}^{new} = \mathbf{b}^c + \rho[\boldsymbol{\varepsilon} + |\boldsymbol{\varepsilon}|] \quad \text{with } \boldsymbol{\varepsilon} = \mathbf{Z}^T \mathbf{w} - \mathbf{b} \quad (40)$$

where $|\cdot|$ denotes the absolute value of the components of the argument vector. A new estimate of $\mathbf{w}$ can now be computed using Eq. (39) and employing the updated margin vector from Eq. (40). This process is iterated until all the components of $\boldsymbol{\varepsilon}$ are zero (or are sufficiently small and positive), which is an indication of the linear separability of the training set, or until $\boldsymbol{\varepsilon} < 0$, which in this case is an indication of the nonlinear separability of the training set (no solution is found). It can be shown (Ho and Kashyap, 1965; Slansky and Wassel, 1981) that the Ho–Kashyap procedure converges in a finite number of steps if the training set is linearly separable. A sufficient condition for convergence is $0 < \rho < 1$. We will refer to the above algorithm as the *direct* Ho–Kashyap (DHK) algorithm.

When the training set is underdetermined ($m < n + 1$), the Ho–Kashyap recording al-gorithm converges in one iteration (Hassoun and Youssef, 1989). That is, Eq. (39) leads to a perfect solution for $\mathbf{w}$ and no margin update is needed. This solution is identical to the LMS solution discussed in Section 1.4.2 if the initial margin vector was chosen to have equal positive components. Therefore, the full benefits of the Ho–Kashyap recording technique are achieved with overdetermined training sets ($m > n + 1$), which leads to optimized weights and margins.

Section 1.5 discusses ways of extending originally underdetermined training sets into overdetermined ones which are well suited for harvesting the full benefits of the Ho–Kashyap recording technique.

The direct synthesis of the **w** estimate in Eq. (39) involves a one-time computation of the pseudoinverse of $\mathbf{Z}^T$. However, such direct computation can be time-consuming, and it requires special treatment when $\mathbf{Z}\mathbf{Z}^T$ (or $\mathbf{Z}^T\mathbf{Z}$, for the underdetermined case $m < n + 1$) is singular. An alternative algorithm that does not require the computation of $(\mathbf{Z}^T)^+$ can be derived based on gradient descent principles.

Starting with the objective function $J(\mathbf{w}, \mathbf{b}) = \|\mathbf{Z}^T\mathbf{w} - \mathbf{b}\|^2$, gradient descent may be performed (Slansky and Wassel, 1981) in **b** and **w** spaces so that J is minimized subject to the constraint $\mathbf{b} > 0$. The gradients of J with respect to **w** and **b** are given by Eqs. (41) and (42), respectively:

$$\nabla_{\mathbf{b}} J(\mathbf{w}, \mathbf{b})|_{\mathbf{w}^c, \mathbf{b}^c} = -2(\mathbf{Z}^T\mathbf{w}^c - \mathbf{b}^c)$$

$$\text{subject to } \mathbf{b} > 0 \quad (41)$$

$$\nabla_{\mathbf{w}} J(\mathbf{w}, \mathbf{b})|_{\mathbf{w}^c, \mathbf{b}^{\text{new}}} = 2\mathbf{Z}(\mathbf{Z}^T\mathbf{w}^c - \mathbf{b}^{\text{new}}) \quad (42)$$

One sufficient analytic method for imposing the constraint $\mathbf{b} > 0$ is to replace $\varepsilon^c = \mathbf{Z}^T\mathbf{w} - \mathbf{b}$ in (41) by $(1/2)(\varepsilon + |\varepsilon|)$. This leads to the following gradient descent formulation of the Ho–Kashyap procedure:

$$\mathbf{b}^{\text{new}} = \mathbf{b}^c + \frac{\rho_1}{2}\left(|\varepsilon^c| + \varepsilon^c\right)$$

$$\text{with } \varepsilon^c = \mathbf{Z}^T\mathbf{w}^c - \mathbf{b}^c \quad (43)$$

$$\mathbf{w}^{\text{new}} = \mathbf{w}^c - \rho_2\mathbf{Z}(\mathbf{Z}^T\mathbf{w}^c - \mathbf{b}^{\text{new}})$$

$$= \mathbf{w}^c + \frac{\rho_1\rho_2}{2}\mathbf{Z}\left[|\varepsilon^c| + \varepsilon^c\left(1 - \frac{2}{\rho_1}\right)\right] \quad (44)$$

We will refer to the above procedure of Eqs. (43) and (44) as the *batch mode* adaptive Ho–Kashyap (AHK) procedure, because of the requirement that all training vectors $\mathbf{z}_k$ (or $\mathbf{x}_k$) must be present and included in **Z**. It can easily be shown that if $\rho_1 = 0$ and $\mathbf{b}^0 = \mathbf{1}$, Eq. (44) reduces to the well-known MSE procedure. Furthermore, convergence is assured (Duda & Hart, 1973) if $\rho_1 = 2$ and $0 < \rho_2 < 2/\lambda_{\max}$, where $\lambda_{\max}$ is the largest eigenvalue of the positive definite matrix $\mathbf{Z}\mathbf{Z}^T$.

A completely adaptive Ho–Kashyap (AHK) procedure for solving Eq. (38) is arrived at by replacing the **Z** in the "batch update" procedure of Eqs. (43) and (44) by $\mathbf{z}_k$, and thus arriving at the following continuous mode update rules:

$$b_k^{\text{new}} = b_k^c + \frac{\rho_1}{2}\left(|\varepsilon^c| + \varepsilon^c\right)$$

$$\text{with } \varepsilon^c = \mathbf{z}_k^T\mathbf{w}^c - b_k^c \quad (45)$$

$$\mathbf{w}^{\text{new}} = \mathbf{w}^c - \rho_2\mathbf{z}_k[\mathbf{z}_k^T\mathbf{w}^c - b_k^{\text{new}}]$$

$$= \mathbf{w}^c + \frac{\rho_1\rho_2}{2}\left[|\varepsilon^c| + \varepsilon^c\left(1 - \frac{2}{\rho_1}\right)\right]\mathbf{z}_k \quad (46)$$

where b_k represents a margin scalar associated with the $\mathbf{x}_k$ input. In all of the above Ho–Kashyap training strategies, the margin values and the perceptron weights are initialized to small positive and zero (or small random) values, respectively. If full margin error correction is assumed in Eq. (45) (i.e., $\rho_1 = 1$), the above AHK procedure reduces to an earlier reported heuristically derived procedure (Hassoun and Song, 1992; Hassoun, 1989b). An alternative way of writing Eqs. (45) and (46) is

$$\Delta b_k = \rho_1\varepsilon^c \quad \text{and} \quad \Delta\mathbf{w} = \rho_2(\rho_1 - 1)\varepsilon^c\mathbf{z}_k$$

$$\text{if } \varepsilon^c > 0 \quad (47)$$

$$\Delta b_k = 0 \quad \text{and} \quad \Delta\mathbf{w} = -\rho_2\varepsilon^c\mathbf{z}_k \quad \text{if } \varepsilon^c \leq 0 \quad (48)$$

where Δb ($\Delta\mathbf{w}$) signifies the difference between the updated and current margin (weight) values. We will refer to this procedure as the AHK I learning rule. For comparison purposes, it may be noted that the α-LMS rule (or the relaxation rule; Duda and Hart, 1973) can be written, employing the notation of this chapter, as $\Delta\mathbf{w} = -\rho_2\varepsilon_k^c\mathbf{z}_k$, where $\rho_2 = \rho_2(k) = \alpha/\|\mathbf{z}_k\|^2$, and b_k is fixed at $+1$. A sufficient condition for the convergence of the α-LMS rule is $0 < \alpha < 2$ (Duda and Hart, 1973; Widrow and Lehr, 1990).

The constraint $b_k > 0$ in Eqs. (45)–(48) was realized by starting with a positive initial margin and restricting the change in Δb to positive real values. An alternative, more flexible way to realize this constraint is to allow both positive and negative changes in Δb, except for cases where a decrease in b_k results in a negative margin. This modification results in the following alternative AHK II learning rule:

$$\Delta b_k = \rho_1\varepsilon^c \quad \text{and} \quad \Delta\mathbf{w} = \rho_2(\rho_1 - 1)\varepsilon^c\mathbf{z}_k$$

$$\text{if } b_k^c + \rho_1\varepsilon^c > 0 \quad (49)$$

$$\Delta b_k = 0 \quad \text{and} \quad \Delta\mathbf{w} = -\rho_2\varepsilon^c\mathbf{z}_k$$

$$\text{if } b_k^c + \rho_1\varepsilon^c \leq 0 \quad (50)$$

Note that in the limit as ρ_1 approaches zero and setting all margins to $+1$, Eqs. (49) and (50) reduce to the α-LMS rule. On the other hand, if all margins are held fixed at $+1$, these equations reduce to

$$\Delta\mathbf{w} = -\rho_2(1 - \rho_1)\varepsilon_k^c \mathbf{z}_k \quad \text{if } \varepsilon_k^c > -\frac{1}{\rho_1} \quad (51)$$

$$\Delta\mathbf{w} = -\rho_2\varepsilon_k^c \mathbf{z}_k \qquad \text{if } \varepsilon_k^c \leq -\frac{1}{\rho_1} \quad (52)$$

If we assume $0 < \rho_1 < 1$, then Eqs. (51) and (52) are identical to the α-LMS rule except that a two-valued effective learning coefficient is employed: $\rho_2(1 - \rho_1)$ in Eq. (51) and ρ_2 in Eq. (52). Hence, the algorithm converges if the larger learning coefficient, ρ_2, satisfies the α-LMS convergence criterion which is equivalent to $0 < \rho_2 < 2/\max\|\mathbf{z}_k\|^2$. In the general case of adaptive margins as in Eqs. (49) and (50), it can be shown (Hassoun and Song, 1992) that a sufficient condition for the convergence of the AHK I and II rules is given by

$$0 < \rho_2 < \frac{2}{\|\mathbf{z}_k\|^2} \quad (53)$$

$$0 < \rho_1 < 2 \quad (54)$$

Finally, we define yet another variation on Eqs. (49) and (50) to arrive at the AHK III learning rule which is appropriate for both linearly separable and non-linearly separable cases. This algorithm is identical to the AHK II except that $\Delta\mathbf{w}$ is set to zero in (49). The advantages of the AHK III rule are that it is capable of adaptively identifying difficult-to-separate class boundaries and that it uses such information to discard nonseparable training vectors and to speed up convergence. These features, among others, are discussed in Hassoun and Song (1992).

In this section we have reviewed several recording/learning algorithms suited for loading memories in simple single-layer (static or dynamic) associative neural memories. Additional extensions and more elaborate learning techniques are presented and analyzed in the later chapters of this book (e.g., see Chapters 5, 6, 7, 16, and 18).

1.5. RECORDING STRATEGIES

The encoding, dimension, and number of stored patterns strongly affect the performance of a DAM. The recording strategy of a given number of associations is also of critical importance in the robustness and dynamics of a DAM. For example, DAM performance will be enhanced by augmenting the training set with an additional set of specialized associations. Here, we present

three examples of recording strategies: (1) training with perfect associations, (2) training with an extended set of noisy/partial associations, and (3) training with the aid of a specialized set of associations.

The first training strategy is employed when only perfect associations are present. The training set consists of the m input/target pairs $\{\mathbf{x}_k, \mathbf{y}_k\}$ and is shown in Fig. 1.7a. This represents the simplest training strategy possible, and relies on intrinsic DAM dynamics to realize the needed basins of attraction around each recorded association. This strategy works if the number of associations is relatively small compared to the smallest of the dimensions n and L, and if the degree of correlation between the m associations is relatively low. With this strategy, the training set is usually underdetermined ($m < n$).

The second recording strategy is employed when the training set consists of a number of clusters with each cluster having a unique label or target. This strategy is common in pattern classification applications and is useful in defining large basins of attraction around each recorded memory, thus increasing DAM error tolerance. Figure 1.7b illustrates this case. In general, the inclusion of the noisy/partial associations increases the size of the training set and may lead to an overdetermined set of associations ($m > n$).

The third recording strategy is employed when specific additional DAM associations must be introduced or eliminated. One possibility of employing this strategy is when a *ground state* must be introduced to act as a *no-decision* state, as depicted in Fig. 1.7c. For example, for associations encoded such that sparse vectors $\mathbf{n}_i$ have low information content, augmenting the original training set with associations of the form $\{\mathbf{n}_i, \mathbf{0}\}$ leads to the creation of a no-decision state $\mathbf{0}$ which attracts highly corrupted or noisy inputs and prevents them from being classified erroneously or mapped into spurious memories. Simulations illustrating the use and benefits of the above first and third recording strategies are presented in the next section, employing various recording algorithms.

1.6. DAM CAPACITY AND PERFORMANCE

In the following, the capacity and performance of autoassociative DAMs are discussed. Capacity is defined as a measure of the ability of a DAM to store a set of *unbiased random* binary patterns $\{\mathbf{x}_k\}$ (with a probability that a 1/0 or $+1/-1$ bit equals 0.5) at a given error correction and recall accuracy levels. Earlier proposed capacity

Input Vectors   Target Vectors

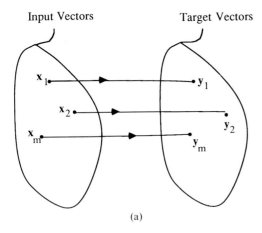

(a)

Input Vectors   Target Vectors

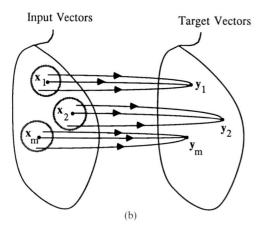

(b)

Input Vectors   Target Vectors

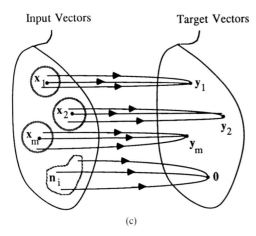

(c)

Fig. 1.7. Examples of DAM recording strategies: (a) simple recording; (b) training with noisy/partial associations; and (c) training with the aid of specialized associations.

measures (Amit et al., 1985; Stiles and Deng, 1987; Meir and Domani, 1987; McEliece et al., 1987; Amari and Maginu, 1988; Newman, 1988; Komlos and Paturi, 1988; Dembo, 1989; Weisbuch and Fogelman, 1985; Personnaz et al., 1985; Venkatesh and Psaltis, 1989) have defined capacity as equivalent to a tight upper bound on the pattern ratio (m/n) for which all stored memories are fixed states, with a probability approaching 1.

The two most-used DAM capacity measures in the literature are the absolute capacity (C_a) and relative capacity (C_r) measures. C_a is defined as the least upper bound on the pattern ratio (m/n) such that the m stored memories are retrievable (with a probability approaching 1) in one pass through the DAM, starting from perfect key patterns; i.e., the DAM is capable of memorizing m patterns as equilibrium points. C_r is a least upper bound on m/n such that the stored memories are approximately retrievable. For correlation-recorded autoassociative DAMs employing hard-clipping activation, the relative capacity is approximately equal to 0.15 (Amari, 1977; Hopfield, 1982; Amit et al., 1985; Meir and Domani, 1987). The requirement for error-free memory recall severely limits the capacity of correlation-recorded DAMs to $C_a = 1/[4 \log(n)]$ (McEliece et al., 1987; Amari and Maginu, 1988; Newman, 1988; Weisbuch and Fogelman, 1985), which approaches zero for large n (log is the natural logarithm). These capacity limits are reviewed next.

Assuming $\mathbf{y}_k = \mathbf{x}_k$ (autoassociative case) in Eq. (20), the condition for perfect recall of the ith bit of the input $\mathbf{x}_h$ (one of the recorded memories) is given by

$$\tilde{x}_i^h = \mathrm{sgn}\left[x_i^h + \frac{1}{n} \sum_{j=1}^n \sum_{k \neq h}^m x_i^k x_j^k x_j^h \right]$$

$$= \mathrm{sgn}[x_i^h + \Delta_i^h]$$

If we now define $C_i^h = -x_i^h \Delta_i^h$, then the one-pass retrieved bit x_i^h is in error if $C_i^h > 1$. If we now assume uniformly distributed random bipolar pattern $\mathbf{x}_k$, then we can estimate the probability of this bit being in error as $P_{\mathrm{error}} = \mathrm{Prob}(C_i^h > 1)$. For large m and n, the C_i^h term is approximately distributed according to a normal distribution $N(\mu, \sigma^2) = N(0, m/n)$ (Amari and Maginu, 1988). Therefore, we can estimate the upper bound on m/n for a given P_{error}. This can be done by solving for m/n in the following inequality:

$$\frac{1}{\sqrt{2\pi(m/n)}} \int_1^\infty \exp\left(-\frac{n\zeta^2}{2m} \right) d\zeta \leq P_{\mathrm{error}}$$

Now, using the fact that for a normal distribution, with zero mean and variance σ^2, $\text{Prob}(x > 3\sigma) = 0.0015$, and if we require that the ith bit be retrievable with $P_{\text{error}} < 0.0015$, then we must have $3\sigma < 1$ or $m/n < 1/9$. Therefore, we must have $m < 0.11n$ for $P_{\text{error}} < 0.0015$. Similarly, the relative capacity $C_r = 0.15$ allows for a P_{error} close to 0.005.

If we require that all stored memories be fixed points of the associative memory with a probability approaching 1, say 0.99, then we can derive an upper bound on m/n by starting from $(1 - P_{\text{error}})^{mn} > 0.99$, thus requiring all bits of all memories to be retrievable with less than a 1 percent error. Employing binomial expansion, we may approximate the above inequality as $P_{\text{error}} < 0.01/mn$. Noting that this stringent error correction requirement results in n/m approaching infinity, we may use this fact to write (Hertz et al., 1991)

$$P_{\text{error}} = \frac{1}{2}\left[1 - \text{erf}\left(\sqrt{\frac{n}{2m}}\right)\right]$$

$$\cong \frac{1}{2}\frac{e^{-n/2m}}{\sqrt{(\pi n/2m)}} < \frac{0.01}{mn}$$

By taking the natural logarithm (log) of both sides of the above inequality and using the fact that $n/m \gg \log(n/m)$, we get (Chapter 12 gives a mathematically more rigorous derivation of the following result)

$$\frac{m}{n} < \frac{1}{2\log mn} \approx \frac{1}{2\log n^2} = \frac{1}{4\log n}$$

Another, more useful, DAM capacity measure gives an upper bound on m/n in terms of attraction radius and memory size (McEliece et al., 1987; Weisbuch and Fogelman, 1985). According to this measure, a correlation-recorded DAM has a maximum pattern ratio

$$\frac{m}{n} = \frac{(1 - 2\rho)^2}{4\log n}$$

which guarantees error-free fundamental memory retrieval (by direct one-pass convergence in the limit of large n) from random key patterns lying inside the Hamming sphere of radius ρn ($\rho < 0.5$). This capacity result is only approximate for the limiting case $\rho = 0$, and it has been shown (Komlos and Paturi, 1988; see also Chapter 9) that, when the number of stored patterns approaches $n/[4\log(n)]$, the fundamental memories have a basin of attraction of radius $0.024 < \rho < 1/8$, and convergence to such

memories is achieved in $O(\log\log n)$ parallel iterations.

Amari (1989) has extended the above capacity measures to the case of a correlation-recorded DAM with sparsely encoded memories, where he shows that C_a is of order $1/\log^2(n)$, which is much larger than that of non-sparsely encoded memory (refer to Chapter 3 for other sparse associative memory architectures and their capacity). In another DAM variation, Yanai and Sawada (1990) derive the expression

$$C_a = \frac{(1 + h)^2}{2\log n}$$

for a correlation-recorded DAM with unbiased random memories and with perceptrons employing a hysteric hard-clipping activation function with a hysteric range $[-h, +h]$. Marcus et al. (1989) have studied the retrieval dynamics of parallel updated continuous DAMs employing correlation and generalized-inverse recording of unbiased random bipolar memories with large n. Phase diagrams depicting various DAM dynamics (recall, oscillatory, spin glass, and ground regions) in the space of pattern vs. activation function gain β (a $\tanh(\beta z)$ activation function was assumed) were derived for the correlation and generalized-inverse DAMs; for instance, it was shown that period-2 limit cycles are eliminated from the DAM for both recording techniques when $\beta < n/m$, and that the origin is the only attractor for the correlation DAM when

$$\beta < \frac{1}{1 + 2\sqrt{(m/n)}}$$

and for the generalized-inverse DAM when

$$\beta < \frac{1}{1 - (m/n)}$$

(The reader is referred to Chapters 8, 10, and 13 for additional results on the capacities of other DAMs.)

Let us turn our attention back to parallel updated autoassociative DAMs employing hard-clipping activation and unbiased random high-dimensional binary memories. It can be shown (Komlos, 1967) that for an underdetermined training set of unbiased random binary vectors, and in the limit as n approaches infinity, the probability of linear independence of the training vectors approaches 1. This makes the single-layer DAM appropriate for the realization of such training sets. Little theoretical work has been done on the capacity and performance of DAMs recorded with generalized-inverse or

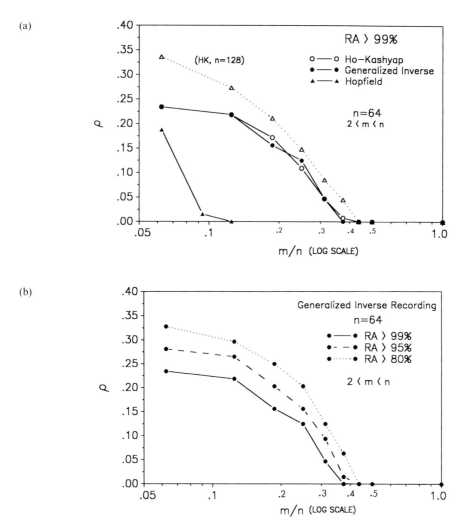

Fig. 1.8. (a) Capacity/performance curves comparing the correlation, generalized-inverse, and Ho–Kashyap DAMs for $n = 64$. Dashed curve represents the cases of GI and HK DAM with $n = 128$. (b) Generalized-inverse performance curves for various values of recall accuracy (RA).

Ho–Kashyap techniques. Youssef and Hassoun (1989) report Monte Carlo simulations for $16 < n < 128$ where the retrieval performance and capacity of various recording/learning techniques, including the generalized-inverse and Ho–Kashyap, were investigated. They propose a capacity/performance measure similar to C_a given above, with one additional DAM performance parameter: recall accuracy (RA). Here, capacity is computed under the strict requirement that all fundamental memories are perfectly retrievable and that *retrieval is not restricted to a single pass.* Figure 1.8a shows plots for ρ vs. m/n where ρ is an upper bound on the normalized basin radius

around fundamental memories, guaranteeing RA > 99 percent. Three curves are shown for the correlation (Hopfield), generalized-inverse (GI), and Ho–Kashyap (HK) DAMs, respectively, with $n = 64$. A fourth curve is shown (dotted line) for the Ho–Kashyap and generalized-inverse DAM, with $n = 128$. These curves clearly depict the superiority of the GI and HK recording techniques over the correlation technique. It can also be concluded that the GI- and HK-recorded DAMs have a relatively large operating region where error correction is possible. This range is defined, roughly, as $0 < m/n < 0.5$. From the simulations in Youssef and Hassoun (1989) it is to be noted that the synthesized weight matrix becomes a diagonal-dominant matrix in the limit as m approaches $n/2$, at which point all fundamental memories lose their basin of attraction (this is consistent with the results of Personnaz et al. (1986) discussed in Section 1.4). At $m = 0.15n$ with large n, the GI and HK algorithms lead to DAMs capable of correcting in excess of 25 percent unbiased random noise. On the other hand, and at a loading level of $m = 0.15n$, the correlation DAM is not capable of retaining the fundamental memories, nor their basin of attraction. Figure 1.8b extends the results just discussed to cases of reduced recall accuracy constraints (RA > 95 percent, 85 percent) for the GI-recorded DAM. The Ho–Kashyap-recorded DAM exhibits similar characteristics.

It is interesting to note the similarity between the GI and HK DAM retrieval performance. This should not be surprising, since the training set assumed above is underdetermined where, according to the discussion in Section 1.4.3, the HK solution is equivalent to the GI solution except for the added bias bits. The effects of the bias bits disappear when high-dimensional unbiased-random memories (associations) are used. However, when nonrandom memories are used, the GI and HK DAMs exhibit substantially different dynamics, as is shown next.

Heteroassociative DAM performance and dynamics are less understood. Kosko (1987, 1988) has analyzed the stability and capacity of correlation-recorded heteroassociative DAM [or bidirectional associative memory (BAM)] and its extension to general Hebbian learning. Hassoun (1989a) employs Monte Carlo simulations in the analysis of the capacity and performance of parallel updated GI- and HK-recorded hetero-associative DAMs (for more details on BAM dynamics, the reader is referred to Chapters 5 and 17).

The dynamics and capacity of continuous or discrete updated DAMs employing multilayer architectures have not received adequate attention in the literature. Multilayer DAMs are important, since single-layer DAMs are limited to the realization of linearly-separable mappings; in practice, many interesting problems are non-linearly-separable. The recording of DAMs also needs to be developed further, in such a way that the dynamical nature of the architecture is taken into account during the learning/recording synthesis phase, which can result in more predictable retrieval dynamics. Also, computationally efficient methods for controlling the shape and size of the basins of attraction of fundamental and spurious memories are desirable [see Venkatesh et al. (1990) for an attempt to address this problem; chapters 6, 7, and 12 also address these questions]. Some of the later chapters of this book deal with interesting extensions and analyses of more general DAMs than those considered in this chapter.

Next, a limited but illustrative simulation comparing the various *direct* and *adaptive* recording techniques, discussed in Section 1.4, is presented. The advantages of employing specialized recording strategies in improving the retrieval characteristics of DAMs is also illustrated. Four 16-dimensional binary patterns are chosen such that any two distinct patterns have a Hamming distance of 8, as shown in Fig. 1.9a, b.

Two sets of simulations were performed; the results are tabulated in Tables 1.1 and 1.2, respectively. The first simulation assumes a direct recording strategy employing the four memories ($m = 4$; $n = 16$). The second set of simulations employs an overdetermined training set consisting of the four memories of Fig. 1.9 and 16 additional heteroassociations representing the mapping of the rows of the 16-dimensional unit matrix into the 16-dimensional zero pattern. In both cases, the DHK, GI, correlation (HOP), LMS, AHK I, and AHK II recording/learning algorithms were employed in DAM synthesis.

The DAMs in these simulations were tested with 10 000 unbiased random vectors. Unipolar encoding was assumed for the Ho–Kashyap and GI DAMs. On the other hand, bipolar encoding

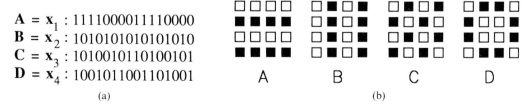

$\mathbf{A} = \mathbf{x}_1 : 1111000011110000$
$\mathbf{B} = \mathbf{x}_2 : 1010101010101010$
$\mathbf{C} = \mathbf{x}_3 : 1010010110100101$
$\mathbf{D} = \mathbf{x}_4 : 1001011001101001$

(a)

(b)

Fig. 1.9. (a) Four memory vectors used to train an autoassociative DAM. (b) 2-D representation of the vectors in (a).

Table 1.1. Retrieval dynamics statistics for various DAMs trained with the memories of Fig. 1.9.

Retrieval Dynamics[b]	Recording/Learning Algorithm[a]					
	DHK	GI	HOP[c]	LMS[c]	AHK I	AHK II
Convergence rate to fundamental memories (%)	54.5	28.3	18.5	34.7	61.7	53.1
Convergence rate to spurious memories (%)	45.5	65.7	23.0	65.3	38.3	46.9
Number of spurious memories detected	22	51	10	79	18	16
Convergence rate to ground memory (%)	0	6.0	0.0	0.0	0.0	0.0
Convergence rate to oscillatory memories (%)	0	0	58.5	0	0	0

[a] Training was performed with the original set of 4 memories.
[b] DAM tested with 1000 input vectors having uniformly random binary bits.
[c] Training and testing were performed assuming bipolar memories.

Table 1.2. Retrieval dynamics statistics for DAMs with specialized associations.

Retrieval Dynamics[b]	Recording/learning Algorithm[a]					
	DHK	GI	HOP[c]	LMS[c]	AHK I	AHK II
Convergence rate to fundamental memories (%)	34.65	24.7	–	12.7	24.8	1.1
Convergence rate to spurious memories (%)	10.2	60.4	–	83.6	70.9	62.8
Number of spurious memories detected	7	44	–	127	38	16
Convergence rate to ground memory (%)	55.1	14.9	–	3.7	4.3	36.1
Convergence rate to oscillatory memories (%)	0.05	0	–	0	0	0

[a] Training was performed using sparse memory-to-ground mapping.
[b] DAM tested with 1000 input vectors having uniformly random binary bits.
[c] Training and testing were performed assuming bipolar memories.

was assumed for the Hopfield and LMS DAMs, for both training and retrieval phases (the LMS DAM also employed bias). In addition, all adaptive learning algorithms (LMS, AHK I, and AHK II) were initialized with zero weight matrices. The learning rates were 0.1 for LMS, 0.1 for direct HK and $\rho_1 = 1$ and $\rho_2 = 0.01$ for AHK I (with margin vector initialized to all $+1$ components), and $\rho_1 = 0.1$ and $\rho_2 = 0.05$ for AHK II (with initial margin vectors of $+0.1$ components). Learning was terminated after 50 iterations for DHK and AHK I. For LMS and AHK II, learning stopped after reducing the error function $J(\mathbf{W})$ to 0.0001. After learning, all weight matrices were normalized and rounded to integer weights in the range $[-99, +99]$.

The three Ho–Kashyap recording algorithms have comparable performances which exceed those of generalized-inverse, Hopfield, and LMS. The first row depicts the formation of large basins of attraction around fundamental memories for the HK algorithms. Noisy inputs which do not converge to fundamental memories are attracted by spurious memories for HK recording. The GI and LMS DAMs resulted in a relatively large number of spurious memories which attracted about 65 percent of the test vectors. The GI DAM was the only DAM with a no-decision state (ground state), which attracted 6 percent of the input. As expected, the worst performance is that of the Hopfield DAM, which has a low convergence rate to fundamental memories and a high convergence rate to period-2 oscillations (these oscillations can be eliminated if stochastic update is employed; however, this will result in an increased convergence rate to spurious states). It is also interesting to note that even though the **W** matrix of the HK-recorded DAM is not symmetric and that parallel updating is used, no oscillations were encountered. This phenomenon is due in part to the weight normalization and rounding employed, and to the low memory loading level ($m = 4$) (Youssef and Hassoun, 1989). The use of sparse memory-to-ground mapping enhances the performance of the DHK DAM, as depicted in the first column of Table 1.2. Here, the number of spurious states is reduced to seven and the convergence of highly noisy inputs is redirected from spurious and fundamental memories to a ground state which attracts about half of the test inputs. In addition, only five noisy inputs (0.05 percent) converged to period-2 oscillations. The GI also shows slight improvement in performance compared to the underdetermined recording case in Table 1.1. On the other hand, this recording strategy does not seem to be adequate for the Hopfield DAM (fundamental memories were not retrievable) or for the adaptive LMS and AHK recording techniques.

1.7. DAM PERFORMANCE WITH CONSTRAINED WEIGHT ACCURACY

This section addresses the question of dynamic associative memory (DAM) performance due to constraints on synaptic weight accuracy, which arise in analog hardware DAM implementations. High-performance recording algorithms (such as generalized-inverse and Ho–Kashyap algorithms) synthesize real-valued DAM weights which may constrain hardware implementations (electronic or optoelectronic) due to the restricted dynamic range/accuracy of analog voltages, currents, and/or optical signals involved (for an analysis of the performance of correlation-recorded DAMs with technology-dependent faulty weights, the reader is referred to Chapter 14). We first introduce a simple normalization technique which converts DAM real-valued weights into finite-accuracy integer weights. Based on this normalization technique, we study, through extensive simulations, the effects of weight accuracy on DAM dynamics and retrieval.

A very interesting and useful result is that the retrieval characteristics of a high-performance DAM, with uniformly distributed memories and moderate memory loading levels (pattern ratio less than 0.2), is minimally affected when implemented with restricted 4-bit or higher accuracy synaptic weights.

In the digital computer simulation of DAM, the weight matrix **W** is used to full accuracy and is only limited by the floating-point accuracy of the computer used. However, in hardware implementations, the weight matrix accuracy is limited by the accuracy of the hardware used. DAM performance degrades relative to the accuracy of its weight matrix. The recall accuracy and dynamics of hardware-implemented artificial neural networks is strongly affected by the accuracy of the implementation technology used (Hecht-Nielsen, 1986; Yu et al., 1989; see also Chapters 15, 16, and 18 for additional discussion on hardware accuracy effects).

DAM's weight matrices have been implemented using various techniques, such as binary transparencies (Psaltis and Farhat, 1985), holograms (Psaltis et al., 1987), resistors (Denker, 1986), and high-resolution TV monitors (Lu et al., 1989) (the reader is referred to Chapters 15–19 for specific examples of DAM implementation using various technologies). The performance of a DAM using any one of the above and other implementation techniques is affected by limited weight accuracy. The question is how much does the weight accuracy affect DAM performance. In this section, this question is answered for a simple parallel updated single-layer DAM architecture with thresholding and feedback (shown in Fig. 1.1) with weight matrix **W** and threshold vector **T**. Here the Ho–Kashyap (HK) recording algorithm, introduced in Section 1.4.3, is employed.

1.7.1. Synaptic Weight Normalization Technique

Weight accuracy is defined as a function of the number of binary bits representing an element of the weight matrix **W**. Ideally, the elements in **W** are real numbers with infinite accuracy. However, when implementing the weight matrix in hardware, the weight accuracy is constrained by the chosen technology/circuitry. The hardware-realized weight matrix is equivalent to a normalized weight matrix **Z** with a restricted weight accuracy in the range $[-2^b, +2^b]$ with integer $b > 0$. Each element in **Z** is a normalized integer of b binary bit accuracy. The normalized restricted-accuracy weight matrix **Z** can have low or high accuracy depending on the value of b. The weight accuracy is defined as a function of the number of binary bits b representing each element.

The performance of a DAM is a function of weight accuracy. DAM performance is measured by means of the performance parameters P_f and P_s, the probabilities of convergence to fundamental and spurious memories, respectively. The range of weight accuracy is bounded by low accuracy 1-bit ($b = 1$) and full accuracy (b very large, which corresponds to the exact real-valued weight matrix). The procedure of encoding the real-valued weight matrix into integers for different numbers of b-bit accuracy is referred to as normalization. For instance, if $\mathbf{W}$ has the three elements $[-0.5, -0.01, 0.5]$ and a bit accuracy $b = 3$ is assumed, each element will be replaced by three digits of 1s and 0s. Therefore, the elements will have the normalized values $[-7, 0, +7]$. This normalization procedure is detailed next.

The weight matrix $\mathbf{W}$ and threshold vector $\mathbf{T}$ have real-valued components. The normalization procedure described here is a mapping which maps $\mathbf{W}$ and $\mathbf{T}$ onto $\mathbf{Z}$ and $\mathbf{V}$, respectively, where $\mathbf{Z}$ and $\mathbf{V}$ have integer values with b-bit accuracy. The mapping used is empirical and other mappings can be employed.

For a bit accuracy equal to 1 ($b = 1$), the following normalization procedure is employed.

1. Clip the weight matrix as follows:

$$z(i,j) = +1 \quad \text{if } w(i,j) \geq 0 \quad \text{and } -1 \text{ otherwise}$$

where $w(i, j)$ is the ijth element of $\mathbf{W}$.
2. Compute the mean of $\mathbf{W}$:

$$\mu = \frac{\sum_{i=1}^{n} \sum_{j=1}^{n} w(i, j)}{n^2}$$

3. Compute the mean of $\mathbf{Z}$:

$$\mu_z = \frac{\sum_{i=1}^{n} \sum_{j=1}^{n} z(i, j)}{n^2}$$

4. Compute a threshold normalization factor τ by dividing μ_z by μ:

$$\tau = \frac{\mu_z}{\mu}$$

5. Normalize the threshold vector $\mathbf{T}$ according to

$$\mathbf{V}' = \tau \mathbf{T}$$

Finally, the components of the vector $\mathbf{V}'$ are rounded off to the nearest integer to produce $\mathbf{V}$.

For a bit accuracy greater than one, the normalization procedure is as follows:

1. Compute the absolute mean of $\mathbf{W}$ as

$$\mu = \frac{\sum_{i=1}^{n} \sum_{j=1}^{n} |w(i, j)|}{n^2}$$

2. Compute a normalization factor β by mapping the mean μ to the middle of the range $[0, +2^b]$. This factor β is given by

$$\beta = \frac{2^b}{2\mu}$$

3. Multiply the weight matrix $\mathbf{W}$ by the normalization factor β to obtain the normalized weight matrix $\mathbf{Z}'$:

$$\mathbf{Z}' = \beta \mathbf{W}$$

4. Make all the tallies of the normalized weight matrix $\mathbf{Z}'$ equal to the possible maximum values, either positive or negative according to

$$\text{if } w(i, j) \geq \mu + 3\sigma \rightarrow z'(i, j) = +(2^b - 1)$$

$$\text{if } w(i, j) \leq \mu - 3\sigma \rightarrow z'(i, j) = -(2^b - 1)$$

where σ is the standard deviation of the distribution of $|w(i, j)|$.

The threshold vector is normalized according to

$$\mathbf{V}' = \beta \mathbf{T}$$

The normalized weight matrix $\mathbf{Z}'$ and threshold vector $\mathbf{V}'$ are rounded off to the nearest integer to give $\mathbf{Z}$ and $\mathbf{V}$. This normalization procedure is used in the following simulations.

1.7.2. Simulations

In order to evaluate the performance of DAMs with normalized (restricted-accuracy) weight matrices, Monte Carlo simulations were performed. Here, we choose to study the performance of an n-dimensional (n neurons) DAMs with HK recording employing the recording strategy, presented in Section 1.5, which augments all unity vectors of dimension n to the training set of fundamental memories (all these unit vectors assume the zero vector as their associated target). This recording strategy enforces the desirable creation of a wide ground (no-decision state) attractor which attracts highly corrupted inputs (as discussed at the end of Section 1.6).

Ten sets of m random unipolar binary patterns ($2 < m < n/2$) were generated for each tested DAM of dimension n, with $n = 32$ or 64. The DAM performance studied here is measured in terms of the probabilities of convergence to fundamental memories (P_f) and to spurious memories (P_s). The fundamental memories are randomly generated uniform n-dimensional unipolar binary patterns, and the test data base is also randomly generated.

We expect the probability of convergence to fundamental memories (P_f) to increase with the weight accuracy b and to reach a plateau near the value obtained using the full-accuracy weight matrix. Each set of recorded memories is tested for the above two convergence probabilities. For any memory size and loading level, the DAM has a poor performance when the weight is normalized with a bit accuracy equal to 1 (refer to Chapter 15 for an optimization learning strategy that enhances the performance of clipped-weight DAMs). The DAM performance increases as the bit accuracy increases. In the following figures, the bit accuracy axis is broken to indicate that the last data point corresponds to full-accuracy weights.

The probability of convergence to fundamental memories for $n = 32$ is shown in Fig. 1.10, as a family of curves for $2 < m < 15$. The data shown in this figure show how P_f increases and saturates as the weight accuracy employed increases. The plateau level approaches P_f of the exact weight matrix. At low bit accuracy, $b = 1$ or 2, the DAM exhibits poor performance. P_f increases to a level near the one obtained using the exact weight matrix for a low pattern ratio (i.e., low loading). For a larger pattern ratio, $m/n = 0.25$, P_f is zero for $b = 1$ and increases to

reach 40 percent at $b = 6$. For the same pattern ratio, the DAM gives a P_f near 50 percent with exact weights. There is a reduction in performance at high loading as expected. The above phenomenon is expected to persist at higher DAM dimension n. In order to have good performance for a large pattern ratio, the bit accuracy must be high. On the other hand, at relatively moderate loading levels, low bit accuracy may be used.

For the above simulations with $n = 32$, the probability of convergence to spurious memories is plotted in Fig. 1.11. Here, the retrieval characteristics of the DAM in terms of bit accuracy are consistent with the above performance trends. We also note here that the consistency of the plots in Figs. 1.10 and 1.11 may be seen more clearly if one looks at the family of curves with even (odd) m independent of the family of curves with odd (even) m. The dynamics of a DAM loaded with m fundamental memories may vary substantially depending on whether m is odd or even (Hassoun and Watta, 1991).

The two probabilities, for $m = 8$ and $n = 32$ (for a pattern ratio $m/n = 8/32 = 0.25$), are shown in Fig. 1.12. The probability P_f changes from 0 to 48 percent for $b = 1$ and 6, respectively, and never reaches the 62 percent obtained using the full-accuracy weight matrix. On the other hand, the probability P_s changes from 95 percent to 50 percent for b changing from 1 to 6 bits. It approaches but never reaches the 30% obtained

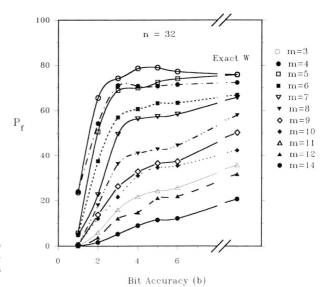

Fig. 1.10. Probability of convergence to fundamental memories as a function of synaptic weight bit accuracy for various numbers of stored random memories ($n = 32$).

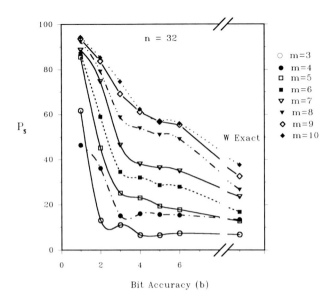

Fig. 1.11. Probability of convergence to spurious memories as a function of synaptic weight bit accuracy, for various numbers of stored random memories ($n = 32$).

using the full-accuracy **W** matrix. Increasing the bit accuracy beyond 6 did not improve performance at such a relatively high loading level. This may be attributed to the statistical nature of the empirical normalization technique employed here, which, at high loading levels, does not guarantee the recovery of the exact weights, even as b approaches large values.

One interesting observation is the change of

P_f versus pattern ratio for several bit accuracies. These data are shown in the family of curves in Fig. 1.13. It is interesting to note that both $b = 1$ and 2 give drastic change in performance as the pattern ratio increases. For other bit accuracies, the performance drops only a maximum of 15 percent for a change in pattern ratio from 0.08 to 0.25. Another observation is that P_f decreases by only about 10 percent for $b = 3$ and pattern

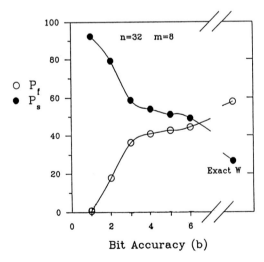

Fig. 1.12. Probability of convergence to fundamental memories and probability of convergence to spurious memories as a function of weight accuracy for $n = 32$ and $m = 8$.

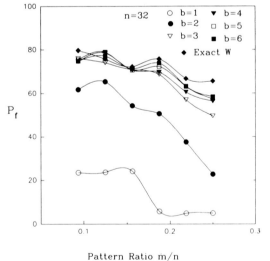

Fig. 1.13. Probability of convergence to fundamental memories as a function of pattern ratio, for various selections of weight bit accuracy b ($n = 32$).

ratio 0.22. Here, the performance obtained using a bit accuracy equal to or larger than 3 is not far from the one obtained using the full-accuracy weight matrix for a pattern ratio of 0.15. Thus, using three bits to implement a Ho–Kashyap-recorded DAM can give rise to a performance close to the one obtained using exact weights as long as m/n is low; at $m/n = 0.15$, almost equivalent performance is obtained. Additional simulations with $n = 64$, 128, and 256 show that for the weight normalization/quantization technique employed in this section, 4-bit accuracy weights consistently led to DAM performance within 10–20 percent (in terms of P_f and P_s) of that of an unrestricted weight accuracy DAM, for a moderately low loading level ($m/n < 0.15$).

1.8. SUMMARY

This chapter presents a review of the fundamental concepts relating to basic DAM architectures. It gives an overview of the various results on stability, capacity, and retrieval capability of simple single-layer nonlinear dynamical associative neural-like memories trained with various learning/recording algorithms. The correlation, generalized inverse (projection), and Ho–Kashyap recording algorithms and their adaptive learning versions are described, and the necessary requirements on fundamental memories for perfect storage are derived for these algorithms. The chapter also presents the main criteria for high-performance associative memories. Such criteria are employed in the later sections of the chapter to compare the performance of DAMs recorded with the above-mentioned learning algorithms.

Necessary conditions on the interconnection matrix for stability are derived for the autoassociative fully interconnected single-layer DAM, and similar results are presented for other simple DAM variations. Some of these variations include the BSB memory and the continuous parallel updated Hopfield memory. The capacity and error correction capabilities of the correlation-recorded DAM are derived mathematically for the case of one-pass retrieval. It is noted that, because of its tractable nature, the correlation-recorded DAM is extensively studied and various performance results are presented in the literature; an attempt is made to point out the significance of such results. The capacity and error-correction capabilities (one-pass and multiple-pass retrieval) of the generalized-inverse and Ho–Kashyap-recorded DAMs are more difficult to analyze mathematically, and Monte

Carlo simulations are employed in the study of these DAMs. The results show the superior performance of these latter DAMs over the correlation-recorded DAMs. This increased performance comes at the expense of higher computational complexity and a need for increased weight accuracy.

The chapter concludes with an investigation of the effects of synaptic weight accuracy on DAM performance for DAMs recorded with the Ho–Kashyap recording algorithm. It concludes that a high-performance DAM may be implemented in hardware having as little as 4-bit synaptic weight accuracy (assuming uniform random fundamental memories) without compromising performance, as long as the pattern ratio (m/n) is low ($m/n < 0.2$). This chapter also introduces some of the terminology and architectures used in later chapters of this book.

REFERENCES

Abu-Mustafa, Y. S. and St Jacques, J. M. (1985). "Information Capacity of the Hopfield Model," *IEEE Trans. Information Theory*, **IT-31**, 461–464.

Albert, A. (1972). Regression and the Moore–Penrose Pseudoinverse. Academic Press, New York.

Amari, S.-I. (1972). "Learning Patterns and Pattern Sequences by Self-Organizing Nets of Threshold Elements," *IEEE Trans. Computers*, **C-21**, 1197–1206.

Amari, S.-I. (1977). "Neural Theory of Association and Concept-Formation," *Biological Cybernetics*, **26**, 175–185.

Amari, S.-I. (1989). "Characteristics of Sparsely Encoded Associative Memory," *Neural Networks*, **2**, 451–457.

Amari, S.-I. (1990). "Mathematical Foundations of Neurocomputing," *Proc. IEEE*, **78**(9), 1443–1463.

Amari, S.-I. and Maginu, K. (1988). "Statistical Neurodynamics of Associative Memory," *Neural Networks*, **1**, 63–73.

Amit, D. J., Gutfreund, H., and Sompolinsky, H. (1985). "Storing Infinite Numbers of Patterns in a Spin-Glass Model of Neural Networks," *Phys. Rev. Lett.*, **55**(14), 1530–1533.

Anderson, J. A. (1972). "A Simple Neural Network Generating Interactive Memory," *Math. Biosci.*, **14**, 197–220.

Anderson, J. A., Silverstein, J. W., Ritz, S. A., and Jones, R. S. (1977). "Distinctive Features, Categorical Perception, and Probability Learning: Some Applications of Neural Model," *Psychol. Rev.*, **84**, 413–451.

Cohen, M. A. (1992). "The Construction of Arbitrary Stable Dynamics in Nonlinear Neural Networks," *Neural Networks*, **5**(1), 83–103.

Cohen, M. A. and Grossberg, S. (1983). "Absolute Stability of Global Pattern Formation and Parallel

NORTHEAST WISCONSIN TECHNICAL COLLEGE
LEARNING RESOURCE CENTER
GREEN BAY WI 54307

Memory Storage by Competitive Neural Networks," *IEEE Trans. Systems, Man, and Cybernetics*, **13**, 815–826.

Dembo, A. (1989). "On the Capacity of the Associative Memories with Linear Threshold Functions," *IEEE Trans. Information Theory*, **IT-35**, 709–720.

Denker, J. S. (1986). "Neural Networks Models of Learning and Adaptation," *Proc. Fifth Annual International Conference Center for Nonlinear Studies*, **21D**, 216–228.

Duda, R. and Hart, P. (1973). *Pattern Classification and Scene Analysis*. Wiley-Interscience, New York.

Fogelman, S. F. (1985). "Contributions à une Théorie de Calcul sur Réseaux," Thése d'Etat, Grenoble.

Golden, R. M. (1986). "The 'Brain-State-in-a-Box' Neural Model is a Gradient Descent Algorithm," *J. Math. Psychol.*, **30**, 73–80.

Greenberg, H. J. (1988). "Equilibria of the Brain-State-in-a-Box (BSB) Neural Model," *Neural Networks*, **1**, 323–324.

Hecht-Nielsen, R. (1986). "Performance Limits of Optical, Electro-Optical, and Electronic Neurocomputers," in *Optical and Hybrid Computing*, H. Szu, ed., *Proc. SPIE*, **634**, 277–306.

Hassoun, M. H. (1988). "Two-Level Neural Network for Deterministic Logic Processing," in *Optical Computing and Nonlinear Materials*, N. Peyghambarian, ed., *Proc. SPIE*, **881**, 258–264.

Hassoun, M. H. (1989a). "Dynamic Heteroassociative Neural Memories," *Neural Networks*, **2**, 275–287.

Hassoun, M. H. (1989b). "Adaptive Dynamic Heteroassociative Neural Memories for Pattern Classification," in *Optical Pattern Recognition*, H-K. Liu, ed., *Proc. SPIE*, **1053**, 75–83.

Hassoun, M. H. and Clark, D. W. (1988). "An Adaptive Attentive Learning Algorithm for Single Layer Neural Networks," *Proc. IEEE Annual Conference on Neural Networks*, San Diego, CA, pp. 431–440.

Hassoun, M. H. and Song, J. (1992). "Adaptive Ho-Kashyap Rules for Optimal Training of Perceptrons," *IEEE Trans. Neural Networks*, **3**(1), 52–61.

Hassoun, M. H. and Watta, P. B. (1991). "Exact Associative Neural Memory Dynamics Utilizing Boolean Matrices," *IEEE Trans. Neural Networks*, **2**(4), 437–448.

Hassoun, M. H. and Youssef, A. M. (1988). "New Recording Algorithm for Hopfield Model Associative Memories," in *Neural Network Models for Optical Computing*, R. Athale and J. Davis, eds., *Proc. SPIE*, **882**, 62–70.

Hassoun, M. H. and Youssef, A. M. (1989). "A High-Performance Recording Algorithm for Hopfield Model Associative Memories," *Opt. Eng.*, **27**, 46–54.

Hassoun, M. H. and Youssef, A. M. (1990). "Autoassociative Neural Memory Capacity and Dynamics," *International Joint Conference on Neural Networks (IJCNN)*, June 17–21, San Diego, Vol. I, pp. 763–769.

Hertz, J., Krogh, A., and Palmer, R. G. (1991). *Introduction to the Theory of Neural Computation*. Addison Wesley, Redwood City, CA.

Ho, Y. and Kashyap, R. L. (1965). "An Algorithm for Linear Inequalities and its Applications," *IEEE Trans. Electronics and Computers*, **EC-14**, 683–688.

Hopfield, J. J. (1982). "Neural Networks and Physical Systems with Emergent Collective Computational Abilities," *Proc. Nat. Acad. Sci. U.S.A.*, **79**, 2445–2558.

Hopfield, J. J. (1984). "Neurons with Graded Response have Collective Computational Properties Like Those of Two-State Neurons," *Proc. Nat. Acad. Sci. U.S.A.*, **81**, 3088–3092.

Horowitz, L. L. and Senne, K. D. (1981). "Performance Advantage of Complex LMS for Controlling Narrow Band Adaptive Arrays," *IEEE Trans. Circuits Systems*, **CAS-28**, 562–576.

Kohonen, T. (1972). "Correlation Matrix Memories," *IEEE Trans. Computers*, **C-21**, 353–359.

Kohonen, T. (1974). "An Adaptive Associative Memory Principle," *IEEE Trans. Computers*, **C-23**, 444–445.

Kohonen, T. (1984). *Self-Organization and Associative Memory*. Springer-Verlag, New York.

Kohonen, T. and Ruohonen, M. (1973). "Representation of Associated Data by Matrix Operators," *IEEE Trans. Computers*, **C-22**, 701–702.

Komlos, J. (1967). "On the Determinant of (0, 1) Matrices," *Studia Scientarum Mathematicarum Hungarica*, **2**, 7–21.

Komlos, J. and Paturi, R. (1988). "Convergence Results in an Associative Memory Model," *Neural Networks*, **3**, 239–250.

Kosko, B. (1987). "Adaptive Bidirectional Associative Memories," *Appl. Opt.*, **26**, 4947–4960.

Kosko, B. (1988). "Bidirectional Associative Memories," *IEEE Trans. Systems, Man, and Cybernetics*, **SMC-18**, 49–60.

Lu, T., Wu, S., Xu, X., and Yu, F. T. (1989). "Optical Implementation of Programmable Neural Networks," in *Optical Pattern Recognition*, H. Liu, ed., *Proc. SPIE*, **1053**, 30–39.

Marcus, C. M. and Westervelt, R. M. (1989). "Dynamics of Iterated-Map Neural Networks," *Phys. Rev. A*, **40**(1), 501–504.

Marcus, C. M., Waugh, F. R., and Westervelt, R. M. (1989). "Associative Memory in an Analog Iterated-Map Neural Network," *Phys. Rev. A*, **41**(6), 3355–3364.

McEliece, R. J., Posner, E. C., Rodemich, E. R., and Venkatesh, S. S. (1987). "The Capacity of the Hopfield Associative Memory," *IEEE Trans. Information Theory*, **IT-33**, 461–482.

Meir, R. and Domany, E. (1987). "Exact Solutions of a Layered Neural Network Memory," *Phys. Rev. Lett.*, **59**, 359–362.

Montgomery, B. L. and Kumar, V. K. (1986). "Evaluation of the Use of the Hopfield Neural Network Model as a Nearest-Neighbor Algorithm," *Appl. Opt.*, **25**(20), 3759–3766.

Nakano, K. (1972). "Associatron: A Model of Associative Memory," *IEEE Trans. Systems, Man, and Cybernetics*, **SMS-2**, 380–388.

Newman, C. (1988). "Memory Capacity in Neural Network Models: Rigorous Lower Bounds," *Neural Networks*, **3**, 223–239.

Okajima, K., Tanaka, S., and Fujiwara, S. (1987). *A Heteroassociative Memory Network with Feedback Connection*, M. Caudill and C. Butler, eds., Proc. IEEE First International Conference on Neural Networks, San Diego, CA, pp. II: 711–718.

Penrose, R. (1955). "A Generalized Inverse for Matrices," *Proc. Cambridge Philosophical Society*, **51**, 406–413.

Personnaz, L., Guyon, I., and Dreyfus, G. (1985). "Information Storage and Retrieval in Spin-Glass like Neural Networks," *J. Physique Lett.*, **46**, L359–L365.

Personnaz, L., Guyon, I., and Dreyfus, G. (1986). "Collective Computational Properties of Neural Networks: New Learning Mechanisms," *Phys. Rev. A*, **34**(5), 4217–4227.

Psaltis, D. and Farhat, N. (1985). "Optical Information Processing Based on an Associative Memory of Neural Nets with Thresholding and Feedback," *Opt. Lett.*, **10**(2), 98–100.

Psaltis, D. and Park, C. H. (1986). "Nonlinear Discriminant Functions and Associative Memories," in *Neural Networks for Computing*, J. S. Denker, ed., *Proc. Am. Inst. Physics*, **151**, 370–375.

Psaltis, D., Wagner, K., and Brady, D. (1987). *Learning in Optical Neural Computers*, M. Caudill and C. Butler, eds., Proceedings of the IEEE First International Conference on Neural Networks, Vol. III, pp. 549–556.

Rao, C. R. and Mitra, S. K. (1971). *Generalized Inverse of Matrices and its Applications*. Wiley, New York.

Rumelhart, D. E., McClelland, J. L., and the PDP Research Group (1986). *Parallel Distributed Processing: Explorations in the Microstructure of Cognition*. MIT Press, Cambridge. Vol. I.

Slansky, J. and Wassel, G. N. (1981). *Pattern Classification and Trainable Machines*. Springer-Verlag, New York.

Stiles, G. S. and Denq, D-L. (1987). "A Quantitative Comparison of Three Discrete Distributed Associative Memory Models," *IEEE Trans. Computers*, **C-36**, 257–263.

Uesaka, G. and Ozeki, K. (1972). "Some Properties of Associative Type Memories," *J. Inst. Electrical and Communication Engineers of Japan*, **55-D**, 323–330.

Venkatesh, S. S. and Psaltis, D. (1985). "Efficient Strategies for Information Storage and Retrieval in Associative Neural Nets," *Workshop on Neural Networks for Computing*, Santa Barbara, CA.

Venkatesh, S. S. and Psaltis, D. (1989). "Linear and Logarithmic Capacities in Associative Neural Networks," *IEEE Trans. Information Theory*, **IT-35**, 558–568.

Venkatesh, S. S., Pancha, G., Psaltis, D., and Sirat, G. (1990). "Shaping Attraction Basins in Neural Networks," *Neural Networks*, **3**(6), 613–623.

Weisbuch, G. and Fogelman, S. (1985). "Scaling Laws for the Attractors of Hopfield Networks," *J. de Physique Lett.*, **46**(14), L-623–L-630.

Widrow, B. and Hoff, M. E. Jr. (1960). "Adaptive Switching Circuits," *IRE WESCON Convention Record*, 96–104.

Widrow, B. and Lehr, M. A. (1990). "30 Years of Adaptive Neural Networks: Perceptron, Madaline, and Backpropagation," *Proc. IEEE*, **78**(9), 1415–1442.

Wigstrom, H. (1973). "A Neuron Model with Learning Capability and its Relation to Mechanism of Association," *Kybernetic*, **12**, 204–215.

Yanai, H. and Sawada, Y. (1990). "Associative Memory Network Composed of Neurons with Hysteric Property," *Neural Networks*, **3**, 223–228.

Youssef, A. M. and Hassoun, M. H. (1989). "Dynamic Autoassociative Neural Memory Performance vs. Capacity," in *Optical Pattern Recognition*, H-K. Liu, ed., *Proc. SPIE*, **1053**, 52–59.

Yu, J., Jonston, A., Psaltis, D., and Brady, D. (1989). "Limitations of Opto-electronic Neural Networks," in *Optical Pattern Recognition*, H-K. Liu, ed., *Proc. SPIE*, **1053**, 40–49.

PART I

BIOLOGICAL AND PSYCHOLOGICAL CONNECTIONS

2

Biological Plausibility of Artificial Neural Networks: Learning by Non-Hebbian Synapses

DANIEL L. ALKON, KIM T. BLACKWELL, THOMAS P. VOGL, and SUSAN A. WERNESS

2.1. INTRODUCTION

Artificial neural networks (ANN) are characterized by a variety of learning rules, computational elements and interconnection strategies (Hopfield, 1982; Rumelhart and McClelland, 1986; Lippmann, 1987; Grossberg, 1987; Kohonen, 1988; Carpenter et al., 1991), most of which rarely approximate the properties of true biological neurons. Furthermore, thus far, none of the artificial networks has demonstrated the pattern recognition capability, the memory capacity, or the robustness of biological networks. While it has been argued that part of the disparity of performance capabilities between biological and artificial networks is due to the disparity in number of neurons, studies of biological neural networks suggest that the basic building blocks of artificial networks and their architecture of synaptic connections are too simple, i.e., artificial networks are not biologically plausible. Close examination of biological networks for important design features has led to biologically plausible learning rules and improved performance of artificial networks (Alkon et al., 1990, 1991; Blackwell et al., 1992; Vogl et al., 1991, 1992, 1993) and may lead to further improvements.

The two biological networks we investigated have been implicated in associative learning (Alkon, 1987). The marine snail, *Hermissenda*, and the rabbit both demonstrate the ability to learn to associate two stimuli, one called the conditioned stimulus (CS) and the other called the unconditioned stimulus (UCS). Naive *Hermissenda* reacts to light by elongating its foot and moving toward the light; it reacts to turbulence by contracting its foot and clinging. However, when the snail experiences repeated pairings of light, which serves as a conditioned stimulus during Pavlovian conditioning, followed promptly by turbulence, which serves as an unconditioned stimulus, it learns the predictive

relationship between the two stimuli and exhibits the learned association by contracting its foot in response to light alone. Similarly, a rabbit may be conditioned to blink its nictitating third eyelid membrane in response to a tone, when the tone is repeatedly paired with and precedes a puff of air to the rabbit's eye.

Electrophysiologic measurements demonstrate that the sensory stimuli used to classically condition the snail elicit action potentials and synaptic potentials along well-defined neuronal pathways (Fig. 2.1). At identified locations within the animal's neural network, e.g., the type B photoreceptor cell, action potentials in response to a light step (CS) are affected by synaptic potentials elicited by rotation (UCS). Synaptic potentials affected by stimulus pairing in turn regulate the action potentials of interneurons and motor neurons that control the learning-induced behavior. Thus it is the interaction of CS- and UCS-elicited signals that is ultimately responsible for learning and memory storage.

We first describe the results of biochemical and biophysical experiments that elucidate the properties of the four neurons of the *Hermissenda* visual-vestibular network, investigate the interactions between the visual (CS) and vestibular (UCS) signals, and illustrate the non-Hebbian nature of learning in the network. This is followed by descriptions of two computer models, each representing a different level of aggregation of the essential features of learning and memory abstracted from *Hermissenda* and rabbit.

One of the computer models is a resistance–capacitance (RC) circuit of the membrane of each of the four relevant neurons in the network of *Hermissenda* (Alkon et al., 1989; Hodgkin and Huxley, 1952; Eccles, 1957; Kuffler and Nicholls, 1976). The model's output, membrane potential, an indirect measure of firing frequency, closely parallels the behavioral and electrophysiologic outputs of *Hermissenda* in response to the same

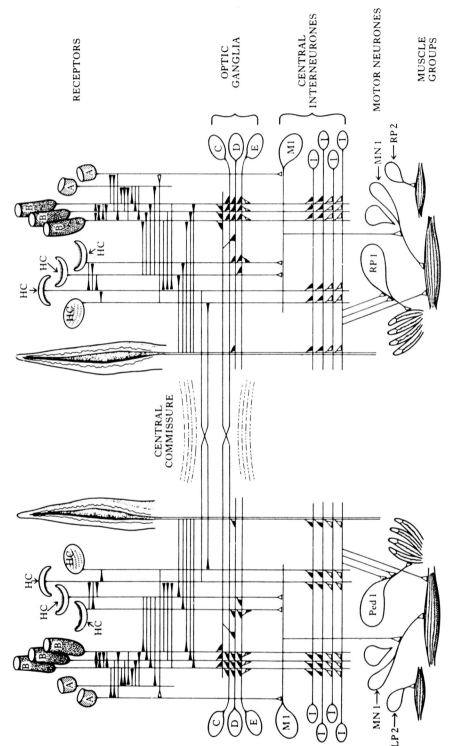

RECEPTORS

OPTIC GANGLIA

CENTRAL INTERNEURONES

MOTOR NEURONES

MUSCLE GROUPS

CENTRAL COMMISSURE

Fig. 2.1. Diagram of *Hermissenda* nervous system. (From Alkon, 1982–3.)

input stimuli presented during and after associative learning. Such correspondence between computer model outputs and biologic outputs suggests that the model captures the features necessary and sufficient for associative learning.

The second model, an artificial neural network called Dystal, is based on the neurobiology of both *Hermissenda* and rabbit and represents a much higher level of aggregation of the principles and mechanisms of associative learning. Biochemical, biophysical, and behavioral experiments on *Hermissenda* and rabbit lead to insights on non-Hebbian learning mechanisms; namely, learning takes place locally, on small areas of the cell membrane in the immediate vicinity of paired (CS and UCS) incoming stimuli. Networks based on these neurobiological insights exhibit desirable mathematical properties and demonstrate excellent performance in face recognition and optical character recognition.

2.2. BIOLOGICAL PLAUSIBILITY OF SYNAPTIC ASSOCIATIVE MEMORY MODELS

2.2.1. Anatomy

Site of Interaction for Associated Signals

For any associative memory process to function, there must exist anatomic sites where signals from two or more temporally associated stimuli interact. In the mammalian brain, the overwhelming preponderance of synapses occur on dendritic branches. A smaller but still large number occur on neuronal somata. Axo-axonal and dendrodendritic synapses do occur but are extremely infrequent. Examples of presynaptic terminals ending on other presynaptic terminals have been identified or inferred in the spinal cord, the olfactory bulb, and the retina, but represent exceptions rather than the rule. The anatomy of brain centers, therefore, dictates that signals from associated stimuli must first interact on and/or within post-synaptic sites, most often the dendritic branches.

Sites of Change

Although interaction must begin postsynaptically, there is no anatomic necessity that the changes caused by that interaction must be confined to postsynaptic sites. The changes might reasonably begin on postsynaptic dendritic branches and/or spines and subsequently affect presynaptic endings. Data from biological networks that change with associative learning have

implicated persistent alteration of postsynaptic K^+ channels (Alkon, 1984; Disterhoft et al., 1986), activation of protein kinase C (Alkon, 1989), and phosphorylation of low-molecular-weight G-protein substrates of protein kinase C (Nelson et al., 1990). Data from models of short-term synaptic change which might occur during memory, such as long-term potentiation (LTP) and long-term depression (LTD), have implicated both postsynaptic and presynaptic changes (Changeux et al., 1987).

2.2.2. Physiology

All data from models of short-term synaptic change (LTP, LTD) have suggested interpretations consistent with either the classical or expanded Hebb synapse models. Whether LTP is produced by single inputs activated at high frequency but in parallel, or by associated inputs, the targets and sites for change have always been conceived of as single synapses.

Recent observations of the most explicitly understood biological associative network, the *Hermissenda* visual-vestibular network (Fig. 2.2), have suggested entirely new interpretations, consistent with more than one site for synaptic change even for a single discrete association. In vitro studies of isolated *Hermissenda* nervous systems allow simulation of in vivo conditions for associative learning. Light steps presented to living animals can also be presented to the visual-vestibular network which, together with the central nervous system, has been isolated from the rest of the animal. Vestibular stimuli can be presented by rotation of the nervous system, vibrating the vestibular organ or statocyst, injection of positive current into statocyst sensory neurons called hair cells, or by pulsed ejection of the hair cells' neurotransmitter, GABA (gamma-amino benzoic acid) onto the terminal receiving branches of the postsynaptic visual sensory neuron called type B cell (Alkon et al., 1992).

Such in vitro studies have recently demonstrated that the signals of the statocyst hair cells interact with the signals of the eye's type B cell by releasing GABA onto terminal branches of the type B cell. When hair cell and type B cell signals are not temporally correlated, i.e., when the light step and vestibular stimulus occur separately, hair cells release GABA to cause inhibition of the type B cells. When the hair cell signals and type B signals are correlated, i.e., when the onset of a light step occurs 0.5–1.0 sec before the onset of a coterminating vestibular stimulus, hair cells initially cause inhibition followed by excitation of the type B cells. If the

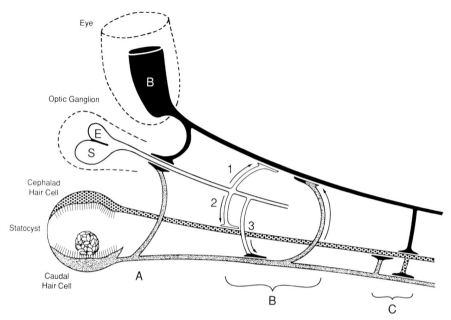

Fig. 2.2. Intersensory integration by the *Hermissenda* nervous system. (A) Convergence of synaptic inhibition from type B and caudal hair cells (part of the statocyst) on the E cell. (B) Positive synaptic feedback onto a type B photoreceptor. 1—direct synaptic excitation; 2—indirect excitation E-cell excites the cephalic hair cell that inhibits the caudal hair cell and thus disinhibits the type B cell; 3—indirect excitation E inhibits the caudal hair cell and thus disinhibits the type B cell. (C) Intra- and intersensory inhibition. Cephalic and caudal hair cells are mutually inhibitory. The type B cell inhibits mainly the cephalic hair cell. All filled endings indicate inhibitory synapses; open endings indicate excitatory synapses; half-filled–half-open endings indicate variable synapse. (From Tabata and Alkon, 1982.)

correlated presentation of visual and vestibular signals is repeated just three times, the release of GABA by hair cells causes pure excitation of the type B cells. Correlated stimulus presentations, therefore, transform classical GABA-mediated synaptic inhibition into persistent GABA-mediated synaptic excitation. With three paired presentations of a 4.0-sec light step with a 3.0-sec GABA application (or current-induced firing of the presynaptic hair cell), this synaptic transformation typically lasts more than 1 hour. Since 90 such pairings occur during a 2-hour training interval on three consecutive days, the transformation can be expected to have long persistence as confirmed by recent observations discussed below.

2.2.3. Mechanisms of Heterosynaptic GABA-Mediated Transformations

Current- and voltage-clamp experiments have helped reveal the mechanisms responsible for the GABA-mediated transformation. Briefly summarized, the hair cells release GABA which combines with postsynaptic GABA-A and

GABA-B receptors on the type B photoreceptor cell. Before pairing, activated GABA-A receptors open Cl^- channels and activated GABA-B receptors open K^+ channels to cause hyperpolarization of the type B cell. When the type B cell is depolarized sufficiently to cause significant elevation of intracellular Ca^{2+} (i.e., at membrane potentials ≥ 20 mV above the resting level of -60 mV), GABA-B receptor activation also causes prolonged reduction of steady-state K^+ conductance. This reduction results when GABA-B receptor activation causes elevation of intracellular Ca^{2+} by releasing it from intracellular stores. Following several pairings of GABA-activated A and B receptors with depolarization-induced elevation of intracellular Ca^{2+}, GABA subsequently no longer activates GABA-A receptors to open Cl^- channels or GABA-B receptors to open K^+ channels. GABA does, however, continue to release Ca^{2+} from intracellular stores. In fact, fura measurements of intracellular Ca^{2+} reveal that, after pairing, Ca^{2+} elevation induced by GABA remains prolonged. This prolonged GABA-mediated release of Ca^{2+} also appears to persist on days following paired training of the living animals but not controls (Ito et al., 1992). Transformation of synaptic inhibition to excitation, then, is due to a shift from a net increase of conductance to a net decrease. Before pairing, GABA increases conductance to Cl^- and K^+ to cause inhibition. After pairing, GABA decreases conductance to K^+ (via elevation of intracellularly released Ca^{2+}) to cause excitation. It is particularly interesting that, to generate the transformed synaptic response, GABA works via a mechanism previously implicated in the production of memory-specific reduction of K^+ currents for the days and weeks during which the animal retains its learned conditioned response (Alkon, 1984, 1989).

2.2.4. Implications of Combinatorial Synaptic Memory

Taken together, the nature of the Ca^{2+}-mediated K^+ currents during memory retention and the Ca^{2+}-mediated transformation of GABA synaptic response suggest an entirely new role for the site of UCS-synaptic input. Heretofore, only one site, that of the CS-synaptic input, was considered to change. Now the UCS site appears to change and this change could, in turn, facilitate the change of the CS site. If during, and for some days after, training, GABA causes depolarization and more prolonged elevation of intracellular Ca^{2+}, it could, indirectly, cause more reduction of K^+ currents at the CS site which is in local proximity. A more prolonged Ca^{2+} elevation might spread via a wave of Ca^{2+}-mediated release of Ca^{2+} from intracellular stores. More K^+ current reduction at the CS site could, in turn, cause more voltage-dependent influx of Ca^{2+}, elevation of intracellular Ca^{2+}, and spread of Ca^{2+} to the UCS site. GABA-induced synaptic transformation, together with learning-induced reduction of voltage-dependent K^+ currents, might functionally link the CS and UCS sites together via prolonged and spreading elevation of intracellular Ca^{2+} and, presumably, the Ca^{2+}-triggered phosphorylation pathways previously identified.

Such CS–UCS linkage, of course, need not be limited to two synaptic sites, i.e., of single CS and UCS inputs. As previously formulated (for pattern recognition), multiple CS sites together with UCS could interact in an analogous manner.

2.3. COMPUTER MODEL

A computer model of the *Hermissenda* visual-vestibular network was developed (Werness et al., 1992, 1993) to further explore the role of synaptic interactions in the type B photoreceptor cell (B cell) and the transformation of the GABA synapse. The computer model included only the essential visual-vestibular network using simplifications derived from previous empirical measurements of the biological networks neurons and their connections. We examined the responses of the network model to the same stimuli used to train the snails. Quantitative comparisons of animal behavior with the output from computer simulations provided validation of the model. The close correspondence between computer network responses, and *Hermissenda* behavioral and electrophysiologic responses (to the same training stimuli) offers strong support for the concept that the essential network modeled is indeed responsible for the associative learning behavior previously observed in the animal (Werness et al., 1992, 1993).

2.3.1. General Neuron Model

Basic Model

A resistance–capacitance (RC) circuit is used to model the membrane of each of the four neurons in the network (Alkon et al., 1989; Hodgkin and Huxley, 1952; Eccles, 1957; Kuffler and Nicholls, 1976). The four neurons were selected as a critical subset of the snail's visual-vestibular network on the basis of earlier electrophysiologic studies. Differential equations describing the RC circuit behavior are approximated by difference equations. Membrane resistance, capacitance, currents, and time constants of the four neurons are represented as lumped parameters (circuit elements) within the equations.

The membrane potential of each neuron is computed from the differential equation describing the RC circuit model of the neuron (illustrated in Fig. 2.3):

$$C\frac{dV_m(t)}{dt} + \frac{V_m(t)}{R} = \sum_{k=1}^{NS} \text{sign}_k\, I_k(t) + I_{\text{reb}}(t)$$
$$+ I_{\text{stim}}(t) + I_{Ca^{2+}}(t) \quad (1)$$

where C is the membrane capacitance, $V_m(t)$ is the membrane potential, $I_{\text{eq}}(t)$ is the equivalent current and R is the membrane resistance; NS is the number of synaptic currents, $I_k(t)$, from other neurons; NS is different for each neuron. $I_{\text{reb}}(t)$ is a time-varying postinhibitory rebound current found in the hair cells and E cell. The B cell has a leakage current but no rebound current; therefore, in the equation for the B cell, $I_{\text{reb}}(t)$ is replaced by $I_{\text{leak}}(t)$. $I_{\text{stim}}(t)$ is a stimulus-induced current and $I_{Ca^{2+}}(t)$ is an inward calcium current which occurs only in the B cell. Note that each

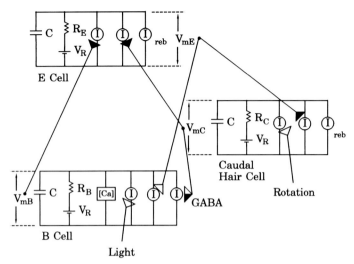

Fig. 2.3. Resistance–capacitance circuit model for three of the neurons in the model: B cell, E cell, and caudal hair cell. The cephalad hair cell is not shown for clarity, but differs from the caudal hair cell only in its synaptic connectivity. Specifically, the cephalad hair cell receives inhibitory synaptic input from, and sends inhibitory synaptic input to, the caudal hair cell and receives excitatory synaptic input from the E cell. Synaptic inputs, stimulus currents, and rebound currents are all represented by current sources; solid or open triangles illustrate synaptic and stimulus currents being driven by stimuli or synaptic input. V_m is membrane potential, V_r is the membrane's resting potential, C is membrane capacitance, and R is membrane resistance. (From Werness et al., 1993.)

neuron is considered to be a single isopotential compartment.

Model Inputs

Inputs to the model are simulated as excitation or a positive increase in membrane potential of the sensory receptors. Light to the B cell is simulated as an increase in $I_{stim}(t)$ in the B cell equation; similarly, orbital rotation to the statocyst hair cells is simulated as an increase in $I_{stim}(t)$ in the caudal and cephalad hair cell equations.

Model Outputs

The output of a neuron, firing frequency, is a function of its membrane potential. In the model, for simplicity, firing frequencies of the network neurons are assumed to be linearly related to membrane potential when it exceeds a threshold: the output of a neuron equals its membrane potential when it exceeds a threshold and is zero otherwise. The threshold was set equal to the biological value of membrane potential at which action potentials of individual neurons were triggered, taking the background firing of each neuron into account. By using membrane potential rather than action potentials in the model,

accurate simulation of the features relevant to learning can be accomplished using relatively long (0.1-sec) time steps. The membrane potential of the B cell is taken as the output of the network as a whole (Alkon, 1987).

Synaptic Connections in the Network

Previously established interactions among the four neurons of the network (cf. Alkon, 1987) are ilustrated in Figs. 2.2 and 2.3:

1. The B cell inhibits the cephalic hair cell and the S-E optic ganglion cell.
2. The caudal hair cell inhibits the cephalic hair cell; caudal hair cell input to the B cell may be either excitatory or inhibitory (cf. Matzel and Alkon, 1991), depending on the membrane potential of the B cell. When the B cell receives GABA-mediated synaptic input from the caudal hair cell and the B cell is *not* depolarized, this caudal synaptic input is exclusively inhibitory. In contrast, depolarized B cells receive an early inhibition followed by excitation from the caudal hair cell. After three presentations of caudal hair cell input while the B cell is depolarized, the synaptic effect of the hair cell becomes purely excitatory (cf. Matzel and Alkon, 1990).

3. The cephalic hair cell inhibits the caudal hair cell.
4. The S-E optic ganglion cell excites the B cell and inhibits the caudal hair cell.
5. Postinhibitory rebound: Synaptic inhibition of all neurons except the B cell is followed by transient excitation within the neuron. This transient excitation is due to rebound depolarization thought to be caused by transient activation of previously inactivated ionic fluxes (Hodgkin, 1938; Hodgkin and Huxley, 1952).

2.3.2. Biophysical Events in the B Cell

Two additional biophysical properties of the B cell are included in the model because the B cell has been causally implicated in the snail's learning (Alkon, 1987) and because here its membrane potential and membrane resistance are considered as the output of the four-neuron network. These properties are a voltage-dependent calcium current and secondary changes of membrane resistance following voltage-dependent Ca^{2+} influx and elevation of intracellular Ca^{2+}.

Voltage-dependent Calcium Current

The inward Ca^{2+} current is modeled in detail because experimental evidence (Alkon, 1987) indicates the Ca^{2+} current critically determines the level of elevation of intracellular Ca^{2+}, which

in turn is directly responsible for the long-lasting depolarization (LLD) of the B cell following the light step. By experimentally determining parameter values of Hodgkin–Huxley activation equations, Alkon and Sakakibara (1985) found that $I_{Ca^{2+}}$ activation kinetics can be modeled by

$$I_{Ca^{2+}}(n) = \frac{g_{Ca}(\max) \cdot (V_m(n) - E_{Ca})}{1 + \exp[(-V_m(n) - 5)/10]} \quad (2)$$

where E_{Ca} is the equilibrium potential for calcium, and $g_{Ca}(\max)$ is the maximum conductance.

B Cell Resistance Changes

We model the combined effect of the remaining, consequential B cell currents (I_A and $I_{Ca^{2+}K^+}$) as resistance changes for computational simplicity. Some resistance changes occur during light stimulation; other resistance changes are observed during paired light-rotation stimulation; a third type of resistance change occurs after pairing.

The three phases of resistance changes (Fig. 2.4) due to light stimulation (Alkon, 1979; Alkon and Sakakibara, 1985) are modeled with three equations. At the beginning of a light step (Phase I in Fig. 2.4), the resistance of the photoreceptor's membrane drops from its resting value of 8×10^7 ohms to R_{shunt}.

During the light stimulus and after the initial drop in resistance (phase II in Fig. 2.4), the

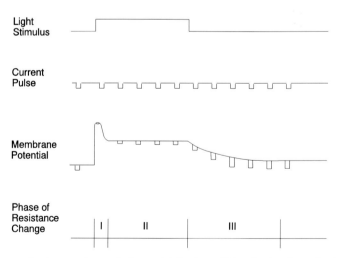

Fig. 2.4. Three phases of resistance change during and following a light stimulus. Immediately after light onset, phase I, the membrane resistance drops to an extremely low value (producing a small membrane potential change in response to a current pulse). During phase II, the membrane resistance increases slowly, hence the large membrane potential changes. During phase III, the membrane resistance increases rapidly to its initial value.

membrane resistance of the B cell gradually increases from the minimum value of R_{shunt}. This increase is modeled by

$$R(n) = R_l - (R_l - R(n-1)) \exp(-T/\tau_{lt}) \quad (3)$$

R_l is the maximum resistance attainable during this phase, and is less than the resting resistance value of 8×10^7 ohms. The rate at which the resistance increases during light is relatively slow and is governed by the time constant, τ_{lt}, whose magnitude is of the order of a few seconds.

For 0.5 sec (5 time steps) after the light stimulus is removed (phase III in Fig. 2.4), the photoreceptor membrane resistance rapidly recovers to a value slightly less than the prestimulus value:

$$R(n) = R_{bl} - [R_{bl} - R(n-1)] \exp(-T/\tau_{rf}) \quad (4)$$

where R_{bl} is the membrane resistance value before the light stimulus, and τ_{rf} is a time constant for the increasing resistance.

When B cell depolarization is paired with GABA stimulation from the caudal hair cell, there is a prolonged elevation of intracellular Ca^{2+} and a consequent decrease in potassium currents (Alkon et al., 1982, 1985; Alkon, 1984, Collin et al., 1988). The decrease in potassium currents is modeled as an increase in membrane resistance using the following equations. After the light stimulus is removed, if the B cell is sufficiently depolarized (due to synaptic inputs arising from paired rotation and light), then the membrane resistance continues to increase slowly during the time the B cell is depolarized:

$$R(n) = R_{max}(n) - [R_{max}(n) - R(n-1)]$$
$$\times \exp(-T/\tau_{rise}) \quad (5)$$

The value of τ_{rise} was set so that the time course of membrane resistance changes agreed with that reported (Matzel et al., 1990; Grover and Farley, 1987). $R_{max}(n)$ is the maximum resistance that the membrane will support at time n:

$$R_{max}(n) = R_r K_{tr}(V_{cm}(n) - V_r) + R_r \quad (6)$$

where K_{tr} is a proportionality constant, R_r is the constant resting membrane resistance (R_{max} is reset to R_r at the beginning of each light step), and $V_{cm}(n)$ is a moving average of membrane potential over a 0.9-sec time interval:

$$V_{cm}(n) = \frac{\sum_{k=k_{start}}^{k_{end}} V_m(n-k)}{k_{end} - k_{start} + 1} \quad (7)$$

2.3.3. Simulation Results

In order to test the computer model, a simulation of the same training regimen as used to train *Hermissenda* is employed: repeated presentations of a simulated light stimulus (of 3 sec), followed by a simulated rotation stimulus (of 2 sec), with an interstimulus interval of 1 sec. The simulated B cell membrane resistance increases with the number of pairings (Fig. 2.5b); this increase is comparable to the increase in B cell membrane resistance measured in *Hermissenda* (Fig. 2.5a). As in *Hermissenda*, the model's membrane resistance in response to light does *not* increase with the number of stimulus presentations when trained with random stimuli.

Following training, the photoreceptor in the model exhibits an enhanced long-lasting depolarization that is quantitatively similar to that observed in *Hermissenda* (Crow and Alkon, 1978; West et al., 1982). Figure 2.6 illustrates the computer model's photoreceptor membrane potential in response to a simulated light stimulus, following training with 150 pairs of light and rotation. As in *Hermissenda*, the model does not exhibit an enhanced LLD when presented with random (unpaired) stimuli, or when no stimuli are presented.

In *Hermissenda*, the degree of learning is a function of the amount of correlation between the signals (Farley and Alkon, 1985a, b). Maximal learning occurs when *Hermissenda* is presented with paired signals (light followed by rotation) within a critical interstimulus interval (ISI); i.e., learning occurs when the signals are temporally correlated. A small amount of learning occurs when the signals are sometimes correlated, as would occur when the training regimen contains both paired and unpaired signal presentations. No learning occurs when the signals are uncorrelated: when the CS (light) or the UCS (rotation) are presented by themselves, or when the CS and UCS are both presented but at random times. Such behavioral results in animals are reported in terms of distances that animals move toward light or the amount by which they contract their foot.

To compare features of the learning-induced behavioral changes with the model's simulation of electrophysiologic changes, learning ratios are used. The learning ratio for the behavioral studies is defined as the ratio of the distance moved before training to the distance moved after training. For the model, this ratio is defined as the membrane resistance after training to the resistance with no training. When no learning occurs, both the model's and *Hermissenda*'s learning ratio are less than or equal to 1; when learning occurs, the ratio is greater than 1.

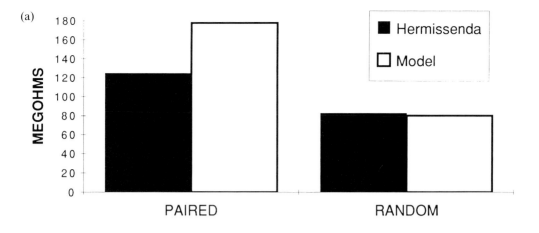

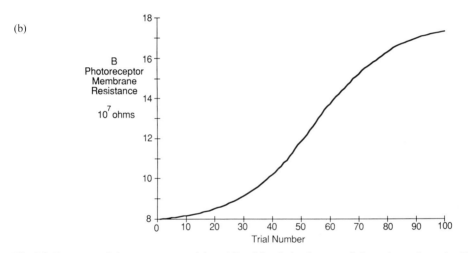

Fig. 2.5. Response of the computer model to 150 trials of simultaneous light and rotation stimuli. In animal experiments, the maximum speed of the rotator was obtained approximately 1 sec following the onset of the light stimulus. Thus, computer experiments were done with the rotation onset following the light onset by 1 sec. (a) Comparison of *Hermissenda* B cell resistance values to those obtained by the computer model. The animal resistance values were obtained by regression of the voltage responses of West et al. (1982) to injections of positive current (0.1, 0.2, 0.3, 0.4, nano-amperes). Resistance values obtained from the conditioning paradigm are compared to two control groups: naive and random light and rotational stimuli. Twelve animals were in each animal group shown. (b) Acquisition curve of the model for the training regimen.

One set of simulation experiments examines the model's sensitivity to ISI (interstimulus interval) and compares it to that of *Hermissenda*. In Lederhendler and Alkon (1989), the CS duration was held constant at 3 sec while the rotation duration was varied to produce ISIs of 1.5, 1.0, 0.5, 0, and −2.0 sec. Fifty pairs were presented with an intertrial interval of 120 sec. Significant learning occurred only for the 1 sec ISI (Fig. 2.7), thus indicating that *Hermissenda*'s associative learning is quite sensitive to ISI.

Simulation of this experiment verifies the computer model's sensitivity to ISI. As illustrated in Fig. 2.7, simulation results are quantitatively comparable to the behavioral results. As *Hermissenda*, the model learns most effectively with a 1 sec ISI and reproduces the steep ISI curve (Fig. 2.7).

(a)

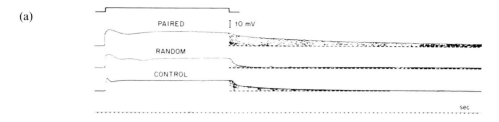

(b)

(c)

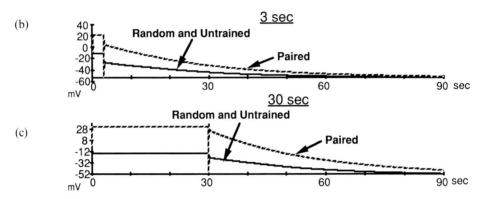

Fig. 2.6. (a) Responses to a light step of type B cell from paired, random, and control groups. Shaded areas indicate LLD (long-lasting depolarization) following the light step (monitored by top trace). Note that the paired LLD (i.e., from conditioned animals) is clearly larger than random and control LLD. Impulse and synaptic activity were eliminated by axotomy. (From West et al., 1982.) (b) The model's B cell membrane potential in response to a 3 sec light stimulus. Solid curve shows response before training with paired stimuli (3 sec light and 2 sec rotation) or after training with unpaired stimuli. Dashed curve shows response after training with 150 paired stimuli. Same training regimen used by Matzel et al. (1990). (c) The model's B cell membrane potential in response to a 30 sec light stimulus. Solid curve shows response before training with paired 30 sec stimuli, or after training with unpaired stimuli. Dashed curve shows response after training with 150 pairs of 30 sec stimuli. Same training regimen used by Crow and Alkon (1978).

Similarly to *Hermissenda*, the computer model does not exhibit learning when presented with a single stimulus or when the light and rotation stimuli are randomly presented in time. Furthermore, when either light or rotation is used repeatedly as the sole stimulus, there is no cumulative change of B cell membrane resistance or membrane potential following stimulation. If extra presentations of either the CS (light) or the UCS (rotation) are interspersed among presentations of paired stimul during training of the animal, the effectiveness of the training trials is reduced.

Figure 2.8 illustrates the degradation of learning due to the addition of extra unpaired stimuli to the training regimen. The first set of bars shows the learning ratio for both *Hermissenda* and the model when trained with stimulus pairs only. The second set of bars shows the learning ratios when extra light trials are presented. Degradation of model learning is quantitatively similar to that in *Hermissenda*.

Thus, we have shown that a lumped-parameter computer model of a four-neuron circuit in the invertebrate marine mollusk, *Hermissenda crassicornis*, reproduces the observed learning-induced changes in behavior and electrophysiology. It is important to note that the model's learning behavior is purely associative; it explicitly lacks any nonassociative components. The model reproduces *Hermissenda*'s synaptic connectivity, but not synaptic properties; it reproduces the dynamic effects of selected currents on membrane resistance and membrane potential, but not biophysical properties of the other currents (see Fig. 2.3). This level of modeling of *Hermissenda*'s associative learning suggests that network organization, independent of synaptic, biophysical, and biochemical properties, is responsible for essential features of Pavlovian conditioning. This model has proven useful for enhancing understanding of the underlying biological system, and for extraction of mechanisms to incorporate into

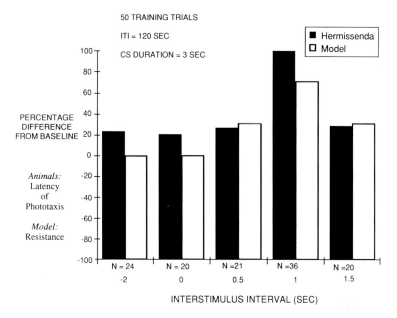

Fig. 2.7. Comparison of model with animal's ISI characteristics in forward-conditioning experiments in which the CS and US coterminate. The light stimulus (CS) was always 3 sec and the duration of the rotation stimulus (US) was varied to create the different ISI conditions. Animal data taken from Lederhendler and Alkon (1989). Each bar represents the percentage change from baseline conditions. In the animal studies, the measured variable was latency before starting locomotion toward a light source. In the model studies, the measured variable was B cell membrane resistance.

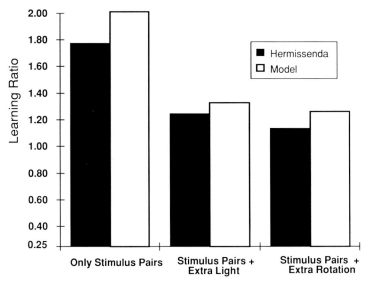

Fig. 2.8. Degradation of associative learning with the addition of extra unpaired stimuli to the training regimen. (30 sec stimuli.) Shown are ratios of post-training membrane resistances to untrained membrane resistance. (1) Pairs only: 50 pairs (animals), 100 pairs (model); (2) extra light trials: 50 pairs + 25 lights (animals), 100 pairs + 50 lights (model); (3) extra rotation trials: 50 pairs + 25 rotations (animals), 100 pairs + 50 rotations (model). Animal resistance measurements from Farley (1987a, b). There were 6 animals in each treatment group. (From Werness et al., 1993)

artificial neural networks (Alkon et al., 1990, 1991; Blackwell et al., 1992; Vogl et al., 1991, 1992, 1993).

2.4. DYSTAL

The results of research on *Hermissenda*, described above, together with results from Olds et al. on the hippocampus of associatively trained rabbits (1989) and on water-maze trained rats (1990), suggest that learning takes place locally, on small areas of the dendritic cell membrane in the immediate vicinity of paired (CS and UCS) incoming stimuli, areas that we refer to as "patches." In marked contrast to earlier theories of neuronal learning, it appears that initial learning depends on the local interaction of a multiplicity of synapses and less, if at all, on the output of the postsynaptic neuron. We have incorporated a non-Hebbian learning rule based on these neurobiologic insights into an ANN that we call Dystal (DYnamically STable Associative Learning), which has a number of advantageous properties.

2.4.1. Non-Hebbian Learning Algorithm

Dystal learns to associate a CS input pattern with a UCS input pattern and demonstrates the learned association by output of the UCS pattern in response to the CS pattern alone. Just as *Hermissenda* and the rabbit have two separate inputs (vestibular and visual for *Hermissenda*, auditory and sensory for rabbit), Dystal has two separate input pathways, named the CS and the UCS. In addition to the two input pathways, a single layer Dystal network [equivalent to a three-layer feed-forward network in the context of Kolmogorov's theorem (Hecht-Nielsen, 1987)] consists of output elements and their patches. Each individual output unit (illustrated in Fig. 2.9) receives input from a receptive field (a subset of CS inputs) and one (scalar) component of the UCS input vector via its patches. All the patterns learned by Dystal are, collectively, stored in the entire set of patches; however, each patch individually stores only an association between a group of similar CS inputs and their associated UCS component. As illustrated, a patch is composed of (1) a patch vector, the running average of CS input patterns that share similar UCS values; (2) the running average of similar UCS values; and (3) a weight that reflects the frequency of utilization of the patch. The number of output units is equal to the number of components in the UCS vector. The UCS vector can be a classification vector, or of the same size as the CS input to permit pattern completion (reconstruction). Note that both the CS and the UCS inputs can be noisy; for pattern completion the UCS can, in fact, be drawn from the same population as the CS inputs.

Prior to training, no patches exist. The number and content of the patches are determined during training and are a function of the content of the

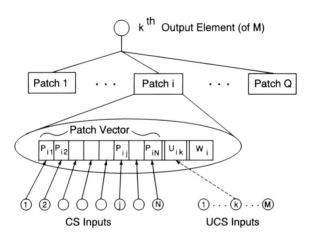

Fig. 2.9. The structure of the Dystal output neuron. There is one output neuron for each component of the UCS vector. The patch structure for one of the patches of an output neuron is shown. In the present implementation, the CS inputs, which may be a subset (receptive field) of the entire input pattern, remain the same for each patch of an output neuron.

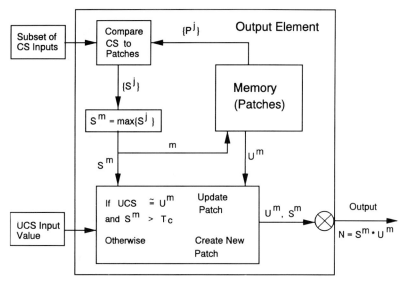

Fig. 2.10. An overview of the Dystal algorithm, which is carried out independently for each output unit. The CS input pattern is compared to the patch vector of each patch on the neuron. (P^j is the patch vector of the jth patch; S^j is the similarity between the CS inputs and P^j). The patch, m, whose patch vector is most similar to the input CS is selected. If the measure of similarity (Pearson's R in the present implementation) is sufficiently high (i.e., $> T_c$) and the stored UCS value, U^m, is close to the input UCS value, the patch is considered as matched and both the patch vector and the UCS value are updated by the use of a running average. If the similarity is too low, a new patch is created using the input CS and UCS values. The output of the neuron is the product of the stored UCS, U^m, and the similarity measure, S^m ($0 > S^m > 1$). A detailed description of the algorithm may be found in Blackwell et al (1992) and Vogl et al. (1992, 1993).

training sets and of global network parameters. The training procedure is described in Fig. 2.10: The first paired CS–UCS presentation causes each output unit to create one patch. At each subsequent presentation, each output unit compares its CS input with its patch vectors and selects the patch whose vector is most similar. At present, Pearson's R correlation is our measure of similarity (Irvine et al., 1993). If the stored UCS is different from the input UCS, or the CS vector is not sufficiently similar to the input CS (less than some threshold), a new patch is created and it becomes the active patch. Otherwise, the selected patch is updated and it becomes the active patch. The output equals the similarity times the stored UCS of the most similar patch. It should be noted that all decisions required in the training process are made on the basis of information available locally at the level of the patches of each output element irrespective of whether the output element fires or not. Consequently, neither local nor global feedback is required and iterations in the learning loop (associated with nonlinear searches) are avoided. Thus, learning is monotonic and rapid. We have not encountered a problem that requires more

than 10 presentations of the training set to reach fully trained equilibrium; for most problems 2–4 presentations suffice, and the number of iterations is independent of the number of patterns in the training set, or the size of the patterns (Barbour et al., 1992). Performance is insensitive to the global parameter (patch matching threshold), and default values provide good performance on a wide variety of problems. The only difference between training and testing is that during testing the CS inputs are *not* accompanied by a UCS input. Therefore, training and testing can be intermingled as required.

2.4.2. Results

Dystal has been applied both to pattern classification (character recognition) and to pattern completion (image restoration) tasks. Dystal is remarkably insensitive to changes in its global parameters. Satisfactory performance on all three of the tasks described below can be obtained using identical global parameters; fine tuning is accomplished by univariate adjustment of a single global parameter, the patch creation threshold.

Pattern Completion

Consider the problem of face recognition and reconstruction. Figure 2.11 shows 16 images, 4 different facial expressions of each of four different persons. Each image consists of 2496 pixels (64 high by 39 wide), 256 shades of gray. The images were chipped from digitized photographs, but no preprocessing was utilized. The four pictures (in Fig. 2.11) of each of the four subjects are used as the training set and provide both CS and UCS input patterns. The testing set, five additional images of each individual, is shown in the first and third rows of Fig. 2.12. Therefore, for the task of image reconstruction, CS input array, UCS input array and output array each contain 2496 elements. Each of the output elements receives input from one UCS component and a 17×17 pixel subfield of the CS. During training, one of the four images of a subject is randomly chosen to be the CS and a second random choice from the set of four is used as the UCS. This is repeated for each of the four images of the four subjects. Four such sets of pairings of each subject constitute the entire training procedure, which requires approximately 50 minutes on a Silicon Graphics Personal Iris workstation. Two experiments were performed to test how well Dystal had learned each individual.

In the first experiment, five expressions of each subject not included in the training set were presented to the trained network. The output from the network for the five images in the testing set is shown in the second and fourth rows of Fig. 2.12. Note that each output image for each individual is a composite image of the four different expressions of that individual. Note that the facial expression of the output is independent of facial expression of the input.

Fig. 2.11. The four images of four individuals used to train the network in the experiments on face recognition. Note the difference in facial expressions across each row. Each image is 64 pixels high by 39 pixels wide. To achieve image completion (rather than classification) the number and arrangement of output neurons is the same as the CS input. The receptive field for each output neuron is 17×17 pixels. (Photograph courtesy of Mont Blackwell.)

Fig. 2.12. The response of the network to the 20 testing images whose expressions are different from those of the training set (Fig. 2.11). Note that the expressions of the output do not reflect the expressions of the test images; rather, they represent a composite of the images in the training set for each individual. Neutral gray areas in the output are areas where the similarity of the input CS to the patch vector is less than the 0 (i.e., there is no correlation between the two). In that case, the output of that neuronal output element is neural gray. (Photograph courtesy of Mont Blackwell.)

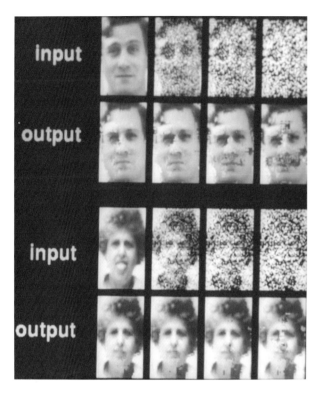

Fig. 2.13. The response of the network to two of the test images with various amounts of added Gaussian noise. In each pair of rows, the top row is the same image with the amount of additive noise increasing from left to right. Note that even in the cases with the most noise, the right-most images, the reconstruction is clearly recognizable despite increasing areas of neutral gray (unrecognized) or other misrecognized areas in the output. (Photograph courtesy of Mont Blackwell.)

In the second experiment (illustrated in Fig. 2.13) Gaussian noise was added to each of the images by adding to each pixel a random value drawn from Gaussian distributions whose variances are 0.02, 0.08, and 0.25 for the 2nd, 3rd, and 4th images from the left, respectively. Figure 2.13 shows that Dystal can reconstruct recognizable images even when the input images are so degraded by noise as to be barely recognizable as faces by human observers. We emphasize that the input images, degraded by nonlinear noise, were derived from images of facial expressions not included in the training set.

Classification

Pattern classification can be considered a special case of pattern completion, in which the UCS pattern is much smaller than the CS pattern. Two series of classification experiments were performed, the first using hand-written ZIP Codes compiled by the U.S. Postal Service and the second using hand-printed Japanese Kanji characters from a compilation by the Electrotechnical Laboratory in Tsukuba, Japan.

Hand-written ZIP Codes. A set of segmented digits from the USPS ZIP Code data set was provided to us, in rectified format, by the staff of the ERIM Post Office project in Ann Arbor, Michigan. The rectified set of about 11 000 digits, scaled to fit into a 22 × 16 pixel array and rotated to a roughly vertical orientation, was divided into a 2000-digit training set (200 of each digit) and an 8964-digit testing set. The CS input to Dystal consisted either of a randomly chosen subset of CS input pixels or of the coefficients of the first 20 principal components derived from the entirety of the training set (Hyman et al., 1991; Rubner and Schulten, 1990). The UCS input is a 10-element vector all of whose elements but one are zero. The ordinal of the single element of the UCS vector whose value is unity corresponds to the ordinal number of the digit. The network was trained by presenting each member of the training set and its corresponding UCS a single time. During testing the input is considered classified if, and only if, one and only one of the 10 output units has an output activity over 0.5. If the ordinal number of the output unit with the activity over 0.5 matches the ordinal number of the input, the network is considered to have

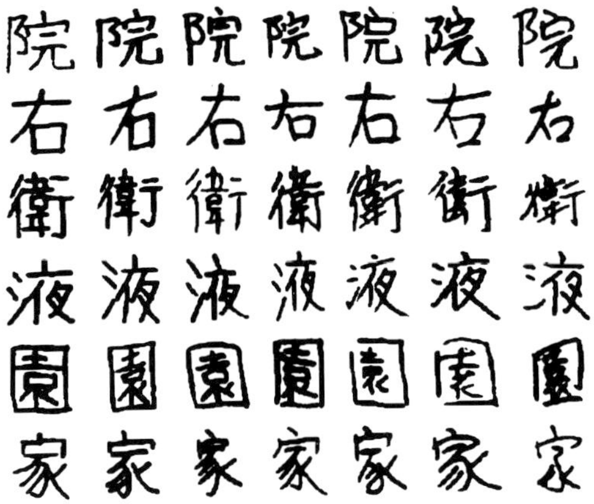

院 院 院 院 院 院 院
右 右 右 右 右 右 右
衛 衛 衛 衛 衛 衛 衛
液 液 液 液 液 液 液
園 園 園 園 園 園 園
家 家 家 家 家 家 家

Fig. 2.14. Examples of Japanese Kanji characters from the database of hand-printed Kanji. Note the variation in handwriting across the rows, in which the columns are different individuals' hand-printing of the same character. These examples are not chosen as extreme cases. Each character is 63 × 64 binary pixels.

correctly classified the input. Testing the network with the 9000-member test set shows that the network classifies in excess of 90% of the inputs and correctly classifies 98% of those inputs that it classified. Using inputs preprocessed by principal component analysis decreases the number of unclassified inputs.

Hand-printed Japanese Kanji. Kanji characters present a significantly more formidable challenge than digits. Not only is each character

much more complicated (see Fig. 2.14) but for a practical system 3000 or more different Kanji must be classified. We review preliminary results from a Dystal-based Kanji recognition system. The database consists of 956 different Kanji characters, 63 × 64 binary pixels in size, with 160 exemplars of each character, each exemplar being written by a different individual. The database, ETL8B2 was provided by the courtesy of Drs. T. Mori and K. Yamamoto of the Electrotechnical Laboratory (ETL), Tsukuba, Japan. The training set consisted of the first 20, 40, 80, or 120 exemplars of each of 20 different characters, and the testing set consisted of the remaining exemplars of those characters. Figure 2.15 presents the classification performance of Dystal on Kanji preprocessed using the PCA pre-processor. Clearly, performance is limited by the size of the training set. When the training set contains 120 exemplars, performance is comparable to that on ZIP code digits.

2.5. SUMMARY

We have developed an overarching description of associative memory formation and retrieval in biological systems derived from neurophysiologic experiments on both invertebrate and vertebrate animals. The data can be explained by a non-Hebbian learning rule in which the inputs to a post-synaptic patch of membrane are modulated by a GABAergic UCS synapse to store associations among inputs that occur in the correct, temporally proximate sequence.

Fig. 2.15. Performance of Dystal on 20 different characters from the Kanji set (crossed), showing the percentage of correctly identified characters as a function of the number of exemplars per character in the training set. The performance of Dystal on ZIP code digits (10 different digits), 200 exemplars per character; circle) is provided for comparison. Dystal's performance on Kanji recognition is comparable to that of digit recognition if trained on enough exemplars, even though twice as many Kanji characters as digits were used.

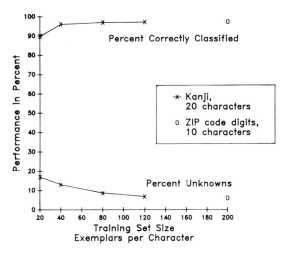

As demonstrated by computed modeling, the specifics of the architecture of the neural network are as important as the intraneuronal mechanisms in facilitating learning and memory storage. Based on these insights, we have constructed an associative neural network, called Dystal, that utilizes these non-Hebbian learning rules and network architecture. The performance, learning speed, and storage capacity of Dystal demonstrate the important contributions that neurobiology can make to the design of artificial neural networks and demonstrate that, even at our present state of knowledge, reverse engineering of principles underlying mechanisms and features of brain function is not only possible but extremely productive. As we learn to concatenate such networks into more elaborate structures, these "third-generation" neural networks [such as Dystal, Faust (Wilson, 1993), and Artmap (Carpenter et al., 1991)] will demonstrate capabilities far beyond those of the currently popular second-generation associative neural network.

ACKNOWLEDGMENTS

The partial support of this research effort by ONR contracts N00014-88-K-0659 and N00014-92-C-0018 and NIH/NINDS contract N01NS02389 is gratefully acknowledged.

REFERENCES

Alkon, D. L. (1979). "Voltage-dependent Calcium and Potassium Ion Conductances, Contingency Mechanism for an Associative Model," *Science*, **205**, 810–816.

Alkon, D. L. (1982–3). "Regenerative Changes of Voltage-Dependent Ca^{2+} and R^+ Currents Encode a Learned Stimulus Association," *J. Physiol. Paris*, **78**, 700–706.

Alkon, D. L. (1984). "Calcium-mediated Reduction of Ionic Currents: a Biophysical Memory Trace," *Science*, **226**, 1037–1045.

Alkon, D. L. (1987). *Memory Traces in the Brain.* Cambridge University Press, Cambridge.

Alkon, D. L. (1989). "Memory Storage and Neural Systems." *Scientific American*, July, pp. 42–50.

Alkon, D. L. and Sakakibara, M. (1985). "Calcium Activates and Inactivates a Photoreceptor Soma Potassium Current," *Biophys. J.*, **48**, 983–995.

Alkon, D. L., Shoukimas, J. J., and Heldman, E. (1982). "Primary Changes of Membrane Currents During Retention of Associative Learning," *Science*, **215**, 693–695.

Alkon, D. L., Sakakibara, M., Forman, R., Harrigan, J., Lederhendler, I., and Farley, J. (1985). "Reduction of Two Voltage Dependent K^+ Currents Mediates

Retention of a Learned Association," *Behav. Neural Biol.*, **44**, 278–300.

Alkon, D. L., Quek, F., and Vogl, T. P. (1989). "Computer Modeling of Associative Learning," in *Advances in Neural Information Processing Systems 1*, D. S. Touretzky, ed. Morgan Kaufman, San Mateo, pp. 419–435.

Alkon, D. L., Blackwell, K. T., Barbour, G. S., Rigler, A. K., and Vogl, T. P. (1990). "Pattern-recognition by an Artificial Network Derived from Biological Neural Systems," *Biological Cybernetics*, **62**, 363–376.

Alkon, D. L., Vogl, T. P., and Blackwell, K. T. (1991). "Artificial Learning Networks Derived from Biological Neural Systems," in *Prentice-Hall Series on Neural Networks*, Part IV, V. Milutinovic, ed. Prentice-Hall, Englewood Cliffs, NJ, pp. 24–46.

Alkon, D. L., Sanchez-Andres, J. V., Ito, P., Oka, K., Yoshioka, T., and Collin, C. (1992). "Long Term Transformation of an Inhibitory into an Excitatory GABA-ergic Synaptic Response," *Proc. Nat. Acad. Sci., USA*, **89**, 11862–11866.

Barbour, G., Blackwell, K., Busse, T., Alkon, D., and Vogl, T. (1992). "Dystal: a Self-organizing ANN with Pattern Independent Training Time," *Proc. IJCNN '92 Baltimore*, Vol. IV, pp. 814–819.

Blackwell, K. T., Vogl, T. P., Hyman, S. D., Barbour, G. S., and Alkon, D. L. (1992). "A New Approach to the Classification of Hand-written Characters," *Pattern Recognition*, **25**, 655–666.

Carpenter, G. A., Grossberg, S., and Reynolds, J. H. (1991). "Artmap: Supervised Real-time Learning and Classification of Nonstationary Data by a Self-organizing Neural Network," *Neural Networks*, **4**, 565–588.

Changeux, J. P., Konishi, M., and Baudry, M. (eds.) (1987). *The Neural and Molecular Bases of Learning: Report of the Dahlem Workshop on the Neural and Molecular Bases of Learning, Berlin 1985.* Wiley, New York.

Collin, C. H., Ikeno, J. F., Harrigan, I., Lederhendler, I. I., and Alkon, D. L. (1988). "Sequential Modification of Membrane Currents with Classical Conditioning," *Biophys. J.*, **54**, 955–960.

Crow, T. J. and Alkon, D. L. (1978). "Retention of an Associative Behavioral Change in *Hermissenda*," *Science*, **201**, 412–414.

Disterhoft, J. F., Coulter, D. A., and Alkon, D. L. (1986). "Conditioning-specific Membrane Changes of Rabbit Hippocampal Neurons Measured in Vitro," *Proc. Nat. Acad. Sci., U.S.A.*, **83**, 2733–2737.

Eccles, J. C. (1957). *The Physiology of Synapses.* Academic Press, New York.

Farley, J. (1987a). "Contingency Learning and Causal Detection in *Hermissenda*. I, Behavior," *Behavioral Neurosci.*, **101**, 13–27.

Farley, J. (1987b). "Contingency Learning and Causal Detection in *Hermissenda*. II, Cellular Mechanisms," *Behavioral Neurosci.*, **101**, 28–56.

Farley, J. and Alkon, D. L. (1985a). "Cellular Analysis of Gastropod Learning," in *Cell Receptors and Cell Communications in Invertebrates*, A. J. Greenberg, ed. Marcel Dekker, New York.

Farley, J. and Alkon, D. L. (1985b). "Cellular Mechanisms of Learning and Information Storage," *Ann. Rev. Psychol.*, **36**, 419–494.

Grover, L. M. and Farley, J. (1987). "Temporal Order Sensitivity of Associative Neural and Behavioral Changes in *Hermissenda*," *Behavioral Neurosci.*, **5**, 658–675.

Grossberg, S. (1987). "Competitive Learning: From Interactive Activation to Adaptive Resonance," *Cognitive Sci.*, **11**, 23–63.

Hebb, D. (1949). *Organization of Behavior*. Wiley, New York.

Hecht-Nielsen, R. (1987). "Kolmogorov's Mapping Neural Network Existence Theorem," *Proc. IJCNN '87*, Vol. III, pp. 11–13.

Hodgkin, A. L. (1938). "The Subthreshold Potentials in a Crustacean Nerve Fiber," *Proc. R. Soc. London Ser. B*, **126**, 87.

Hodgkin, A. L. and Huxley, A. F. (1952). "A Quantitative Description of Membrane Current and its Application to Conduction and Excitation in Nerve," *J. Physiol.*, **117**, 500–544.

Hopfield, J. J. (1982). "Neural Networks and Physical Systems with Emergent Collective Computational Abilities," *Proc. Nat. Acad. Sci., U.S.A.*, **79**, 2254–2258.

Hyman, S. D., Vogl, T. P., Blackwell, K. T., Barbour, G. S., Irvine, J., and Alkon, D. L. (1991). "Classification of Japanese Kanji Using Principal Component Analysis as a Preprocessor to an Artificial Neural Network," *Proc. IJCNN '91*, Vol. I, pp. 233–238.

Irvine, J. M., Blackwell, K. T., Alkon, D. L., and Vogl, T. P. (1993). "Angular Separation in Neural Networks," *J. Artificial Neural Networks*, in press.

Ito, E., Oka, K., Schreurs, B. G., Collin, C., and Alkon, D. L. (1992). "Associative Learning Prolongs GABA-mediated Ca^{2+} Responses of the B Photoreceptors of *Hermissenda*," *Soc. Neurosci. Abstr.*, **18**, 15.

Kohonen, T. (1988). *Self-organization and Associative Memory*. Springer-Verlag, Berlin.

Kuffler, S. W. and Nicholls, J. G. (1976). *From Neuron to Brain*. Sinauer Associates, Inc., Sunderland, MA.

Lederhendler, I. I. and Alkon, D. L. (1989). "The Interstimulus Interval and Classical Conditioning in the Marine Snail *Hermissenda crassicornis*," *Behavioral Brain Res.*, **35**, 75–80.

Lippmann, R. P. (1987). "An Introduction to Computing with Neural Nets," *IEEE ASSP Magazine*, Vol. 4, pp. 4–22, April.

Matzel, L. D. and Alkon, D. L. (1990). "GABA-induced Protein Kinase Activation is Enhanced When Paired with Post-synaptic Depolarization," *Soc. Neurosci. Abstr.*, **16**, 918.

Matzel, L. D. and Alkon, D. L. (1991). "GABA-induced Potentiation of Neuronal Excitability is Enhanced by Post-synaptic Calcium: a Mechanism for Temporal Contiguity in Associative Learning," *Brain Res.*, **554**, 77–84.

Matzel, L. D., Schreurs, B. G., Lederhendler, I. I., and

Alkon, D. L. (1990). "Acquisition of Conditioned Associations in *Hermissenda*: Additive Effects of Contiguity and the Forward Interstimulus Interval," *Behavioral Neurosci.*, **104**, 597–606.

Nelson, T., Collin, C., and Alkon, D. L. (1990). "Isolation of a G Protein That is Modified by Learning and Reduces Potassium Currents in *Hermissenda*," *Science*, **247**, 1479–1483.

Olds, J. L., Anderson, M. L., McPhie, D. L., Staten, L. D., and Alkon, D. L. (1989). "Imaging Memory-specific Changes in the Distribution of Protein Kinase C Within the Hippocampus," *Science*, **245**, 866–869.

Olds, J. L., Golski, S., McPhie, D. L., Olton, D., Mishkin, M., and Alkon, D. L. (1990). "Learning-specific Changes in Rat Hippocampal Protein Kinase C Distribution for Two Water Maze Discrimination Tasks," *J. Neurosci.*, **10**, 3707–3713.

Rubner, J. and Schulten, K. (1990). "Development of Feature Detectors by Self-organization. A Network Model," *Biological Cybernetics*, **62**, 193–199.

Rumelhart, D. E. and McClelland, J. L. (eds.) (1986). *Parallel Distributed Processing: Explorations in the Microstructure of Cognition*. MIT Press, Cambridge.

Tabata, M. and Alkon, D. L. (1982). "Positive Synaptic Feedback in Visual System of Nudibranch Mollusk *Hermissenda crassicornis*," *J. Neurophysiol.*, **48**, 174–191.

Vogl, T. P., Blackwell, K. T., Hyman, S. D., Barbour, G. S., and Alkon, D. L. (1991). "Classification of Hand-written Digits and Japanese Kanji," *Proc. IJCNN '91*, Vol. I, pp. 97–102.

Vogl, T. P., Blackwell, K. T., Barbour, G. S., and Alkon, D. L. (1992). "Dynamically Stable Associative Learning (DYSTAL): A Neurobiologically Based ANN and its Applications," in *Science of Artificial Neural Networks*, D. W. Ruck, ed., *Proc. SPIE*, **1710**, Bellingham, WA, pp. 165–176.

Vogl, T. P., Blackwell, K. T., Irvine, J. M., Barbour, G. S., Hyman, S. D., and Alkon, D. L. (1993). "Dystal: A Neural Network Architecture Based on Biological Associative Learning," in *Progress in Neural Networks*, Vol. III, C. L. Wilson and O. M. Omidvar, eds., in press.

Werness, S. A., Fay, S. D., Blackwell, K. T., Vogl, T. P., and Alkon, D. L. (1992). "Associative Learning in a Model of *Hermissenda crassicornis*. I. Theory," *Biological Cybernetics*, **68**, 125–133.

Werness, S. A., Fay, S. D., Blackwell, K. T., Vogl, T. P., and Alkon, D. L. (1993). "Associative Learning in a Model of *Hermissenda crassicornis*. II. Experiments," *Biological Cybernetics*, in press.

West, A., Barnes, E., and Alkon, D. L. (1982). "Primary Changes of Voltage Responses During Retention of Associative Learning," *J. Neurophysiol.*, **48**, 1243–1255.

Wilson, C. L. (1993). "A New Self-organizing Neural Network Architecture for Parallel Multi-map Pattern Recognition—Faust," in *Progress in Neural Networks*, Vol. III, C. L. Wilson, O. M. Omidvar, eds. Ablex Publishing Co., Norwood, NJ, in press.

3

Sparse Distributed Memory and Related Models

PENTTI KANERVA

3.1. INTRODUCTION

This chapter describes one basic model of associative memory, called the sparse distributed memory, and relates it to other models and circuits: to ordinary computer memory, to correlation-matrix memories, to feed-forward artificial neural nets, to neural circuits in the brain, and to associative-memory models of the cerebellum. Presenting the various designs within one framework will hopefully help the reader see the similarities and the differences in designs that are often described in different ways.

3.1.1. Sparse Distributed Memory as a Model of Human Long-term Memory

Sparse distributed memory (SDM) was developed as a mathematical model of human long-term memory (Kanerva, 1988). The pursuit of a simple idea led to the discovery of the model, namely, that the distances between concepts in our minds correspond to the distances between points of a high-dimensional space. In what follows, "high-dimensional" means that the number of dimensions is at least in the hundreds, although smaller numbers of dimensions are often found in examples.

If a concept, or a percept, or a moment of experience, or a piece of information in memory—a point of interest—is represented by a high-dimensional (or "long") vector, the representation need not be exact. This follows from the distribution of points of a high-dimensional space: Any point of the space that might be a point of interest is relatively far from most of the space and from other points of interest. Therefore, a point of interest can be represented with considerable slop before it is confused with other points of interest. In this sense, long vectors are fault-tolerant or robust, and a device based on them can take advantage of the robustness.

This corresponds beautifully to how humans and animals with advanced sensory systems and brains work. The signals received by us at two different times are hardly ever identical, and yet we can identify the source of the signal as a *specific* individual, object, place, scene, thing. The representations used by the brain must allow for such identification; in fact, they must make the identification nearly automatic, and high-dimensional vectors as internal representations of things do that.

Another property of high-dimensional spaces also has to do with the distances between points. If we take two points (of interest) at random, they are relatively far from each other, on the average: they are uncorrelated. However, there are many points between the two that are close to both, in the sense that the amount of space around an intermediate point—in a hypersphere—that contains both of the two original points is very small. This corresponds to the relative ease with which we can find a concept that links two unrelated concepts.

Strictly speaking, a mathematical space need not be a high-dimensional vector space to have the desired properties; it needs to be a huge space, with an appropriate similarity measure for pairs of points, but the measure need not define a metric on the space.

The important properties of high-dimensional spaces are evident even with the simplest of such spaces—that is, when the dimensions are binary. Therefore, the sparse distributed memory model was developed using long (i.e., high-dimensional) binary vectors or words. The memory is addressed by such words, and such words are stored and retrieved as data.

The following two examples demonstrate the memory's robustness in dealing with approximate data. The memory works with 256-bit words: it is addressed by them, and it stores and retrieves them. At the top of Fig. 3.1 are nine similar (20% noisy) 256-bit words. To help us compare long words, their 256 bits are laid on a 16-by-16 grid, with 1s shown in black. The noise-free prototype word was designed in the shape of a circle within the grid. (This example

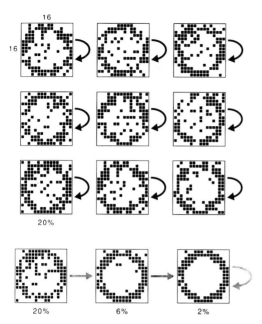

Fig. 3.1. Nine noisy words (20% noise) are stored, and the tenth is used as a retrieval cue.

address, a relatively noise-free eleventh word was retrieved (bottom middle), and with that as the address, a nearly noise-free twelfth word was retrieved (bottom right), which in turn retrieved itself. This example demonstrates the memory's tendency to construct a prototype from noisy data.

Figure 3.2 demonstrates sequence storage and recall. Six words, shaped as Roman numerals, are stored in a linked list: I is used as the address to store II, II is used as the address to store III, and so forth. Any of the words I to V can then be used to recall the rest of the sequence. For example, III will retrieve IV will retrieve V will retrieve VI. The retrieval cue for the sequence can be noisy, as demonstrated at the bottom of the figure. As the retrieval progresses, a retrieved word, which then serves as the next address, is less and less noisy. This example resembles human ability to find a familiar tune by hearing a piece of it in the middle, and to recall the rest. This kind of recall applies to a multitude of human and animal skills.

is confusing in that it might be taken to imply that humans recognize circles based on stored retinal images of circles. No such claim is intended.) The nine noisy words were stored in a sparse distributed memory autoassociatively, meaning that each word was stored with itself as the address. When a tenth noisy word (bottom left), different from the nine, was used as the

3.2. SDM AS A RANDOM-ACCESS MEMORY

Except for the lengths of the address and data words, the memory resembles ordinary computer memory. It is a generalized random-access memory for long words, as will be seen shortly, and its construction and operation can be explained in terms of an ordinary random-access memory. We will start by describing an ordinary random-access memory.

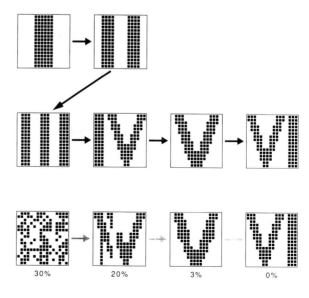

Fig. 3.2. Recalling a stored sequence with a noisy (30% noise) retrieval cue.

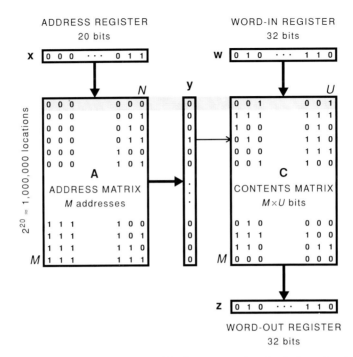

Fig. 3.3. Organization of a random-access memory. The first memory location is shown by shading.

3.2.1. Random-Access Memory

A random-access memory (RAM) is an array of M addressable storage registers or memory locations of fixed capacity. A location's place in the memory array is called the location's *address*, and the value stored in the register is called the location's *contents*. Figure 3.3 represents such a memory, and a horizontal row through the figure represents one memory location. The first location is shown shaded. The addresses of the locations are on the left, in matrix **A**, and the contents are on the right, in matrix **C**.

A memory with a million locations ($M = 2^{20}$) is addressed by 20-bit words. The length of the address will be denoted by N ($N = 20$ in Fig. 3.3). The capacity of a location is referred to as the memory's *word size*, U ($U = 32$ bits in Fig. 3.3), and the capacity of the entire memory is defined conventionally as the word size multiplied by the number of memory locations (i.e., $M \times U$ bits).

Storage and retrieval happen one word at a time through three special registers: the *address register*, for an N-bit address into the memory array; the *word-in register*, for a U-bit word that is to be stored in memory; and the *word-out register*, for a U-bit word retrieved from memory. To store the word **w** in location **x** (the location's address **x** is used as a name for the location), **x**

is placed in the address register, **w** is placed in the word-in register, and a write-into-memory command is issued. Consequently, **w** replaces the old contents of location **x**, while all other locations remain unchanged. To retrieve the word **w** that was last stored in location **x**, **x** is placed in the address register and a read-from-memory command is issued. The result **w** appears in the word-out register. The figure shows (a possible) state of the memory after $\mathbf{w} = 010\ldots110$ has been stored in location $\mathbf{x} = 000\ldots011$ (the word-in register holds **w**) and then retrieved from the same location (the address register holds **x**).

Between matrices **A** and **C** in the figure is an *activation vector*, **y**. Its components are 0s except for one 1, which indicates the memory location that is selected for reading or writing (i.e., the location's address matches the address register). In a hardware realization of a random-access memory, a location's activation is determined by an address-decoder circuit, so that the address matrix **A** is implicit. However, the contents matrix **C** is an explicit array of $2^{20} \times 32$ one-bit registers or flip-flops.

3.2.2. Sparse Distributed Memory

Figure 3.4 represents a sparse distributed memory. From the outside, it is like a random-access

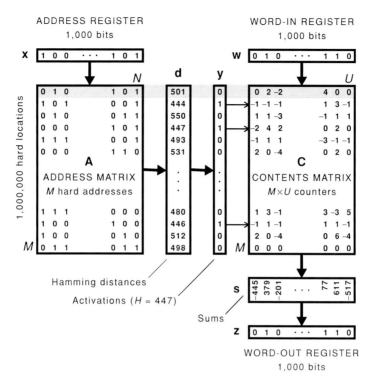

Fig. 3.4. Organization of a sparse distributed memory. The first memory location is shown by shading.

memory: it has the same three special registers—address, word-in, and word-out—and they are used in the same way when words are stored and retrieved, except that these registers are large (e.g., $N = U = 1000$).

Construction. The internal organization of sparse distributed memory, likewise, is an array of addressable storage locations of fixed capacity. However, since the addresses are long, it is impossible to build a hardware location—a *hard location*, for short—for each of the 2^N addresses. (Neither is it necessary, considering the enormous capacity that such a memory would have.)

A memory of reasonable size and capacity can be built by taking a reasonably large sample of the 2^N addresses and by building a hard location for each address in the sample. Let M be the size of the sample: we want a memory with M locations ($M = 1\,000\,000$ in Fig. 3.4). The sample can be chosen in many ways, and only some will be considered here.

A good choice of addresses for the hard locations depends on the data to be stored in the memory. The data consist of the words to be stored and of the addresses used in storing them. For simplicity, we assume in the basic model that

the data are distributed randomly and uniformly (i.e., bits are independent of each other, and 0s and 1s are equally likely, both in the words being stored and in the addresses used for storing them). Then the M hard locations can be picked at random; that is to say, we can take a uniform random sample, of size M, of all N-bit addresses. Such a choice of locations is shown in Fig. 3.4, where the addresses of the locations are given in matrix **A** and the contents are given in matrix **C**, and where a row through the figure represents a hard location, just as in Fig. 3.3 (row $\mathbf{A}_m$ of matrix **A** is the mth hard address, and $\mathbf{C}_m$ is the contents of location $\mathbf{A}_m$; as with RAM, we use $\mathbf{A}_m$ to name the mth location).

Activation. In a random-access memory, to store or retrieve a word with **x** as the address, **x** is placed in the (20-bit) address register, which activates location **x**. We say that the address register *points* to location **x**, and that whatever location the address register points to is activated. This does not work with a sparse distributed memory because its (1000-bit) address register never—practically never—points to a hard location because the hard locations are so few compared to the number of possible addresses (e.g., 1 000 000 hard addresses vs. 2^{1000} possible addresses; matrix **A** is an exceedingly sparse sampling of the address space).

To compensate for the extreme sparseness of the memory, a *set* of *nearby* locations is activated at once, for example, all the locations that are within a certain *distance* from **x**. Since the addresses are binary, we can use Hamming distance, which is the number of places at which two binary vectors differ. Thus, in a sparse distributed memory, the mth location is activated by **x** (which is in the address register) if the Hamming distance between **x** and the location's address A_m is below or equal to a threshold value H [H stands for a (Hamming) radius of activation]. The threshold is chosen so that only a small fraction of the hard locations are activated by any given **x**. When the hard addresses **A** are a uniform random sample of the N-dimensional address space, the binomial distribution with parameters N and $1/2$ can be used to find the activation radius H that corresponds to a given probability p of activating a location. Notice that, in a random-access memory, a location is activated only if its address matches **x**, meaning that $H = 0$.

Vectors **d** and **y** in Fig. 3.4 show the activation of locations by addresses **x**. The distance vector **d** gives the Hamming distances from the address register to each of the hard locations, and the 1s of the activation vector **y** mark the locations that are close enough to **x** to be activated by it: $y_m = 1$ if $d_m \leq H$, and $y_m = 0$ otherwise, where $d_m = h(\mathbf{x}, A_m)$ is the Hamming distance from **x** to location A_m. The number of 1s in **y** therefore equals the size of the set activated by **x**.

Figure 3.5 is another way of representing the activation of locations. The large circle represents the space of 2^N addresses. Each tiny square is a hard location, and its position within the large circle represents the location's addresses. The small circle around **x** includes the locations that are within H bits of **x** and that therefore are activated by **x**.

Storage. To store U-bit words, a hard location has U up–down counters. The range of a counter can be small, for example, the integers from -15 to 15. The U counters for each of the M hard locations constitute the $M \times U$ contents matrix, **C**, shown on the right in Fig. 3.4, and they correspond to the $M \times U$ flip-flops of Fig. 3.3. We will assume that all counters are initially set to zero.

When **x** is used as the storage address for the word **w**, **w** is stored in each of the locations activated by **x**. Thus, multiple copies of **w** are

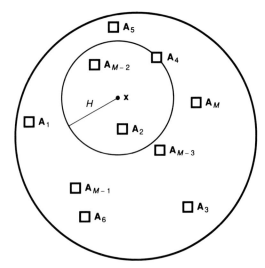

Fig. 3.5. Address space, hard locations, and the set activated by **x**. H is the (Hamming) radius of activation.

stored; in other words, **w** is distributed over a (small) number of locations. The word **w** is stored in, or written into, an active location as follows: Each 1-bit of **w** increments, and each 0-bit of **w** decrements the corresponding counter of the location. This is equivalent to saying that the word **w**′ of -1s and 1s is added (vector addition) to the contents of each active location, where **w**′ is gotten from **w** by replacing 0s with -1s. Furthermore, the counters in **C** are not incremented or decremented past their limits (i.e., overflow and underflow are lost).

Figure 3.4 depicts the memory after the word $\mathbf{w} = 010 \ldots 110$ (in the word-in register) has been stored with $\mathbf{x} = 100 \ldots 101$ as the address (in the address register). Several locations are shown as selected, and the vector $\mathbf{w}' = (-1, 1, -1, \ldots, 1, 1, -1)$ has been added to their contents. The figure also shows that many locations have been selected for writing in the past (e.g., the first location has nonzero counters), that the last location appears never to have been selected, and **w** appears to be the first word written into the selected location near the bottom of the memory (the location contains **w**′). Notice that a positive value of a counter, $+5$, say, tells that five more 1s than 0s have been stored in it; similarly, -5 tells that five more 0s than 1s have been stored in it (provided that the capacity of the counter has never been exceeded).

Retrieval. When **x** is used as the retrieval address, the locations activated by **x** are pooled as follows: Their contents are accumulated (vector addition) into a vector of U sums, $\mathbf{s}$, and the sums are compared to a threshold value 0 to get an output vector, $\mathbf{z}$, which then appears in the word-out register ($z_u = 1$ iff $s_u > 0$; $\mathbf{s}$ and $\mathbf{z}$ are below matrix $\mathbf{C}$ in Fig. 3.4). This pooling constitutes a majority rule, in the sense that the uth output bit is 1 if, and only if, more 1s than 0s have been stored in the uth counters of the activated locations; otherwise, the output bit is 0.

In Fig. 3.4 the word retrieved, $\mathbf{z}$, is the same as, or very similar to, the word $\mathbf{w}$ that was stored, for the following reason: The same $\mathbf{x}$ is used as both storage and retrieval address, so that the same set of locations is activated both times. In storing, each active location receives one copy of $\mathbf{w}'$, as described above; in retrieving, we get back *all* of them, plus a few copies of many other words that have been stored. This biases the sums, $\mathbf{s}$, in the direction of $\mathbf{w}'$, so that $\mathbf{w}$ is a likely result after thresholding. This principle holds even when the retrieval address is not exactly $\mathbf{x}$ but is close to it. Then we get back *most* of the copies of $\mathbf{w}'$.

The ideas of storing multiple copies of target words in memory, and of retrieving the most likely target word based on the majority rule, are found already in the *redundant hash addressing* method of Kohonen and Reuhkala (1978; Kohonen, 1980). The method of realizing these ideas in redundant hash addressing is very different from their realization in a sparse distributed memory.

Retrieval and memory capacity will be analyzed statistically at the end of the next section, after a uniform set of symbols and conventions for the remainder of this chapter has been established. We will note here, however, that the intersections of activation sets play a key role in the analysis, for they appear as weights for the words stored in the memory when the sum vector $\mathbf{s}$ is evaluated.

Random-Access Memory as a Special Case. One more comment about a random-access memory: Proper choice of parameters for a sparse distributed memory yields an ordinary random-access memory. First, the address matrix $\mathbf{A}$ must contain all 2^N addresses; second, the activation radius H must be zero; and, third, the capacity of each counter in $\mathbf{C}$ must be one bit. The first condition guarantees that every possible address $\mathbf{x}$ points to at least one hard location. The second condition guarantees that only a location that is pointed to is activated. The third condition guarantees that when a word is written into a location, it replaces the location's old contents, because overflow and underflow are lost. In

memory retrieval, the contents of all active locations are added together; in this case, the sum is over one or more locations with hard address $\mathbf{x}$. Any particular coordinate of the sum is zero if the word last written (with address $\mathbf{x}$) has a 0 in that position; and it is positive if the word has a 1, so that after thresholding we get the word last written with address $\mathbf{x}$. Therefore, the sparse distributed memory is a generalization of the random-access memory.

Parallel Realization. Storing a word, or retrieving a word, in a sparse distributed memory involves massive computation. The contents of the address register are compared to each hard address, to determine which locations to activate. For the model memory with a million locations, this means computing one million Hamming distances involving 1000 bits each, and comparing the distances to a threshold. This is very time-consuming if done serially. However, the activations of the hard locations are independent of each other so that they can be computed in parallel; once the address is broadcast to all the locations, millionfold parallelism is possible. The addressing computation that determines the set of active locations corresponds to address decoding by the address-decoder circuit in a random-access memory.

In storing a word, each column of counters in matrix $\mathbf{C}$ (see Fig. 3.4) can be updated independently of all other columns, so that there is an opportunity for thousandfold parallelism when 1000-bit words are stored. Similarly, in retrieving a 1000-bit word, there is an opportunity for thousand-fold parallelism. Further parallelism is achieved by updating many locations at once when a word is stored, and by accumulating many partial sums at once when a word is retrieved. It appears that neural circuits in the brain are wired for these kinds of parallelism.

3.3. SDM AS A MATRIX MEMORY

The construction of the memory was described above in terms of vectors and matrices. We will now see that its operation is described naturally in vector–matrix notation. Such description is convenient in relating the sparse distributed memory to the correlation-matrix memories described by Anderson (1968) and Kohonen (1972)—see also Hopfield (1982), Kohonen (1984), Willshaw (1981), and Chapter 1 by Hassoun (this volume)—and in relating it to many other kinds of artificial neural networks. The notation will also be used for describing variations and generalizations of the basic sparse distributed memory model.

3.3.1. Notation

In comparing the memory to a random-access memory, it is convenient to express binary addresses and words in 0s and 1s. In comparing it to a matrix memory, however, it is more convenient to express them in -1s and 1s (also called *bipolar* representation). This transformation is already implicit in the storage algorithm described above: a binary word $\mathbf{w}$ of 0s and 1s is stored by adding the corresponding word $\mathbf{w}'$ of -1s and 1s into (the contents of) the active locations. From here on, we assume that the binary components of $\mathbf{A}$ and $\mathbf{x}$ (and of $\mathbf{w}$ and $\mathbf{z}$) are -1s and 1s, and whether *bit* refers to 0 and 1 or to -1 and 1 will depend on the context.

How is the activation of a location determined after this transformation? In the same way as before, provided that Hamming distance is defined as the number of places at which two vectors differ. However, we can also use the inner product (scalar product, dot product) of the hard address $\mathbf{A}_m$ and the address register $\mathbf{x}$ to measure their similarity: $d = d(\mathbf{A}_m, \mathbf{x}) = \mathbf{A}_m \cdot \mathbf{x}$. It ranges from $-N$ to N ($d = N$ means that the two addresses are most similar—they are identical), and it relates linearly to the Hamming distance, which ranges from 0 to N (0 means identical). Therefore, Hamming distance $h(\mathbf{A}_m, \mathbf{x}) \leq H$ if, and only if, $\mathbf{A}_m \cdot \mathbf{x} \geq N - 2H \, (=D)$. In a computer simulation of the memory, however, the exclusive-or (XOR) operation on addresses of 0s and 1s usually results in the most efficient computation of distances and of the activation vector $\mathbf{y}$.

The following typographic conventions will be used:

s	italic lowercase for a scalar or a function name
S	italic uppercase for a scalar upper bound or a threshold
$\mathbf{v}$	bold lowercase for a (column) vector
v_i	ith component of a vector, a scalar
$\mathbf{M}$	bold uppercase for a matrix
$\mathbf{M}_i$	ith row of a matrix, a (column) vector
$\mathbf{M}_{.,j}$	jth column of a matrix, a (column) vector
$M_{i,j}$	scalar component of a matrix

$\mathbf{M}^{\mathrm{T}}$	transpose of a matrix (or of a vector)
$\cdot$	scalar (inner) product of two vectors: $\mathbf{u} \cdot \mathbf{v} = \mathbf{u}^{\mathrm{T}} \mathbf{v}$
$\square$	matrix (outer) product of two vectors: $\mathbf{u} \square \mathbf{v} = \mathbf{u}\mathbf{v}^{\mathrm{T}}$
$n = 1, 2, 3, \ldots, N$	index into the bits of an address
$u = 1, 2, 3, \ldots, U$	index into the bits of a word
$t = 1, 2, 3, \ldots, T$	index into the data
$m = 1, 2, 3, \ldots, M$	index into the hard locations

3.3.2. Memory Parameters

The sparse distributed memory, as a matrix memory, is described below in terms of its parameters, progressing with the information flow from upper left to lower right in Fig. 3.5. *Sample memory* refers to a memory whose parameter values appear in parentheses in the descriptions below, as in "(e.g., $N = 1000$)."

The *external dimensions* of the memory are given by

N — Address length; dimension of the address space; input dimension (e.g., $N = 1000$). Small demonstrations can be made with N as small as 25, but $N > 100$ is recommended, as the properties of high-dimensional spaces will then be evident.

U — Word length; the number of bits (-1s and 1s) in the words stored; output dimension (e.g., $U = 1000$). The minimum, $U = 1$, corresponds to classifying the data into two classes. If $U = N$, it is possible to store words autoassociatively and to store sequences of words as pointer chains, as demonstrated in Figs. 3.1 and 3.2.

The *data set* to be stored—the *training set* $(\mathbf{X}, \mathbf{W})$—is given by

T — Training-set size; number of elements in the data set (e.g., $T = 10\,000$).

$\mathbf{X}$ — Data-address matrix; T training addresses; $T \times N$ -1s and 1s (e.g., uniform random).

$\mathbf{W}$ — Data-word matrix; T training words; $T \times U$ -1s and 1s (e.g., uniform random). Autoassociative data (self-addressing) means that $\mathbf{X} = \mathbf{W}$, and sequence data means that $\mathbf{X}_t = \mathbf{W}_{t-1}$ ($t > 1$).

The memory's *internal parameters* are

M — Memory size; number of hard locations (e.g., $M = 1\,000\,000$). Memory needs to be sufficient for the data being stored and for the amount of noise to be tolerated in retrieval.

Memory capacity is low, so that T should be 1–5 percent of M [T is the number of stored patterns; storing many noisy versions of the same pattern (cf. Fig. 3.1) counts as storing one pattern, or as storing few].

A Hard-address matrix; M hard addresses; $M \times N$ −1s and 1s (e.g., uniform random). This matrix is fixed. Efficient use of memory requires that **A** correspond to the set of data addresses **X** (see Section 3.8 on SDM research).

p Probability of activation (e.g., $p = 0.000\,445$; "ideally," $p = 0.000\,368$). This important parameter determines the number of hard locations that are activated, on the average, by an address, which, in turn, determines how well stored words are retrieved. The best p maximizes the signal (due to the target word that is being retrieved) relative to the noise (due to all other stored words) in the sum, **s**, and is approximately $(2MT)^{-1/3}$ (see end of this section, where signal, noise, and memory capacity are discussed).

H Radius of activation (e.g., $H = 447$ bits). The binomial distribution or its normal approximation can be used to find the (Hamming) radius for a given probability. For the sample memory, optimal p is $0.000\,368$, so that about 368 locations should be activated at a time. Radius $H = 446$ captures 354 locations, and $H = 447$ captures 445 locations, on the average. We choose the latter.

D Activation threshold on similarity (e.g., $D = 106$). This threshold is related to the radius of activation by $D = N - 2H$, so that $D = 108$ and $D = 106$ correspond to the two values of H given above.

c Range of a counter in the $M \times U$ contents matrix **C** (e.g., $c = \{-15, -14, -13, \ldots, 14, 15\}$). If the range is one bit ($c = \{0, 1\}$), the contents of a location are determined wholly by the most-recent word written into the location. An 8-bit byte, an integer variable, and a floating-point variable are convenient counters in computer simulations of the memory.

The following variables describe the *memory's state and operation*:

x Storage or retrieval address; contents of the address register; N −1s and 1s (e.g., $\mathbf{x} = \mathbf{X}_t$).
d Similarity vector; M integers in $\{-N, -N+2, -N+4, \ldots, N-2, N\}$. Since the similarity between the mth hard address and the address register is given by their inner product $\mathbf{A}_m \cdot \mathbf{x}$ (see Section 3.3.1 on

Notation), the similarity vector can be expressed as $\mathbf{d} = \mathbf{Ax}$.

y Activation vector; M 0s and 1s. The similarity vector **d** is converted into the activation vector **y** by the (nonlinear) threshold function y defined by $y(\mathbf{d}) = \mathbf{y}$, where $y_m = 1$ if $d_m \geq D$, and $y_m = 0$ otherwise. The number of 1s in **y**, $|\mathbf{y}|$, is small compared to the number of 0s ($|\mathbf{y}| \approx pM$); the activation vector is a very sparse vector in a very high-dimensional space. Notice that this is the only vector of 0s and 1s; all other binary vectors consist of −1s and 1s.

w Input word; U −1s and 1s (e.g., $\mathbf{w} = \mathbf{W}_t$).
C Contents matrix; $U \times M$ up–down counters with range c, initial value usually assumed to be 0. Since the word **w** is stored in active location $\mathbf{A}_m$ (i.e., when $y_m = 1$) by adding **w** into the location's contents $\mathbf{C}_m$, it is stored in *all* active locations indicated by **y** by adding the (outer-product) matrix $\mathbf{y} \square \mathbf{w}$ (most of whose rows are 0) into **C**, so that $\mathbf{C} := \mathbf{C} + \mathbf{y} \square \mathbf{w}$, where $:=$ means substitution, and where addition beyond the range of a counter is ignored. This is known as the outer-product, or Hebbian, learning rule.

s Sum vector; U sums [each sum has (at most) $|\mathbf{y}|$ nonzero terms]. Because the sum vector is made up of the contents of the active locations, it can be expressed as $\mathbf{s} = \mathbf{C}^T \mathbf{y}$. The U sums give us the final output word **z**, but they also tell us how reliable each of the output bits is. The further a sum is from the threshold, the stronger is the memory's evidence for the corresponding output bit.

z Output word; U −1s and 1s. The sum vector **s** is converted into the output vector **z** by the (nonlinear) threshold function z defined by $z(\mathbf{s}) = \mathbf{z}$, where $z_u = 1$ if $s_u > 0$, and $z_u = -1$ otherwise.

In summary, storing the word **w** into the memory with **x** as the address can be expressed as

$$\mathbf{C} := \mathbf{C} + y(\mathbf{Ax}) \square \mathbf{w}$$

and retrieving the word **z** corresponding to the address **x** can be expressed as

$$\mathbf{z} = z(\mathbf{C}^T y(\mathbf{Ax}))$$

3.3.3. Summary Specification

The following matrices describe the memory's operation on the data set—the training set $(\mathbf{X}, \mathbf{W})$—as a whole:

D $T \times M$ matrix of similarities corresponding to the data addresses **X**: $\mathbf{D} = (\mathbf{AX}^T)^T = \mathbf{XA}^T$.

Y Corresponding $T \times M$ matrix of activations: $\mathbf{Y} = y(\mathbf{D})$.

S $T \times U$ matrix of sums for the data set: $\mathbf{S} = \mathbf{YC}$.

Z Corresponding $T \times U$ matrix of output words: $\mathbf{Z} = z(\mathbf{S}) = z(\mathbf{YC})$.

If the initial contents of the memory are 0, and if the capacities of the counters are never exceeded, storing the T-element data set yields memory contents

$$\mathbf{C} = \sum_{t=1}^{T} \mathbf{Y}_t \square \mathbf{W}_t = \sum_{t=1}^{T} y(\mathbf{AX}_t) \square \mathbf{W}_t$$

This expression for **C** follows from the outer-product learning rule (see **C** above), as it is the sum of T matrices, each of which represents an item in the data set. However, **C** can be viewed equivalently as a matrix of $M \times U$ inner products $C_{m,u}$ of pairs of vectors of length T. One set of these vectors is the M columns of **Y**, and the other set is the U columns of **W**, so that $C_{m,u} = \mathbf{Y}_{\cdot,m} \cdot \mathbf{W}_{\cdot,u}$, and

$$\mathbf{C} = \mathbf{Y}^{\mathrm{T}}\mathbf{W} = y(\mathbf{AX}^{\mathrm{T}})\mathbf{W}$$

The accuracy of recall of the training set after it has been stored in memory is then given by

$$\mathbf{Z} - \mathbf{W} = z(\mathbf{YC}) - \mathbf{W}$$
$$= z(\mathbf{YY}^{\mathrm{T}}\mathbf{W}) - \mathbf{W}$$

This form is convenient in the mathematical analysis of the memory. For example, it is readily seen that if the T rows of **Y** are orthogonal to one another, $\mathbf{YY}^{\mathrm{T}}$ is a diagonal matrix approximately equal to $p M \mathbf{I}$ (**I** is the identity matrix), so that $z(\mathbf{YY}^{\mathrm{T}}\mathbf{W}) = \mathbf{W}$ and recall is perfect. Notice that the rows of **Y** for the sample memory are nearly orthogonal to one another, and that the purpose of addressing through **A** is to produce (nearly) orthogonal activation vectors for most pairs of addresses, which is a way of saying that the sets of locations activated by dissimilar addresses overlap as little as possible.

3.3.4. Relation to Correlation-Matrix Memories

The $M \times U$ inner products that make up **C** are correlations of a sort: they are unnormalized correlations that reflect agreement between the M variables represented by the columns of **Y**, and the U variables represented by the columns of **W**. If the columns were normalized to zero mean and to unit length, their inner products would equal the correlation coefficients used commonly in statistics. Furthermore, the inner products of activation vectors (i.e., unnormalized

correlations) $\mathbf{Y}_t \cdot \mathbf{y}$ serve as weights for the training words in memory retrieval, further justifying the term correlation-matrix memory.

The **Y**-variables are derived from the **X**-variables (each **Y**-variable compares the data addresses **X** to a specific hard address), whereas in the original correlation-matrix memories (Anderson, 1968; Kohonen, 1972) the **X**-variables are used directly and the variables are continuous. Changing from the **X**-variables to the **Y**-variables means, mathematically, that the input dimension is blown way up (from a thousand to a million); in practice it means that the memory can be made arbitrarily large, rendering its capacity independent of the input dimension N. The idea of expanding the input dimension goes back at least to Rosenblatt's (1962) α-perceptron network.

3.3.5. Recall Fidelity

We will now look at the retrieval of words stored in memory, that is, how faithfully the stored words are reconstructed by the retrieval procedure. The asymptotic behavior of the memory, as the input dimension N grows without bound, has been analyzed in depth by Chou (1989). Specific dimension N is assumed here, and the analysis is simple but approximate. The analysis follows one given by Jaeckel (1989a) and uses some of the same symbols.

What happens when we use one of the addresses, say, the last data address $\mathbf{X}_T$, to retrieve a word from memory; how close to the stored word $\mathbf{W}_T$ is the retrieved word $\mathbf{Z}_T$? The output word $\mathbf{Z}_T$ is gotten from the sum vector $\mathbf{S}_T$ by comparing its U sums to zero. Therefore, we need to find out how likely a sum in $\mathbf{S}_T$ will be on the correct side of zero. Since the data are uniform random, all columns of **C** have the same statistics, and all sums in $\mathbf{S}_T$ have the same statistics. So it suffices to look at a single coordinate of the data words, say, the last, and to assume that the last bit of the last data word, $W_{T,U}$, is 1. How likely is $S_{T,U} > 0$ if $W_{T,U} = 1$? This likelihood is called the *fidelity* for a single bit, denoted here by φ (phi for "fidelity"), and we now proceed to estimate it.

The sum vector $\mathbf{S}_T$ retrieved by the address $\mathbf{X}_T$ is a sum over the locations activated by $\mathbf{X}_T$. The locations are indicated by the 1s of the activation vector $\mathbf{Y}_T$, so that $\mathbf{S}_T = \mathbf{Y}_T{}^{\mathrm{T}}\mathbf{C}$, which equals $\mathbf{Y}_T{}^{\mathrm{T}}\mathbf{Y}^{\mathrm{T}}\mathbf{W}$ (that $\mathbf{C} = \mathbf{Y}^{\mathrm{T}}\mathbf{W}$ was shown above). The last coordinate of the sum vector is then

$$S_{T,U} = \mathbf{Y}_T^{\mathrm{T}}\mathbf{C}_{\cdot,U} = \mathbf{Y}_T^{\mathrm{T}}\mathbf{Y}^{\mathrm{T}}\mathbf{W}_{\cdot,U} = (\mathbf{YY}_T)^{\mathrm{T}}\mathbf{W}_{\cdot,U}$$
$$= (\mathbf{YY}_T) \cdot \mathbf{W}_{\cdot,U},$$

which shows that only the last bits of the data

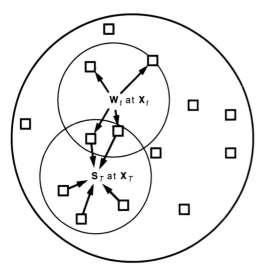

Fig. 3.6. Activation overlaps as weights for stored words. When reading at X_T, the sum S_T includes one copy of the word W_t from each hard location in the activation overlap (two copies in the figure).

words contribute to it. Thus, the Uth bit-sum is the (inner) product of two vectors, YY_T and $W_{.,U}$, where the T-vector $W_{.,U}$ consists of the stored bits (the last bit of each stored word), and the T components of YY_T act as weights for the stored bits.

The weights YY_T have a clear interpretation in terms of activation sets and their intersections or overlaps: they equal the sizes of the overlaps. This is illustrated in Fig. 3.6 (cf. Fig. 3.5). For example, since the 1s of Y_t and Y_T mark the locations activated by X_t and X_T, respectively, the weight $Y_t \cdot T_T$ for the tth data word in the sum S_T equals the number of locations activated by both X_t and X_T. Because the addresses are uniform random, this overlap is $p^2 M$ locations on the average, where p is the probability of activating a location, except that for $t = T$ the two activation sets are the same and the overlap is complete, covering pM locations on the average.

In computing fidelity, we will abbreviate notation as follows: Let B_t $(= W_{t,U})$ be the last bit of the tth data word, let $L_t = Y_t \cdot Y_T$ be its weight in the sum $S_{T,U}$, and let Σ $(= S_{T,U})$ be the last bit sum. Regard the bits B_t and their weights L_t as two sets of T random variables, and recall our assumption that addresses and data are uniform random. Then the bits B_t are independent -1s and 1s with equal probability (i.e., mean $E\{B_t\} = 0$), and they are also independent of the

weights. The weights L_t, being sizes of activation overlaps, are nonnegative integers. When activation is low, as it is in the sample memory ($p = 0.000\,445$), the weights resemble independent Poisson variables: the first $T-1$ of them have a mean (and variance $\mathrm{Var}\{L_t\} \approx$) $E\{L_t\} = \lambda_t = \lambda = p^2 M$ and the last has a mean (and variance $\mathrm{Var}\{L_T\} \approx$) $E\{L_T\} = \lambda_T = \Lambda = pM$ (i.e., complete overlap). For the sample memory these values are: mean activation $\Lambda = pM = 445$ locations (out of a million), and mean activation overlap $\lambda = p^2 M = 0.2$ location ($t < T$). We will proceed as if the weights L_t were independent Poisson variables, and hence our result will be approximate.

We are assuming that the bit we are trying to recover equals 1 (i.e., $B_T = W_{T,U} = 1$); by symmetry, the analysis of $B_T = -1$ is equivalent. The sum Σ is then the sum of T products $L_t B_t$, and its mean, or expectation, is

$$\mu = E\{\Sigma\} = \sum_{t=1}^{T-1} E\{L_t B_t\} + E\{L_T \cdot 1\}$$

$$= E\{L_T\}$$

$$= \Lambda$$

because independence and $E\{B_t\} = 0$ yield $E\{L_t B_t\} = 0$ when $t < T$. The mean sum can be interpreted as follows: It contains all Λ ($=445$) copies of the target bit B_T that have been stored and they reinforce each other, while the other $(T-1)\lambda$ ($=2000$) bits in Σ tend to cancel out each other.

Retrieval is correct when the sum Σ is greater than 0. However, random variation can make $\Sigma \le 0$. The likelihood of that happening, depends on the variance σ^2 of the sum, which variance we will now estimate. When the terms are approximately independent, their variances are approximately additive, so that

$$\sigma^2 = \mathrm{Var}\{\Sigma\} \approx (T-1)\,\mathrm{Var}\{L_1 B_1\}$$
$$+ \mathrm{Var}\{L_T \cdot 1\}$$

The second variance is simply $\mathrm{Var}\{L_T\} \approx \Lambda$. The first variance can be rewritten as

$$\mathrm{Var}\{L_1 B_1\} \equiv E\{L_1^2 B_1^2\} - (E\{L_1 B_1\})^2$$
$$= E\{L_1^2\}$$

because $B_1^2 = 1$, and because $E\{L_1 B_1\} = 0$ as above. It can be rewritten further as

$$\equiv \mathrm{Var}\{L_1\} + (E\{L_1\})^2$$
$$\approx \lambda + \lambda^2$$

and we get, for the variance of the sum,

$$\sigma^2 \approx (T - 1)(\lambda + \lambda^2) + \Lambda$$

Substituting p^2M for λ and pM for Λ, approximating $T - 1$ with T, and rearranging finally yields

$$\sigma^2 = \text{Var}\{\Sigma\} \approx pM[1 + pT(1 + p^2M)]$$

We can now estimate the probability of incorrect recall, that is, the probability that $\Sigma \leq 0$ when $B_T = 1$. We will use the fact that, if the products $L_t B_t$ are T independent random variables, their sum Σ tends to the normal (Gaussian) distribution with mean and variance equal to those of Σ. We then get, for the probability of a single-bit failure,

$$\Pr\{\Sigma \leq 0 | \mu, \sigma\} \approx \Phi(-\mu/\sigma)$$

where Φ is the normal distribution function; and for the probability of recalling a bit correctly, or bit-fidelity φ, we get $1 - \Phi(-\mu/\sigma)$, which equals $\Phi(\mu/\sigma)$.

3.3.6. Signal, Noise, and Probability of Activation

We can regard the mean value μ $(=pM)$ of the sum Σ as signal, and the variance σ^2 $(\approx pM[1 + pT(1 + p^2M)])$ of the sum as noise. The standard quantity $\rho = \mu/\sigma$ is then a *signal-to-noise ratio* (rho for "ratio") that can be compared to the normal distribution, to estimate bit-fidelity, as was done above:

$$\varphi = \Pr\{\text{bit recalled correctly}\} = \Phi(\rho)$$

The higher the signal-to-noise ratio, the more likely are stored words recalled correctly. This points to a way to find good values for the probability p of activating locations and, hence, for the activation radius H: We want p that maximizes ρ. To find this value of p, it is convenient to start with the expression for ρ^2 and to reduce it to

$$\rho^2 = \frac{\mu^2}{\sigma^2} \approx \frac{pM}{1 + pT(1 + p^2M)}$$

Taking the derivative with respect to p, setting it to 0, and solving for p gives

$$p = \frac{1}{\sqrt[3]{2MT}}$$

as the best probability of activation. This value of p was mentioned earlier, and it was used to set parameters for the sample memory.

The probability $p = (2MT)^{-1/3}$ of activating a location is optimal only when exact storage addresses are used for retrieval. When a retrieval address is approximate (i.e., when it equals a storage address plus some noise), both the signal and the noise are reduced, and also their ratio is reduced. Analysis of this is more complicated than the one above, and it is not carried out here. The result is that, for maximum recovery of stored words with approximate retrieval addresses, p should be somewhat larger than $(2MT)^{-1/3}$ (typically, less than twice as large); however, when the data are clustered rather than uniform random, optimum p tends to be smaller than $(2MT)^{-1/3}$.

In a case yet more general, the training set is not "clean" but contains many noisy copies of each word to be stored, and the data addresses are noisy (cf. Fig. 3.1). Then it makes sense to store words within a smaller radius and to retrieve them within a larger. To allow such memories to be analyzed, Avery Wang (unpublished) and Jaeckel (1988) have derived formulas for the size of the overlap of activation sets with different radii of activation. As a rule, the overlap decreases rapidly with increasing distance between the centers of activation.

3.3.7. Memory Capacity

Storage and retrieval in a standard random-access memory are deterministic. Therefore, its capacity (in words) can be expressed simply as the number of memory locations. In a sparse distributed memory, retrieval of words is statistical. However, its capacity, too, can be defined as a limit on the number T of words that can be stored and retrieved successfully, although the limit depends on what we mean by success.

A simple criterion of success is that a stored bit is retrieved correctly with high probability φ (e.g., $0.99 \leq \varphi \leq 1$). Other criteria can be derived from it or are related to it. Specifically, capacity here is the maximum T, $T_{\max}$, such that $\Pr\{Z_{t,u} = W_{t,u}\} \geq \varphi$; we are assuming that exact storage addresses are used to retrieve the words. It is convenient to relate capacity to memory size M and to define it as $\tau = T_{\max}/M$. As fidelity φ approaches 1, capacity τ approaches 0, and the values of τ that concern us here are smaller than 1. We will now proceed to estimate τ.

In Section 3.3.5 on Recall Fidelity we saw that the bit-recall probability φ is approximated by $\Phi(\rho)$, where ρ is the signal-to-noise ratio as defined above. By writing out ρ and substituting

τM for T we get

$$\varphi \approx \Phi(\rho) \approx \Phi\left(\frac{pM}{1 + p\tau M(1 + p^2 M)}\right)^{1/2}$$

which leads to

$$[\Phi^{-1}(\varphi)]^2 \approx \rho^2 \approx \frac{pM}{1 + p\tau M(1 + p^2 M)}$$

where Φ^{-1} is the inverse of the normal distribution function. Dividing by pM in the numerator and the denominator gives

$$[\Phi^{-1}(\varphi)]^2 \approx \frac{1}{(1/pM) + \tau(1 + p^2 M)}$$

The right side goes to $1/\tau$ as the memory size M grows without bound, giving us a simple expression for the asymptotic capacity:

$$\tau \approx \frac{1}{[\Phi^{-1}(\varphi)]^2}$$

To verify this limit, we use the optimal probability of activation, taking note that it depends on both M and τ: $p = (2MT)^{-1/3} = (2\tau M^2)^{-1/3}$. Then, in the expression above, $1/(pM) = (2\tau/M)^{1/3}$ and goes to zero as M goes to infinity, because $\tau < 1$. Similarly, $\tau(1 + p^2 M) = \tau + (\frac{1}{4}\tau/M)^{1/3}$ and goes to τ.

To compare this asymptotic capacity to the capacity of a finite memory, consider $\varphi = 0.999$, meaning that about one bit in a thousand is retrieved incorrectly. Then the asymptotic capacity is $\tau \approx 0.105$, and the capacity of the million-location sample memory is 0.096. Keeler (1988) has shown that the sparse distributed memory and the binary Hopfield net trained with the outer-product leaning rule, which is equivalent to a correlation-matrix memory, have the same capacity per storage element or counter. The $0.15N$ capacity of the Hopfield net ($\tau = 0.15$) corresponds to fidelity $\varphi = 0.995$, meaning that about one bit in 200 is retrieved incorrectly. The practical significance of the sparse distributed memory design is that, by virtue of the hard locations, the number of storage elements is independent of the input and output dimensions. Doubling the hardware doubles the number of words of a given size that can be stored, whereas the capacity of the Hopfield net is limited by the word size.

A very simple notion of capacity has been used here, and it results in capacities of about 10 percent of memory size. However, the assumption has been that exact storage addresses are used in retrieval. If approximate addresses are used, and if less error is tolerated in the words

retrieved than in the addresses used for retrieving them, the capacity goes down. The most complete analysis of capacity under such general conditions has been given by Chou (1989). Expressing capacity in absolute terms, for example, as Shannon's information capacity, is perhaps the most satisfying. This approach has been taken by Keeler (1988). Allocating the capacity is then a separate issue: whether to store many words or to correct many errors. A practical guide is that the number of stored words should be from 1 to 5 percent of memory size (i.e., of the number of hard locations).

3.4. SDM AS AN ARTIFICIAL NEURAL NETWORK

The sparse distributed memory, as an artificial neural network, is a synchronous, fully connected, three-layer (or two-layer, see below), feed-forward net illustrated by Fig. 3.7. The flow of information in the figure is from left to right. The column of N circles on the left is called the *input* layer, the column of M circles in the middle is called the *hidden* layer, and the column of U circles on the right is called the *output* layer, and the circles in the three columns are called input units, hidden units, and output units, respectively.

The hidden units and the output units are bona fide artificial neurons, so that, in fact, there are only two layers of "neurons." The input units merely represent the outputs of some other neurons. The inputs x_n to the hidden units label the input layer, the input coefficients $A_{m,n}$ of the hidden units label the lines leading into the hidden units, and the outputs y_m of the hidden units label the hidden layer. If y is the activation

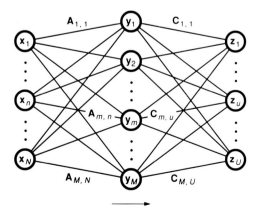

Fig. 3.7. Feed-forward artificial neural network.

function of the hidden units (e.g., $y(d) = 1$ if $d \geq D$, and $y(d) = 0$ otherwise), the output of the mth hidden unit is given by

$$y_m = y\left(\sum_{n=1}^{N} A_{m,n} x_n \right)$$

which, in vector notation, is $y_m = y(\mathbf{A}_m \cdot \mathbf{x})$, where $\mathbf{x}$ is the vector of inputs to, and $\mathbf{A}_m$ is the vector of input coefficients of, the mth hidden unit.

A similar description applies to the output units, with the outputs of the hidden units serving as their inputs, so that the output of the uth output unit is given by

$$z_u = z\left(\sum_{m=1}^{M} C_{m,u} y_m \right)$$

or, in vector notation, $z_u = z(\mathbf{C}_{\cdot,u} \cdot \mathbf{y})$. Here $\mathbf{C}_{\cdot,u}$ is the vector of input coefficients of the uth output unit, and z is the activation function.

From the equations above it is clear that the input coefficients of the hidden units form the address matrix $\mathbf{A}$, and those of the output units form the contents matrix $\mathbf{C}$, of a sparse distributed memory. In the terminology of artificial neural nets, these are the matrices of connection strengths (synaptic strengths) for the two layers. "Fully connected" means that all elements of these matrices can assume nonzero values. Later we will see sparsely connected variations of the model.

Correspondance between Figs. 3.7 and 3.4 is now demonstrated by transforming Fig. 3.7 according to Fig. 3.8, which shows four ways of drawing artificial neurons. View (A) shows how they appear in Fig. 3.7. View (B) is laid out similarly, but all labels now appear in boxes and circles. In view (C), the diamond and the circle that represent the inner product and the output, respectively, appear below the column of input coefficients, so that these units are easily stacked side by side. View (D) is essentially the same as view (C), for stacking units on top of each other. We will now redraw Fig. 3.7 with units of type (D) in the hidden layer and with units of type (C) in the output layer. An input (a circle) that is shared by many units is drawn only once. The result is Fig. 3.9. Its correspondence to Fig. 3.4 is immediate, the vectors and the matrices implied by Fig. 3.7 are explicit, and the cobwebs of Fig. 3.7 are gone.

In describing the memory, the term "synchronous" means that all computations are completed in what could be called a machine cycle, after which the network is ready to perform another cycle. The term is superfluous if the net is used as a feed-forward net akin to a

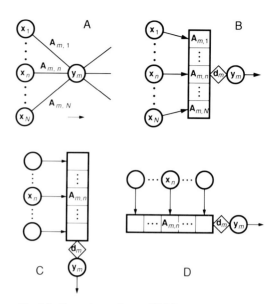

Fig. 3.8. Four views of an artificial neuron.

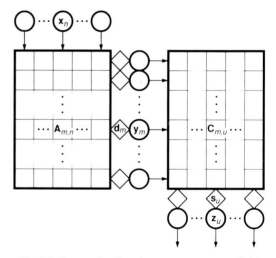

Fig. 3.9. Sparse distributed memory as an artificial neural network (Fig. 3.7 redrawn in the style of Fig. 3.4).

random-access memory. However, it is meaningful if the network's output is fed back as input: the network is allowed to settle with each input so that a completely updated output is available as the next input.

As a multilayer feed-forward net, the sparse distributed memory is akin to the nets trained with the error back-propagation algorithm (Rumelhart and McClelland, 1986). How are the two different? In a broad sense they are not: we try to find matrices **A** and **C**, and activation functions y and z, that fit the source of our data. In practice, many things are done differently.

In error back-propagation, the matrices **A** and **C** and the activation vector **y** are usually real-valued, the components of **y** usually range over the interval $[-1, 1]$ or $[0, 1]$, the activation function y and its inverse are differentiable, and the data are stored over a uniform algorithm to change both **A** and **C**. In sparse distributed memory, the address matrix **A** is usually binary, and various methods are used for choosing it, but once a location's address has been set, it is not changed as the data are stored (**A** is constant); furthermore, the activation function y is a step function that yields an activation vector **y** that is mostly 0s, with a few 1s. Another major difference is in the size of the hidden layer. In back-propagation nets, the number of hidden units is usually smaller than the number of input units or the number of items in the training set; in a sparse distributed memory, it is much larger.

The differences imply that, relative to back-propagation nets, the training of a sparse distributed memory is fast (it is easy to demonstrate single-trial learning), but applying it to a new problem is less automatic (it requires choosing an appropriate data representation, as discussed in the section on SDM research below).

3.5. SDM AS A MODEL OF THE CEREBELLUM

3.5.1. Modeling Biology with Artificial Neural Networks

Biological neurons are cells that process signals in animals and humans, allowing them to respond rapidly to the environment. To achieve speed, neurons use electrochemical mechanisms to generate a signal (a voltage level or electrical pulses) and to transmit it to nearby and distant sites.

Biological neurons come in many varieties. The peripheral neurons couple the organism to the world. They include the sensory neurons that convert an external stimulus into an electrical signal, the motor neurons whose electrical pulses cause muscle fibers to contract, and other effector neurons that regulate the secretion of glands. However, most neurons in highly evolved animals are interneurons that connect directly to other neurons rather than to sensors or to effectors. Interneurons also come in many varieties and they are organized into a multitude of neural circuits.

A typical interneuron has a cell body and two kinds of arborizations: a dendrite tree that receives signals from other neurons, and an axon tree that transmits the neuron's signal to other neurons. Transmission-contact points between neurons are called synapses. They are either excitatory (positive synaptic weight) or inhibitory (negative synaptic weight) according to whether a signal received through the synapse facilitates or hinders the activation of the receiving neuron. The axon of one neuron can make synaptic contact with the dendrites and cell bodies of many other neurons. Thus, a neuron receives multiple inputs, it integrates them, and it transmits the result to other neurons.

Artificial neural networks are networks of simple, interconnected processing units, called (*artificial*) *neurons*. The most common artificial neuron in the literature has multiple (N) inputs and one output and is defined by a set of input coefficients—a vector of N reals, standing for the synaptic weights—and a nonlinear scalar activation function. The value of this function is the neuron's output, and it serves as input to other neurons. A linear threshold function is an example of an artificial neuron, and the simplest kind—one with binary inputs and output—is used in the sparse distributed memory.

It may seem strange to model brain activity with binary neurons when real neurons are very complex in comparison. However, the brain is organized in large circuits of neurons working in parallel, and the mathematical study of neural nets is aimed more at understanding the behavior of circuits than of individual neurons. An important fact—perhaps the most important—is that the states of a large circuit can be mapped onto the points of a high-dimensional space, so that although a binary neuron is a grossly simplified model of a biological neuron, a large circuit of binary neurons, by virtue of its high dimension, can be a useful model of a circuit of biological neurons.

The sparse distributed memory's connection to biology is made in the standard way. Each row through **A**, **d**, **y**, and **C** in Fig. 3.9—each hidden unit—is an artificial neuron that represents a biological neuron. Vector **x** represents the N signals coming to these neurons as inputs from N other neurons (along their axons), vector $\mathbf{A}_m$ represents the weights of the synapses through which the input signals enter the mth neuron (at its dendrites), d_m represents the integration of the input signals by the mth neuron, and y_m represents the output signal, which is passed along the neuron's axon to U other neurons through synapses with strengths $\mathbf{C}_m$.

We will call these (the hidden units) the *address-decoder neurons* because they are like the address-decoder circuit of a random-access memory: they select locations for reading and writing. The address that the mth address-decoder neuron decodes is given by the input coefficients $\mathbf{A}_m$; location $\mathbf{A}_m$ is activated by inputs $\mathbf{x}$ that equal or are sufficiently similar to $\mathbf{A}_m$. How similar depends on the radius of activation H. It is interesting that a linear threshold function with N inputs, which is perhaps the oldest mathematical model of a neuron, is ideal for address decoding in the sparse distributed memory, and that a proper choice of a single parameter, the threshold, makes it into an address decoder for a location of an ordinary random-access memory.

Likewise, in Fig. 3.9, each column through $\mathbf{C}$, $\mathbf{s}$, and $\mathbf{z}$ is an artificial neuron that represents a biological neuron. Since these U neurons provide the output of the circuit, they are called the output neurons. The synapses made by the axons of the address-decoder neurons with the dendrites of the output neurons are represented by matrix $\mathbf{C}$, and they are modifiable; they are the sites of information storage in the circuit.

We now look at how these synapses are modified; specifically, what neural structures are implied by the memory's storage algorithm (cf. Figs. 3.4 and 3.9). The word $\mathbf{w}$ is stored by adding it into the counters of the active locations, that is, into the axonal synapses of active address-decoder neurons. This means that if a location is activated for writing, its counters are adjusted upward and downward; if it is not activated, its counters stay unchanged.

Since the output neurons are independent of each other, it suffices to look at just one of them, say, the uth output neuron; see Fig. 3.10 center. The uth output neuron produces the uth output bit, which is affected only by the uth bits of the words that have been stored in the memory. Let us assume that we are storing the word $\mathbf{w}$. Its uth bit is w_u. To add w_u into all the active synapses in the uth column of $\mathbf{C}$, it must be made physically present at the active synaptic sites of the column. Since different sites in a column are active at different times, it must be made present at all synaptic sites of the column. A neuron's way of presenting a signal is by passing it along the axon. This suggests that the uth bit w_u of the word-in register should be represented by a neuron that corresponds to the uth output neuron z_u, and that its output signal should be available at each synapse in column u, although it is "captured" only by synapses that have just been activated by address-decoder neurons $\mathbf{y}$. Such an

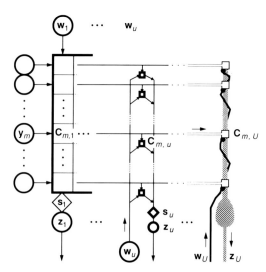

Fig. 3.10. Connections to an output neuron. Three output units are shown. The first unit is drawn as a column through the contents matrix $\mathbf{C}$, the middle unit shows the connections explicitly, and the last unit corresponds to Fig. 3.11.

arrangement is shown in Fig. 3.10. It suggests that word-in neurons are paired with the output neurons, with the axon tree of a word-in neuron possibly meshing with the dendrite tree of the corresponding output neuron, as that would help carry the signal to all synaptic sites of a column. This kind of pairing, when found in a brain circuit, can help us interpret the circuit (Fig. 3.10, on the right).

3.5.2. The Cortex of the Cerebellum

Of the neural circuits in the brain, the cortex of the cerebellum most resembles the sparse distributed memory. The cerebellar cortex of mammals is a fairly large and highly regular structure with an enormous number of neurons of only five major kinds, and with two major kinds of input. Its morphology has been studied extensively since the early 1900s, its role in fine motor control has been established, and its physiology is still studied intensively (Ito, 1984).

The cortex of the cerebellum is sketched in Fig. 3.11 after Llinás (1975). Figure 3.12 is Fig. 3.9 redrawn in an orientation that corresponds to the sketch of the cerebellar cortex.

Within the cortex are the cell bodies of the granule cells, the Golgi cells, the stellate cells, the basket cells, and the Purkinje cells. Figure 3.11 shows the climbing fibers and the mossy fibers entering and the axons of the Purkinje cells leaving

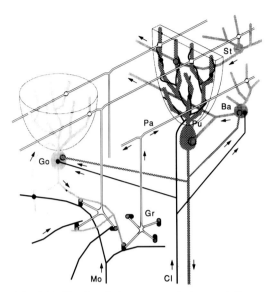

Fig. 3.11. Sketch of the cortex of the cerebellum. Ba = basket cell, Cl = climbing fiber (black), Go = Golgi cell, Gr = granule cell, Mo = mossy fiber (black), Pa = parallel fiber, Pu = Purkinje cell (cross-hatched), St = stellate cell. Synapses are shown with small circles and squares of the axon's "color." Excitatory synapses are black or white, inhibitory synapses are cross-hatched or gray.

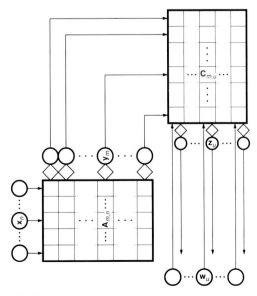

Fig. 3.12. Sparse distributed memory's resemblance to the cerebellum (Fig. 3.9 redrawn in the style of Fig. 3.11; see also Fig. 3.10).

the cortex. This agrees with the two inputs into and the one output from a sparse distributed memory. The correspondence goes deeper: The Purkinje cells that provide the output are paired with the climbing fibers that provide input. A climbing fiber, which is an axon of an olivary cell that resides in the brain stem, could thus have the same role in the cerebellum that the line from a word-in cell through a column of counters has in a sparse distributed memory (see Fig. 3.10), namely, to make a bit of a data word available at a bit-storage site when words are stored.

The other set of inputs enters along the mossy fibers, which are axons of cells outside the cerebellum. They would then be like an address into a sparse distributed memory. The mossy fibers feed into the granule cells, which thus would correspond to the hidden units of Fig. 3.12 (they appear as rows across Fig. 3.9) and would perform address decoding. The firing of a granule cell would constitute activating a location for reading or writing. Therefore, the counters of a location would be found among the synapses of a granule cell's axon; these axons are called parallel fibers. A parallel fiber makes synapses with Golgi cells, stellate cells, basket cells, and Purkinje cells. Since the Purkinje cells provide the output, it is natural to assume that their synapses with the parallel fibers are the storage sites or the memory's counters.

In addition to the "circuit diagram," other things suggest that the cortex of the cerebellum in an associative memory reminiscent of the sparse distributed memory. The numbers are reasonable. The numbers quoted below were compiled by Loebner (1989) in a review of the literature and they refer to the cerebellum of a cat. Several million mossy fibers enter the cerebellum, suggesting that the dimension of the address space is several million. The granule cells are the most numerous—in the billions—implying a memory with billions of hard locations, and only a small fraction of them is active at once, which agrees with the model. Each parallel fiber intersects the flat dendritic trees of several Purkinje cells, implying that a hard location has several hundred counters. The number of parallel fibers that pass through the dendritic tree of a single Purkinje cell is around a hundred thousand, implying that a single "bit" of output is computed from about a hundred thousand counters (only few of which are active at once). The number of Purkinje cells is around a million, implying that the dimension of the data words is around a million. However, a single olivary cell sends about ten climbing fibers to that many Purkinje cells, and if, indeed, the climbing fibers train the Purkinje cells, the output dimension is more like a hundred thousand than a million. All these numbers mean, of course, that the cerebellar cortex is far from fully connected: every granule cell does not reach every Purkinje cell (nor does every mossy fiber reach every granule cell; more on that below).

This interpretation of the cortex of the cerebellum as an associative memory, akin to the sparse distributed memory, is but an outline, and it contains discrepancies that are evident even at the level of cell morphology. According to the model, an address decoder (a hidden unit) should receive all address bits, but a granule cell receives input from three to five mossy fibers only and, for a granule cell to fire, most or all of its inputs must be firing (the number of active inputs required for firing appears to be controlled by the Golgi cells that provide the other major input to the granule cells; the Golgi cells could control the number of locations that are active at once). The very small number of inputs to a granule cell means that activation is not based on Hamming distance from an address but on certain address bits being on in the address register. Activation of locations of a sparse distributed memory under such conditions has been treated specifically by Jaeckel, and the idea is presented already in the cerebellar models of Marr and of Albus. These will be discussed in the next two sections.

Many details of the cerebellar circuit are not addressed by this comparison with the sparse distributed memory. The basket cells connect to the Purkinje cells in a special way, the stellate cells make synapses with the Purkinje cells, and signals from the Purkinje cells and climbing fibers go to the basket cells and Golgi cells. The nature of synapses and signals—the neurophysiology of the cerebellum—has not been considered. Some of these things are addressed by the mathematical models of Marr and of Albus. The point here has been to demonstrate some of the variety in a real neural circuit, to show how a mathematical model can be used to interpret such a circuit, and to suggest that the cortex of the cerebellum constitutes an associative memory. Because its mossy-fiber input comes from all over the cerebral cortex—from many sensory areas—the cerebellum is well located for correlating action that it regulates with information about the environment.

3.6. VARIATIONS OF THE SDM MODEL

The basic sparse distributed memory model is fully connected. This means that every input unit

(address bit) is seen by every hidden unit (hard location), and that every hidden unit is seen by every output unit. Furthermore, all addresses and words are binary. If -1 and 1 are used as the binary components, "fully connected" means that none of the elements of the address and contents matrices $\mathbf{A}$ and $\mathbf{C}$ is (identically) zero. Partially— and sparsely—connected models have zeros in one or both of the matrices, as a missing connection is marked by a weight that is zero.

Jaeckel has studied designs with sparse address matrices and binary data. In the selected-coordinate design (1989a), -1s and 1s are assumed to be equally likely in the data addresses; in the hyperplane design (1989b), the data-address bits are assumed to be mostly (e.g., 90%) -1s. Jaeckel's papers are written in terms of binary 0s and 1s, but here we will use -1s and 1s, and will let a 0 stand for a missing connection or a "don't care" bit (for which Jaeckel uses the value $1/2$). Jaeckel uses one-million-location memories ($M = 1\,000\,000$) with a 1000-dimensional address space ($N = 1000$) to demonstrate the designs.

3.6.1. Jaeckel's Selected-Coordinate Design

In the selected-coordinate design, the hard-address matrix $\mathbf{A}$ has a million rows with ten -1s and 1s ($k = 10$) in each row. The -1s and 1s are chosen with probability $1/2$ and they are placed randomly within the row and independently of other rows; the remaining 990 coordinates of a row are 0s. This is equivalent to taking a uniform random $\mathbf{A}$ of -1s and 1s and setting a random 990 coordinates in each row to zero (different 990 for different rows). A location is activated if the values of all ten of its selected coordinates match the address register $\mathbf{x}$: $y_m = 1$ iff $\mathbf{A}_m \cdot \mathbf{x} = k$. The probability of activating a hard location is related to the number of nonzero coordinates in a hard address by $p = 0.5^k$. Here, $k = 10$ and $p = 0.001$.

3.6.2. Jaeckel's Hyperplane Design

The hyperplane design deals with data where the addresses are skewed (e.g., 100 1s and 900 -1s). Each row of the hard-address matrix $\mathbf{A}$ has three 1s ($k = 3$), placed at random, and the remaining 997 places have 0s (there are no -1s). A location is activated if the address register has 1s at those same three places: $y_m = 1$ iff $\mathbf{A}_m \cdot \mathbf{x} = k$. The probability of activating a location is related to the number of 1s in its address by $p \approx (L/N)^k$, where L is the number of 1s in the data addresses $\mathbf{x}$. Here, $N = 1000$, $L = 100$, $k = 3$, and $p \approx 0.001$.

Jaeckel has shown that both of these designs are better than the basic design in recovering previously stored words, as judged by signal-to-noise ratios. They are also easier to realize physically—in hardware—because they require far fewer connections and much less computation in the address-decoder unit that determines the set of active locations.

The region of the address space that activates a hard location in the three designs can be interpreted geometrically as follows: A location of the basic sparse distributed memory is activated by all addresses that are within H Hamming units of the location's address, so that the exciting part of the address space is a hypersphere around the hard address. In the selected-coordinate design, a hard location is activated by all addresses in a subspace of the address space defined by the k selected coordinates—that is, by the vertices of an $(N - k)$-dimensional hypercube. In the hyperplane design, the address space is a hyperplane defined by the number of 1s in an address, L (which is constant over all data addresses), and a hard location is activated by the intersection of the address space with the $(N - k)$-dimensional hypercube defined by the k 1s of the hard address.

The regions have a spherical interpretation also in the latter two designs, as suggested by the activation condition $A_m \cdot x = k$ (same formula for both designs; see above). It tells us that the exciting points of the address space lie on the surface of a hypersphere in Euclidean N-space, with center coordinates A_m (the hard address) and with Euclidean radius $(N - k)^{1/2}$ (no points of the address space lie inside the sphere). This gives rise to *intermediate designs*, as suggested by Jaeckel (1989b): let the hard addresses be defined in -1s, 0s, and 1s as above, and let the mth hard location be activated by all addresses x within a suitably large hypersphere centered at the hard address. Specifically, $y_m = 1$ if, and only if, $A_m \cdot x \geq G$. The parameters G and k (and L) have to be chosen so that the probability of activating a location is reasonable.

The optimum probability of activation p for the various sparse distributed memory designs is about the same—it is in the vicinity of $(2MT)^{-1/3}$—and the reason is that, in all these designs, the sets of locations activated by two addresses, x and x', overlap minimally unless x and x' are very similar to each other. The sets behave in the manner of random sets of approximately pM hard locations each, with two such sets overlapping by p^2M locations, on the average (unless x and x' are very similar to each other). This is a consequence of the high dimension of the address space.

In the preceding section on the cerebellum we saw that the hard-address matrix A, as implied by the few inputs (3–5 mossy fibers) to each granule cell, is very sparse, and that the number of active inputs required for a granule cell to fire can be modulated by the Golgi cells. This means that the activation of granule cells in the cerebellum resembles the activation of locations in an intermediate design that is close to the hyperplane design.

Not only are the mossy-fiber connections to a granule cell few (3–5 out of several million), but also the granule-cell connections to a Purkinje cell are few (hundred thousand out of billions), so that also the contents matrix C is very sparse. This aspect of the cerebellum has not been modeled mathematically.

3.6.3. Hassoun's Pseudorandom Associative Neural Memory

Independently of the above developments, Hassoun (1988) has proposed a model with a random, fixed address matrix A and variable contents matrix C. This model allows us to extend the concepts of this chapter to data with short addresses (e.g., $N = 4$ bits), and it introduces ideas about storing the data (i.e., training) that can be applied to associative memories at large.

The data addresses X and words W in Hassoun's examples are binary vectors in 0s and 1s. The elements of the hard-address matrix A are small integers; they are chosen at uniform random from the symmetric interval $\{-L, -L + 1, -L + 2, \ldots, L\}$, where L is a small positive integer (e.g., $L = 3$). Each hard location has its own activation threshold D_m, which is chosen so that approximately half of all possible N-bit addresses x activate the location: $y_m = 1$ if $A_m \cdot x \geq D_m$, and $y_m = 0$ otherwise. The effect of such addressing through A is to convert the matrix X of N-bit data addresses into the matrix Y of M-bit activation vectors, where $M \gg N$ and where each activation vector y_m is about half 0s and half 1s (probability of activation p is around 0.5).

Geometric interpretation of addressing through A is as follows. The space of hard addresses is an N-dimensional hypercube with sides of length $2L + 1$. The unit cubes or cells of this space are potential hard locations. The M hard addresses A_m are chosen uniformly at random from within this space. The space of data addresses is an N-cube with sides of length 2; it is at the center of the hard-address space, with the cell $000 \ldots 0$

at the very center. The data addresses that activate the location $\mathbf{A}_m$ are the ones closest to $\mathbf{A}_m$ and they can be visualized as follows: A straight line is drawn from $\mathbf{A}_m$ through $000 \ldots 0$. Each setting of the threshold D_m then corresponds to an $N - 1$ dimensional hyperplane perpendicular to this line, at some distance from $\mathbf{A}_m$. The cells $\mathbf{x}$ of the data-address space that are on the $\mathbf{A}_m$ side of the plane will activate location $\mathbf{A}_m$. The threshold D_m is chosen so that the plane cuts the data-addresses space into two nearly equal parts.

The hard addresses $\mathbf{A}_m$ correspond naturally to points (and subspaces) $\mathbf{A}'_m$ of the data-address space $\{0, 1\}^N$ gotten by replacing the negative components of $\mathbf{A}_m$ by 0s, the positive components by 1s, and the 0s by either (a "don't care"). The absolute values of the components of $\mathbf{A}_m$ then serve as weights, and the mth location is activated by $\mathbf{x}$ if the *weighted* distance between $\mathbf{A}'_m$ and $\mathbf{x}$ is below a threshold (cf. Kanerva, 1988, pp. 46–48).

High probability of activation (≈ 0.5) works poorly with the outer-product learning rule. However, it is appropriate for an analytic solution to storage by the Ho–Kashyap recording algorithm (Hassoun and Youssef, 1989; also refer to Chapter 1). This algorithm finds a contents matrix $\mathbf{C}$ that solves the linear inequalities implied by $\mathbf{Z} = \mathbf{W}$, where $\mathbf{W}$ is the matrix of data words to be stored, and $\mathbf{Z} = z(\mathbf{S}) = z(\mathbf{YC})$ is the matrix of words retrieved by the rows of $\mathbf{X}$. The inequalities follow from the definition of the threshold function z, as $W_{t,u} = 1$ implies that $S_{t,u} > 0$, and $W_{t,u} = 0$ implies that $S_{t,u} < 0$. Hassoun and Youssef have shown that this storage algorithm results in large basins of attraction around the data addresses, and that if data are stored autoassociatively, false attractors (i.e., spurious stable patterns and limit cycles) will be relatively few.

of the bits in a label is a 1, and its position corresponds to the vowel in question. This is a standard setup for classification by artificial neural nets.

For processing on a computer, the input variables are discretized into 513 integers in the range 16 127–16 639. The memory is constructed by choosing (2000) hard addresses at uniform random from a 128-dimensional hypercube with sides of length 32 768. The cells of this outer space are addressed naturally by 128-place integers to base 32 768 (i.e., these are the vectors $\mathbf{A}_m$), and the data addresses $\mathbf{x}$ then occupy a small hypercube at the center of the hard-address space; the data-address space is a 128-dimensional cube with sides of length 513. Activation is based on distance. Address $\mathbf{x}$ activates the mth hard location if the maximum coordinate separation (i.e., L_∞ distance) between $\mathbf{x}$ and $\mathbf{A}_m$ is at most 16 091. About 10 percent of the hard locations will be activated. Experiments with connected speech deal similarly with 896-dimensional real vectors. In other experiments with the same data, the use of Euclidean distance and other distance measures in place of the L_∞ distance resulted in only minor changes in the outcome. See also Clarke et al. (1991) for a further analysis of the model and an example of its use.

Prager and Fallside train the contents matrix $\mathbf{C}$ iteratively by correcting errors so as to solve the inequalities implied by $\mathbf{Z} = \mathbf{W}$ (see the last paragraph of Section 3.6.3).

This design is similar to Hassoun's design discussed in Section 3.6.3, in that both have a large space of hard addresses that includes, at the center, a small space of data addresses, and in that the hard locations are placed at random within the hard-address space. The designs are in contrast with Albus's CMAC (discussed in the next section), where the placement of the hard locations is systematic.

3.6.4. Adaptation to Continuous Variables by Prager and Fallside

All the models discussed so far have had binary vectors as inputs and outputs. Prager and Fallside (1989) consider several ways of extending the sparse distributed memory model into real-valued inputs. The following experiment with spoken English illustrates their approach.

Eleven vowels were spoken several times by different people. Each spoken instance of a vowel is represented by a 128-dimensional vector of reals that serves as an address or cue. The corresponding data word is an 11-bit label. One

3.7. RELATION TO THE CEREBELLAR MODELS OF MARR AND OF ALBUS

The first comprehensive mathematical models of the cerebellum as an associative memory are by Marr (1969) and by Albus (1971), developed independently in their doctoral dissertations, and they still are the most complete of any such models. They were developed specifically as models of the cerebellar cortex, whereas the sparse distributed memory's resemblance to the cerebellum was noticed only after the model had been fully developed.

Marr's and Albus's models attend to many of the details of the cerebellar circuit. The models are based mostly on connectivity but also on the nature of the synapses. Albus (1989) has made a comparison of the two models. The models will be described here to show their relation to the sparse distributed memory.

3.7.1. Marr's Model of the Cerebellum

The main circuit in Marr's model—in Marr's vocabulary and in our symbols—consists of $(N =)$ 7000 input fibers that feed into $(M =)$ 200 000 codon cells that feed into a single output cell. The input fibers activate codon cells, and codon-cell connections with the output cell store information. The correspondence to the cerebellum is straightforward: the input fibers model mossy fibers, the codon cells model granule cells, and the output cell models a Purkinje cell.

Marr discusses at length the activation of codon cells by the input fibers. Since the input fibers represent mossy fibers and the codon cells represent granule cells, each codon cell receives input from 3–5 fibers in Marr's model. The model assumes discrete time intervals. During an interval an input fiber is either inactive (-1) or active $(+1)$, and at the end of the interval a codon cell is either inactive (0) or active $(+1)$ according to the activity of its inputs during the interval; the codon-cell output is a linear threshold function of its inputs, with $+1$ weights.

The overall pattern of activity of the N input fibers during an interval is called the input pattern (an N-vector of -1s and 1s), and the resulting pattern of activity of the M codon cells at the end of the interval is called a codon representation of the input pattern (an M-vector of 0s and 1s). These correspond, respectively, to the address register x, and to the activation vector y, of a sparse distributed memory.

Essential to the model is that M is much larger than N, and that the number of 1s in a codon representation is small compared to M and relatively constant; conditions that hold also for the sparse distributed memory. Then the codon representation amplifies differences between input patterns. To make differences in N-bit patterns commensurate with differences in M-bit patterns, Marr uses a relative measure defined as the number of 1s that two patterns have in common, divided by the number of places where either pattern has a 1 (i.e., the size of the intersection of 1s relative to the size of their union).

Marr's model's relation to artificial neural networks is simple. The input fibers correspond to input units, the codon cells correspond to hidden units, and the output cell corresponds to an output unit. Each hidden unit has only 3–5 inputs, chosen at random from the N input units, and the input coefficients are fixed at $+1$. Obviously, the net is far from fully connected, but all hidden units are connected to the output unit, and these connections are modifiable. The hidden units are activated by a linear threshold function, and the threshold varies. However, it varies not as the result of training but dynamically so as to keep the number of active hidden units within desired limits (500–5000). Therefore, to what first looks like a feed-forward net must be added feedback connections that adjust dynamically the thresholds of the hidden units. The Golgi cells are assumed to provide this feedback.

In relating Marr's model to the sparse distributed memory, the codon cells correspond to hard locations, and the hard-address matrix A is very sparse, as each row has k_m 1s ($k_m = 3, 4, 5$), placed at random, and $N - k_m$ 0s (there are no -1s in A). A codon cell fires if most of its 3–5 inputs are active, and the Golgi cells set the firing threshold so that 500–5000 codon cells (out of the 200 000) are active at any one time, regardless of the number of active input lines. Thus, the activation function y_m for hard location A_m is a threshold function with value 1 (the codon cell fires) when most—but not necessarily all—of the k_m 1s of A_m are matched by 1s in the address x. The exact condition of activation in the examples developed by Marr is that $A_m \cdot x \geq R$, where the threshold R is between 1 and 5 and depends on x. Thus, the codon cells are activated in Marr's model in a way that resembles the activation of hard locations in an intermediate design of sparse distributed memory that is close to the hyperplane design (in the hyperplane design, all inputs must be active for a cell to fire).

One of the conditions of the hyperplane design is far from being satisfied—namely, that the number of 1s in the address is about constant (hence the name hyperplane design). In Marr's model it is allowed to vary widely (between 20 and 1000 out of 7000), and this creates the need for adjusting the threshold dynamically. In the sparse distributed memory variations discussed so far, the threshold is fixed, but later in this chapter we will refer to experiments in which the thresholds are adjusted either dynamically or by training with data.

Marr estimates the capacity of his model under the most conservative of assumptions, namely, that (0s and) 1s are added to one-bit counters that are initially 0. Under this assumption, all counters eventually saturated and all information is lost, as pointed out by Albus (1989).

3.7.2. Albus's Cerebellar Model Arithmetic Computer (CMAC)

This description of CMAC is based on the one in Albus's book *Brains, Behavior, and Robotics* (1981) and uses its symbols. The purpose here is to describe it sufficiently to allow its comparison to the sparse distributed memory.

CMAC is an associative memory with a large number of addressable storage locations, just as the sparse distributed memory is, and the address space is multidimensional. However, the number of dimensions, N, is usually small (e.g., $N = 14$), while each dimension, rather than being binary, spans a discrete range of values $\{0, 1, 2, \ldots, R - 1\}$. The dimensions are also called input variables, and an input variable might represent a joint angle of a robot arm (0–180 degrees) discretized in five-degree increments (resolution $R = 36$), and a 14-dimensional address might represent the angular positions and velocities in a seven-jointed robot arm. Different dimensions can have different resolutions, but we assume here, for simplicity, that all have the same resolution R.

An N-dimensional address in this space can be represented by an N-dimensional unit cube, or *cell*, and the entire address space is then represented by R^N of these cells packed into an N-dimensional cube with sides of length R. The cells are addressed naturally by N-place integers to base R.

A storage location is activated by some addresses and not by others. In the sparse distributed memory, these exciting addresses occupy an N-dimensional sphere with Hamming radius H, centred at the location's address. The exciting region of the address space in Albus's CMAC is an N-dimensional cube with sides of length K ($1 < K < R$); it is a cubicle of K^N cells (near the edge of the space it is the intersection of such a cubicle with the address space and thus contains fewer than K^N cells). The center coordinates of the cubicle can be thought of as the location's address (the center coordinates are integers if K is odd and half-way between two integers if K is even, and the center can lie outside the R^N cube).

The hard locations of a sparse distributed memory are placed randomly in the address space; those of CMAC—the cubicles—are arranged systematically as follows: First, the R^N cube is packed with the K^N cubicles starting from the corner cell at the origin—the cell addressed

by $(0, 0, 0, \ldots, 0)$. This defines a set of $\lceil R/K \rceil^N$ hard locations (the ceiling of the fraction means that the space is covered completely). The next set of $(1 + \lceil (R - 1)/K \rceil)^N$ hard locations is defined by moving the entire package of cubicles up by one cell along the principal diagonal of the R^N cube—a translation. To cover the entire address space, cubicles are added next to the existing ones at this stage. This is repeated until K sets of hard locations have been defined (K translations take the cubicles to the starting position), resulting in a total of at least $K \lceil R/K \rceil^N$ hard locations. Since each set of hard locations covers the entire R^N address space, and since the locations in a set do not overlap, each address activates exactly one location in each set and so it activates K locations overall. Conversely, each location is activated by the K^N addresses in its defining cubicle (by fewer if the cubicle spills over the edge of the space). The systematic placement of the hard locations allows addresses to be converted into activation vectors very efficiently in a hardware realization or in a computer simulation (Albus, 1980).

Correspondence of the hard locations to the granule cells of the cerebellum is natural in Albus's model. To make the model lifelike, each input variable (i.e., each coordinate of the address) is encoded in $R + K - 1$ bits. A bit in the encoding represents a mossy fiber, so that a vector of N input variables (an address) is presented to CMAC as binary inputs on $N(R + K - 1)$ mossy fibers. In the model, each granule cell receives input from N mossy fibers, and each mossy fiber provides input to at least $\lceil R/K \rceil^N$ granule cells.

The 20-bit code for an input variable s_n with range $R = 17$ and with $K = 4$ is given in Table 3.1. It corresponds to the encoding of the variables s_1 and s_2 in figure 6.8 in Albus's book (1981, p. 149). The bits are labeled with letters above the code in Table 3.1, and the same letters appear below the code in four rows. Bit A, for example, is on ($+$) when the input variable is at most 3, bit B is on when the input variable falls between 4 and 7, and so forth.

This encoding mimics nature. Many receptor neurons respond maximally to a specific value of an input variable and to values near it. An address bit (a mossy fiber) represents such a receptor, and it is ($+$)1 when the input variable is near this specific value. For example, this "central" value for bit B is 5.5.

The four rows of labels below the code in Table 3.1 correspond to the four sets of cubicles ($K = 4$) that define the hard locations (the granule cells) of CMAC. The first set depends only on the input bits labeled by the first row. If

Table 3.1. Encoding a 17-level Input Variable s_n in 20 Bits ($K = 4$).

	Input bit																			
s_n	F	M	S	A	G	N	T	B	H	P	V	C	J	Q	W	D	K	R	X	E
0	+	+	+	+	−	−	−	−	−	−	−	−	−	−	−	−	−	−	−	−
1	−	+	+	+	+	−	−	−	−	−	−	−	−	−	−	−	−	−	−	−
2	−	−	+	+	+	+	−	−	−	−	−	−	−	−	−	−	−	−	−	−
3	−	−	−	+	+	+	+	−	−	−	−	−	−	−	−	−	−	−	−	−
4	−	−	−	−	+	+	+	+	−	−	−	−	−	−	−	−	−	−	−	−
5	−	−	−	−	−	+	+	+	+	−	−	−	−	−	−	−	−	−	−	−
6	−	−	−	−	−	−	+	+	+	+	−	−	−	−	−	−	−	−	−	−
7	−	−	−	−	−	−	−	+	+	+	+	−	−	−	−	−	−	−	−	−
8	−	−	−	−	−	−	−	−	+	+	+	+	−	−	−	−	−	−	−	−
9	−	−	−	−	−	−	−	−	−	+	+	+	+	−	−	−	−	−	−	−
10	−	−	−	−	−	−	−	−	−	−	+	+	+	+	−	−	−	−	−	−
11	−	−	−	−	−	−	−	−	−	−	−	+	+	+	+	−	−	−	−	−
12	−	−	−	−	−	−	−	−	−	−	−	−	+	+	+	+	−	−	−	−
13	−	−	−	−	−	−	−	−	−	−	−	−	−	+	+	+	+	−	−	−
14	−	−	−	−	−	−	−	−	−	−	−	−	−	−	+	+	+	+	−	−
15	−	−	−	−	−	−	−	−	−	−	−	−	−	−	−	+	+	+	+	−
16	−	−	−	−	−	−	−	−	−	−	−	−	−	−	−	−	+	+	+	+

```
        A             B             C             D             E
  F           G             H             J             K
     M             N             P             Q             R
        S             T             V             W             X
```

the code for an input variable s_n has Q_1 first-row bits ($Q_1 = 5$ in Table 3.1), then the NQ_1 first-row bits of the N input variables define Q_1^N hard locations by assigning a location to each set of N inputs that combines one first-row bit from each input variable. The second set of Q_2^N hard locations is defined similarly by the NQ_2 second-row bits, and so forth with the rest.

We are now ready to describe Albus's CMAC design as a special case of Jaeckel's hyperplane design. The N input variables s_n are encoded and concatenated into an $N(R + K - 1)$-bit address $\mathbf{x}$, which will have NK 1s and $N(R - 1) - 1$s. The address matrix $\mathbf{A}$ will have $\sum_k Q_k^N$ rows, and each row will have N 1s, arranged according to the description in the preceding paragraph. The rest of $\mathbf{A}$ will be 0s (for "don't care"; there will be no -1s in $\mathbf{A}$). The activation vector $\mathbf{y}$ can then be computed as in the hyperplane design: the mth location is activated by $\mathbf{x}$ if the 1s of the hard address $\mathbf{A}_m$ are matched by 1s in $\mathbf{x}$ (i.e., iff $\mathbf{A}_m \cdot \mathbf{x} = N$).

If the number of input variables is large enough (e.g., $N > 20$), the number of rows in the address matrix $\mathbf{A}$, as given above, will be so large that building a hard location for each address in

$\mathbf{A}$ is impractical. To handle such cases, many addresses in $\mathbf{A}$ will use a single hard location. The contributions into a location's contents from disparate parts of the address space will then act as noise with respect to each other. The mapping of the addresses in $\mathbf{A}$ to the hard locations is pseudorandom and is effected by a hashing function. Multiple assignment of memory locations in this manner has been described also by Kohonen and Reuhkala (1978; Kohonen, 1980) in a method called redundant hash addressing.

After a set of locations has been activated, CMAC is ready to transfer data. Here, as with the sparse distributed memory, we can look at a single coordinate of a data words only, say, the uth coordinate. Since CMAC data are continuous or graded rather than binary, the storage and retrieval rules cannot be identical to those of a sparse distributed memory, but they are similar. Retrieval is simpler: we use the sum s_u as output and we omit the final thresholding. From the regularity of CMAC it follows that the sum is over K active locations.

From this is derived a storage (learning) rule for CMAC: Before storing the desired output value $\hat{p}_u$ at $\mathbf{x}$, retrieve s_u using $\mathbf{x}$ as the address and compute the error $s_u - \hat{p}_u$. If the error is acceptable, do nothing. If the error is too large, correct the K active counters (elements of the matrix $\mathbf{C}$) by adding $g(\hat{p}_u - s_u)/K$ to each, where g $(0 < g \leq 1)$ is a gain factor that affects the rate of learning. This storage rule implies that the counters in $\mathbf{C}$ count at intervals no greater than one Kth of the maximum allowable error (the counting interval in the basic sparse distributed memory is 1).

In summary, multidimensional input to CMAC can be encoded into a long binary vector that serves as an address to a hyperplane-design sparse distributed memory. The address bits and the hard-address decoders correspond very naturally to the mossy fibers and the granule cells of the cerebellum, respectively, and the activation of a hard location corresponds to the firing of a granule cell. The synapses of the parallel fibers with the Purkinje cells are the storage sites suggested by the model, and the value of an output variable is represented by the firing frequency of a Purkinje cell. Training of CMAC is by error-correction, which presumably is the function of the climbing fibers in the cerebellum.

3.8. SDM RESEARCH

So far in this chapter we have assumed that the hard addresses and the data are a uniform random sample of their respective spaces (the distribution of the hard locations in CMAC is uniform systematic). This has allowed us to establish a baseline: we have estimated signal, noise, fidelity, and memory capacity, and we have suggested reasonable values for various memory parameters. However, data from real processes tend to occur in clusters, and large regions of the address space are empty. When such data are stored in a uniformly distributed memory, large numbers of locations are never activated and hence are wasted, and many of the active locations are activated repeatedly so that they, too, are mostly wasted as their contents turn into noise.

There are many ways to counter this tendency of data to cluster. Let us look at the clustering of data addresses first. Several studies have used the memory efficiently by distributing the hard addresses $\mathbf{A}$ according to the distribution of the data addresses $\mathbf{X}$. Keeler (1988) observed that when the two distributions are the same and the activation radius H is adjusted for each storage and retrieval operation so that nearly an optimal number of locations are activated, the statistical properties of the memory are close to those of the basic memory with uniformly random hard

addresses. In agreement with that, Joglekar (1989) experimented with NETtalk data and got his best results by using a subset of the data addresses as hard addresses (NETtalk transcribes English text into phonemes; Sejnowski and Rosenberg, 1986). In a series of experiments by Danforth (1990), recognition of spoken digits, encoded in 240 bits, improved dramatically when uniformly random hard addresses were replaced by addresses that represented spoken words, but the selected-coordinate design with three coordinates performed the best. In yet another experiment, Saarinen et al. (1991b) improved memory utilization by distributing the hard addresses with Kohonen's self-organizing algorithm.

Two studies have shown that uniform random hard addresses can be used with clustered data if the rule for activating locations is adjusted appropriately. In Kanerva (1991), storage and retrieval require two steps: the first to determine a vector of N positive weights for each data address $\mathbf{X}_t$, and the second to activate locations according to a weighted Hamming distance between $\mathbf{X}_t$ and the hard addresses $\mathbf{A}$. In Pohja and Kaski (1992), each hard location has its own radius of activation H_m, which is chosen based on the data addresses $\mathbf{X}$ so that the probability of activating a location is nearly optimal.

It is equally important to deal with clustering in the stored words. For example, some of their bits may be mostly on, some may be mostly off, and some may depend on others. It is possible to analyze the data $(\mathbf{X}, \mathbf{Z})$ and the hard addresses $\mathbf{A}$ and to determine optimal storage and retrieval algorithms (Danforth, 1991), but we can also use iterative training by error correction, as described above for Albus's CMAC. This was done by Joglekar and by Danforth in their above-mentioned experiments. When error correction is used, it compensates for the clustering of addresses as well, but it also introduces the possibility of overfitting the model to the training set.

Two studies by Rogers (1989a, 1990a) deal specifically with the interactions of the data with the hard addresses $\mathbf{A}$. In the first of these he concludes that, in computing the sum vector $\mathbf{s}$, the active locations should be weighted according to the words stored in them—in fact, each active counter $C_{m,u}$ might be weighted individually. This would take into account at once the number of words stored in a hard location and the uniformity of those words, so as to give relatively little weight to locations or counters that record mostly noise. In the second study he uses a genetic algorithm to arrive at a set of hard addresses that would store the most information about a variable in weather data.

Other research issues include the storage of sequences (Manevitz, 1991) and the hierarchical storage of data (Manevitz and Zemach, work in progress).

Most studies of sparse distributed memory have used binary data and have dealt with multivalued variables by encoding them according to an appropriate binary code. Table 3.1 is an example of such a code. Important about the code is that the Hamming distance between codewords corresponds to the difference between the values being encoded (it grows with the difference until a maximum of $2k$ is reached, after which the Hamming distance stays at the maximum). Jorgensen (1990) proposes the Radial Basis Sparse Distributed Memory that uses ideas from radial-basis functions and probabilistic neural networks to deal with continuous variables; the paper also introduces the Infolding Net for working with nonstationary data. The use of continuous variables by Prager and Fallside has been discussed in Section 6.4.

Sparse distributed memory has been simulated on many computers (Rogers, 1990b), including the highly parallel Connection Machine (Rogers, 1989b) and special-purpose neural-network computers (Nordström, 1991). Hardware implementations have used standard logic circuits and memory chips (Flynn et al., 1987) and programmable gate arrays (Saarinen et al., 1991a). A systolic-array implementation of sparse distributed memory and a resistor circuit for computing the Hamming distances have been described by Keeler and Denning (1986).

3.9. ASSOCIATIVE MEMORY AS A COMPONENT OF A SYSTEM

In practical systems, an associative memory plays but a part. It can store and recall large numbers of large patterns (high-dimensional vectors) based on other large patterns that serve as memory cues, and it can store and recall long sequences of such patterns, doing it all in the presence of noise. In addition to generating output patterns, the memory provides an estimate of their reliability based on the data it has stored. But that is all; the memory assigns no meaning to the data beyond the reliability estimate. The meaning is determined by other parts of the system, which are also responsible for processing data into forms that are appropriate for an associative memory. Sometimes these other tasks are called preprocessing and

postprocessing, but the terms are misleading inasmuch as they imply that preprocessing and postprocessing are minor peripheral functions. They are major functions—at least in the nervous systems of animals—and feedback from memory is integral to these "peripheral" functions.

For an example of what a sensor processor must do in producing patterns for an associative memory, consider identifying objects by sight, and assume that the memory is trained to respond with the name of an object, in some suitable code, when presented with an object (i.e., when addressed by the encoding for the object). In what features should objects be encoded? To make efficient use of the memory, all views of an object—past, present, and future—should get the same encoding, and any two different objects should get different encodings. The name, as an encoding, satisfies this condition and so it is an ideal encoding, except that it is arbitrary. What we ask of the visual system is to produce an encoding that reflects physical reality and that can serve as an input to an associative memory, which then outputs the name.

For this final naming step to be successful—even with views as yet unseen—different views of an object should produce encodings that are similar to each other as measured by something like the Hamming distance, but that are dissimilar to the encodings of other objects. A raw retinal image (a pixel map) is a poor encoding, because the retinal cells excited by an object vary drastically with viewing distance and with gaze relative to the object. It is simple for us to fix the gaze—to look directly at the object—but it is impractical to bring objects to a standard viewing distance in order to recognize them. Therefore, the visual system needs to compensate for changes in viewing distance by encoding—by expressing images in features that are relatively insensitive to viewing distance. Orientations of lines in the retinal image satisfy this condition, making them good features for vision. This may explain the abundance of orientation-sensitive neurons in the visual cortex, and why the human visual system is much more sensitive to rotation than to scale (we are poor at recognizing objects in new orientations; we must resort to mental rotation). Encoding shapes in long vectors of bits for an associative memory, where a bit encodes orientation at a location, has been described by Kanerva (1990).

What about the claim that "peripheral" processing, particularly sensory processing, is a major activity in the brain? Large areas of the brain are specific to one sensory modality or another.

In robots that learn, an associative memory stores a world model that relates sensory input to action. The flow of events in the world is presented to the memory as a sequence of large patterns. These patterns encode sensor data, internal-state variables, and commands to the actuators. The memory's ability to store these sequences and to recall them under conditions that resemble the past allows its use for predicting and planning. Albus (1981, 1991) argues that intelligent behavior of animals and robots in complex environments requires not just one associative memory but a large hierarchy of them, with the sensors and the actuators at the bottom of the hierarchy.

3.10. SUMMARY

In this chapter we have explored a number of related designs for an associative memory. Common to them is a feed-forward architecture through two layers of input coefficients or weights represented by the matrices $\mathbf{A}$ and $\mathbf{C}$. The matrix $\mathbf{A}$ is constant, and the matrix $\mathbf{C}$ is variable. The M rows of $\mathbf{A}$ are interpreted as the addresses of M hard locations, and the M rows of $\mathbf{C}$ are interpreted as the contents of those locations. The rows of $\mathbf{A}$ are a random sample of the hard-address space in all but Albus's CMAC model, in which the sample is systematic. When the sample is random, it should allow for the distribution of the data.

The matrix $\mathbf{A}$ and the threshold function y transform N-dimensional input vectors into M-dimensional activation vectors of 0s and 1s. Since M is much larger than N, the effect is a tremendous increase over the input dimension and a corresponding increase in the separation of patterns and in memory capacity. This simplifies the storage of words by matrix $\mathbf{C}$. The training of $\mathbf{C}$ can be by the outer-product learning rule, by error correction (delta rule), by an analytic solution of a set of linear inequalities, or by a combination of the above. Training, by and large, is fast. These memories require much hardware per stored pattern, but the resolution of the components can be low.

The high fan-out and subsequent fan-in (divergence and convergence) implied by these designs are found also in many neural circuits in the brain. The correspondence is most striking in the cortex of the cerebellum, suggesting that the cerebellum could function as an associative memory with billions of hard locations, each one capable of storing several-hundred-bit words.

The properties of these associative memories imply that if such memory devices, indeed, play an important part in the brain, the brain must also include devices that are dedicated to the sensory systems and that transform sensory signals into forms appropriate for an associative memory.

Pattern Computing. The nervous system offers us a new model of computing, to be contrasted with traditional numeric computing and symbolic computing. It deals with large patterns as computational units and therefore it might be called *pattern computing.* The main units in numeric computing are numbers, say, 32-bit integers or 64-bit floating-point numbers, and we think of them as data; in symbolic computing they are pointers of fewer than 32 bits, and we can think of them as names (very compact, "ideal" encodings; see discussion on sensory encoding in Section 3.9). In contrast, the units in pattern computing have hundreds or thousands of bits, they serve both as pointers and as data, and they need not be precise. Nature has found a way to compute with such units, and we are barely beginning to understand how it is done. It appears that much of the power of pattern computing derives from the geometry of very-high-dimensional spaces and from the parallelism in computing that it allows.

ACKNOWLEDGMENTS

This work was supported by the National Aeronautics and Space Administration (NASA) Cooperative Agreement NC2-387 with the Universities Space Research Association (USRA). Computers for the work were a gift from Apple Computer Company. Many of the ideas came from the SDM Research Group of RIACS at the NASA–Ames Research Center. We are indebted to Dr. Michael Raugh for organizing and directing the group.

Copyright © 1993 Pentti Kanerva

REFERENCES

Albus, J. S. (1971). "A Theory of Cerebellar Functions," *Math. Biosci.,* **10**, 25–61.

Albus, J. S. (1980). "Method and Apparatus for Implementation of the CMAC Mapping Algorithm," U.S. Patent No. 4,193,115.

Albus, J. S. (1981). *Brains, Behavior, and Robotics.* BYTE/McGraw-Hill, Peterborough, N.H.

Albus, J. S. (1989). "The Marr and Albus Theories of the Cerebellum: Two Early Models of Associative Memory," in *Proc. COMPCON Spring '89* (34th IEEE Computer Society International Conference). IEEE Computer Society Press, Washington, D.C., pp. 577–582.

Albus, J. S. (1991). "Outline for a Theory of Intelligence," *IEEE Trans. Systems, Man, and Cybernetics*, **21**(3), 473–509.

Anderson, J. A. (1968). "A Memory Storage Module Utilizing Spatial Correlation Functions," *Kybernetik*, **5**(3), 113–119.

Chou, P. A. (1989). "The Capacity of the Kanerva Associative Memory," *IEEE Trans. Information Theory*, **35**(2), 281–298.

Clarke, T. J. W., Prager, R. W., and Fallside, F. (1991). "The Modified Kanerva Model: Theory and Results for Real-time Word Recognition," *IEE Proceedings—F*, **138**(1), 25–31.

Danforth, D. (1990). "An Empirical Investigation of Sparse Distributed Memory Using Discrete Speech Recognition," in *Proc. Int. Neural Network Conference (Paris)*, Vol. 1, pp. 183–186. Kluwer Academic, Norwell, Mass.). (Complete report, with the same title, in RIACS TR 90.18, Research Institute for Advanced Computer Science, NASA Ames Research Center.)

Danforth, D. (1991). *Total Recall in Distributed Associative Memories*. Report RIACS TR 91.3, Research Institute for Advanced Computer Science, NASA Ames Research Center.

Flynn, M. J., Kanerva, P., Ahanin, B., Bhadkamkar, N., Flaherty, P., and Hinkley, P. (1987). *Sparse Distributed Memory Prototype: Principles of Operation*. Report CSL-TR78-338, Computer Systems Laboratory, Stanford University.

Hassoun, M. H. (1988). "Two-level Neural Network for Deterministic Logic Processing," in *Optical Computing and Nonlinear Materials*, N. Peyghambarian, ed. *Proc. SPIE*, **881**, 258–264.

Hassoun, M. H. and Youssef, A. M. (1989). "High Performance Recording Algorithm for Hopfield Model Associative Memories," *Opt. Eng.*, **28**(1), 46–54.

Hopfield, J. J. (1982). "Neural Networks and Physical Systems with Emergent Collective Computational Abilities," *Proc. Nat. Acad. Sci. U.S.A. (Biophysics)*, **79**(8), 2554–2558. (Reprinted in Anderson, J. A. and Rosenfeld, E., eds. (1988). *Neurocomputing: Foundations of Research*. MIT Press, Cambridge, MA, pp. 460–464.)

Ito, M. (1984). *The Cerebellum and Neuronal Control*. Raven Press, New York.

Jaeckel, L. A. (1988). *Two Alternate Proofs of Wang's Lune Formula for Sparse Distributed Memory and an Integral Approximation*. Report RIACS TR 88.5, Research Institute for Advanced Computer Science, NASA Ames Research Center.

Jaeckel, L. A. (1989a). *An Alternative Design for a Sparse Distributed Memory*. Report RIACS TR 89.28, Research Institute for Advanced Computer Science, NASA Ames Research Center.

Jaeckel, L. A. (1989b). *A Class of Designs for a Sparse Distributed Memory*. Report RIACS TR 89.30, Research Institute for Advanced Computer Science, NASA Ames Research Center.

Joglekar, U. D. (1989). "Learning to Read Aloud: A Neural Network Approach Using Sparse Distributed Memory," Master's thesis, Computer Science, UC Santa Barbara. (Reprinted as report RIACS TR 89.27, Research Institute for Advanced Computer Science, NASA Ames Research Center.)

Jorgensen, C. C. (1990). *Distributed Memory Approaches for Robotic Neural Controllers*. Report RIACS TR 90.29, Research Institute for Advanced Computer Science, NASA Ames Research Center.

Kanerva, P. (1988). *Sparse Distributed Memory*. Bradford/MIT Press, Cambridge, MA.

Kanerva, P. (1990). "Contour-map Encoding of Shape for Early Vision," in *Proc. NIPS-89*, D. S. Touretzky, ed., *Neural Information Processing Systems*, Vol. 2, pp. 282–289. Kaufmann, San Mateo, CA.

Kanerva, P. (1991). "Effective Packing of Patterns in Sparse Distributed Memory by Selective Weighting of Input Bits," in *Proc. ICANN-91*, T. Kohonen, K. Mäkisara, O. Simula, and J. Kangas, eds., *Artificial Neural Networks*, Vol. 1, pp. 279–284. Elsevier/North-Holland, Amsterdam.

Keeler, J. D. (1988). "Comparison Between Kanerva's SDM and Hopfield-type Neural Networks," *Cognitive Sci.*, **12**, 299–329.

Keeler, J. D. and Denning, P. J. (1986). *Notes on Implementation of Sparse Distributed Memory*. Report RIACS TR 86.15, Research Institute for Advanced Computer Science, NASA Ames Research Center.

Kohonen, T. (1972). "Correlation Matrix Memories," *IEEE Trans. Computers*, **C21**(4), 353–359. (Reprinted in Anderson, J. A. and Rosenfeld, E., Eds. (1988). *Neurocomputing: Foundations of Research*, MIT Press, Cambridge, MA, pp. 174–180.)

Kohonen, T. (1980). *Content-Addressable Memories*. Springer-Verlag, New York.

Kohonen, T. (1984). *Self-Organization and Associative Memory*, 2nd edn. Springer-Verlag, New York.

Kohonen, T. and Reuhkala, E. (1978). "A Very Fast Associative Method for the Recognition and Correction of Misspelt Words, Based on Redundant Hash Addressing," in *Proc. Fourth Int. Joint Conference on Pattern Recognition*, Kyoto, Japan, pp. 807–809.

Llinás, R. R. (1975). "The Cortex of the Cerebellum," *Scientific American*, **232**(1), 56–71.

Loebner, E. E. (1989). "Intelligent Network Management and Functional Cerebellum Synthesis," in *Proc. COMPCON Spring '89* (34th IEEE Computer Society International Conference), pp. 583–588. IEEE Computer Society Press, Washington, D.C. (Reprinted in *The Selected Papers of Egon Loebner*, Hewlett-Packard Laboratories, Palo Alto, CA, 1991, pp. 205–209.)

Manevitz, L. M. (1991). "Implementing a 'Sense of Time' via Entropy in Associative Memories," in *Proc. ICANN-91*, T. Kohonen, K. Mäkisara, O. Simula, and J. Kangas, eds., *Artificial Neural Networks*, **2**, 1211–1214. Elsevier/North-Holland, Amsterdam.

Manevitz, L. M. and Zemach, Y. "Assigning Meaning to Data: Multilevel Information Processing in Kanerva's SDM," work in progress.

Marr, D. (1969). "A Theory of Cerebellar Cortex," *J. Physiol.* (*London*), **202**, 437–470.

Nordström, T. (1991). "Designing and Using Massively Parallel Computers for Artificial Neural Networks," Licentiate thesis 1991: 12L, Luleå University of Technology, Sweden.

Pohja, S. and Kaski, K. (1992). *Kanerva's Sparse Distributed Memory with Multiple Hamming Thresholds*. Report RIACS TR 92.06, Research Institute for Advanced Computer Science, NASA Ames Research Center.

Prager, R. W. and Fallside, F. (1989). "The Modified Kanerva Model for Automatic Speech Recognition," *Computer Speech and Language*, **3**(1), 61–81.

Rogers, D. (1989a). "Statistical Prediction with Kanerva's Sparse Distributed Memory," in *Proc. NIPS-88*, D. S. Touretzky, ed., *Neural Information Processing Systems*, **1**, 586–593. Kaufmann, San Mateo, CA.

Rogers, D. (1989b). "Kanerva's Sparse Distributed Memory: An Associative Memory Algorithm Well-suited to the Connection Machine," *Int. J. High Speed Computing*, **1**(2), 349–365.

Rogers, D. (1990a). "Predicting Weather using a Genetic Memory: A Combination of Kanerva's Sparse Distributed Memory and Holland's Genetic Algorithms," in *Proc. NIPS-89*, D. S. Touretzky, ed., *Neural Information Processing Systems*, **2**, 455–464. Kaufmann, San Mateo, CA.

Rogers, D. (1990b). *BIRD: A General Interface for Sparse Distributed Memory Simulators*. Report

RIACS TR 90.3, Research Institute for Advanced Computer Science, NASA Ames Research Center.

Rosenblatt, F. (1962). *Principles of Neurodynamics*. Spartan, Washington, D.C.

Rumelhart, D. E. and McClelland, J. L., eds. (1986). *Parallel Distributed Processing*, Vols. 1 and 2. Bradford/MIT Press, Cambridge, MA.

Saarinen, J., Lindell, M., Kotilainen, P., Tomberg, J., Kanerva, P. and Kaski, K. (1991a). "Highly Parallel Hardware Implementation of Sparse Distributed Memory," in *Proc. ICANN-91*, T. Kohonen, K. Mäkisara, O. Simula, and J. Kangas, eds., *Artificial Neural Networks*, **1**, 673–678. Elsevier/North-Holland, Amsterdam.

Saarinen, J., Pohja, S., and Kaski, K. (1991b). "Self-organization with Kanerva's Sparse Distributed Memory," in *Proc. ICANN-91*, T. Kohonen, K. Mäkisara, O. Simula, and J. Kangas, eds., *Artificial Neural Networks*, **1**, 285–290. Elsevier/North-Holland, Amsterdam.

Sejnowski, T. J. and Rosenberg, C. R. (1986). *NETtalk: A Parallel Network that Learns to Read Aloud*. Report JHU/EECS-86/01, Department of Electrical Engineering and Computer Science, Johns Hopkins University. (Reprinted in Anderson, J. A. and Rosenfeld, E., eds. (1988). *Neurocomputing: Foundations of Research*. MIT Press, Cambridge, MA, pp. 663–672.)

Willshaw, D. (1981). "Holography, Associative Memory, and Inductive Generalization," in *Parallel Models of Associative Memory*, G. E. Hinton and J. A. Anderson, eds. Erlbaum, Hillsdale, N.J., pp. 83–104.

4

The BSB Model: A Simple Nonlinear Autoassociative Neural Network[1]

JAMES A. ANDERSON

It is, sir, as I have said, a simple network,
and yet there are those who love it.

—Modified from Daniel Webster (1818)

4.1. INTRODUCTION

I have no particular interest in the details of neural networks. However, I have a great deal of interest in what neural networks compute. If we are to take neural networks seriously as models of human computation then it is necessary to get some idea of their computational powers and limitations. The BSB model, which I will discuss in this chapter, was proposed in 1977 (Anderson, Silverstein, Ritz, and Jones) as a simple, nonlinear network that was rich in behavior but easy to simulate and to analyze formally. One reason it is easy to work with is that it is two networks combined: a simple associative network coupled to a nonlinear dynamics. It is a weak model, in that it does not have the intrinsic power to form correct input–output relationships shown by more powerful networks using complex learning rules like back propagation.

Since it is a weak network, we cannot count on the BSB network to learn difficult discriminations for us. It is not a magic network that can theoretically solve any problem. We can make a virtue of this weakness and say that since there are so few places to tweak the network, we must worry instead about externals: the data representation, what we want to compute, and what we want the answer to look like. We also must focus on network size. A small weak network can solve very little, but a large weak network can be very powerful if it scales well. It is hard not to be impressed by the huge size of biological networks, especially in structures like mammalian cerebral cortex.

A useful cognitive network must be at least adequate for a very wide range of tasks, though not necessarily outstanding at any of them. Versatility is a major virtue in a biological system. My wistful hope is that BSB captures enough of the essentials of mental computation in its parallelism, associativity, nonlinearity, and temporal dynamics to be useful as a low-order approximation to a broad range of human mental operations. In any case, focusing on models for human cognitive computation strikes me as a more important direction in which to look than designing yet another network to solve X-OR or parity.

Papers describing uses of BSB in cognitive modeling are: (1) storing information in a knowledge base, simple qualitative physics, and realizing a semantic network (Anderson, 1986); (2) disambiguation and multistable perception (Kawamoto and Anderson, 1985; Kawamoto, 1986; Anderson and Murphy, 1986); (3) probability learning and categorical perception in speech (Anderson et al., 1977; (4) concept and category formation (Anderson and Mozer, 1981; Anderson and Rosenfeld, 1988; Begin, 1991); (5) clustering and organizing a complex radar environment (Anderson et al., 1990a); (6) learning the multiplication tables (Viscuso, 1989; Anderson et al., in press); (7) for part of a speech recognition system (Rossen, 1989; Anderson et al., 1990b); (8) as a model for psychological response time (Anderson, 1991); and (9) as a tool for flexible programming of a neural network (Anderson et al., in press).

In one sentence, a BSB network is an autoassociative, energy-minimizing nonlinear dynamical system. This chapter is a description of the theory behind BSB and an attempt to

[1] This material is adapted from a book to appear from MIT Press, *Practical Neural Modelling* by James A. Anderson.

describe informally some of its mathematical properties. We have provided one "psychological" simulation at the end as a worked-out example to show how BSB can be used to model psychological response time. The simulation is of some independent interest as an indication of how long it takes a dynamical system neural network to settle on an answer.

It is worth remembering that states vectors in a neural network are "real" in that they correspond in some way to unit discharges. They are also real in that they contain the "meaning" in the system and the results of the network computation. Therefore we have to be careful when we discuss operations on the state vectors in terms of abstract mathematical operations: these operations are describing information processing by a real entity.

4.2. THE BASIC STRUCTURE OF THE BSB MODEL

4.2.1. Feedback Models

The BSB model is a feedback model. The connectivity is shown in Fig. 4.1. The nervous system is full of recurrent systems of various kinds. Examples occur at all levels. For example, the recurrent collateral system of cortical pyramids provides widely distributed local excitatory feedback over a few millimeters. At a higher level, cortical areas have strong downward as well as upward projections. As Van Essen puts it in a review article on visual cortex, "A particularly significant feature of cortical connectivity is a strong, perhaps universal tendency for pathways to occur as reciprocal pairs.... there are as yet no convincing counterexamples to what appears to be a basic organizational principle in macaque visual cor-

tex." (van Essen, 1985, p. 284.) There is little evidence suggesting that cortical information processing is done only by simple feedforward networks. Actual anatomy suggests the substrate for powerful feedback loops at many levels and that computation may be done by many richly interconnected functional modules settling on a consensus answer. The BSB model is a simple example of a formal model which tries to capture some of this flavor.

4.2.2. Autoassociative Reconstruction

Consider what Kohonen calls an "autoassociative" system, which is a variant of the standard associative structure. We can use autoassociative feedback to reconstruct the missing part of a state vector. Suppose we have a normalized state vector $\mathbf{f}$, which is composed of two parts, say $\mathbf{f}'$ and $\mathbf{f}''$, i.e., $\mathbf{f} = \mathbf{f}' + \mathbf{f}''$. Suppose $\mathbf{f}'$ and $\mathbf{f}''$ are orthogonal. The easiest way to accomplish this would be to have $\mathbf{f}'$ and $\mathbf{f}''$ be subvectors that occupy different sets of elements—say $\mathbf{f}'$ is nonzero only for elements $[1 \ldots n]$ and $\mathbf{f}''$ is nonzero only for elements $[(n + 1) \ldots \text{Dimen-}$ sionality$]$. Then consider a matrix $\mathbf{A}$ storing only the single autoassociation of $\mathbf{f}$ with itself using outer product learning, that is,

$$\mathbf{A} = (\mathbf{f}' + \mathbf{f}'')(\mathbf{f}' + \mathbf{f}'')^{\mathrm{T}}$$

Suppose that at some future time part of the complete state vector $\mathbf{f}$, say $\mathbf{f}'$, is presented at the input to the system. The output is then given by

$$(\text{output}) = \mathbf{A}\mathbf{f}'$$
$$= (\mathbf{f}'\mathbf{f}'^{\mathrm{T}} + \mathbf{f}'\mathbf{f}''^{\mathrm{T}} + \mathbf{f}''\mathbf{f}'^{\mathrm{T}} + \mathbf{f}''\mathbf{f}''^{\mathrm{T}})\mathbf{f}'$$

Since $\mathbf{f}'$ and $\mathbf{f}''$ are orthogonal

$$(\text{output}) = (\mathbf{f}' + \mathbf{f}'')[\mathbf{f}', \mathbf{f}']$$
$$= \alpha\mathbf{f}$$

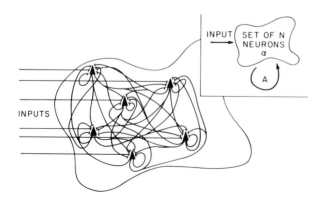

INPUT → SET OF N NEURONS α

A

INPUTS

Fig. 4.1. A group of neurons feeds back on itself. The basis of autoassociative feedback in the BSB model. (From Anderson et al., 1977. Reprinted by permission. © American Psychological Association.

where α is a constant since the inner product $[\mathbf{f}', \mathbf{f}']$ is simply a number. Autoassociation can reconstruct the missing part of the state vector. This property forms the basis for many of the uses of autoassociation.

The easiest way to analyze an autoassociative feedback system is in terms of the eigenvectors and eigenvalues of the connection matrix, $\mathbf{A}$. Let us assume we have multiple items stored in the connection matrix $\mathbf{A}$. Let us now assume that $\mathbf{f}$ is an *eigenvector* of the connection matrix, $\mathbf{A}$, with eigenvalue λ, that is,

$$\mathbf{A}\mathbf{f} = \lambda\mathbf{f} \qquad (1)$$

Suppose a set of units with autoassociative feedback is receiving an input, say, $\mathbf{f}$, and also feedback through the connection matrix, $\mathbf{A}\mathbf{f}$, and adds them. Let us assume the feedback pathway operates with a time delay, say one time step, shown in Fig. 4.2. Then, if we assume that the input is constant and always present, at $t = 0$, the activity on the set of elements is $\mathbf{f}$.

at $t = 1$, activity $= \mathbf{f} + \lambda\mathbf{f} = (1 + \lambda)\mathbf{f}$

at $t = 2$, activity $= (1 + \lambda)\mathbf{f} + \lambda(1 + \lambda)\mathbf{f}$

$\qquad\qquad\qquad = (1 + \lambda)^2\mathbf{f}$

at $t = 3$, activity $= (1 + \lambda)^2\mathbf{f} + \lambda(1 + \lambda)^2\mathbf{f}$

$\qquad\qquad\qquad = (1 + \lambda)^3\mathbf{f} \ldots$

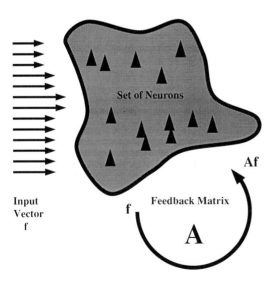

Fig. 4.2. Schematic diagram of the vector feedback used in the BSB model. For a single iteration, the previous state vector, $\mathbf{f}$, is fed back through the connection matrix, $\mathbf{A}$, and adds to the previous state of the system.

It can be seen that as long as λ is positive, activity keeps on growing geometrically.

The enhancement is proportional to the magnitude of the eigenvalue, which becomes a kind of "strength of self-association," effectively a gain parameter. Here we have a simple example of a system where a very large and complex pattern of unit activities, in this case an eigenvector, behaves in a very simple way under the control of a single scalar variable, the eigenvalue. For someone familiar with the classical mathematical psychology tradition, it is hard to avoid the suggestion that the eigenvalues of the connection matrix in this system have similarities to the concept of "trace strength" as used in many formal psychological models. Trace strength is usually conceived of as a scalar value that is related to the retrievability and overall importance of an item stored in memory.

Recent interest in nonlinear dynamical systems in physics and mathematics has led to productive attempts to predict and work with global system properties, rather than focusing on the behavior of the small parts of the system. As one of many examples, Haken and the well known "synergetics" group in Germany have followed this approach for years. Recently, they have turned their attention to complex biological and psychological phenomena (see Haken and Stadler, 1990, for example). In one of Haken's analogies, a laser is best understood by considering the behavior of the entire system: "In a laser there may be 10^{16} degrees of freedom of the atoms and 1 degree of freedom of the field mode. ... at the lasing threshold, the whole system is governed by a single degree of freedom" (Haken, 1985, p. 5). Similarly, in the feedback systems we will describe, the behavior of a large, nonlinear, and complex system can be understood to some degree by use of well-defined large-scale entities like the eigenvectors of the system. Therefore this approach suggests a simple way of linking the behavior of very many low-level units with overall system behavior. The most desirable and useful kind of psychological model would be one where the richness of the world could be represented by extremely complex patterns of unit activation but overall system behavior would be governed by a small number of global parameters.

If an input is composed of the sum of multiple eigenvectors, the eigenvector with largest magnitude eigenvalue will receive the greatest relative enhancement. This is the basis of the *power method* of eigenvector computation, which can be used to pick out the eigenvector with the largest-magnitude eigenvalue of a matrix, $\mathbf{A}$, by repeatedly passing a random starting vector through the matrix. After a number of iterations, the eigenvector with largest magnitude eigenvalue will dominate.

4.2.3. An Experiment with the Power Method

Let us do a numerical experiment to see if we can get some feeling for how matrix feedback behaves. More general discussions of the power method can be found in many books on numerical analysis (Householder, 1964; Wilkinson, 1965; Hornbeck, 1975; Young and Gregory, 1973). The unmodified power method is most often used for computing a few eigenvectors, the eigenvectors with the largest-magnitude eigenvalues. Variants such as the inverse power method are capable of computing eigenvectors with eigenvalues near any value. The more widely spaced the eigenvalues, the better and faster the system works.

The idea is simple. Suppose we have a matrix, $\mathbf{A}$. Let us assume the eigenvectors are orthogonal. Let us start with a random vector, $\mathbf{v}(0)$. We will repeatedly pass this vector through $\mathbf{A}$. At the ith iteration

$$\mathbf{v}(i + 1) = \mathbf{A}\mathbf{v}(i)$$

But $\mathbf{v}(0)$ can be represented as a weighted sum of eigenvectors, $\mathbf{e}_i$

$$\mathbf{v}(0) = \sum a_i \mathbf{e}_i \qquad (2)$$

where a_i are weighting coefficients.

After the first iteration, $\mathbf{v}(1)$ is given by

$$\mathbf{v}(1) = \sum \lambda_i a_i \mathbf{e}_i$$

Consider the ratio of two terms, i and j. After a single iteration, the ratio will be given by $a_i \lambda_i / a_j \lambda_j$ and after n iterations the ratio will be $(a_i/a_j)(\lambda_i/\lambda_j)^n$. If $|\lambda|_i > |\lambda|_j$ the magnitude of this ratio will grow without bound.

The eigenvector with the eigenvalue with largest magnitude is called the *dominant* eigenvector. After a large number of iterations, the vector $\mathbf{v}(n)$ will be almost entirely in the direction of the dominant eigenvector. The only case where this would not happen would be in the improbable event that the initial random vector happened to have no component in the direction of the dominant eigenvector.

By successive iterations we can find the direction of the dominant eigenvector to whatever precision we want. We can also get an estimate of the eigenvalue. Suppose we have waited long enough so that the vector $\mathbf{v}(n)$ is in the direction of the eigenvector $\mathbf{e}_i$. Then,

$$\mathbf{v}(n + 1) = \mathbf{A}\mathbf{v}(n) = \lambda\mathbf{v}(n)$$

By looking at the ratio of lengths of $\mathbf{v}(n + 1)$ and $\mathbf{v}(n)$ we can estimate the eigenvalue, λ_i.

We can easily demonstrate this process. First, we need to know how to construct a matrix $\mathbf{A}$ with known eigenvectors and eigenvalues. An outer product matrix has this property. Consider a set of orthogonal normalized vectors, $\{\mathbf{e}_i\}$ with eigenvalues $\{\lambda_i\}$. Then we can construct a matrix $\mathbf{A}$ with desired eigenvectors and eigenvalues by

$$\mathbf{A} = \sum \lambda_i \mathbf{e}_i \mathbf{e}_i^{\mathrm{T}} \qquad (3)$$

It is easy to check that $\mathbf{e}_i$ is an eigenvector and that λ_i is its eigenvalue.

For the first simulation, the vectors used are given in Table 4.1. In Table 4.1, the "+s" correspond to +1 and the "−s" to −1. These vectors are examples of Walsh functions, a useful set of orthogonal functions often used in image processing. A matrix with known eigenvectors and eigenvalues was constructed from these 32-dimensional vectors by normalizing each vector and forming the autoassociative outer product of the vector with itself. The resulting outer product matrix for each vector was multiplied by the

Table 4.1. Vectors Used to Construct a Matrix with Known Eigenvectors. Vectors are Walsh Functions.

Eigenvalue	State Vectors[a]			
1.1	+ + + + + + + +	+ + + + + + + +	− − − − − − − −	− − − − − − − −
1.0	+ + + + + + + +	− − − − − − − −	+ + + + + + + +	− − − − − − − −
0.9	+ + + + − − − −	+ + + + − − − −	+ + + + − − − −	+ + + + − − − −
0.8	+ + − − + + − −	+ + − − + + − −	+ + − − + + − −	+ + − − + + − −
0.7	+ − − + + − − +	+ − − + + − − +	+ − − + + − − +	+ − − + + − − +
0.6	+ − + − + − + −	+ − + − + − + −	+ − + − + − + −	+ − + − + − + −

[a] The "+s" correspond to vector elements of +1 and the "−s" to −1.

Table 4.2. State Vector after 100 Iterations Starting from Random Vector.

| Estimated Eigenvalue | Cosines Between True Eigenvector and Final State Vector After 100 Iterations | | | | | |
	1	2	3	4	5	6
1.099 9998	0.999 9999	0.000 4442	0.000 0000	0.000 0000	0.000 0000	0,000 0000

The estimate of the eigenvalue is the ratio of lengths of input vector and vector after an iteration through the matrix.

Table 4.3. Power Method Demonstration: Computation of Successive Eigenvalues and Eigenvectors.

| Pass | Estimated Eigenvalue | Cosines Between True Eigenvector and Final State Vector After 100 Iterations | | | | | |
		1	2	3	4	5	6
1.	1.099 9998	0.999 9999	0.000 4442	0.000 0000	0.000 0000	0.000 0000	0.000 0000
2.	1.000 0001	−0.000 4887	0.999 9999	0.000 0134	0.000 0000	0.000 0000	0.000 0000
3.	0.900 0000	0.000 0000	0.000 0149	−1.000 0000	0.000 0557	0.000 0000	0.000 0000
4.	0.800 0001	0.000 0000	0.000 0000	−0.000 0626	−0.999 9999	0.000 0052	0.000 0000
5.	0.700 0000	0.000 0000	0.000 0000	−0.000 0001	−0.000 0060	−1.000 0000	−0.000 0002
6.	0.600 0001	0.000 0000	0.000 0000	0.000 0000	0.000 0000	0.000 0002	−0.999 9999

eigenvalue, and the six outer product matrices were summed to form the final matrix.

We choose a random starting vector. For the first simulation, the state vector was renormalized after each iteration. After 100 passes through the matrix, Table 4.2 gives the cosine between the six eigenvectors and the resulting vector. The final vector is pointing in the direction of e_1, the eigenvector with largest eigenvalue. The ratio of lengths between the input and output at the 100th iteration is a very accurate (6 places) estimate of the eigenvalue of the dominant eigenvector. There are no components from any eigenvectors except the two with the largest-magnitude eigenvalues, and the contribution from the eigenvector with the second largest eigenvalue is very small, and decreasing in magnitude with each iteration.

We need not stop with computing only the dominant eigenvector. We can modify our matrix construction method, which used the sum of outer products, in an obvious way, and a natural way for a neural network, to find the other eigenvectors. (For use of this technique in a cognitive simulation, see Kawamoto and Anderson, 1985.)

We have just estimated the eigenvector and eigenvalue by the power method. Suppose we form the outer product of the estimated normalized eigenvector, multiply it by the estimated eigenvalues and *subtract* it from the matrix. (To a neural network model, the outer product is

simple Hebb learning, in this case with a negative sign.) If e' and λ' are the estimates of eigenvector and eigenvalue, then we can modify A, forming matrix A', so that,

$$A' = \sum \lambda_i e_i e_i^T - \lambda' e' e'^T \qquad (4)$$

Assuming that e' and λ' are good estimates then the new matrix, A', will have reduced the eigenvalue of the dominant eigenvector, e_i, say, to

$$(\lambda_i - \lambda') \simeq 0$$

By repeating this process for each eigenvector as it is computed by the power method, it is possible to compute the eigenvectors in order of the magnitude of their eigenvalue. Using anti-Hebb adaptation is an extremely robust technique. The estimate of the dominant eigenvalue and eigenvector can actually be quite poor. Adaptation merely has to reduce the magnitude of the dominant eigenvalue enough so that it is no longer dominant.

Table 4.3 shows the results of this process applied to the matrix we constructed earlier. The estimates of eigenvectors and eigenvalues are accurate to five or six decimal places. These estimates will contain small errors. If this process is continued indefinitely, the estimates will become inaccurate enough to be useless.

The values in Table 4.3 give the estimated eigenvalue and the cosine between the true eigenvector and the eigenvector computed by this process. These can be compared with the actual eigenvalues used in constructing the matrix in Table 4.1. One hundred iterations were used for each estimation. Eigenvectors are only determined to a multiplicative constant. A cosine of -1 simply means that the eigenvector is pointing in the opposite direction to the reference direction; it is still a correct solution.

4.2.4. Shifting Eigenvalues

Because the ratio of eigenvalues dominates how rapidly the computation converges, it is possible to speed up the computation and improve accuracy if we have some idea of the value of the eigenvalues. Sometimes this eigenvalue can be estimated from other information in the problem. For example, when we make the connection with neural networks, we often use a learning rule that will produce eigenvalues with maximum values near one and minimum values near zero.

The complete set of eigenvalues is called the *eigenvalue spectrum*. It is easy to translate the eigenvalue spectrum up or down by subtracting a multiple of the identity matrix, $\mathbf{I}$, from the matrix $\mathbf{A}$. For any vector, $\mathbf{Ix} = \mathbf{x}$. The identity matrix is all zeros except for ones down the main diagonal. Suppose we form the matrix $\mathbf{A}'$ so that

$$\mathbf{A}' = \mathbf{A} + \gamma\mathbf{I}$$

Consider an eigenvector $\mathbf{e}$ of $\mathbf{A}$, and its associated eigenvalue, λ. Then we know that $\mathbf{Ae} = \lambda\mathbf{e}$. But $\mathbf{e}$ is also an eigenvector of $\mathbf{A}'$ because

$$\mathbf{A}'\mathbf{e} = \mathbf{Ae} + \gamma\mathbf{Ie}$$

$$= \lambda\mathbf{e} + \gamma\mathbf{e}$$

$$= (\lambda + \gamma)\mathbf{e}$$

The new eigenvalue of $\mathbf{e}$ is $(\lambda + \gamma)$. The entire eigenvalue spectrum has been translated upward by an amount γ. Consider the ratio between two eigenvalues, a quantity which dominates the power method computation. The larger this ratio is, the faster is convergence. By using a proper γ this ratio can often be made larger.

As an aside, in numerical analysis, it is often more convenient to compute the eigenvectors of the inverse matrix, which will be the same as for the uninverted matrix. The eigenvalues of the inverse matrix are the reciprocal of the eigenvalues of the uninverted matrix. By shifting the eigenvalue spectrum, the dominant eigenvector of the inverse matrix can be placed at any point by moving its uninverted eigenvalue closest to zero. Eigenvectors in any region of the eigenvalue spectrum can be computed using this method,

which is fast and accurate (see Wilkinson, 1965). It is difficult to see how to apply this trick to neural networks, however.

Although the pure power method simply picks out the dominant eigenvalue, positive and negative eigenvalues behave differently in the feedback model. A negative eigenvalue will cause the sign of the eigenvector component to change after each iteration. When we discussed neural network feedback models, we assumed that the previous state vector remained present. Suppose the connectivity matrix is $\mathbf{A}$. If the current state vector is the eigenvector $\mathbf{e}(t)$ then the next iteration gives

$$\mathbf{x}(t + 1) = \mathbf{e}(t) + \mathbf{Ae}(t) = (\lambda + 1)\mathbf{e}(t)$$

This process has shifted the eigenvalue by $+1$, increasing the magnitude of positive eigenvalues and decreasing the magnitude of negative eigenvalues.

4.2.5. Learning

In the simplest version of the BSB model, the matrix is formed using a Hebbian rule, giving a matrix that is a sum of outer products. If a set $\{\mathbf{f}_i\}$ is to be learned, then

$$\mathbf{A} = \eta \sum_i \mathbf{f}_i\mathbf{f}_i \qquad (5)$$

This matrix (except for the constant) has the form of the covariance matrix of statistics. The eigenvectors of this matrix are used to form the *principal components* of statistics. This matrix is symmetric, and therefore has orthogonal eigenvectors and real eigenvalues. The eigenvalues are zero or positive. The principal components, ordered in terms of size of eigenvalue, are the components that contain the most variance; the first principal component contains the largest amount of the variance of the system, the second principal component contains the next largest amount of the variance, and so on.

The eigenvectors that get relatively enhanced by matrix feedback are therefore the ones that are the best to use, if it is desired to make discriminations among members of the input set. We once suggested that such a system when combined with simple Hebbian learning was acting like analysis in terms of what are called *distinctive features* in psychology, i.e., critical features that are particularly useful in making discriminations. (Anderson et al., 1977). Many other neural networks turn out to have close connections to principal components. Cottrell, Munro, and Zipser (1988) showed this experimentally in an application to image compression.

Baldi and Hornik (1989) suggested a close connection between encoder networks in back propagation and principal components. Linsker (1988) analyzed an intriguing theory for the organization of visual cortex in terms of principal components (also see MacKay and Miller, 1990). Neural network techniques have been used by several groups to compute principal components directly (Oja, 1982, 1983; Foldiak, 1989; Krogh and Hertz, 1990). The similarity between the rule for Hebbian synaptic modification and correlation seems to make connection between principal components and neural networks natural and inevitable. In any case, matrix feedback combined with a variant of Hebbian learning can serve as valuable and statistically valid preprocessing.

At the present time, besides simple Hebbian learning, another useful associative learning algorithm to construct the connection matrix is the well-known least-mean-squares algorithm (Widrow and Hoff, 1960). [See Proulx and Begin (1990) for further comments about learning.] This technique is fast and robust, and does gradient descent in weight space in a simple error surface with no local minima. Simple vector Widrow–Hoff learning can be viewed as Hebbian learning of the error vector. That is, if $\mathbf{g}$ is the desired output to an input $\mathbf{f}$, and $\mathbf{g}'$ is what is obtained, then learning adds to the connectivity matrix a term

$$\Delta\mathbf{A} = \eta(\mathbf{g} - \mathbf{g}')\mathbf{f}^{\mathrm{T}} \qquad (6)$$

that is, the error vector is learned, not the associated vector. In the context of auto-associative matrix feedback (i.e., $\mathbf{g} = \mathbf{f}$), if the Widrow–Hoff technique works perfectly, we will develop a system that has eigenvectors corresponding to the learned set and with eigenvalue 1, i.e.

$$\mathbf{Af}_i = \mathbf{f}_i$$

Perfect simple Widrow–Hoff learning, ultimately, aims to produce a degenerate system, with multiple eigenvectors having eigenvalue 1. This means that any linear combination of eigenvectors will also be an eigenvector.

As learning progresses, the eigenvalue spectrum will contract so that the differences between the largest eigenvalues will decrease. This makes vector feedback and the power method enhancement less efficient. However, because of the nonlinearities in BSB, this is not a major problem in practice. There are a number of ways to deal with it. One possibility, with plausible biological support, would be to assume the presence of adaptive processes that constantly shift the

eigenvalues of a particular eigenvector up and down, depending on past history. One could also use Widrow–Hoff learning where the error measure is not simple error but error computed from a constant multiple of the desired vector, for example,

$$\Delta\mathbf{A} = \eta(c\mathbf{g} - \mathbf{g}')\mathbf{f}^{\mathrm{T}} \qquad (7)$$

The target eigenvalue will then become the constant, c, instead of 1.

There is also a connection between principal components and error correction learning. Widrow and Stearns (1985) point out that the eigenvectors of the covariance matrix define the principal axes of the error surface. Corrections will be largest to the principal components with the largest eigenvalue because the error surface is steepest there.

4.2.6. Introducing the BSB Nonlinearity

Since positive feedback may make the output vector grow without bounds, the simplest way to contain the vector is to put upper and lower limits on the allowable values of elements of the vectors. This is physiologically reasonable since neurons have a limited dynamic range. Hard limits are a simple special case of the "sigmoid" or "squashing function" used in the generic neuron model of neural networks (Anderson and Rosenfeld, 1988). It has the effect of limiting the state vector to restricted region of space, hence the name "Brain-State-in-a-Box" or BSB model (Anderson et al., 1977; Anderson and Mozer, 1981; Golden, 1986; Hui and Zak, 1992).

We will describe here the simplest BSB variant. The activity pattern in a set of neurons feeding back on itself is assumed to be composed of (1) the input to the neurons, (2) the feedback coming back through recurrent connections from past states of the neurons, and (3) the persistence of activity in the state, that is, activity does not drop to zero immediately but decays with some kind of time constant. We assume that time is quantized.

Suppose $\mathbf{x}(t + 1)$ is the state vector at the $t + 1$ time step. In general we will assume the input to be presented at $t = 0$, and then the system "processes" that input. The feedback is given by $\alpha\mathbf{Ax}(t)$, where α is a feedback constant to control strength of feedback. The decay of the previous state is given by some constant, γ, multiplying the previous state, that is $\gamma\mathbf{x}(t)$. Sometimes it is desirable to keep the initial input, $\mathbf{f}(0)$, present, perhaps multiplied by a constant δ, usually either 0 (not present) or 1 (present). So we have the next state, $\mathbf{x}(t + 1)$, given as

$$\mathbf{x}(t + 1) = \text{LIMIT}(\gamma\mathbf{x}(t) + \alpha\mathbf{A}\mathbf{x}(t) + \delta\mathbf{f}(0)) \quad (8)$$

The LIMIT operation clips values that exceed upper or lower limits. In most BSB simulations the limits are symmetrical, that is, the upper and lower limits are equal.

Constants are chosen so that feedback is positive. This equation produces a simple dynamical system. The positive feedback causes the vector to lengthen, just as we have seen for the power method in the linear case. However, in BSB the lengthening slows when the state vector components start to limit and ceases when all components are at a limit. In the terminology of dynamical system theory, the final stable states are called *attractors*. BSB dynamics can be shown (Golden, 1986) to be minimizing an "energy" (Liapunov) function, so that the stable states are points of minimum energy. Some corners tend to be stable attractor states because feedback will always be pushing them outward. All the points in a region will end in a final stable attractor state. These points are referred to as the *basin of attraction* of the attractor. Figure 4.3 shows a

very simple example of a two-dimensional BSB model in action. (Anderson and Silverstein, 1978). There are other dynamical system models in the neural net literature. Among them are the models of Hopfield (1982, 1984), the ART models (Carpenter and Grossberg, 1987), and the BAM model of Kosko (1988). Many important results on this class of models, especially the Hopfield networks, are found in a valuable book by Amit (1989). The qualitative behavior of these different networks is often quite similar and they partake of a number of formal similarities (see Cohen and Grossberg, 1983).

For the BSB network, it is easy to show that there are three classes of attractors in interesting cases, where there is more than one attractor. The origin is an unstable attractor, that is, the origin itself is stable in that a state vector there will not change. However, any shift, no matter how small, away from the origin will cause the state vector to shift to a stable attractor. The boundaries between basins are also unstable attractors. Finally, there can be stable point attractors, which are generally corners of the hypercube of limits. Simple BSB models cannot

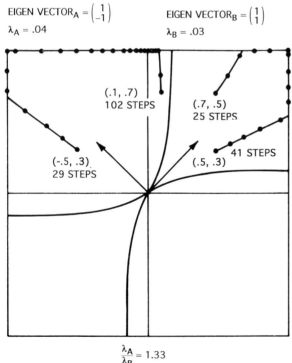

EIGEN VECTOR$_A = \begin{pmatrix} 1 \\ -1 \end{pmatrix}$ $\qquad$ EIGEN VECTOR$_B = \begin{pmatrix} 1 \\ 1 \end{pmatrix}$

$\lambda_A = .04$ $\qquad\qquad\qquad$ $\lambda_B = .03$

(.1, .7)
102 STEPS

(.7, .5)
25 STEPS

(-.5, .3)
29 STEPS

(.5, .3)

41 STEPS

$$\frac{\lambda_A}{\lambda_B} = 1.33$$

FEEDBACK MATRIX $= \begin{pmatrix} .035 & -.005 \\ -.005 & .035 \end{pmatrix}$

Fig. 4.3. A two-dimensional BSB simulation taken from an early paper. The basin boundaries are curved lines. Differing starting points take varying numbers of iterations to reach stable attractor states, in corners of the box of limits. The arrows indicate the direction of the eigenvectors; the larger basin is associated with an eigenvalue of 0.04 and the smaller with an eigenvalue of 0.03. (From Anderson and Silverstein, 1978. Reprinted by permission. © American Psychological Association.).

normally generate the limit cycle attractors found in more complex dynamical systems (see Greenberg, 1988; Hui and Zak, 1992).

4.3. IMPORTANT SYSTEM PROPERTIES AND PARAMETERS

4.3.1. Corner Stability

If all the eigenvalues are positive, and if we add the previous state to the current state (that is, if $\gamma = 1$), then all corners are attractors. However, most applications of BSB incorporate decay of the previous state. If we are interested in BSB as a classification and categorizing algorithm, we are interested in having only a few stable attractors with large basins of attraction.

Consider the simple BSB model

$$x(k + 1) = \text{LIMIT}(\gamma x(k) + \alpha A x(k)) \qquad (9)$$

where A is the weight matrix. Suppose we teach this network one state vector, c, which is a corner of the box of limits. We will assume the limits, and therefore, all elements of c are $+1$ or -1. Let us assume that A is formed by simple outer product learning of c,

$$A = cc^T$$

Note that corner c is an eigenvector of A since

$$Ac = c(c^T c)$$
$$= nc$$

where n is the dimensionality of the space, since the inner product $[c, c] = c^T c = n$. If the sum of the off-diagonal elements in a row is less than the diagonal elements, the matrix is referred to as "row dominant" and it is possible to show that then all corners are stable (Hui and Zak, 1992). The identity matrix, an extremely uninteresting matrix for information processing, is row-dominant. (Uninteresting for most, except possibly for the spiritually evolved. Perhaps the identity matrix says one should accept the world exactly as it is.) The identity matrix can arise with Hebbian learning when large amounts of random noise are learned, providing a natural neural network senility mechanism, if one is needed.

For the outer product matrix with a single learned binary vector, the diagonal elements are all 1. The other elements are either $+1$ or -1. Therefore, the sum of the other elements may not be less than the diagonal elements, the matrix is not row dominant, and all the corners may not be stable. We can see the conditions for the stability of corner c. If the state at the kth

iteration, $x(k)$, equals corner c then

$$x(k + 1) = \text{LIMIT}(\gamma c + \alpha Ac)$$
$$= \text{LIMIT}(\gamma c + \alpha nc)$$
$$= \text{LIMIT}(\gamma + \alpha n)c$$

If $(\gamma + \alpha n) \geq 1$ then c will be stable; if $(\gamma + \alpha n) < 1$, then the corner will destabilize and decay to the origin.

Now, consider a corner c' which is equal to c except that one element, the jth element of c' is the opposite sign to the jth element of c. If the inner product $[c, c]$ is n, the dimensionality, the inner product of $[c, c']$ will be $(n - 2)$.

Consider the stability of corner c':

$$x(k + 1) = \text{LIMIT}(\gamma c' + \alpha Ac')$$

A is the outer product cc^T. It is easy to compute the product Ac'. For all the elements i where c and c' are equal, then the ith element of Ac',

$$(Ac')_i = (cc^T c')_i = c_i[c, c'] = c_i(n - 2)$$
$$= c_i'(n - 2) \quad \text{since } c_i = c_i'$$

For the jth element, where $c_j = -c_j'$ we have

$$(Ac')_j = c_j[c, c'] = c_j(n - 2) = -c_j'(n - 2)$$

Then, the jth element of $x(k + 1)$, if $x(k)$ is c', will be

$$x_j(k + 1) = \text{LIMIT}(\gamma c_j' + \alpha(-c_j'(n - 2)))$$
$$= \text{LIMIT}(\gamma c_j' - \alpha c_j'(n - 2))$$
$$= \text{LIMIT}(c_j'(\gamma - \alpha(n - 2)))$$

If $(\gamma - \alpha(n - 2)) \geq 1$ then c' will be stable, otherwise it will shift toward the value of c_j.

This value can be very negative for many BSB implementations. The corner corresponding to the eigenvector c can be stable, while a corner with only one element differing will be very unstable and a state vector there will be attracted toward the stable corner.

4.3.2. Attractor Basins and Stability

The basic mode of processing done by BSB is to turn an unlimited number of possible input vectors into a small number of stable attractors. The basins of attraction of the BSB model are, in general, well behaved and, for most interesting situations, only a few attractors are present. The qualitative behavior of the basin boundaries is intuitive. For a simple two-dimensional case (Anderson et al., 1977) the boundary has an exact solution: segments of a logarithmic spiral. Although we have no exact

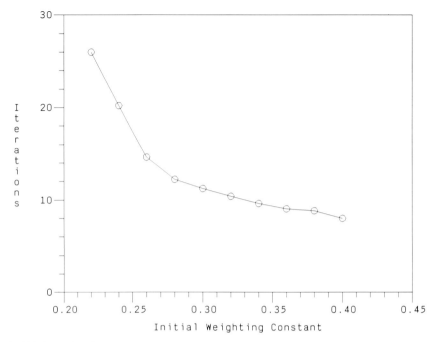

Fig. 4.4. A 1000-dimensional system learned 10 random binary vectors using outer product learning. As expected, the initial vectors with larger weighting constants (approximately eigenvalues) reached an attractor more quickly than vectors with smaller weighting constants.

solutions for higher-dimensionality boundaries, we would expect similar simplicity.

Let us do some simulations to check boundary behavior. Suppose we start with 10 binary vectors, constructed with values +1 and −1 equally probable, and then normalize them. We assume a high-dimensional space; between 250 and 1000 dimensions were used for the following simulations. In the first simulation, in a 1000-dimensional system, we constructed a connection matrix using outer product (Hebbian) learning from these 10 vectors using a weighting constant ranging from 0.4 for the first vector to 0.22 for the tenth, with weighting constants spaced 0.02 apart. As we would expect, the initial vectors are approximately orthogonal, and the initial vectors are within a few degrees of eigenvectors with eigenvalues spaced approximately from 0.4 to 0.22, that is, the eigenvalues were very close to the initial weighting constant and the 10 eigenvectors with largest positive eigenvectors were approximately in the direction of the original 10 vectors.

When tested, each input vector was associated with a different BSB attractor. The number of iterations required to reach an attractor was a function of the eigenvalue associated with that

vector. Figure 4.4 shows the number of iterations required in one simulation to reach an attractor for various initial weighting constants (approximately eigenvalues).

The next simulation also used a 1000-dimensional system. Suppose we next consider the plane between two of the initial vectors. Suppose we construct a normalized starting vector for BSB that lies in this plane. The simulation used 100 starting positions at roughly one-degree intervals. If the system is well behaved, then we would expect that starting points close to one starting vector would end in the attractor associated with that vector, and when close to the second starting vector would end in the attractor associated with the second starting vector.

Figures 4.5 and 4.6 show typical results for two pairs of vectors. There is a slow increase in number of iterations to the attractor as the starting point moves away from the initial vector. Despite the different starting points, however, the final state of the system is in either the first or the second attractor. An attractor is reached when all the vector elements are fully limited. There is a large region where the attractor is reached in a roughly constant number of

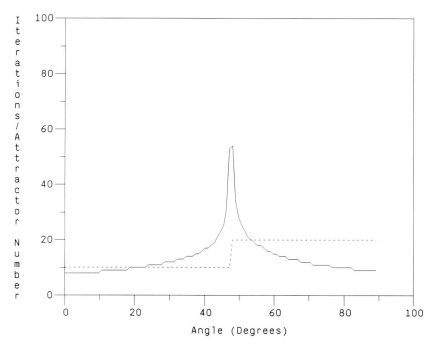

Fig. 4.5. Number of iterations required to reach an attractor for starting points lying in the plane between two learned random vectors. A 1000-dimensional system learned 10 random binary vectors using outer product learning. Normalized test vectors were constructed falling in the plane between vector 1, the initial vector with the highest weighting constant (0.4, approximately the eigenvalue) and vector 2, with the second highest weighting constant (0.38). The dashed line shows the number of the attractor corresponding to the final state of the system. (The *y*-axis value divided by 10 gives the attractor number.) All starting points in the plane are attracted into the attractor associated with vector 1 or with vector 2, though the number of iterations to reach that final state varies depending on starting point.

iterations, though the number of iterations required increases slowly as the boundary is approached. As the final state shifts from one attractor basin to the other, very near the boundary, many iterations are required to reach the attractor.

Occasionally, the boundary has a slightly more complex structure, as shown in Fig. 4.7. The starting points used here were in the plane between two initial vectors with relatively small associated eigenvalues. The two initial vectors were numbers 5 and 9. Starting points in the plane between the two initial vectors gave rise to a small interpolated region associated with vector 3. There was also a single point on the plane that gave rise to a new attractor that was not associated with any of the learned vectors (shown as attractor 0). There are *spurious* attractors in the BSB system, though the term spurious is highly misleading. The spurious attractors are what can give the system some ability to generalize in cognitively oriented simulations.

As we expect, larger eigenvalues have larger attractor basins. If we look at the boundary between two basins, we can see that it shifts toward the initial vector with the smaller associated eigenvalue. (This shift toward the vector with smaller eigenvalues can be seen clearly in Figs. 4.5 and 4.6). Figure 4.8 shows the location of the boundary as the starting vector moves from the plane between vectors 1 and 2, where the initial weighting constants (close to the eigenvalues) are 0.4 and 0.38, to the location of the boundary on the plane between vectors 1 and 10 (with eigenvalues approximately 0.4 and 0.22). When the eigenvalues of the two vectors forming the plane are nearly the same, the boundary is nearly 45 degrees. A 2:1 ratio of eigenvalues shifts the boundary to only 10 degrees away from the smaller eigenvalue.

4.3.3. Effects of Limits

Basin size is affected by the size of the box of limits. For example, the linear power method

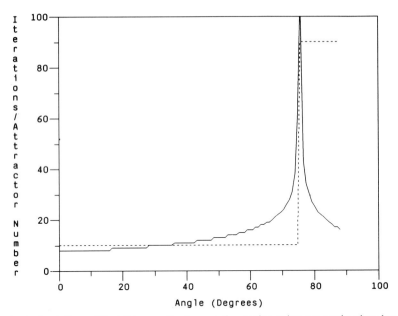

Fig. 4.6. Same basic simulation as Fig. 4.5 (see caption) except the starting points are on the plane between vector 1, with initial weighting constant 0.4, and vector 9, with weighting constant 0.24. The dashed line shows the number of the attractor corresponding to the final state of the system. (The y-axis value divided by 10 gives the attractor number.) All starting points are attracted into the attractor associated with vector 1 or with vector 9. Note that the boundary between the attractor basins has shifted away from vector 1, as compared to Fig. 4.5.

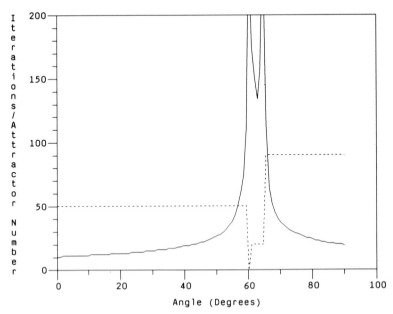

Fig. 4.7. A more complex boundary structure. Same basic simulation as Fig. 4.5 (see caption) except the starting points are on the plane between vector 5, with initial weighting constant 0.32, and vector 9, with weighting constant 0.24. The dashed line shows the number of the attractor corresponding to the final state of the system. (The y-axis value divided by 10 gives the attractor number.) There is a small region near the boundary which contains points which are attracted into the final state associated with learned vector 3, and there is a "spurious" attractor, which is not associated with any of the learned vectors, around 60 degrees.

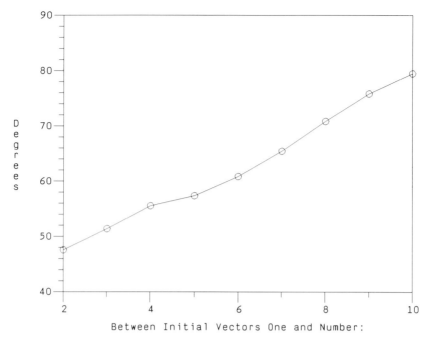

Fig. 4.8. Same basic simulation as Figs. 4.5 through 4.7. This graph gives the location of the attractor basin boundaries in the planes between vector 1, the initial vector with the largest weighting constant, and the other nine initial vectors. Notice the nearly linear shift of the angle of the basin boundary away from the vector, vector 1, with the largest initial weighting constant as the weighting constants decrease linearly from 0.4 to 0.22.

could be considered to realize a nonlinear BSB system where the limits are extremely far from the origin. In the power method, the dominant eigenvector becomes the attractor, in that every starting point eventually comes asymptotically close to it. Therefore, we should expect that the basins associated with large eigenvalues should grow relative to small ones as the limits increase. This is what is seen in simulations; however, the effect is usually rather small. Figure 4.9 shows the results of a 250-dimensional simulation where 1000 different normalized random binary vectors were used as initial starting points. In this system there were five learned normalized binary vectors. Each of the five learned vectors had an associated attractor. Almost all the random vectors finished in one of these five attractors; there were a small number of new attractors. (For this simulation, an attractor and its anti-attractor, that is, the corner diagonally across the box, were considered the same attractor.) The limits were increased from one times an element magnitude to 35 times an element magnitude. In the first case, the starting point was actually at a corner; in the last, it was small and far away from a corner. It can be seen that as the box increases in size, the basin

associated with the larger eigenvalues increases to some degree, but the effect is not large. If the box is asymmetrical, so that the positive and negative limits differ, behavior is more complex.

4.3.4. Decay Rate

The general BSB equation contains a term γ which multiplies the previous state vector, that is, it governs the decay of the state vector in the absence of feedback. (If the connection matrix were zero, then the state vector would decay exponentially with a time constant of γ.) If γ is less than 1, we would expect that attractors involving the smallest eigenvalues would be most strongly affected since the eigenvalue spectrum is shifted downward. Qualitatively, as the state vector decays more and more quickly, there are fewer and fewer attractors; γ less than 1 simplifies the attractor structure. If the sum of the largest positive eigenvalue and the decay term is less than 1, there will be no stable corners at all, and the only stable attractor will be at zero; that is, all starting points will end at the origin. In general, if we have a system (as is often found in applications) where there are a small number of

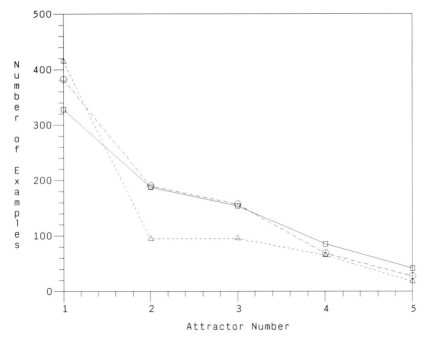

Fig. 4.9. Change in attractor basin structure for different sizes of the box of limits. This 250-dimensional system learned, using the outer product rule, five binary random vectors. Attractors 1 through 5 are the attractors associated with the five learned vectors. Attractor 1 had the largest weighting constant during learning (approximately the associated eigenvalue) and attractor 5 had the smallest. Attractors were computed for 1000 normalized binary random starting vectors. (An attractor and its "anti-attractor" were counted as the same attractor.) As the limits increase from one times the starting element size (solid line), to 10 times the starting element size (dashed line) to 50 times the starting element size (dotted line), the attractor basins changed in relative size. The attractor basins associated with the largest eigenvalue (number 1) grow at the expense of the basins associated with the smaller eigenvalues.

eigenvectors with large eigenvalues and a number of other eigenvectors with small eigenvalues, produced, perhaps, by random noise processes, reducing γ will have a disproportionate effect on the small eigenvalues and will tend to eliminate attractors associated with them. For example, if the sum of γ and a small eigenvalue is less than 1, the related eigenvector cannot support a stable attractor. This effect can act as a noise-suppression technique.

Figure 4.10 shows the number of different attractors seen in a 500-dimensional system, where there were 10 learned starting vectors given weighting constants from 0.4 to 0.22 when the connection matrix was constructed, as before. One thousand random normalized binary vectors were used as starting points. The same random number seed was used for all the conditions, so the same set of random starting vectors was used for each value of γ. The attractor corner and the anti-corner were considered to be the same attractor. The simplest BSB model, with γ of 1.00

showed about 60 different attractors from 1000 random starting vectors. When γ was lowered to 0.70 there was only one stable attractor that attracted almost all the initial vectors. (There were a few starting vectors that did not reach an attractor at all in 400 iterations.) Small decreases in γ were effective at eliminating large numbers of attractors with small basins of attraction. This observation is consistent with other simulations.

4.3.5. Connectivity

Many neural networks assume that the network is fully connected, so that every unit in a group connects to every other unit in the same or a different group. Examples of networks where this is generally assumed are Hopfield networks, projections between layers in back propagation, the ART models, as well as many others. In fact, connectivity in the real nervous system is small when compared with the number of cells involved. A "typical" central nervous system unit

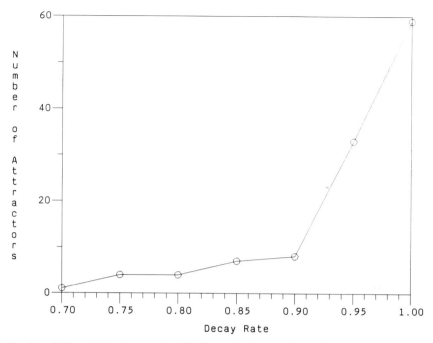

Fig. 4.10. Number of different attractors seen in a 500-dimensional system for varying rates of decay, γ. The network learned, using the outer product rule, 10 binary random vectors. Decreasing γ simplifies the attractor structure by eliminating attractors associated with eigenvalues near zero. (An attractor and its "anti-attractor" were counted as the same attractor.)

might receive a thousand inputs, give or take an order of magnitude. This is a very small number compared to the total number of neurons in a mammalian brain, which is probably greater than 10 billion, or even the number of neurons in a single cortical area such as primary visual cortex, which has over 100 million cells in primates.

One reason for this low connectivity is physical: even though they are small, connections between neurons still take up space. It has been calculated that if each of our neurons connected to every other one, our heads would be kilometers across. Therefore, network models that depend for their analysis and operation on full connectivity—and many do, though it is rarely emphasized—are in trouble if they are to be taken seriously as brain models.

Because BSB is a feedback model with point attractors, serving primarily as an associator and categorizer, the BSB model works well with partial connectivity. All the qualitative results described above, and the cognitive simulations described in the references to applications, hold over a wide range of connectivities. (In this paper, the simulations presented in Figs. 4.6, 4.7, and 4.8 gave identical qualitative results for all

connectivities tested between 30 percent and 100 percent.) Cognitive simulations done at Brown University routinely use 50 percent random connectivity or less. This is done partly to save computer time and partly in the interests of (a little) realism. Qualitative effects observed when connectivity is incomplete are increase in time to reach attractors, somewhat lower accuracy, lowered storage capacity, and, with the lowest connectivity, occasional elements that change so slowly that they do not reach limit values in reasonable numbers of iterations.

The qualitative behavior of the attractor structure is remarkably resistant to changes in connectivity. However, when connectivity is changed, there are some obvious "gain" changes that have to be taken into account. Consider, for example, the weighted outer product we used to construct matrices with known or approximately known eigenvectors and eigenvalues. With simple Hebbian (outer product) learning of normalized input vectors, partial connectivity reduces the eigenvalues proportionally (subject to some statistical assumptions). When many weights are set to zero the learned pattern is effectively no longer normalized.

In the real nervous system, the limited allowable connectivity is under tight biological control and is of the utmost importance for function. Neuroscientists have an idea of the outlines of connectivity in much of the brain, but it is still sketchy. One major task of modelers in the next decade—it is already starting to occur—is to incorporate biologically realistic connectivity patterns into all network models. A large part of the biological computation is done by the initial wiring and biological learning modifies and tunes up this rather small number of preexisting connections. The brain seems to operate in a regime where there are huge numbers of units, a small number of connections per unit, and where units are arranged in groups (nuclei and sub-regions), each group subserving one or a small number of simple computational steps. Study of networks of this type has not been popular with other than biologically motivated theorists. Part of the reason for this is that the large number of units required makes them difficult to simulate with existing computer technology and difficult to realize for practical devices. However, if they can be made, there is reason to think that such large modular networks may be reliable, robust, and versatile.

4.4. SEQUENCES OF STATES AND UNSUPERVISED CLUSTERING

4.4.1. Corner Shifting and Adaptation

Many problems require the learning of sequences of state vectors. Since BSB, like many simple dynamical system models, has stable point attractors, the network must be modified somehow to produce temporal sequences of attractors. There are *many* ways to do this for attractor networks and many published variants in the literature. Amit (1989) gives a number of examples in his book. One straightforward way to change the state vector from one attractor to another is to manipulate the matrix so that the attractor's local energy minimum is raised and the state vector "spills" into a nearby minimum.

As we have seen in our power method demonstration, we can manipulate eigenvalues of the connection matrix easily using outer product learning with a negative sign. If we assume that an adaptive process of an anti-Hebbian type exists, then the longer the state vector stays in an attractor the weaker the eigenvalue(s) associated with that attractor become. At some critical value, the state vector shifts to a new attractor. There have been several applications

of this technique; the one we shall discuss is from Kawamoto and Anderson (1985).

4.4.2. Necker Cube Reversals

The Necker Cube is the best known example of what is called *multistable perception* (Necker, 1832). Figure 4.11 shows a Necker cube. This two-dimensional drawing is ambiguous, in that it is consistent with a three-dimensional perception that goes into the paper or comes out of it. If it is looked at steadily for a period, first one interpretation will be seen, then the other, then the first again, and so on. The perception will oscillate between interpretations. Ambiguity is a common perceptual problem. For example, many words are ambiguous as to meaning, pronunciation, or part of speech. Multistability can be demonstrated for many types of stimuli.

Kawamoto and Anderson (1985) list some of the major experimental findings related to multistability in vision:

1. There are two or more possible configurations of the stimulus. (That is, the stimulus is ambiguous.)
2. Once a particular configuration is perceived, it remains stable for a while before spontaneously changing to an alternative configuration.
3. The rate of fluctuation increases during the first 3 minutes of viewing before reaching an asymptote.

One natural model for such an effect is an attractor model with adaptation. Suppose we postulate adaptation in a BSB network in the form of Hebbian anti-learning to let a state vector shift from attractor to attractor. Assume a set of initial conditions, $\mathbf{x}(0)$, the initial state vector, and $\mathbf{A}(0)$, the long-term values of the connection

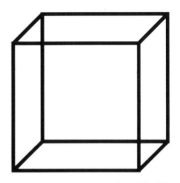

Fig. 4.11. The classic example of multistable perception: The Necker Cube.

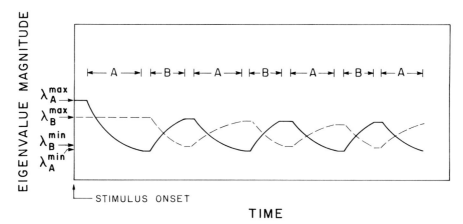

Fig. 4.12. Behavior of the eigenvalues of the system during bistable perception. The system state will cycle between corners A and B and as the eigenvalues cycle up and down. Note that the first reversal takes the longest time. (From Kawamoto and Anderson, 1985. © Elsevier Science Publishers, B.V. Reprinted by permission.)

matrix, **A**. These long-term values represent the stable component of the strength of the trace. We assume that the time constant for permanent modification of these values is slow relative to the dynamic modifications.

For simplicity, changes in matrix **A** were not allowed until the activity of the system reached a stable state. Once the activity of the network reaches an attractor, adaptation starts and continues as the activity remains in the corner. We assume that there are two eigenvectors corresponding to the two perceptions of the cube, the eigenvectors point toward corners, and are orthogonal.

If the system is in a corner **f**, change in eigenvalue magnitude is accomplished simply by Hebbian synaptic "anti-learning"

$$\mathbf{A}(t+1) = \mathbf{A}(t) - (1-\sigma)\mathbf{f}\mathbf{f}^{\mathrm{T}}$$

where σ is a constant determining amount of adaptation. This implies that

$$\lambda(t+1) = \sigma\lambda(t) \qquad (10)$$

where λ is the eigenvalue associated with eigenvector **f**. If λ drops too much, the eigenvector will leave the corner. For a computer simulation to test this, simple stimuli were used which lay between the eigenstates, which were low-dimensionality Walsh functions.

With these relationships governing the decay and recovery of the eigenvalues, the values of λ_1 and λ_2 as a function of time are plotted in Fig. 4.12. The system state oscillated between the two eigenvectors. In conformance with the psychological result, the first reversal took a long time; the oscillations speeded up until a constant

frequency of perceptual reversal was attained. Several other effects that correspond to the psychological data could also be demonstrated in this simple model, including hysteresis and adaptation.

4.4.3. Vector Adaptation

Perhaps the cleanest and most obvious way to shift states is Hebbian anti-learning because it is theoretically well founded and works effectively in both the linear and nonlinear systems. However, as a demonstration of just how robust a nonlinear system can be in practice, let us demonstrate an adapting technique that might be easy to implement, works very poorly in a linear system, but is quite robust in the nonlinear network.

Hebbian anti-learning requires detailed changes at synapses. Suppose we assume a much cruder form of adaptation: when a unit is active, after a while, it becomes less responsive. This can be accomplished in a number of ways. The unit can become "tired" or any one of a number of "habituation" processes might occur. Or an active inhibitory process might occur, producing a "negative afterimage" of the cell's discharge. The exact details do not matter much.

Let us consider a system where the activity of the system for a period of time has been an attractor, a state vector **c**. We will use the BSB model, but, again, most attractor networks should show similar behavior. We will assume that adaptation occurs, so the attractor, **c**, times a constant, κ, is simply subtracted from the state vector. The BSB equation then becomes, with

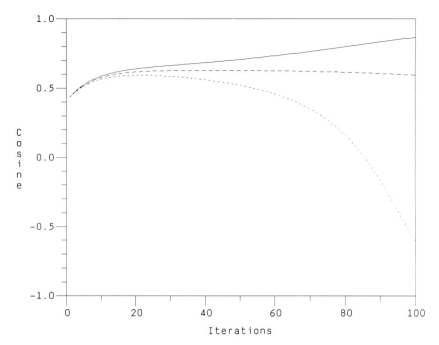

Fig. 4.13. Vector adaptation with linear dynamics does not work well. The graph gives the cosine between the first (dominant) eigenvector and the system state vector after subtractive vector adaptation. Three values for the adapting constant, κ, are plotted, corresponding to about a 5 percent range of values. Eventually every value of κ approaches a system state composed of one sign or the other of the dominant eigenvector. Near the transition point from positive to negative final system state, the approach to the dominant eigenvector may be slow.

$\delta = 0$ for simplicity,

$$\mathbf{x}(t + 1) = \text{LIMIT}(\gamma\mathbf{x}(t) + \alpha\mathbf{A}\mathbf{x}(t) - \kappa\mathbf{c}) \ (11)$$

The qualitative result is that vector $\mathbf{c}$ will be less likely to occur. Unfortunately, and what makes this attractively simple idea so hard to use in practice, the constant subtraction of $-\kappa\mathbf{c}$ tends to force the system strongly into the stable anti-corner, $-\mathbf{c}$.

When this model is simulated in the linear system, it works very poorly. The simulation used the 32-dimensional Walsh functions used before. Figure 4.13 shows the size of the components of the first eigenvector in the state vector for different values of κ in the linear system. The eigenvector with the second largest eigenvalue grows a little, but the eigenvector with largest positive eigenvalue, or its anti-vector rapidly dominates the system. Small differences in κ make large differences in the dynamics of the system; the three qualitatively different curves are from κ values that vary over a total range of only 5%. Such extreme parameter sensitivity is undesirable in a practical neural network model.

When the BSB model is applied to the same

system, the qualitative behavior becomes very different. The eigenvector with second largest eigenvalue becomes a stable state over a wide range of κ. The nonlinearity keeps the system in a stable state corresponding to this eigenvector and the size of the basin makes the model much less parameter sensitive. Figure 4.14 shows the dynamics of the first eigenvector component in the case where simple BSB is applied. Notice that this component drops to zero; the stable system state vector at this point was entirely the second eigenvector, the one that we wanted to retrieve. The three curves represent a variation in κ of about 50 percent of the center of the range. Simple, but unstable, subtractive vector adaptation becomes a potential usable technique that is rather insensitive to parameters.

4.4.4. Clustering

A natural application of attractor networks in general is clustering, because the basins of attraction divide up state space into regions, each associated with a particular attractor. The simulations we have presented have indicated

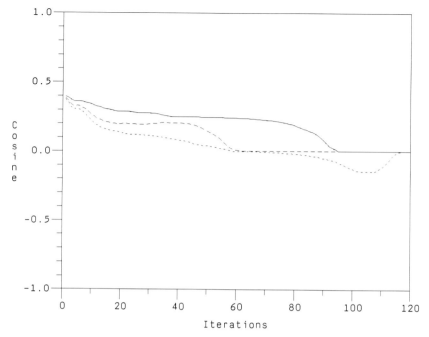

Fig. 4.14. Vector adaptation with nonlinear BSB dynamics. The graph gives the cosine between the first (dominant) eigenvector and the system state vector after subtractive vector adaptation. Three values for the adapting constant, κ, are plotted, corresponding to about a 50 percent range of values. Note that the final system state vector for the three values is exactly the eigenvector with second largest eigenvalue. This is the desired result of adaptation. Time to reach the attractor varies with the fastest times in the middle of the range (the dashes). The positive and negative extremes of the range (solid line and dots) take longer to reach the attractor.

that the attractor basins in BSB are well behaved, and respond to important global parameters like eigenvalues in appropriate ways. If we assume that a set of data is generated from underlying discrete events or processes, the attractors might reflect this, that is, a basin might come to correspond to one particular class of events.

One specific BSB application to radar analysis made use of this ability to cluster (Anderson et al., 1990). The basic problem addressed was what is called in radar electronics the "de-interleaving problem." A region of the microwave spectrum can have a great many radar pulses coming from a number of different emitters. These pulses can overlap and there is sufficient noise in the system and uncertainty in measurement of pulse properties so that it is difficult to tell easily how many emitters are present. It is an important practical problem to know how many emitters are present and to characterize them it terms of emitter properties.

For this problem, it is assumed that there are a small number of discrete emitters, significant noise, and a very large number of samples of each

emitter. When a pulse was received, its coding was learned using the Widrow–Hoff learning algorithm with a very small learning constant. Although Widrow–Hoff is generally considered to be a supervised algorithm, in this unsupervised situation, the "correct" answer was assumed to be the input pulse being learned. It is easy to show that learning pulse codings using this rule will tend to develop a BSB attractor at the average pulse in the cluster of pulse codings coming from a single emitter. The distinct attractors, each associated with a different emitter, are highly discriminable and are noise free. Counting and characterizing them becomes easy. Further, once a single attractor attracts all examples of pulses from a single emitter, it then becomes possible to do detailed average measurements of the properties of the emitter, as well as looking for additional structure, for example, the pattern of pulses shown by a single emitter. [See Knapp and Anderson (1984) for a pscyhological application of a related linear model.] Hassoun and co-workers have applied a related multilayer autoassociative attractor model to clustering biomedical data, with considerable success (Hassoun et al., 1993; Spitzer et al., 1990).

The BSB model turned out to be remarkably successful at performing correct and robust clustering of realistic simulated radar signals using this technique. Of at least equal importance in this success was the data representation used for single pulses, which used bar codes to represent measured pulse parameters. (Bar codes use a moving area of activation in a state vector to represent a parameter. Topographic maps of sensory data are common in the nervous system.) This data representation greatly increased the dimensionality of the system, but was easily able to integrate data from all the measurements into a single state vector, as well as provide a simple and effective method for generalization. It would be of some interest to see if it would be possible to relate attractor network clustering to more traditional techniques. The computational parallelism obtainable from a neural network is often desirable, but it is not clear how well the accuracy of network clustering compares to better-analyzed traditional methods.

4.5. A SIMULATION OF NETWORK RESPONSE TIME

4.5.1. Response Time

Most neural network research emphasizes the accuracy of the computation: whether or not the final system state is the correct answer to the question posed. However, the time required to reach that answer is sometimes of great importance. For example, from the point of view of an experimental psychologist the *response time* or the *reaction time* to perform a task is a common experimental measure. There is an immense literature on psychological response times. (See Luce, 1986, for a review.) It is also important to know the time required for a network computation to be completed for many practical applications.

Let us consider the time for a response to be generated. In feed-forward neural models such as back propagation, during *operation*, when an input state vector is presented at the input layer, the output state vector is generated in a few parallel steps. The function of various learning algorithms is to generate weights so that the right input–output relations are produced, but the actual computation is not strongly time dependent without additional assumptions.

Dynamical system models such as BSB have intrinsic time domain behavior due to their dynamics. Some early cognitive applications of neural networks, in particular, the interactive activation model for letter perception (McClelland and Rumelhart, 1981; Rumelhart and McClelland, 1982), a complex dynamical system, showed time domain behavior which often displayed similarities to human reaction-time data.

We will assume that response time—time for a computation to be completed—is related to dynamical system behavior. The key assumption is that *simple response time is related to the time required for an input vector to move to a point attractor*. Classical reaction-time models that are most similar formally to this approach are also the ones that are the most successful in explaining experimental reaction-time data: random walk and diffusion models (see Luce, 1986, chapters 8 and 9).

4.5.2. Theory: Reaction-Time Computation

Suppose that response time in experiments is related to the time required to reach an attractor in simulations. Computation of the time to reach an attractor can be a difficult calculation in the BSB model if the state vector starts to limit in one region of state space and not in others, because large discontinuous changes in rate of change of the state vector can occur. However, it is possible to compute easily the time required to reach an attractor for one interesting case.

For simplicity in analysis, we will assume that the connection matrix has only two eigenvectors and two attractors, which correspond to two possible responses. Because the dynamics of the BSB model are controlled by an equation with three constants (α, γ, δ), let us assume that the dynamics of the system are represented by a single matrix, $\mathbf{B}$, with two eigenvectors and the associated eigenvalues λ_1 and λ_2. (If $\mathbf{f}$ is an eigenvector, and $\mathbf{A}$ is the connection matrix, and if $\lambda_{1\mathbf{A}}$ and $\lambda_{2\mathbf{A}}$ are the eigenvalues of $\mathbf{A}$ then the eigenvalues of $\mathbf{B}$ are

$$\lambda_1 = \alpha\lambda_{1\mathbf{A}} + \gamma + \delta$$
$$\lambda_2 = \alpha\lambda_{2\mathbf{A}} + \gamma + \delta$$

The addition of a multiple of the identity matrix simply translates all the eigenvalues up or down together.

Suppose we start the system at some point P which lies along the direction of the eigenvector, at a distance r_0 from the origin. We want to compute the time it takes to get from the point P into the attractor, which is assumed to be a distance d from the origin. (The magnitude of d

depends on the dimensionality and the limits of the box.) We assume that the eigenvector is exactly aligned with the attractor, so that all the element values limit simultaneously when the attractor is reached.

Qualitatively, the calculation below will *under-estimate* the time required for starting points that are not multiples of the eigenvector to reach the attractor because, if some elements limit before others, there are large discontinuous changes in the rate of approach to the attractor. However, the simulations presented earlier on basin boundaries suggest that this effect is not large when there are a small number of learned vectors in a high-dimensional space. Overall, there will be more long times in the tails of the distribution than the calculation indicates, and there exists an obvious mechanism for producing a few extremely long response times.

Suppose we are a distance r from the origin along one eigenvector. Then, the next position of the point after one iteration will be given by λr, where λ is the eigenvalue of **B** associated with the eigenvector. The change in position, Δr, is given by

$$\Delta r = \lambda r - r = (\lambda - 1)r$$

If each iteration takes time Δt, then the "velocity" of the state vector is

$$\frac{\Delta r}{\Delta t} = (\lambda - 1)r \qquad (12)$$

or

$$\frac{\Delta r}{(\lambda - 1)r} = \Delta t$$

When the distance to be traveled equals the sum of the Δr's, the state vector will have reached the attractor. The time will be given by integrating the above expression from the starting point, r_0, to the attractor, d, so

$$t = \int_{r_0}^{d} \frac{dr}{(\lambda - 1)r}$$

Assuming we start at $t = 0$, then

$$t = \frac{1}{(\lambda - 1)} (\ln(d) - \ln(r_0))$$

$$= \frac{1}{(\lambda - 1)} \ln\left(\frac{d}{r_0}\right) \qquad (13)$$

A graph of this function is given in Fig. 4.15. Note

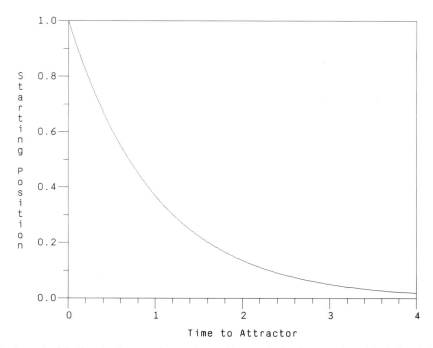

Fig. 4.15. A graph of the function between the starting position on the line between the origin (0.0) and the corner attractor (1.0) and the time required to reach the attractor. (Reprinted from Anderson, 1991, with permission.)

that this function is the negative exponential function, turned on its side.

Luce points out that many (for example, Ratcliff, 1978) have suggested that reaction time data is experimentally found to be well fitted by an unusual distribution called the "ex-Gaussian," which is the convolution of an exponential distribution and a Gaussian. Such a distribution is skewed, with a long tail and a rapid rise. Several examples of the ex-Gaussian distribution fitted to data taken from Ratcliff (1978) are presented in Fig. 4.16.

Suppose we have a Gaussian distribution of starting points along the eigenvector. The values of different regions of the input distribution are multiplied by a negative exponential, and then spread out in time to generate the actual distribution of times along the horizontal axis. This function is not the convolution of a Gaussian and an exponential, but an exponential scaling of the Gaussian. The resulting time distribution looks identical, though moments will differ. Theoretical reaction-time distributions computed for several differing mean starting points and input distribution standard deviations are presented in Fig. 4.17.

4.5.3. Simulation

We can simulate this model as well. Let us consider a "same–different" task, that is, reporting whether two items presented to a subject are the same or different. Besides being a popular psychological experiment, this task is significant for neural networks for historical reasons. It realizes the truth table of Exclusive-OR (X-OR) which is particularly difficult for neural networks to learn (Minsky and Papert, 1969; see also Hinton, 1981).

We assume that the problem is solved by the data representation. Suppose we have a nonlinear mechanism that compares subject and object, say, with element-by-element matching. If the two items match, we have one coding; if they do not match, we have another coding. The additional component that violates the superposition rule for linear systems can be produced by nonlinearities, or by an extension to the vector coding used to represent the input data.

4.5.4. Data Representation

Our bias has always been that small neural nets are uninteresting. But the outputs of large neural

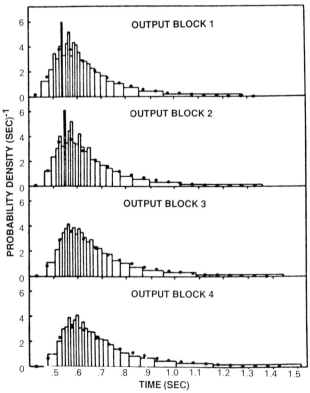

Fig. 4.16. Ratcliff has suggested that reaction time data is experimentally found to be well fitted by a distribution called the "ex-Gaussian" (Luce, 1986). Such a distribution is skewed, with a rapid rise and a long tail. These data from Ratcliff (1989) contain both experimental reaction time distributions and the best-fitting distribution. (From Ratcliff, 1978. Reprinted by permission. © American Psychological Association.)

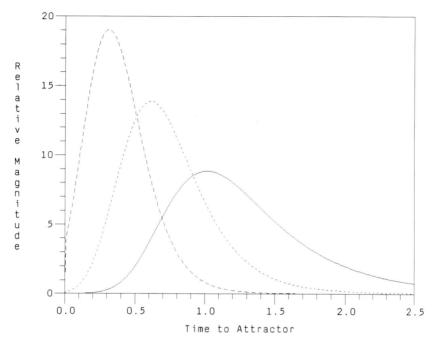

Fig. 4.17. Theoretical reaction-time distributions computed for several differing mean starting points. The solid line corresponds to a mean starting position of 0.3, dots to 0.5, and dashes to 0.7. The input distribution is Gaussian with standard deviation of 0.15. (Modified from Anderson, 1991.)

nets can be difficult to interpret. Strictly for convenience, we have often used in preliminary simulations a set of programs that construct state vectors from ASCII characters, so a character such as "s" is represented by eight vector elements. For example,

$$\text{"s"} = 1 \quad 1 \quad 1 \quad 1 \quad -1 \quad -1 \quad 1 \quad 1$$

We can then concatenate characters, so that 25 characters are represented by a 200-dimensional state vector. Although such a simulation is anything but efficient, it is easy to work with and captures some of the profligate way the nervous system uses the computing power of the computing elements. Such a coding was used for many of the cognitive simulations described in the references.

We generated random character strings of two characters to represent one or another value of a binary feature. For example, if "+" represents one value of a feature and "−" the other, we can easily construct a number of random pairs of strings.

Figure 4.18 depicts several sets of character strings. In Fig. 4.18, strings of 12 characters were constructed. If the strings are identical, the first four characters in the string are "SAME," if the

strings are different, the first four characters are "DIFF." The strings themselves are given in the next fields, separated (for our reading convenience) by the character "|".

To indicate a match or mismatch between the strings, a "+" is used when the characters in that position in the first and second strings match, otherwise a "−" is used. This is a nonlinear function that allows the network to compute "same–different." The computation is local, based on match or mismatch of pairs of characters. The characters at the right end of the string were not used in the simulations we will describe. Both the same and different strings were generated randomly.

The resulting system used 384 elements, or 48 eight-element bytes, each representing a character. All the simulations used 50 percent connectivity, that is, each element connected randomly to half of the other elements. The Widrow–Hoff error correction procedure was used to form the connection matrix. The training set consisted of 100 examples of "Same" and 120 examples of "Different" in the 12 character comparisons. Each of the levels of difference (one element different, two elements different, and so on) was presented 10 times. During learning, 1000

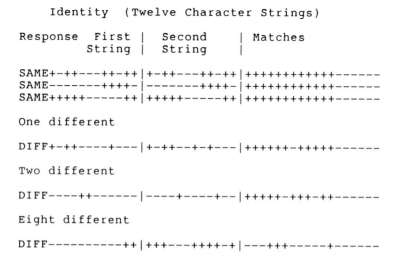

Fig. 4.18. Character strings used in the simulations. These strings have 12 elements to test for identity. When the strings are identical, the first four characters in the string are "SAME," if the strings are different, the first four characters are "DIFF." A "+" is used when the characters in that position in the first and second strings match, otherwise a "−" is used. The resulting state vector contained 384 elements, or 48 eight-element bytes, each representing a character. (Modified from Anderson, 1991.)

learning trials were used, with a state vector presented randomly. During testing, the connection matrix did not learn. Two hundred and fifty newly generated random stimuli were used at all levels of difference, as well as 250 new "same" state vectors. The testing procedure used input state vectors with the initial four characters replaced with zeros. BSB dynamics were applied and the resulting strings were checked to see, first, if the correct four characters were reconstructed, and second, to note how long it took the system to reach the stable attractor where all elements were fully limited. The constants used in the simulation were $\alpha = 0.225$, $\gamma = 0.9$ and $\delta = 1$. Choice of constants was not critical.

The first thing we looked at from the simulations was the distribution of the number of iterations required to reach an attractor, defined here as number of iterations required to reach a corner of the hypercube of limits. Three typical patterns are shown in Fig. 4.19. Note the rapid rise and extended tail of this distribution, very like the distributions seen in human data for this and similar tasks and also predicted by the theory. "Same" responses were fast. Although "different" response times where almost every element differed were faster than "sames," the rest of "different" response times were slower than the "same" response times.

It is possible to fit other aspects of the experimental data with this simple response-time

model. For example, both experimental data and theory find that the response time for a "same" response does not increase much as the number of characters to be compared increases but the response times for "different" responses are strong functions of the number of differences and the overall number of characters. A canonical set of experimental data for this task is from Bamber (1969; reprinted in Luce). More detailed discussion is found in Anderson (1991).

4.5.5. Connection with Theory

We should make one comment on the relation between the simulation, where there were 32 zero elements out of 384 in the state vector, and this simple calculation. If we assume the nonzero part of the initial input is reasonably close to an eigenvector, and if δ is 1, that is, the input is constantly present, the 352 initially nonzero elements limit almost immediately, which is what is seen in the simulations. Consider the structure of the connection matrix in relation to the two pieces, initially zero (0) and initially nonzero (NZ) regions, of the input state vector. We wish to reconstruct the response in the zero block:

$$\left(\frac{\mathbf{0}}{\mathbf{NZ}}\right) \quad \text{and} \quad \left(\frac{\mathbf{A} \mid \mathbf{B}}{\mathbf{C} \mid \mathbf{D}}\right) \qquad (14)$$

In terms of its action on the state vector, the

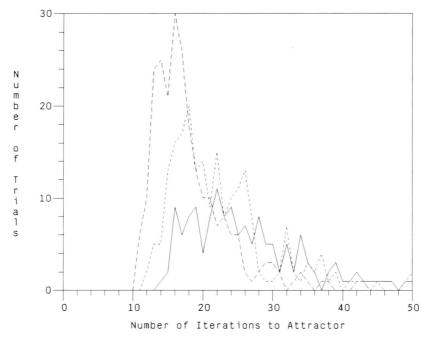

Fig. 4.19. Three typical distributions of correct "response time" in a computer simulation. Two hundred and fifty random input "different" patterns were used and the number of iterations required to each an attractor was recorded. The solid line has 4 of 12 characters different, the dotted line 6 of 12, and the dashed line 8 of 12. (See Figure 4.18 for the coding used.) (Modified from Anderson, 1991.)

matrix has been partitioned into four parts. If the nonzero part is fully limited, it does not change. Therefore, block **D**, which connects the nonzero part of the state vector to itself has no effect. Block **B**, which connects the reconstructing, initially zero, part to the nonzero part also has no effect. Block **C**, connecting the unchanging limited region to the reconstructing region acts as a constant input to the reconstructing region. Block **A** acts as the feedback matrix driving the reconstruction. The dynamics of the system are being driven by the eigenvectors of this smaller block, which will follow the analysis given above, and where the constant input is provided by block **D** acting on the nonzero inputs. Many simulations have similar structure, so that this reaction-time computation is actually reasonably general.

Historical note: A description of the genesis of the BSB model can be found in the introduction to the paper which is reprinted in the collection *Neurocomputing* (Anderson and Rosenfeld, 1988).

4.6. SUMMARY

The BSB (Brain-State-in-a-Box) model is a simple, autoassociative, nonlinear, energy-minimizing

neural network. It is easy to simulate and a number of special cases of its behavior can be analyzed mathematically. It is a relatively weak neural network model in terms of its information-processing power. However, it captures some of the flavor of network computation in its dynamics, associativity, and parallelism. It is especially well suited to simulations using many model neurons. It has primarily been used to model effects and mechanisms seen in psychology and cognitive science.

This chapter discusses a number of aspects of the BSB model.

1. First, the properties of autoassociative neural networks with feedback are discussed. Matrix feedback models have strong similarities to the "power method" of eigenvector computation, and some numerical experiments with the power method are described.
2. The close relationship between many network learning algorithms and principal component analysis is mentioned next.
3. The basic BSB equations use a simple associator with a limiting nonlinearity in the model neurons. An interconnected network of such units is an energy-minimizing dynamical

system. Stable attractors in BSB are usually corners of a box of limits in hyperspace. Some corners are stable and others are not. Criteria for corner stability are computed. In BSB, boundaries between attractor basins are usually well behaved; some simulations are presented designed to illustrate basin structure.

4. The simplest version of BSB only has point attractors. However, several techniques can be used to generate sequences of output vectors with Hebbian "anti-learning," or with vector adaptation. BSB can be used for unsupervised clustering.

5. In the last section, the time it takes the system state to reach an attractor in one simple case is computed. This computation is interesting because the model can then be used to make contact with the extensive psychological literature on reaction times. A simulation of a simple psychological reaction-time experiment using BSB is presented.

ACKNOWLEDGMENTS

We acknowledge indispensable support from the National Science Foundation for work on the cognitive applications of neural networks, most recently in grant BNS-90-23283. Special thanks to Dr. Joseph L. Young of the National Science Foundation, who supported work on neural networks long before it became popular to do so. We received valuable support as well from the Office of Naval Research, currently under grant N-00014-91-J-4032. We also gratefully acknowledge assistance from the Digital Equipment Corporation, who provided valued support to let us improve our computing facilities.

Copyright © 1993 James A. Anderson.

REFERENCES

Amit, D. J. (1989). *Modelling Brain Function: The World of Attractor Neural Networks.* Cambridge University Press, Cambridge, UK.

Anderson, J. A. (1986). "Cognitive Capabilities of a Parallel System," in *Disordered Systems and Biological Organization*, E. Bienenstock, F. Fogelman-Soulie and G. Weisbuch, eds. Springer-Verlag, Berlin.

Anderson, J. A. (1991). "Why, Having So Many Neurons, Do We Have So Few Thoughts," in *Relating Theory to Data: Essays on Human Memory*, W. E. Hockley and S. Lewandowsky, eds. Erlbaum, Hillsdale, NJ.

Anderson, J. A. and Mozer, M. C. (1981). "Categorization and Selective Neurons," in *Parallel Models of Associative Memory*, G. E. Hinton and J. A. Anderson, eds. Erlbaum, Hillsdale, NJ.

Anderson, J. A. and Murphy, G. M. (1986). "Psychological Concepts in a Parallel System," in *Evolution, Games, and Learning*, D. Farmer, A. Lapedes, N. Packard, and B. Wendroff, eds. North-Holland, Amsterdam.

Anderson, J. A. and Rosenfeld, E. (1988). *Neurocomputing: Foundations of Research.* MIT Press, Cambridge, MA.

Anderson, J. A. and Silverstein, J. W. (1978). "Reply to Grossberg," *Psychol. Rev.*, **85**, 597–603.

Anderson, J. A., Silverstein, J. W., Ritz, S. A., and Jones, R. S. (1977). "Distinctive Features, Categorical Perception, and Probability Learning: Some Applications of a Neural Model," *Psychol. Rev.*, **84**, 413–451.

Anderson, J. A., Gateley, M. T., Penz, P. A., and Collins, D. R. (1990a). "Radar Signal Categorization Using a Neural Network," *IEEE Proc.*, **78**, 1646–1657.

Anderson, J. A., Rossen, M. L., Viscuso, S. R., and Sereno, M. E. (1990b). "Experiments with Representation in Neural Networks: Object Motion, Speech, and Arithmetic," in *Synergetics of Cognition*, H. Haken and M. Stadler, eds. Springer-Verlag, Berlin.

Anderson, J. A., Spoehr, K. T., and Bennett, D. (in press). "A Study in Numerical Perversity: Teaching Arithmetic in a Neural Network," in *Neural Networks for Knowledge Representation and Inference*, D. S. Levine and M. Aparacio, eds. Erlbaum, Hillsdale, NJ.

Baldi, P. and Hornik, E. (1989). "Neural Networks and Principal Component Analysis: Learning From Examples Without Local Minima," *Neural Networks*, **2**, 53–58.

Bamber, D. (1969). "Reaction Times and Error Rates for 'Same'–'Different' Judgments of Multidimensional Stimuli," *Percept. Psychophys.*, **6**, 169–174.

Bégin, J. (1991). "Élaboration d'un nouveau modèle de mémoire associative appliqué au problème de la catégorisation," Ph.D. Thesis, Départment of Psychology, Université du Québec à Montréal.

Carpenter, G. A. and Grossberg, S. (1987). "ART 2: Self Organization of Stable Category Recognition Codes for Analog Input Patterns," *Appl. Opt.*, **26**, 4919–4930.

Cohen, M. A. and Grossberg, S. (1983). "Absolute Stability of Global Pattern Formation and Parallel Memory Storage by Competitive Neural Networks," *IEEE Proc. Systems, Man, and Cybernetics*, **SMC-13**, 815–825.

Cottrell, G. W., Munro, P., and Zipser, D. (1988). "Image Compression by Back Propagation: An Example of Extensional Programming," in *Advances in Cognitive Science*, Vol. 3, N. E. Sharkey, ed. Ablex, Norwood, NJ.

Foldiak, P. (1989). "Adaptive Network for Optimal Linear Feature Extraction," *Proc. IJCNN*, Washington, DC, June, 1989, pp. 401–406.

Golden, R. M. (1986). "The 'Brain-State-in-a-Box' Neural Model is a Gradient Descent Algorithm," *J. Math. Psychol.*, **30**, 73–80.

Greenberg, H. J. (1988). "Equilibria of the Brain-State-in-a-Box (BSB) Neural Model," *Neural Networks*, **1**, 323–324.

Haken, H. (1985). "Operational Approaches to Complex Systems: An Introduction," in *Complex Systems—Operational Approaches in Neurobiology, Physics, and Computers*, H. Haken, ed. Springer-Verlag, Berlin.

Haken, H. and Stadler, M. (1990). *Synergetics of Cognition*, Springer-Verlag, Berlin.

Hassoun, M. H., Wang, C., and Spitzer, A. R. (1993). "NNERVE: Neural Network Extraction of Repetitive Vectors for Electromyography; Part I: Algorithm," *IEEE Trans. Biomedical Engineering*, to appear.

Hinton, G. E. (1981). "Implementing Semantic Networks in Parallel Hardware," in *Parallel Models of Associative Memory*, G. E. Hinton and J. A. Anderson, eds. (1981, rev. edn., 1989). Erlbaum, Hillsdale, NJ.

Hopfield, J. J. (1982). "Neural Networks and Physical Systems with Emergent Collective Computational Abilities," *Proc. Nat. Acad. Sci., U.S.A.*, **79**, 2554–2558.

Hopfield, J. J. (1984). "Neurons with Graded Response have Collective Computational Properties Like Those of Two-state Neurons," *Proc. Nat. Acad. Sci. U.S.A.*, **81**, 3088–3092.

Hornbeck, R. W. (1975). *Numerical Methods*. Quantum, New York.

Householder, A. S. (1964). *The Theory of Matrices in Numerical Analysis*. Blaisdel, New York. (Reprinted, 1975, by Dover Publications, New York.)

Hui, S. and Zak, S. H. (1992). "Qualitative Analysis of the BSB Neural Model," *IEEE Neural Network Trans.*, **3**, 86–94.

Kawamoto, A. H. (1985). "The (Re)Solution of Semantic Ambiguity," Ph.D. Thesis, Department of Psychology, Brown University.

Kawamoto, A. H. and Anderson, J. A. (1985). "A Neural Network Model of Multistable Perception," *Acta Psychol.*, **59**, 35–65.

Knapp, A. G. and Anderson, J. A. (1984). "Theory of Categorization Based on Distributed Memory Storage," *J. Exp. Psychol.: Learning, Memory and Cognition*, **10**, 616–637.

Kosko, B. (1988). "Bidirectional Associative Memories," *IEEE Trans. Systems, Man, and Cybernetics*, **SMC-18**, 49–60.

Krogh, A. and Hertz, J. A. (1990). "Hebbian Learning of Principal Components," in *Parallel Processing in Neural Systems and Computers*, R. Eckmiller, G. Hartman, and G. Hauske, eds. Elsevier, Amsterdam.

Linsker, R. (1988). "Self-organization in a Perceptual Network," *Computer*, **21**, 105–117.

Luce, R. D. (1986). *Response Times*. Oxford University Press, New York.

MacKay, D. J. C. and Miller, K. D. (1990). "Analysis of Linsker's Application of Hebbian Rules to Linear Networks," *Network*, **1**, 257–297.

McClelland, J. L. and Rumelhart, D. E. (1981). "An Interactive Activation Model of Context Effects in Letter Perception: Part I. An Account of Basic Findings," *Psychol. Rev.*, **88**, 375–407.

Minsky, M. and Papert, S. (1969). *Perceptrons*. MIT Press, Cambridge, MA.

Necker, L. A. (1832). "Observations on Some Remarkable Phenomena Seen in Switzerland; and an Optical Phenomenon Which Occurs on Viewing of a Crystal or Geometrical Solid," *Philosophical Magazine, Ser. 3*, **1**, 329–343.

Oja, E. (1982). "A Simplified Neuron Model as a Principal Component Analyzer," *J. Math. Biol.*, **15**, 267–273.

Oja, E. (1983). *Subspace Methods of Pattern Recognition*. Research Studies Press, Letchworth, UK.

Proulx, R. and Begin, J. (1990). "A New Learning Algorithm for the BSB Model," in *Proceedings of IJCNN-90*, Washington, D.C., M. Caudill, ed., Erlbaum, Hillsdale, NJ, Vol. I, pp. 703–705.

Ratcliff, R. (1978). "A Theory of Memory Retrieval," *Psychol. Rev.*, **85**, 59–108.

Rossen, M. L. (1989). "Speech Syllable Recognition with a Neural Network," Ph.D. Thesis, Department of Psychology, Brown University.

Rumelhart, D. E. and McClelland, J. L. (1982). "An Interactive Activation Model of Context Effects in Letter Perception: Part 2. The Contextual Enhancement Effect and Some Tests and Extensions of the Model," *Psychol. Rev.*, **89**, 60–94.

Spitzer, A. R., Hassoun, M. H., Wang, C., and Bearden, F. (1990). "Signal Decomposition and Diagnostic Classification of the Electromyogram Using a Novel Neural Network Technique," *Proc. 14th Annual Symposium on Computer Applications in Medical Care*, Washington, DC, pp. 552–556.

Van Essen, D. C. (1985). "Functional Organization of Primate Visual Cortex," in *Cerebral Cortex: Vol. 3, Visual Cortex*, A. Peters and E. G. Jones, eds. Plenum, New York.

Viscuso, S. R. (1989). "Memory for Arithmetic Facts: A Perspective Gained from Two Methodologies," Ph.D. Thesis, Department of Psychology, Brown University.

Young, D. M. and Gregory, R. T. (1973). *A Survey of Numerical Mathematics*. Addison-Wesley, Reading, MA.

Widrow, B. and Hoff, M. (1960). "Adaptive Switching Circuits," *1960 IRE WESCON Convention Record*. IRE, New York, pp. 96–104.

Widrow, B. and Stearns, S. D. (1985). *Adaptive Signal Processing*. Prentice-Hall, Englewood Cliffs, NJ.

Wilkinson, J. H. (1965). *The Algebraic Eigenvalue Problem*. Oxford University Press, Oxford, UK.

PART II

ARTIFICIAL ASSOCIATIVE NEURAL MEMORY MODELS

5

Bidirectional Associative Memories

YEOU-FANG WANG, JOSÉ B. CRUZ, Jr., and JAMES H. MULLIGAN, Jr.

5.1. INTRODUCTION

Bidirectional associative memory (BAM) is a two-layer network for heteroassociation. Through simple iterations of forward and backward operations in a simple structure, this network can reach a steady state which achieves the memory function. This network is an extension of Hopfield's single-layer auto-associator. For this reason, we will start from a brief study of the Hopfield models, then move on to the main topic of this chapter—discrete BAM. Convergence analysis as well as both advantages and disadvantages of the discrete BAM are discussed, followed by an introduction to several other BAM models.

Small capacity and no guarantee of recall are two of the major problems of the original BAM encoding strategy. In this chapter, several methods for improvement in the encoding phase will be discussed. Some of the encoding methods may be applicable to other types of associative memories. Examples and computer simulations are provided for illustration.

Since *bidirectional associative memories* (BAMs) are closely related to Hopfield's models, it is useful to briefly review the latter models first. In this section, both discrete and continuous models of the Hopfield neural networks will be discussed. Hopfield models (Hopfield, 1982, 1984) are one-layer neural networks. Basically, there are two aspects of the Hopfield models: one is the energy function; the other one is the retrieval procedure. The idea is to find a set of connection weights such that the local minimum of the energy function can be reached through the retrieval procedure. The concept is used not only in associative memories but also in optimization problems (Hopfield and Tank, 1985). This will be clarified when we discuss the models themselves.

5.1.1. Discrete Model

Figure 5.1 depicts the block diagram of a discrete Hopfield model. In this model, each neuron has only two types of output, ON and OFF. Therefore, it is referred to as a discrete model. From the structure, the state of neuron i (the input to that neuron) is

$$x_i = \sum_j T_{ij} v_j + I_i \tag{1}$$

where x_i is the input to neuron i
v_j is the output of neuron j
T_{ij} is the weight on the output of neuron j contributing to the input of neuron i
I_i is the external input (bias) to neuron i

and

$$v_i(k+1) = \begin{cases} 1 & \text{if } x_i(k) > \theta_i \\ v_i(k) & \text{if } x_i(k) = \theta_i \\ -1 & \text{if } x_i(k) < \theta_i \end{cases} \tag{2}$$

where θ_i is the threshold value for neuron i. This is the basic retrieval procedure of the discrete model. The energy function is defined to be

$$E_h = -\frac{1}{2} \sum_i \sum_j T_{ij} v_i v_j - \sum_i I_i v_i + \sum_i \theta_i v_i \tag{3}$$

When $T_{ii} = 0$ and $T_{ij} = T_{ji}$, it can be proved that the energy E_h will not increase for any v_i changes in a random asynchronous fashion through the processes of (1) and (2). Therefore, if we can find a set of T_{ij} such that the steady states of v_i of the network are the desired values in a local minimum of E_h, then this network can be used as an associative memory.

5.1.2. Continuous Model

Figure 5.2 depicts a continuous Hopfield model. In this model, an input capacitor C_i and an input resistor ρ_i are associated with each neuron. It can be shown that the state equation of each neuron is

$$C_i \frac{dx_i}{dt} = \sum_j T_{ij} v_j - \frac{x_i}{R_i} + I_i \tag{4}$$

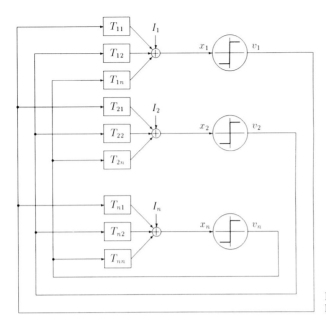

Fig. 5.1. The block diagram of a discrete Hopfield network.

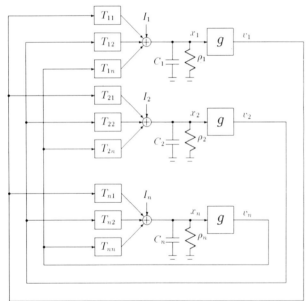

Fig. 5.2. The continuous Hopfield network structure.

where

$$\frac{1}{R_i} = \frac{1}{\rho_i} + \sum_j T_{ij}$$

T_{ij} is the connection weight (reciprocal of resistance) between neurons i and j, I_i is the direct input to neuron i, and $v_i = g(x_i)$ where g is a sigmoidal function with positive first order derivative. (4) and $g(\cdot)$ are the processing operations of the neurons. The energy function is defined to be

$$E_h = -\frac{1}{2}\sum_i \sum_j T_{ij} v_i v_j - \sum_i I_i v_i$$
$$+ \sum_i \left(\frac{1}{R_i}\right) \int_0^{v_i} g^{-1}(v)\, dv \qquad (5)$$

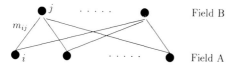

Fig. 5.3. Structure for a Bidirectional Associative Memory. Data in field A are intended to retrieve data in field B. Moreover, data in field B are intended to retrieve data in field A.

It can be shown that $dE_h/dt \leq 0$, when $T_{ij} = T_{ji}$. Therefore, if the output produced by the steady state of (4) is one of the desired memories, then this model can be used as an associative memory. This concept can be extended to optimization problems. If T_{ij} can be chosen to construct an energy function which matches the objective function of an optimization problem which we want to solve, then the steady states of (4) are local optimal solutions to this problem.

5.2. THE DISCRETE BIDIRECTIONAL ASSOCIATIVE MEMORY[1]

A disadvantage of the Hopfield models is the need for a large connection matrix. An extension by Kosko (1987a, b, 1988) of the Hopfield models resulted in a two-layer and two-directional structure which is identified as the bidirectional associative memory (BAM). This arrangement can yield heteroassociation using a smaller correlation matrix. A typical BAM structure is given in Fig. 5.3. The discussion of this section will be limited to the discrete BAM. Other BAM models will be considered in the next section.

The discrete BAM is intended to retrieve the pair $(\mathbf{A}_i, \mathbf{B}_i)$ once an initial pair (α_i, β_i) has been presented, and once the BAM has been trained using a set of training pairs $\{\mathbf{P}_i = (\mathbf{A}_i, \mathbf{B}_i) | i = 1, \ldots, N\}$. The quantities $\mathbf{A}_i$ and $\mathbf{B}_i$ are defined as

$$\mathbf{A}_i = (a_{i1}, a_{i2}, \ldots, a_{in}) \quad \mathbf{B}_i = (b_{i1}, b_{i2}, \ldots, b_{ip})$$

The quantities a_{ij}, b_{ij} represent either ON or OFF states. In the binary mode, ON corresponds to 1 and OFF corresponds to 0. In the bipolar mode, ON and OFF corresponds to $+1$ and -1, respectively.

[1] The material in this section as well as that in Sections 5.4.1, 5.4.2, 5.5.1, and 5.2 is drawn largely from Wang et al. (1990a, b, c, 1991) to which the reader is referred for further details.

5.2.1. Original Coding Strategy

As in the Hopfield model, there are processing equations associated with the energy function of the BAM. Furthermore, it is desired that the steady states be the desired memories. The energy function E for pair (α, β) and correlation matrix $\mathbf{M} = [m_{ij}]_{n \times p}$ (the weight matrix for the connections) is

$$E = -\alpha \mathbf{M} \beta^{\mathrm{T}} \tag{6}$$

The correlation matrix $\mathbf{M}$ plays the same role as $\mathbf{T} = [T_{ij}]$ in Hopfield's model. Equation (6) can be rewritten as

$$E = -\tfrac{1}{2}(\alpha \mathbf{M} \beta^{\mathrm{T}} + \beta \mathbf{M}^{\mathrm{T}} \alpha^{\mathrm{T}})$$

This is seen to be a bidirectional version of (3) with $\theta_i = I_i = 0$, that is zero threshold and no bias to the neurons.

The decoding process associated with the memory begins by specifying an initial condition (α, β). From this information and $\mathbf{M}$, the elements of a finite sequence $(\alpha', \beta'), (\alpha'', \beta''), \ldots$ are determined until an equilibrium point (α_f, β_f) is reached. The terms in the sequence are computed from the relations

$$\beta' = \phi(\alpha \mathbf{M})$$

with $\phi(\mathbf{F}) = \mathbf{G} = (g_1, g_2, \ldots, g_r)$, $\mathbf{F} = (f_1, f_2, \ldots, f_r)$, and

$$g_i = \begin{cases} 1 & \text{if } f_i > 0 \\ \text{previous } g_i & \text{if } f_i = 0 \\ \begin{cases} -1 & \text{(bipolar)} \\ 0 & \text{(binary)} \end{cases} & \text{if } f_i < 0, \end{cases} \tag{7}$$

and

$$\alpha' = \phi(\beta' \mathbf{M}^{\mathrm{T}})$$
$$\beta'' = \phi(\alpha' \mathbf{M})$$
$$\alpha'' = \phi(\beta'' \mathbf{M}^{\mathrm{T}}), \text{ etc.}$$

Note that (7) is a zero-threshold version of (2); the decoding procedure is a bidirectional version of the Hopfield models.

The procedure shown above is a synchronous decoding process. An asynchronous process which updates the elements of α and β randomly can also be used. In this chapter, even though our discussions are not limited to synchronous processes, synchronous processes are assumed for convenience in discussion.

One of the most important aspects of the use of the associative memory is to define a correlation matrix, $\mathbf{M}$, such that the above procedures will produce a desired output that is one of the items the memory is to remember. The process

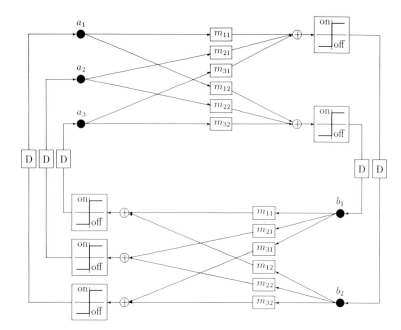

Fig. 5.4. A block diagram for a three by two BAM. The m_{ij} are constant gains and D represents a unit time delay.

of forming a correlation matrix is called an encoding (recording) phase.

Kosko defined the original correlation matrix for the BAM as

$$\mathbf{M}_o = \sum_{i=1}^{N} \mathbf{X}_i^\mathsf{T} \mathbf{Y}_i \qquad (8)$$

with

$$\mathbf{X}_i = (x_{i1}, x_{i2}, \ldots, x_{in})$$
$$\mathbf{Y}_i = (y_{i1}, y_{i2}, \ldots, y_{ip})$$

$x_{ij}(y_{ij})$ is the bipolar form of $a_{ij}(b_{ij})$.

A three-by-two discrete BAM is shown in Fig. 5.4. The behavior of this network is considered in Example 1.

Example 1. Bidirectional associative memory. Assume one training pair (given in binary form) is to be used to establish the correlation matrix, i.e., train the network. Let the pair be

$$\mathbf{A}_1 = [1 \quad 0 \quad 1] \qquad \mathbf{B}_1 = [1 \quad 0]$$

The quantities $\mathbf{X}_1$ and $\mathbf{Y}_1$ are then found to be

$$\mathbf{X}_1 = [1 \quad -1 \quad 1] \qquad \mathbf{Y}_1 = [1 \quad -1]$$

The elements of the correlation matrix can then

be evaluated as

$$\mathbf{M} = \begin{bmatrix} 1 \\ -1 \\ 1 \end{bmatrix} [1 \quad -1] = \begin{bmatrix} 1 & -1 \\ -1 & 1 \\ 1 & -1 \end{bmatrix}$$

$$= \begin{bmatrix} m_{11} & m_{12} \\ m_{21} & m_{22} \\ m_{31} & m_{32} \end{bmatrix}$$

Consider now the recall of $(\mathbf{A}_1, \mathbf{B}_1)$ using the process described above:

$$\alpha\mathbf{M} = \mathbf{A}_1\mathbf{M} = [1 \quad 0 \quad 1] \begin{bmatrix} 1 & -1 \\ -1 & 1 \\ 1 & -1 \end{bmatrix}$$

$$= [2 \quad -2] = \mathbf{F}$$

$$\mathbf{G} = \phi(\mathbf{F}) = [1 \quad 0] = \boldsymbol{\beta}' = \mathbf{B}_1$$

$$\boldsymbol{\beta}'\mathbf{M}^\mathsf{T} = \mathbf{B}_1\mathbf{M}^\mathsf{T} = [1 \quad 0] \begin{bmatrix} 1 & -1 & 1 \\ -1 & 1 & -1 \end{bmatrix}$$

$$= [1 \quad -1 \quad 1] = \mathbf{F}$$

$$\mathbf{G} = \phi(\mathbf{F}) = [1 \quad 0 \quad 1] = \boldsymbol{\alpha}' = \mathbf{A}_1$$

5.2.2. Convergence Analysis

It has been shown by Kosko (1988) for the BAM that after a finite number of steps the decoding process will converge using *any* correlation matrix $\mathbf{M}$. Furthermore, a pair $\mathbf{P}_i$ can be recalled if and only if this pair is a local minimum of the energy surface. It is important to note, however, that the correlation matrix introduced by Kosko (1988) cannot guarantee recall of any particular pair in a training set with more than two training pairs, because it cannot be guaranteed that the energy of any training pair is a local minimum. Example 2 below illustrates this situation. The following two theorems and a corollary deal with the question of convergence in the decoding process and some implications.

THEOREM 1. *The $(\boldsymbol{\alpha}_f, \boldsymbol{\beta}_f)$ determined by the completion of the BAM set of decoding cycles occurs at a local minimum of the energy function.*

PROOF. Note first that "completion of the BAM set of decoding cycles" implies that further decoding iterations cause no change in the pair $(\boldsymbol{\alpha}_f, \boldsymbol{\beta}_f)$.

Assume first that $(\boldsymbol{\alpha}_f, \boldsymbol{\beta}_f)$ is not a local minimum of the energy surface. If so, there exists an $\boldsymbol{\alpha}'$, which differs from $\boldsymbol{\alpha}_f$ only in the sth component, i.e., $a'_s \neq a_s$,

$$\boldsymbol{\alpha}' = [a_1, \ldots, a'_s, \ldots, a_n]$$

such that

$$-\boldsymbol{\alpha}_f \mathbf{M} \boldsymbol{\beta}_f^\mathrm{T} = E > E' = -\boldsymbol{\alpha}' \mathbf{M} \boldsymbol{\beta}^\mathrm{T}$$

Therefore,

$$(\boldsymbol{\alpha}' - \boldsymbol{\alpha}_f) \mathbf{M} \boldsymbol{\beta}_f^\mathrm{T} > 0$$

$$[0, \ldots, 0, (a'_s - a_s), 0, \ldots, 0] \mathbf{M} \boldsymbol{\beta}_f^\mathrm{T} > 0$$

and

$$(a'_s - a_s) \sum_{j=1}^{p} m_{sj} b_j > 0$$

where

$$\boldsymbol{\beta}_f = [b_1, \ldots, b_j, \ldots, b_p]$$

If $a_s = 1$ then $a'_s = 0$ and

$$a'_s - a_s < 0 \qquad \sum_{j=1}^{p} m_{sj} b_j < 0$$

Therefore, on the next cycle, the value of a_s is

$$a_s = \begin{cases} -1 & \text{for bipolar} \\ 0 & \text{for binary} \end{cases}$$

If a_s is -1 or 0, then

$$a'_s - a_s > 0, \qquad \sum_{j=1}^{p} m_{sj} b_j > 0$$

Recall that the next value is determined by

$$\phi\left(\sum_{j=1}^{p} b_j m_{sj} \right)$$

Therefore, the next value of a_s is 1. This means that $(\boldsymbol{\alpha}_f, \boldsymbol{\beta}_f)$ will be changed. This contradicts the assumption that BAM set of decoding cycles has been completed. $\square$

THEOREM 2. *If a training pair $(\mathbf{A}_i, \mathbf{B}_i)$ is a local minimum, then $\phi(\mathbf{A}_i \mathbf{M}) = \mathbf{B}_i$ and $\phi(\mathbf{B}_i \mathbf{M}^\mathrm{T}) = \mathbf{A}_i$.*

PROOF. Define $\mathbf{A}_i$ and $\mathbf{A}'_i$ as

$$\mathbf{A}_i = [a_1, \ldots, a_s, \ldots, a_n]$$
$$\mathbf{A}'_i = [a_1, \ldots, a'_s, \ldots, a_n]$$
$$a_s \neq a'_s$$

Then by the use of Theorem 1 and the expression for the energy function and because $(\mathbf{A}_i, \mathbf{B}_i)$ is a local minimum,

$$-\mathbf{A}_i \mathbf{M} \mathbf{B}_i^\mathrm{T} < -\mathbf{A}'_i \mathbf{M} \mathbf{B}_i^\mathrm{T}$$

That is,

$$(a_s - a'_s) \sum_{j=1}^{p} m_{sj} b_j > 0$$

If $a_s = 1$, then $(a_s - a'_s) > 0$ and

$$\sum_{j=1}^{p} m_{sj} b_j > 0$$

Hence, the next a_s is 1.

If a_s is -1 or 0, then $(a_s - a'_s) < 0$ and

$$\sum_{j=1}^{p} m_{sj} b_j < 0$$

Therefore, the next a_s will be either -1 or 0. This implies that no change occurs in a_s. A similar procedure can be used to show that b_s does not change. $\square$

COROLLARY 1. $(\boldsymbol{\alpha}, \boldsymbol{\beta})$ *is a local minimum if and only if* $\boldsymbol{\beta} = \phi(\boldsymbol{\alpha} \mathbf{M})$ *and* $\boldsymbol{\alpha} = \phi(\boldsymbol{\beta} \mathbf{M}^\mathrm{T})$. $\square$

Example 2. Recall of training pairs—failure. Consider that three training pairs are given as

$A_1 = [1 \ 0 \ 0 \ 1 \ 1 \ 1 \ 0 \ 0 \ 0]$

$B_1 = [1 \ 1 \ 1 \ 0 \ 0 \ 0 \ 0 \ 1 \ 0]$,

$A_2 = [0 \ 1 \ 1 \ 1 \ 0 \ 0 \ 1 \ 1 \ 1]$

$B_2 = [1 \ 0 \ 0 \ 0 \ 0 \ 0 \ 0 \ 0 \ 1]$,

$A_3 = [1 \ 0 \ 1 \ 0 \ 1 \ 1 \ 0 \ 1 \ 1]$

$B_3 = [0 \ 1 \ 0 \ 1 \ 0 \ 0 \ 1 \ 0 \ 1]$,

The correlation matrix **M** is then found to be

M =

$$\begin{bmatrix}
-1 & 3 & 1 & 1 & -1 & -1 & 1 & 1 & -1 \\
1 & -3 & -1 & -1 & 1 & 1 & -1 & -1 & 1 \\
-1 & -1 & -3 & 1 & -1 & -1 & 1 & -3 & 3 \\
3 & -1 & 1 & -3 & -1 & -1 & -3 & 1 & -1 \\
-1 & 3 & 1 & 1 & -1 & -1 & 1 & 1 & -1 \\
-1 & 3 & 1 & 1 & -1 & -1 & 1 & 1 & -1 \\
1 & -3 & -1 & -1 & 1 & 1 & -1 & -1 & 1 \\
-1 & -1 & -3 & 1 & -1 & -1 & 1 & -3 & 3 \\
-1 & -1 & -3 & 1 & -1 & -1 & 1 & -3 & 3
\end{bmatrix}$$

Assume that bipolar form of representation is used. The decoding process then yields the following steps:

$X_2M = [5 \ -19 \ -13 \ -5 \ 1 \ 1 \ -5 \ -13 \ 13]$

$\beta = \phi(X_2M)$

$\quad = [1 \ -1 \ -1 \ -1 \ 1 \ 1 \ -1 \ -1 \ 1] \neq Y_2$

$\beta M^T = [-11 \ 11 \ 5 \ 5 \ -11 \ -11 \ 11 \ 5 \ 5]$

$\alpha = \phi(\beta M^T)$

$\quad = [-1 \ 1 \ 1 \ 1 \ -1 \ -1 \ 1 \ 1 \ 1] = X_2$

The next decoding cycle using the value found for $\alpha (= X_2)$ will produce β again. Thus, the cycle has been completed. However, the value to which β converges is not Y_2. The significance of this is that the correlation matrix **M** cannot recover the original pair (A_2, B_2).

Computation of the energy function E_2 for (X_2, Y_2) and E_f for (α, β) provides an interesting insight regarding the process. By direct substitution of the necessary values, it is found that

$$E_2 = -X_2MY_2^T = -71$$
$$E_f = -\alpha M\beta^T = -75$$

To demonstrate that (X_2, Y_2) is not a local minimum, the energy can be evaluated at a point which is one Hamming distance removed from Y_2. Assume that the fifth component of Y_2 is changed from -1 to $+1$ to obtain Y_2',

$$Y_2' = [1 \ -1 \ -1 \ -1 \ 1 \ -1 \ -1 \ -1 \ 1]$$

The new energy has the value

$$E = -X_2MY_2'^T = -73$$

This is lower than E_2, which confirms the hypothesis that (A_2, B_2) is not a local minimum of E. □

For some applications, it may be necessary to guarantee recall of one or more particular training pairs. Several methods for improving BAM performance will be discussed in Section 5.4. In Section 5.5, the results of computer experiments are reported to illustrate the use and capacity of BAM.

5.2.3. Advantages and Disadvantages of Discrete BAM

A discussion of some advantages and disadvantages of the discrete BAM model provides a perspective from which one can better appreciate some improvements in coding methods to be presented subsequently.

Advantages

1. *BAM needs a smaller correlation matrix than the Hopfield model.* To store the pair $([1, 0, 1], [1, 0])$, for example, BAM needs a 3×2 matrix while the Hopfield model needs a 5×5 matrix. The matrix

$$\begin{bmatrix} 0 & M \\ M^T & 0 \end{bmatrix}$$

where **M** is the correlation matrix for BAM, can be used to convert a bidirectional hetero-associative procedure to a large-scale unidirectional autoassociative procedure. Therefore, in terms of the size of the correlation matrix, it is more economical to use BAM.

2. *BAM has no oscillatory states.* Assume that it is desired to use the Hopfield network to store and recall the vectors

$$X_1 = [1 \ -1 \ 1 \ -1 \ -1 \ 1]$$
$$X_2 = [-1 \ -1 \ 1 \ -1 \ 1 \ 1]$$

The Hopfield correlation matrix is

$$\mathbf{T} = \sum_{i=1}^{N} \mathbf{X}_i^T \mathbf{X}_i - N \cdot \mathbf{I}$$

where $\mathbf{I}$ is the identity matrix. Evaluation of $\mathbf{T}$ yields,

$$\mathbf{T} = \mathbf{X}_1^T \mathbf{X}_1 + \mathbf{X}_2^T \mathbf{X}_2 - 2\mathbf{I}$$

$$\mathbf{T} = \begin{bmatrix}
0 & 0 & 0 & 0 & -2 & 0 \\
0 & 0 & -2 & 2 & 0 & -2 \\
0 & -2 & 0 & -2 & 0 & 2 \\
0 & 2 & -2 & 0 & 0 & -2 \\
-2 & 0 & 0 & 0 & 0 & 0 \\
0 & -2 & 2 & -2 & 0 & 0
\end{bmatrix}$$

If one starts with $\mathbf{X}_{01} = [-1\ 1\ 1\ 1\ 1\ 1]$, then

$$\mathbf{X}_{01}\mathbf{T} = [-2\ \ -2\ \ -2\ \ -2\ \ 2\ \ -2]$$

Applying (2) to $\mathbf{X}_{01}\mathbf{T}$ yields

$$[-1\ \ -1\ \ -1\ \ -1\ \ 1\ \ -1] = \mathbf{X}_{02}$$

Also,

$$\mathbf{X}_{02}\mathbf{T} = [-2\ \ 2\ \ 2\ \ 2\ \ 2\ \ 2]$$

Again, applying (2) to $\mathbf{X}_{02}\mathbf{T}$ yields

$$[-1\ \ 1\ \ 1\ \ 1\ \ 1\ \ 1] = \mathbf{X}_{01}$$

It is seen that the final state is cyclic with terms $\mathbf{X}_{01} \to \mathbf{X}_{02} \to \mathbf{X}_{01} \to \cdots$. The use of (3) with $\theta_i = 0$ and $I_i = 0$, yields

$$E_h(\mathbf{X}_{01}) = -\tfrac{1}{2}\mathbf{X}_{01}\mathbf{T}\mathbf{X}_{01}^T = 2$$

$$E_h(\mathbf{X}_{02}) = -\tfrac{1}{2}\mathbf{X}_{02}\mathbf{T}\mathbf{X}_{02}^T = 2$$

Furthermore, it is shown below that neither $\mathbf{X}_{01}$ nor $\mathbf{X}_{02}$ is a local minimum. Choose

$$\mathbf{X}_{01}' = [-1\ \ 1\ \ 1\ \ 1\ \ 1\ \ -1]$$

$$\mathbf{X}_{02}' = [-1\ \ -1\ \ 1\ \ -1\ \ 1\ \ -1]$$

where it is seen that $\mathbf{X}_{01}'$ is one Hamming distance away from $\mathbf{X}_{01}$ and $\mathbf{X}_{02}'$ is one Hamming distance away from $\mathbf{X}_{02}$. Calculation then yields $E_h(\mathbf{X}_{01}') = E_h(\mathbf{X}_{02}') = -2$. Thus, if a synchronous decoding process is used, the Hopfield model may yield states which are not local minima.

In contrast with the Hopfield model, BAM has no oscillatory states. This is so because $\Delta E < 0$ in BAM if the neuron states change. This may be shown by direct calculation using the energy function. No generality is lost, if one assumes the initial state to be (α, β) and the next state to be

(α, γ), where $\gamma = \phi(\alpha\mathbf{M})$. Define E_1 and E_2 as

$$E_1 = -\alpha\mathbf{M}\beta^T = -\sum_i \sum_j \alpha_i m_{ij} \beta_j$$

$$E_2 = -\alpha\mathbf{M}\gamma^T = -\sum_i \sum_j \alpha_i m_{ij} \gamma_j$$

the change in energy is given by

$$\Delta E = E_2 - E_1 = -\sum_j \sum_i \alpha_i m_{ij} (\gamma_j - \beta_j)$$

$$= -\sum_{j \in S} \Psi_j$$

with

$$S = \{k \mid \gamma_k \neq \beta_k\}$$

If $\gamma_k = 1$, then

$$\sum_i \alpha_i m_{ik} > 0, \quad \gamma_k - \beta_k > 0 \Rightarrow \Psi_k > 0$$

whereas if $\gamma_k = -1$,

$$\sum_i \alpha_i m_{ik} < 0, \quad \gamma_k - \beta_k < 0 \Rightarrow \Psi_k > 0$$

Therefore, $\Delta E < 0$ if $\beta \neq \gamma$. Because α and β are of finite dimension, ΔE will eventually become zero. This implies that the next state will be the same as the current one. From Corollary 1, (α_f, β_f) is a final state if and only if (α_f, β_f) is a local minimum. Thus, regardless of the initial condition at which the process begins, synchronous BAM will always converge to a local minimum. On the other hand, it cannot be guaranteed that the synchronous Hopfield model will converge to a local minimum.

3. *BAM is relatively easy to implement.* It is seen from Fig. 5.4 that BAM requires only the processes of summation and multiplication, plus a threshold function. It is therefore a relatively simple matter to implement it. Kosko (1987b) has suggested an optical method for its implementation.

Disadvantages

1. *BAM can only implement one-to-one mapping.* Since BAM processes data in two directions and each direction has to be a *function*, BAM can only implement one-to-one mappings. This restricts some of the applications of the network. However, it is an advantage to provide backward verification in some applications.

2. *Small capacity.* Small capacity (*capacity* is defined in Chapter 1) is a common problem of those networks based on Hebbian types of correlation matrices. A famous result of the

capacity for the discrete Hopfield model has been derived by McEliece et al. (1987). Basically, their result states that the maximum asymptotic value of the number of training pairs, N, in order that most of the N original memories are exactly recoverable is $n/(2 \log n)$, where n is the number of neurons in the Hopfield network. This suggests that for large n, the pattern ratio N/n required for perfect storage approaches zero which reflects the inefficiency of the Hopfield model. It is seen from Simpson's computer simulation results (1990) that some of the Hebbian type networks, including Hopfield's model and BAM, have low capacities.

5.3. OTHER BAM MODELS

For completeness three other kinds of BAM models are described below.

5.3.1. Continuous BAM

This model, proposed by Kosko (1987b), is an extension of the Hopfield continuous model. Basically, dynamic equations are associated with the neurons in Fig. 5.3. A *Liapunov* function is used as an energy function, and the value of the Liapunov function decreases along the paths of the states of the dynamic equations. The steady states of the dynamic equations are the memories of the BAM.

Assume a_i is the state of neuron i in field A and b_j is the state of neuron j in field B. The dynamic equations associated with the neurons are

$$\dot{a}_i = -a_i + \sum_j S(b_j)m_{ij} + I_i \qquad (9)$$

$$\dot{b}_j = -b_j = \sum_i S(a_i)m_{ij} + J_j \qquad (10)$$

where S is a sigmoid signal function and the derivative of S, S', is assumed to be positive and $S(\cdot)$ is defined to be either in $[0, 1]$ or in $[-1, 1]$. Notice that (9) and (10) are bidirectional versions of (4).

Kosko (1987b) has shown that the energy function

$$E_c(\mathbf{A}, \mathbf{B})$$

$$= \sum_i \int_0^{a_i} S'(x_i)x_i \, dx_i + \sum_j \int_0^{b_j} S'(y_j) y_j \, dy_j$$

$$- \sum_i S(a_i)I_i - \sum_j S(b_j)J_j - \sum_i \sum_j S(a_i)S(b_j)m_{ij} \qquad (11)$$

is an appropriate bounded Liapunov function. From (9), (10), (11), and the condition $S'(\cdot) > 0$,

$$\dot{E}_c = -\sum_i S'(a_i)\dot{a}_i^2 - \sum_j S'(b_j)\dot{b}_j^2$$

$$\leq 0$$

Thus, for the continuous BAM structure, any $\mathbf{M}$ matrix provides a stable system. Therefore, if one can find an $\mathbf{M}$ matrix such that all desired training pairs are located in the local minima of E_c, then such an $\mathbf{M}$ matrix can be used to form a continuous BAM.

5.3.2. Adaptive BAM

Usually, when one determines the correlation matrix $\mathbf{M}$, fixed values are assigned to its elements. The adaptive BAM has the property of adjusting the $\mathbf{M}$ matrix during operation.

Kosko (1987b) suggested the following learning law to adjust the $\mathbf{M}$ matrix,

$$\dot{m}_{ij} = -m_{ij} + S(a_i)S(b_j) \qquad (12)$$

The new energy function is defined to be

$$E_a(\mathbf{A}, \mathbf{B}, \mathbf{M}) = E_c(\mathbf{A}, \mathbf{B}) + \frac{1}{2}\sum_i \sum_j m_{ij}^2 \qquad (13)$$

It is shown in Kosko (1987b) that

$$\dot{E}_a = -\sum_i \sum_j \dot{m}_{ij}[S(a_i)S(b_j) - m_{ij}]$$

$$- \sum_i S'\dot{a}_i^2 - \sum_j S'\dot{b}_j^2 \leq 0$$

Therefore, the adaptive BAM is stable. Obviously, this $\mathbf{M}$ matrix can be used for the continuous BAM.

5.3.3. Nonhomogeneous BAM

Nonhomogeneous BAM was proposed by Haines and Hecht-Nielsen (1988). Basically, this is a discrete BAM with nonzero thresholds. For this structure, (7) is modified to the form

$$\alpha_i = \begin{cases} 1 & \text{if } \sum_{j=1}^p \beta_j m_{ij} > \theta_i^a \\ \text{previous } \alpha_i & \text{if } \sum_{j=1}^p \beta_j m_{ij} = \theta_i^a \\ \begin{cases} -1 & \text{(bipolar)} \\ 0 & \text{(binary)} \end{cases} & \text{if } \sum_{j=1}^p \beta_j m_{ij} < \theta_i^a \end{cases}$$

and

$$
\beta_j = \begin{cases}
1 & \text{if } \sum_{i=1}^{n} \alpha_i m_{ij} > \theta_j^b \\
\text{previous } \beta_j & \text{if } \sum_{i=1}^{n} \alpha_i m_{ij} = \theta_j^b \\
\begin{cases} -1 & \text{(bipolar)} \\ 0 & \text{(binary)} \end{cases} & \text{if } \sum_{i=1}^{n} \alpha_i m_{ij} < \theta_j^b
\end{cases}
$$

These equations are similar to (2) in the discrete Hopfield model. The energy function is defined to be

$$
E_n(\mathbf{A}, \mathbf{B}, \mathbf{\Theta}_a, \mathbf{\Theta}_b) = -\mathbf{A}\mathbf{M}\mathbf{B}^{\mathrm{T}} + \mathbf{A}\mathbf{\Theta}_a^{\mathrm{T}} + \mathbf{B}\mathbf{\Theta}_b^{\mathrm{T}}
\tag{14}
$$

where

$$
\mathbf{\Theta}_a = [\theta_1^a, \ldots, \theta_n^a]
$$
$$
\mathbf{\Theta}_b = [\theta_1^b, \ldots, \theta_p^b]
$$

Notice that E_n in (14) is a bidirectional version of E_h in (3) with no direct input ($I_i = 0$). Haines and Hecht-Nielsen (1988) have shown the convergence of the nonhomogeneous BAM.

5.4. OTHER ENCODING/TRAINING STRATEGIES FOR THE DISCRETE BAM

Since it is known that the correlation matrix used by Kosko in the discrete BAM cannot guarantee any training pair to be located at a local minimum when there are more than two training pairs, it is desirable to modify the encoding process of the discrete BAM to eliminate this difficulty. In this section, some of the techniques to form the correlation matrix $\mathbf{M}$ for the discrete BAM will be discussed that provide improvement in the recall process.

5.4.1. Multiple Training

Multiple training, proposed by Wang, Cruz, and Mulligan (1989, 1990a, b, c, 1991), is a straightforward method to emphasize those unlearned pairs by introducing them to the correlation matrix repeatedly until all training pairs are learned. This condition can be achieved if there exists a Hebbian type correlation matrix which can remember all training pairs. In this case, the correlation matrix becomes a *weighted* correlation matrix with different weights on different training pairs. To facilitate exposition the discussion starts with a single pair case, proceeds to the two-pair condition, and subsequently to consideration of multiple pairs.

Multiple Training Method for a Single Data Pair

To guarantee a specific pair to be recallable, the concept of multiple training can be applied to that pair. The concept is that one puts that specific pair to the $\mathbf{M}$ matrix many times until it is learned. Multiple training of order q directed to the recovery of pair $(\mathbf{A}_i, \mathbf{B}_i)$ consists of adding to the $\mathbf{M}$ matrix a matrix $\mathbf{P}$, defined as

$$
\mathbf{P} = (q - 1)\mathbf{X}_i^{\mathrm{T}}\mathbf{Y}_i
$$

Evaluation of the new energy function E' at the point $(\mathbf{A}_i, \mathbf{B}_i)$ yields

$$
E'(\mathbf{A}_i, \mathbf{B}_i) = -\mathbf{A}_i\mathbf{M}\mathbf{B}_i^{\mathrm{T}} - (q - 1)\mathbf{A}_i\mathbf{X}_i^{\mathrm{T}}\mathbf{Y}_i\mathbf{B}_i^{\mathrm{T}}
$$

The quantity $\mathbf{A}_i\mathbf{X}_i^{\mathrm{T}}\mathbf{Y}_i\mathbf{B}_i^{\mathrm{T}}$ is positive. Therefore, by suitable choice of q one can reduce the energy E' to an arbitrarily low value. To achieve the desired learning result, however, it must also be possible to achieve a local minimum of energy at $(\mathbf{A}_i, \mathbf{B}_i)$. Thus, one must ensure that the energy at $(\mathbf{A}_i, \mathbf{B}_i)$ does not exceed that at points one Hamming distance away.

Consider the contribution of the term $(q - 1)\mathbf{A}_i'\mathbf{X}_i^{\mathrm{T}}\mathbf{Y}_i\mathbf{B}_i^{\mathrm{T}}$ where it is assumed that $\mathbf{A}_i'$ is one Hamming distance away from $\mathbf{A}_i$ (a_{is}' is taken as one unit removed from a_{is}). Two possible values of a_{is} and a_{is}' for both binary and bipolar representations will be used. The following results are obtained:

- Binary representation:

  For $a_{is} = 1$, $x_{is} = 1$, and $a_{is}' = 0$;
  $$a_{is}'x_{is} < a_{is}x_{is}$$

  For $a_{is} = 0$, $x_{is} = -1$, and $a_{is}' = 1$;
  $$a_{is}'x_{is} < a_{is}x_{is}$$

- Bipolar representation:

  For $a_{is} = 1$, $x_{is} = 1$, and $a_{is}' = -1$;
  $$a_{is}'x_{is} < a_{is}x_{is}$$

  For $a_{is} = -1$, $x_{is} = -1$, and $a_{is}' = 1$;
  $$a_{is}'x_{is} < a_{is}x_{is}$$

Because of the inequalities above,

$$
\mathbf{A}_i\mathbf{X}_i^{\mathrm{T}}\mathbf{Y}_i\mathbf{B}_i^{\mathrm{T}} > \mathbf{A}_i'\mathbf{X}_i^{\mathrm{T}}\mathbf{Y}_i\mathbf{B}_i^{\mathrm{T}}
$$

This result demonstrates that a suitable value of q can be selected to reduce the value of the energy function at $(\mathbf{A}_i, \mathbf{B}_i)$ below that at any point in the neighborhood of the point. This process can thus induce a local minimum of E at $(\mathbf{A}_i, \mathbf{B}_i)$.

The addition of $(q-1)$ pairs at $(\mathbf{A}_i, \mathbf{B}_i)$ to the original N trained pairs affects the value of E at the locations of the other trained pairs. A value of q should be used which produces the smallest effect at the other locations. Further discussion of interactive effects will be described under *sequential multiple training* later.

This development is continued now to investigate the size of q necessary to achieve the result which has just been presented. To start the process, define

$$\eta_{ij} = \mathbf{A}_i \mathbf{X}_j^T \mathbf{Y}_j \mathbf{B}_i^T - \mathbf{A}_i' \mathbf{X}_j^T \mathbf{Y}_j \mathbf{B}_i^T \qquad (15)$$

$(\mathbf{A}_i', \mathbf{B}_i') = \mathbf{P}_i'$ is assumed to be one Hamming distance away from $(\mathbf{A}_i, \mathbf{B}_i) = \mathbf{P}_i$. Then

$$\eta_{ij} = \left(\sum_{k_1=1}^{n} a_{ik_1} x_{jk_1} \right) \left(\sum_{k_2=1}^{p} b_{ik_2} y_{jk_2} \right)$$
$$- \left(\sum_{k_3=1}^{n} a_{ik_3}' x_{jk_3} \right) \left(\sum_{k_4=1}^{p} b_{ik_4}' y_{jk_4} \right) \qquad (16)$$

where

$$\mathbf{A}_i' = (a_{i1}', a_{i2}', \ldots, a_{in}')$$
$$\mathbf{B}_i' = (b_{i1}', b_{i2}', \ldots, b_{ip}')$$

η_{ij} has a physical interpretation. It is the energy difference between $\mathbf{P}_i'$ and $\mathbf{P}_i$ due to $\mathbf{P}_j$. If $\eta_{ij} < 0$ then the neighbor $\mathbf{P}_i'$ achieves more negative energy than $\mathbf{P}_i$ from $\mathbf{P}^j$, and $\mathbf{P}_i$ is taken to be relatively more unstable; if $\eta_{ij} > 0$ then the neighbor $\mathbf{P}_i'$ achieves more negative energy than $\mathbf{P}_i'$ from $\mathbf{P}_j$, and $\mathbf{P}_i$ is viewed as relatively more stable. Define η_{ij}^{Fs} to be η_{ij} with respect to the neighbor which has the sth bit different in field F. From (16),

$$\eta_{ij}^{As} = (a_{is} x_{js} - a_{is}' x_{js}) \left(\sum_{k=1}^{p} b_{ik} y_{jk} \right)$$

Assume that $\mathbf{P}_i$ and $\mathbf{P}_j$ have d_{ij}^B different bits in field B, in bipolar form

$$\sum_{k=1}^{p} b_{ik} y_{jk} = p - 2d_{ij}^B \qquad (17)$$

Therefore,

for $a_{is} = a_{js}$: $\quad \eta_{ij}^{As} = 2(p - 2d_{ij}^B) \qquad (18)$

for $a_{is} \neq a_{js}$: $\quad \eta_{ij}^{As} = -2(p - 2d_{ij}^B) \qquad (19)$

Similarly, for field B

for $b_{is} = b_{js}$: $\quad \eta_{ij}^{Bs} = 2(n - 2d_{ij}^A) \qquad (20)$

for $b_{is} \neq b_{js}$: $\quad \eta_{ij}^{Bs} = -2(n - 2d_{ij}^A) \qquad (21)$

Note that the process of training a pair is equivalent to the process of training its complement, since $\mathbf{X}_j^T \mathbf{Y}_j = (-\mathbf{X}_j)^T(-\mathbf{Y}_j)$. Furthermore,

$$\eta_{ij} = \mathbf{A}_i \mathbf{X}_j^T \mathbf{Y}_j \mathbf{B}_i^T - \mathbf{A}_i' \mathbf{X}_j^T \mathbf{Y}_j \mathbf{B}_i'^T$$
$$\eta_{ij} = (-\mathbf{A}_i) \mathbf{X}_j^T \mathbf{Y}_j (-\mathbf{B}_i)^T - (-\mathbf{A}_i') \mathbf{X}_j^T \mathbf{Y}_j (-\mathbf{B}_i')^T$$

The energy difference η_{ij} for $\mathbf{P}_i$ will be the same as that of the complement of $\mathbf{P}_i$.

The quantity ε_{oi} is defined to be the energy difference between $\mathbf{P}_i$ and $\mathbf{P}_i'$ for $\mathbf{M}_o$, the original BAM correlation matrix. Thus, ε_{oi} is the sum of all energy differences contributed from the pairs in the training set. η_{ij} is the energy difference between $\mathbf{P}_i'$ and $\mathbf{P}_i$ due to $\mathbf{P}_j$, and therefore $\varepsilon_{oi} = -\sum_{j=1}^{N} \eta_{ij}$. Let E_{oi} denote the energy of $\mathbf{P}_i$ for the original $\mathbf{M}_o$ matrix and E_{oi}' denote that of $\mathbf{P}_i'$, then

$$\left. \begin{aligned} \varepsilon_{oi} &= E_{oi} - E_{oi}' \\ \varepsilon_{oi} &= -\mathbf{A}_i \mathbf{M}_o \mathbf{B}_i^T - (-\mathbf{A}_i' \mathbf{M}_o \mathbf{B}_i'^T) \\ \varepsilon_{oi} &= -\sum_{j=1}^{N} (\mathbf{A}_i \mathbf{X}_j^T \mathbf{Y}_j \mathbf{B}_i^T - \mathbf{A}_i' \mathbf{X}_j^T \mathbf{Y}_j \mathbf{B}_i'^T) \\ \varepsilon_{oi} &= -\sum_{j=1}^{N} \eta_{ij} \end{aligned} \right\} \quad (22)$$

The question of the number of multiple trainings needed to recall a desired pair is now addressed. Given that E_i is the energy of $\mathbf{P}_i$ for the $\mathbf{M}$ matrix and E_i' is that of $\mathbf{P}_i'$, and ε_i, defined as $\varepsilon_i = E_i - E_i'$, is to be equal to or less than zero, one proceeds by writing

$$\varepsilon_i = -\mathbf{A}_i \mathbf{M} \mathbf{B}_i^T - (-\mathbf{A}_i' \mathbf{M} \mathbf{B}_i'^T)$$

Whence

$$\varepsilon_i = (q-1)(-\mathbf{A}_i \mathbf{X}_i^T \mathbf{Y}_i \mathbf{B}_i^T + \mathbf{A}_i' \mathbf{X}_i^T \mathbf{Y}_i \mathbf{B}_i'^T) + \varepsilon_{oi} \leq 0 \qquad (23)$$

Thus, $(q-1)\eta_{ii} \geq \varepsilon_{oi}$. From (18) and (20), since $d_{ii}^B = 0$ and $d_{ii}^A = 0$, η_{ii} is either $2p$ or $2n$ in the bipolar case. Since $\eta_{ii} > 0$, it is necessary that

$$q \geq \frac{\varepsilon_{oi}}{\eta_{ii}} + 1 \qquad (24)$$

If one defines

$\varepsilon_{oi}^A = $ maximal $(E_{oi} - E_{oi}')$ among all neighbors of $\mathbf{P}_i$ in field A,

and

$\varepsilon_{oi}^B = $ maximal $(E_{oi} - E_{oi}')$ among all neighbors of $\mathbf{P}_i$ in field B, then it follows that

$$q \geq \max\left(1, \frac{\varepsilon_{oi}^A}{2p} + 1, \frac{\varepsilon_{oi}^B}{2n} + 1\right) \quad (25)$$

will ensure that $(\mathbf{A}_i, \mathbf{B}_i)$ can be recalled. If $p = n$ and $\varepsilon_0 = $ maximal $(E_{oi} - E'_{oi})$ among all neighbors, then any

$$q \geq \max\left(1, \frac{\varepsilon_o}{2p} + 1\right) \quad (26)$$

will guarantee recall of $(\mathbf{A}_i, \mathbf{B}_i)$.

Multiple Training Method for Two Data Pairs

It has been demonstrated that multiple training can be used to recall one desired pair. Consider now the possibility of using this method to ensure recall of two pairs among several pairs, whether in a training set or not.

Suppose $\mathbf{P}_i = (\mathbf{A}_i, \mathbf{B}_i)$ and $\mathbf{P}_j = (\mathbf{A}_j, \mathbf{B}_j)$ are the two pairs whose recall is desired. Suppose $\mathbf{P}_i$ has been trained for $t_i = (q_i - 1)$ additional times and $\mathbf{P}_j$ for $t_j = (q_j - 1)$ additional times. The energy functions of $\mathbf{P}_i$ and $\mathbf{P}_j$ have the form

$$\left.\begin{array}{l}E_i = -\mathbf{A}_i(t_i\mathbf{X}_i^T\mathbf{Y}_i + t_j\mathbf{X}_j^T\mathbf{Y}_j + \mathbf{M}_o)\mathbf{B}_i^T \\ E_j = -\mathbf{A}_j(t_i\mathbf{X}_i^T\mathbf{Y}_i + t_j\mathbf{X}_j^T\mathbf{Y}_j + \mathbf{M}_o)\mathbf{B}_j^T\end{array}\right\} \quad (27)$$

Let E'_i and E'_j be the energy functions of the neighbors of $\mathbf{P}_i$ and $\mathbf{P}_j$ one Hamming distance away. For recall one needs $E_i \leq E'_i$ and $E_j \leq E'_j$. This imposes the conditions

$$-\mathbf{A}_i(t_i\mathbf{X}_i^T\mathbf{Y}_i + t_j\mathbf{X}_j^T\mathbf{Y}_j + \mathbf{M}_o)\mathbf{B}_i^T$$
$$\leq -\mathbf{A}'_i(t_i\mathbf{X}_i^T\mathbf{Y}_i + t_j\mathbf{X}_j^T\mathbf{Y}_j + \mathbf{M}_o)\mathbf{B}_i'^T$$

and

$$-\mathbf{A}_j(t_i\mathbf{X}_i^T\mathbf{Y}_i + t_j\mathbf{X}_j^T\mathbf{Y}_j + \mathbf{M}_o)\mathbf{B}_j^T$$
$$\leq -\mathbf{A}'_j(t_i\mathbf{X}_i^T\mathbf{Y}_i + t_j\mathbf{X}_j^T\mathbf{Y}_j + \mathbf{M}_o)\mathbf{B}_j'^T$$

for any $\mathbf{P}'_i$ and $\mathbf{P}'_j$, where $(\mathbf{A}'_i, \mathbf{B}'_i)$ is a neighbor one Hamming distance away from $(\mathbf{A}_i, \mathbf{B}_i)$. Therefore,

$$t_i(\mathbf{A}_i\mathbf{X}_i^T\mathbf{Y}_i\mathbf{B}_i^T - \mathbf{A}'_i\mathbf{X}_i^T\mathbf{Y}_i\mathbf{B}_i'^T)$$
$$+ t_j(\mathbf{A}_i\mathbf{X}_j^T\mathbf{Y}_j\mathbf{B}_i^T - \mathbf{A}'_i\mathbf{X}_j^T\mathbf{Y}_j\mathbf{B}_i'^T) - \varepsilon_{oi} \geq 0$$

$$t_i(\mathbf{A}_j\mathbf{X}_i^T\mathbf{Y}_i\mathbf{B}_j^T - \mathbf{A}'_j\mathbf{X}_i^T\mathbf{Y}_i\mathbf{B}_j'^T)$$
$$+ t_j(\mathbf{A}_j\mathbf{X}_j^T\mathbf{Y}_j\mathbf{B}_j^T - \mathbf{A}'_j\mathbf{X}_j^T\mathbf{Y}_j\mathbf{B}_j'^T) - \varepsilon_{oj} \geq 0$$

This may be expressed as

$$t_i\eta_{ii} + t_j\eta_{ij} - \varepsilon_{oi} \geq 0$$
$$t_i\eta_{ji} + t_j\eta_{jj} - \varepsilon_{oj} \geq 0$$

Denote ε_{oi}^{Fr} to be the energy difference ε_{oi} when the difference bit is the rth bit in field F. For field A,

$$t_i\eta_{ii}^{Ar} + t_j\eta_{ij}^{Ar} - \varepsilon_{oi}^{Ar} \geq 0$$
$$t_i\eta_{ji}^{Ar} + t_j\eta_{jj}^{Ar} - \varepsilon_{oj}^{Ar} \geq 0 \quad \text{for } r = 1, \ldots, n$$

which may be written as

$$t_i\eta_{ii}^{Ar} + t_j\eta_{ij}^{Ar} + \sum_{k=1}^{N} \eta_{ik}^{Ar} \geq 0$$
$$\text{for } r = 1, \ldots, n$$
$$t_i\eta_{ji}^{Ar} + t_j\eta_{jj}^{Ar} + \sum_{k=1}^{N} \eta_{jk}^{Ar} \geq 0 \quad (28)$$

For field B, one finds

$$t_i\eta_{ii}^{Br} + t_j\eta_{ij}^{Br} + \sum_{k=1}^{N} \eta_{ik}^{Br} \geq 0$$
$$\text{for } r = 1, \ldots, p$$
$$t_i\eta_{ji}^{Br} + t_j\eta_{jj}^{Br} + \sum_{k=1}^{N} \eta_{jk}^{Br} \geq 0 \quad (29)$$

Let

$$d_{ij}^B = \frac{p}{2} - \delta_{ij}^B, \text{ if } d_{ij}^B < \frac{p}{2}$$
$$\text{where } 0 \leq \delta_{ij}^B \leq \frac{p}{2}$$
$$d_{ij}^B = \frac{p}{2} + \delta_{ij}^B, \text{ if } d_{ij}^B \geq \frac{p}{2} \quad (30)$$

and

$$d_{ij}^A = \frac{n}{2} - \delta_{ij}^A, \text{ if } d_{ij}^A < \frac{n}{2}$$
$$\text{where } 0 \leq \delta_{ij}^A \leq \frac{n}{2}$$
$$d_{ij}^A = \frac{n}{2} + \delta_{ij}^A, \text{ if } d_{ij}^A \geq \frac{n}{2} \quad (31)$$

Then, for field A

$$2d_{ij}^B = p \mp 2\delta_{ij}^B.$$

The use of (18) and (19) yields

$$\eta_{ij}^{Ar} = \pm 2(p - 2d_{ij}^B)$$
$$= \pm 2(p - (p \mp 2\delta_{ij}^B))$$
$$= \pm 4\delta_{ij}^B, \text{ for } r = 1, \ldots, n \quad (32)$$

Proceeding similarly for field B, one obtains

$$\eta_{ij}^{Br} = \pm 4\delta_{ij}^A, \text{ for } r = 1, \ldots, p \quad (33)$$

By definition, $d_{ij} = d_{ji}$ and $\delta_{ij} = \delta_{ji}$. The left-hand sides of (28) can then be written as,

$$t_i \eta_{ii}^{Ar} + t_j \eta_{ij}^{Ar} + \sum_{k=1}^{N} \eta_{ik}^{Ar}$$

$$= t_i(2p) \pm t_j(4\delta_{ij}^{B}) + \sum_{k=1}^{N} \eta_{ik}^{Ar}$$

$$t_i \eta_{ji}^{Ar} + t_j \eta_{jj}^{Ar} + \sum_{k=1}^{N} \eta_{jk}^{Ar}$$

$$= \pm t_i(4\delta_{ij}^{B}) + t_j(2p) + \sum_{k=1}^{N} \eta_{jk}^{Ar}$$

A sufficient condition for (28) is now established as

$$t_i(2p) - t_j(4\delta_{ij}^{B}) + \sum_{k=1}^{N} \eta_{ik}^{Ar} \geq 0$$

$$-t_i(4\delta_{ij}^{B}) + t_j(2p) + \sum_{k=1}^{N} \eta_{jk}^{A} \geq 0$$

which can be expressed in the form

$$\left. \begin{array}{l} \delta_{ij}^{B} \leq \dfrac{t_i}{t_j} \dfrac{p}{2} + \dfrac{1}{4t_j} \displaystyle\sum_{k=1}^{N} \eta_{ik}^{Ar} \\[3mm] \delta_{ij}^{B} \leq \dfrac{t_j}{t_i} \dfrac{p}{2} + \dfrac{1}{4t_i} \displaystyle\sum_{k=1}^{N} \eta_{jk}^{Ar} \end{array} \right\} \quad (34)$$

If the deviations of the Hamming distances of $\mathbf{B}_i$ and $\mathbf{B}_j$ from $p/2$, δ_{ij}^{B}, satisfy the condition

$$\delta_{ij}^{B} \leq \min \left\{ \dfrac{t_i}{t_j} \left(\dfrac{p}{2} \right) + \dfrac{1}{4t_j} \sum_{k=1}^{N} \eta_{ik}^{Ar}, \right.$$
$$\left. \dfrac{t_j}{t_i} \left(\dfrac{p}{2} \right) + \dfrac{1}{4t_i} \sum_{k=1}^{N} \eta_{jk}^{Ar} \right\} \quad (35)$$

for $r = 1, \ldots, n$, then satisfaction of (28) is assured. If one assumes that $\mathbf{P}_i$ and $\mathbf{P}_j$ have been trained for a sufficiently large number of times to justify assuming $t_i \to \infty$, $t_j \to \infty$, and $t_i = t_j$, then (35) becomes

$$\delta_{ij}^{B} \leq \frac{p}{2} \quad (36)$$

A similar development for field B leads to the results

$$\delta_{ij}^{A} \leq \min \left\{ \dfrac{t_i}{t_j} \left(\dfrac{n}{2} \right) + \dfrac{1}{4t_j} \sum_{k=1}^{N} \eta_{ik}^{Br}, \right.$$
$$\left. \dfrac{t_j}{t_i} \left(\dfrac{n}{2} \right) + \dfrac{1}{4t_i} \sum_{k=1}^{N} \eta_{jk}^{Br} \right\} \quad (37)$$

$$\delta_{ij}^{A} \leq \frac{n}{2}, \text{ as } t_i \to \infty, t_j \to \infty, \text{ and } t_i = t_j \quad (38)$$

For the stated conditions, (36) and (38) are always true. This implies that one can always find weights to guarantee recall of any two pairs in a training set. It is of interest to note that for a training set consisting of only two training pairs, the original correlation matrix given by Kosko can guarantee recall of both. This can be seen from (35) and (37) using $t_i = t_j = 1$ and $\mathbf{M}_o = \mathbf{0}$. It is shown in Wang et al. (1991) that $\mathbf{P}_i$ and $\mathbf{P}_j$ will have the same energy level if they are one Hamming distance away from each other.

Multiple Training for Multiple Pairs— Complete Recall Theorem

If multiple training is to be used for all pairs in a training set, then the correlation matrix will have the form

$$\mathbf{M} = \sum_{i=1}^{N} q_i \mathbf{X}_i^{T} \mathbf{Y}_i \quad (39)$$

Dividing the element of $\mathbf{M}$ by an integer will not affect the recall of training pairs of BAM. By such division, one may obtained a modified correlation matrix $\mathbf{M}$ where all the q_i are positive rational numbers. The q_i are further generalized to be positive real numbers. This modified correlation matrix will be called a *generalized correlation matrix*. In this section, the necessary and sufficient conditions will be derived for the weights q_i such that all pairs can be recalled.

LEMMA 1. *A pair* $\mathbf{P}_i = (\mathbf{A}_i, \mathbf{B}_i)$ *can be recalled, using the generalized correlation matrix in* (39) *if and only if*

$$\sum_{j=1}^{N} q_j \eta_{ij} \geq 0 \quad (40)$$

for every neighbor in both field A and B, where $q_i > 0$ *for all* $i = 1, \ldots, N$, *and* η_{ij} *is defined in* (15).

PROOF. The energies corresponding to the pairs $\mathbf{P}_i$ and $\mathbf{P}'_i$ are

$$E_i = -\mathbf{A}_i \left(\sum_{j=1}^{N} q_j \mathbf{X}_j^{T} \mathbf{Y}_j \right) \mathbf{B}_i^{T}$$

$$E'_i = -\mathbf{A}'_i \left(\sum_{j=1}^{N} q_j \mathbf{X}_j^{T} \mathbf{Y}_j \right) \mathbf{B}_i'^{T}$$

$\mathbf{P}_i$ can be recalled if and only if it represents a local minimum of energy. This requires that $E_i \leq E'_i$ for every neighbor. Accordingly, it is necessary that

$$\mathbf{A}_i \left(\sum_{j=1}^{N} q_j \mathbf{X}_j^{T} \mathbf{Y}_j \right) \mathbf{B}_i^{T} - \mathbf{A}'_i \left(\sum_{j=1}^{N} q_j \mathbf{X}_j^{T} \mathbf{Y}_j \right) \mathbf{B}_i'^{T} \geq 0$$

Table 5.1. Table of η for illustrative Example 3. Left "bit" column indicates the position of the different bit. (Adapted from Wang et al., 1991.)

	Field A							Field B					
Bit	η_{12}	η_{13}	η_{21}	η_{23}	η_{31}	η_{32}	Bit	η_{12}	η_{13}	η_{21}	η_{23}	η_{31}	η_{32}
1	-2	-6	-2	-2	-6	-2	1	-14	-2	-14	6	-2	6
2	-2	-6	-2	-2	-6	-2	2	14	2	14	6	2	6
3	-2	6	-2	2	6	2	3	14	-2	14	-6	-2	-6
4	2	6	2	-2	6	-2	4	-14	-2	-14	6	-2	6
5	-2	-6	-2	-2	-6	-2	5	-14	2	-14	-6	2	-6
6	-2	-6	-2	-2	-6	-2	6	-14	2	-14	-6	2	-6
7	-2	-6	-2	-2	-6	-2	7	-14	-2	-14	6	-2	6
8	-2	-6	-2	-2	-6	-2	8	14	-2	14	-6	-2	-6
9	-2	6	-2	2	6	2	9	14	-2	14	-6	-2	-6

From (15), one notes

$$\sum_{j=1}^{N} q_j \eta_{ij} \geq 0$$

for every neighbor in both field A and B.

COROLLARY 2. *Suppose* $\mathbf{T} = \{\mathbf{P}_i | i = 1, \dots, N\}$ *is a training set and* $\mathbf{T}_s = \{\mathbf{P}_{\zeta_i} | i = 1, \dots, N_s,$ $\zeta_i \in \{1, \dots, N\}\} \subset \mathbf{T}$. *Then all pairs* $\mathbf{P}_{\zeta_i} \in \mathbf{T}_s$ *can be recalled, using a generalized correlation matrix if and only if the positive real weights* q_j *satisfy*

$$\sum_{j=1}^{N} q_j \eta_{\zeta_i j}^{Fr} \geq 0$$

for all

$$\begin{cases} r = 1, \dots, n, F = A \\ r = 1, \dots, p, F = B \end{cases}$$

and all $\mathrm{P}_{\zeta_i} \in \mathbf{T}_s$.

THEOREM 3. Complete Recall Theorem. *Let*

$$\Gamma_i = \begin{bmatrix} \eta_{i1}^{A1} & \cdots & \eta_{iN}^{A1} \\ \vdots & & \vdots \\ \eta_{i1}^{An} & \cdots & \eta_{iN}^{An} \\ \eta_{i1}^{B1} & \cdots & \eta_{iN}^{B1} \\ \vdots & & \vdots \\ \eta_{i1}^{Bp} & \cdots & \eta_{iN}^{Bp} \end{bmatrix} \quad and \quad \mathbf{Q} = \begin{bmatrix} q_1 \\ \vdots \\ q_N \end{bmatrix}$$

All training pairs in $\mathbf{T} = \{\mathbf{P}_i | i = 1, \dots, N\}$ *can be recalled, using a generalized correlation matrix, if*

and only if the positive real weights satisfy

$$\begin{bmatrix} \Gamma_1 \\ \vdots \\ \Gamma^N \end{bmatrix} \mathbf{Q} \geq 0 \qquad (41)$$

PROOF. By Corollary 2, the theorem is true when $\mathbf{T}_s = \mathbf{T}$. $\square$

Example 3. Multiple training for BAM.
Consider the training pairs used in Example 2 and the effect of using the multiple training approach to the problem of their recall. In bipolar mode, the quantities defined in the previous example have the representation

$$A_1 = [\quad 1 \quad -1 \quad -1 \quad 1 \quad 1 \quad 1 \quad -1 \quad -1 \quad -1]$$
$$B_1 = [\quad 1 \quad 1 \quad 1 \quad -1 \quad -1 \quad -1 \quad -1 \quad 1 \quad -1]$$
$$A_2 = [-1 \quad 1 \quad 1 \quad 1 \quad -1 \quad -1 \quad 1 \quad 1 \quad 1]$$
$$B_2 = [\quad 1 \quad -1 \quad -1 \quad -1 \quad -1 \quad -1 \quad -1 \quad -1 \quad 1]$$
$$A_3 = [\quad 1 \quad -1 \quad 1 \quad -1 \quad 1 \quad 1 \quad -1 \quad 1 \quad 1]$$
$$B_3 = [-1 \quad 1 \quad -1 \quad 1 \quad -1 \quad -1 \quad 1 \quad -1 \quad 1]$$

$\eta_{ii} = 2p = 18$ for both fields A and B. The remaining values of η_{ij} are listed in Table 5.1. The left column, "bit", indicates the position of the different bit. If one checks $\mathbf{P}_2$ with its neighbor which has a different bit at bit 6 in field B, (40) becomes $q_1 \eta_{21}^{B6} + q_2 \eta_{22}^{B6} + q_3 \eta_{23}^{B6} \geq 0$. If the original BAM coding strategy is used, $q_1 = q_2 = q_3 = 1$. From Table 5.1,

$$\eta_{21}^{B6} + \eta_{22}^{B6} + \eta_{23}^{B6} = -14 + 18 - 6$$
$$= -2 < 0$$

This implies $\mathbf{P}_2$ cannot be recalled according to Lemma 1. Further confirmation of this is given by the value of the energy levels. The energy of $\mathbf{P}_2$, $E_{o2} = -\mathbf{A}_2\mathbf{M}_o\mathbf{B}_2^T = -71$; the corresponding value for its neighbor (with bit 6 different) is -73. Thus, $\mathbf{P}_2$ cannot be recalled. Next, let $q_1 = q_2 = 2$ and $q_3 = 1$; then $q_1\eta_{21}^{Fj} + q_2\eta_{22}^{Fj} + q_3\eta_{23}^{Fj} \geq 0$, for every position j in both $F = A$ and B. This is also true for $\mathbf{P}_1$ and $\mathbf{P}_3$. Because of this, all three pairs can be recalled for the condition $q_1 = q_2 = 2$ and $q_3 = 1$.

Two Methods for Multiple Training Encoding

1. Linear Programming/Multiple Training.
Equation (40) is a linear inequality. Linear programming methods (Varaiya, 1972; Gass, 1985) provide one means to obtain a feasible solution for $\{q_i | j = 1, \ldots, N\}$. There exists a weight vector $\mathbf{Q}$ such that pair $\mathbf{P}_i$ can be recalled if the set of (40) for all neighbors of $\mathbf{P}_i$ in both field A and B has feasible solutions. Furthermore, if this is true for all training pairs, then there exists a feasible solution for (41). And all training pairs can be recalled (*complete recall theorem*). The process of obtaining $\{q_j | j = 1, \ldots, N\}$ for (41) by using linear programming techniques is called *linear programming/multiple training* (LP/MT). The use of LP/MT will enable the determination of the $\mathbf{Q}$ vector (which guarantees recall of all pairs) or indicate that no such $\mathbf{Q}$ exists.

2. Sequential Multiple Training. Sequential multiple training (SMT) is a dynamic way to determine the $\mathbf{Q}$ vector by the use of multiple training sequentially on those pairs which cannot be recalled correctly. A flowchart for this process is given in Fig. 5.5. Equation (25) can be used as a replacement for the loop test in the dashed box of the figure.

With reference to Example 2, recall that the pair $(\mathbf{X}_2, \mathbf{Y}_2)$ could not be recalled. Consider the application of the multiple training concept to overcome this limitation. Let $q = 2$ for this pair so that $\mathbf{P} = \mathbf{X}_2^T\mathbf{Y}_2$. The value of $\mathbf{M}$ then becomes

$$\mathbf{M} = \mathbf{X}_1^T\mathbf{Y}_1 + 2\mathbf{X}_2^T\mathbf{Y}_2 + \mathbf{X}_3^T\mathbf{Y}_3$$

With this $\mathbf{M}$ matrix recall of $(\mathbf{X}_2, \mathbf{Y}_2)$ becomes possible. But accomplishing this, however, yields the concomitant result that $(\mathbf{X}_1, \mathbf{Y}_1)$ cannot now be recalled under the same augmentation of $\mathbf{M}$. To remedy this condition, multiple training of $(\mathbf{X}_1, \mathbf{Y}_1)$ is also necessary. This produces an additional change in $\mathbf{M}$ to the value

$$\mathbf{M} = 2\mathbf{X}_1^T\mathbf{Y}_1 + 2\mathbf{X}_2^T\mathbf{Y}_2 + \mathbf{X}_3^T\mathbf{Y}_3$$

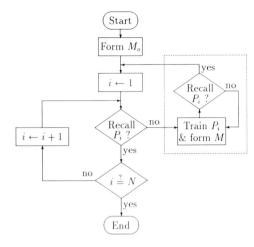

Fig. 5.5. Flow chart for the sequential multiple training algorithm. (Adapted from Wang et al., 1991.)

Explicitly, the matrix is now

$$\mathbf{M} = \begin{bmatrix} -1 & 5 & 3 & 1 & -1 & -1 & 1 & 3 & -3 \\ 1 & -5 & -3 & -1 & 1 & 1 & -1 & -3 & 3 \\ -1 & -3 & -5 & 1 & -1 & -1 & 1 & -5 & 5 \\ 5 & -1 & 1 & -5 & -3 & -3 & -5 & 1 & -1 \\ -1 & 5 & 3 & 1 & -1 & -1 & 1 & 3 & -3 \\ -1 & 5 & 3 & 1 & -1 & -1 & 1 & 3 & -3 \\ 1 & -5 & -3 & -1 & 1 & 1 & -1 & -3 & 3 \\ -1 & -3 & -5 & 1 & -1 & -1 & 1 & -5 & 5 \\ -1 & -3 & -5 & 1 & -1 & -1 & 1 & -5 & 5 \end{bmatrix}$$

One can verify this value of $\mathbf{M}$ enables one to recall all three pairs: $(\mathbf{X}_1, \mathbf{Y}_1)$, $(\mathbf{X}_2, \mathbf{Y}_2)$, and $(\mathbf{X}_3, \mathbf{Y}_3)$.

In Wang et al. (1991), related to these results, the theorem is proved that a set of training pairs, which can be guaranteed to be recalled by using SMT, is guaranteed to be recalled by using LP/MT in a discrete BAM. Multiple training adjusts the weight associated with each pair to guarantee recall of a training set. LP/MT provides a straightforward method of determining the weights, while SMT is a dynamic approach to this problem.

5.4.2. Dummy Augmentation

Multiple training guarantees recall if and only if (41) is satisfied. For an arbitrary number of training pairs, multiple training still cannot guarantee recall of all training pairs (due to the capacity). The dummy augmentation encoding strategy, which has been reported by Wang, Cruz, and Mulligan (1990a), guarantees recall of all training pairs providing one is permitted to utilize augmented vectors in the particular application.

To begin the discussion of the method, define a set of training pairs as *strictly noise free*, if $A_i X_j^T = 0$ and $B_i Y_j^T = 0$ for every $i, j, i \neq j$. The following two theorems describe a method (Wang et al., 1990a) to find a set of strictly noise-free pairs for both bipolar and binary forms.

THEOREM 4. *Given a set of training pairs* $\{(A_i, B_i) | i = 1, \ldots, N\}$. *If*

$$H(X_i, X_j) = n/2 \quad i \neq j \quad \forall i, j$$

$$H(Y_l, Y_k) = p/2 \quad l \neq k \quad \forall l, k$$

where n is the dimension of X_i, *p is the dimension of* Y_i, $H(X_i, X_j)$ *and* $H(Y_l, Y_k)$ *are Hamming distances, then the set* $\{(A_i, B_i) | i = 1, \ldots, N\}$ *is strictly noise-free in the bipolar mode.*

THEOREM 5. *In binary mode, a set of* $n(n + 1)$-*tuple data with ith element*

$$[a_{i1}, \ldots, a_{i,n(n+1)}], a_{ij} = 0 \text{ except } a_{i1} = a_{i,i+1} = 1$$

$\forall i \in 1, \ldots, n$ *is a strictly noise-free set.*

It is known that

$$A_i M = A_i X_i^T Y_i + \sum_{j \neq i} A_i X_j^T Y_j \quad (42)$$

Thus, for strictly noise-free pairs, $A_i M = A_i X_i^T Y_i$. $\phi(A_i M) = \phi(s Y_i) = B_i$, where s is a positive integer. Similarly, $\phi(B_i M^T) = A_i$. Therefore, for a strictly noise-free set, any training pair can be recalled.

Consider N training pairs $\{(A_i, B_i) | i = 1, \ldots, N\}$ and a strictly noise-free set with N elements, $\{D_i | i = 1, \ldots, N\}$, where $D_i = [d_{i1}, \ldots, d_{im}]$. Construct a new set of training pairs

$$\{([A_i | D_i | \ldots | D_i], [B_i | D_i | \ldots | D_i]) | i = 1, \ldots, N\}$$

There are K D_i appended to A_i and K' D_i appended to B_i. Then

$$M = \sum_{i=1}^{N} [X_i | \Delta_i | \ldots | \Delta_i]^T [Y_i | \Delta_i | \ldots | \Delta_i] \quad (43)$$

where Δ_i is a bipolar form of D_i and

$$[A_i | D_i | \ldots | D_i] M$$

$$= [A_i D_i | \ldots | D_i][X_i | \Delta_i \ldots | \Delta_i]^T [Y_i | \Delta_i | \ldots | \Delta_i]$$

$$+ \sum_{j \neq i} [A_i | D_i | \ldots | D_i]$$

$$\times [X_j | \Delta_j | \ldots | \Delta_j]^T [Y_j | \Delta_j | \ldots | \Delta_j] \quad (44)$$

Notice that $\{D_i, i = 1, \ldots, N\}$ is a strictly noise-free set, so that $D_i \Delta_j^T = 0$, for all $i \neq j$. Thus,

$$[A_i | D_i | \ldots | D_i][X_j | \Delta_j | \ldots | \Delta_j]^T$$

$$= A_i X_j^T + K D_i \Delta_j^T = A_i X_j^T \quad (45)$$

From (44),

$$[A_i | D_i | \ldots | D_i] M$$

$$= (A_i X_i^T + K D_i \Delta_i^T)[Y_i | \Delta_i | \ldots | \Delta_i]$$

$$+ \sum_{j \neq i} (A_i X_j^T)[Y_j | \Delta_j | \ldots | \Delta_j] \quad (46)$$

In bipolar form, $D_i \Delta_i^T = m$. In binary form, $D_i \Delta_i^T = s =$ number of 1s in D_i, so (46) becomes

$$[A_i | D_i | \ldots | D_i] M = (A_i X_i^T + q)[Y_i | \Delta_i | \ldots | \Delta_i]$$

$$+ \sum_{j \neq i} (A_i X_j^T)[Y_j | \Delta_j | \ldots | \Delta_j] \quad (47)$$

where q is a positive integer.

By appropriate choice of D_i and K, q can be made large enough, for example $q \geq \sum_{j \neq i} |A_i X_j^T|$, so that

$$\phi([A_i | D_i | \ldots | D_i] M) = [B_i | D_i | \ldots | D_i]$$

This result is evident from (47); a detailed proof is given in Wang et al. (1990a).

The same process is applicable to field B. The implication of this development is that the original training pairs can be augmented by strictly noise-free sets to guarantee recall. The increased number of stored pairs is due to the added redundancy in the BAM state space. One achieves this condition of recall, however, by having the users generate dummy elements for the augmentation during the decoding phase.

5.4.3. Global Optimization Method

This method was proposed by T. Wang et al. (1991). Basically, a cost function is formed such that the global minimum of the function satisfies the necessary and sufficient conditions of the steady states. From Corollary 1, it is known that (A_i, B_i) is a steady state if and only if

$$\phi(A_i M) = B_i \quad (48)$$

$$\phi(B_i M^T) = A_i \quad (49)$$

where ϕ is defined in (7). Then from (48) and (49), one obtains

$$\phi\left(\sum_{j=1}^{n} a_{ij} m_{jk}\right) = b_{ik} \quad \text{for } k = 1, \ldots, p$$

$$\phi\left(\sum_{j=1}^{p} b_{ij} m_{lj}\right) = a_{il} \quad \text{for } l = 1, \ldots, n$$

where $\mathbf{A}_i = [a_{i1}, \ldots, a_{in}]$ and $\mathbf{B}_i = [b_{i1}, \ldots, b_{ip}]$. This yields

$$b_{ik} \sum_{j=1}^{n} a_{ij} m_{jk} \geq 0 \quad \text{and} \quad a_{il} \sum_{j=1}^{p} b_{ij} m_{lj} \geq 0$$

for $k = 1, \ldots, p$ and $l = 1, \ldots, n$. Denote $a_{il} \sum_{j=1}^{p} b_{ij} m_{lj}$ by ρ_{il}^a and $b_{ik} \sum_{j=1}^{n} a_{ij} m_{jk}$ by ρ_{ik}^b. Furthermore, define a function $F(\cdot)$ with the property that $F(x) = 0$ if $x > 0$ and $F(x) = 1$ if $x \leq 0$. A cost function can then be constructed having the form

$$C(\mathbf{M}) = -\sum_{i=1}^{N} \left(\sum_{l=1}^{n} \rho_{il}^a \cdot F(\rho_{il}^a) + \sum_{k=1}^{p} \rho_{ik}^b \cdot F(\rho_{ik}^b) \right) \tag{50}$$

If $\rho_{il}^a \geq 0$, $\rho_{il}^a \cdot F(\rho_{il}^a) = 0$, otherwise $\rho_{il}^a \cdot F(\rho_{il}^a) < 0$ which creates a positive term in $C(\mathbf{M})$. The same argument applies to ρ_{ik}^b. Since $\rho_{il}^a \geq 0$ and $\rho_{ik}^b \geq 0$ are desired and $C(\cdot) \geq 0$, $C(\mathbf{M})$ should be minimized to 0. If an $\mathbf{M}$ matrix can be found such that $C(\mathbf{M}) = 0$ which is the global minimum of $C(\cdot)$, then this $\mathbf{M}$ can guarantee recall of all training pairs.

5.4.4. Unlearning

Spurious memories, which are the undesired local minima in the energy surface, are the traps to the use of the network. Therefore, ensuring training pairs to be local minima is not enough; eliminating spurious memories is also important. Srinivasan and Chia (1991) propose to add *unlearning* to *multiple training*. This results in a matrix $\mathbf{M}$ having the form

$$\mathbf{M} = \sum_{i=1}^{N} q_i \mathbf{X}_i^{\mathrm{T}} \mathbf{Y}_i - \sum_{j} q_{js} \mathbf{X}_{js}^{\mathrm{T}} \mathbf{Y}_{js} \tag{51}$$

where $(\mathbf{X}_i, \mathbf{Y}_i)$ are the desired training pairs and $(\mathbf{X}_{js}, \mathbf{Y}_{js})$ are the spurious memories. However, present knowledge requires that the q_i and q_j have to be determined experimentally. There is no presently available systematic way to find q_i and q_j. Through computer simulations, Srinivasan and Chia (1991) have shown that their unlearning technique can increase the capacity and noise tolerance.

5.4.5. Householder Transformation Method

In (48) and (49), suppose the correlation matrix, $\mathbf{M}$, is permitted to be different in different directions; thus, one defines two correlation matrices $\mathbf{M}_1$ and $\mathbf{M}_2$. Then (48) and (49) can be

rewritten as

$$\phi(\mathbf{A}_i \mathbf{M}_1) = \mathbf{B}_i \tag{52}$$

$$\phi(\mathbf{B}_i \mathbf{M}_2) = \mathbf{A}_i \tag{53}$$

The structure, shown in Fig. 5.6, was proposed by Okajima et al. (1987). A sufficient condition for (52) and (53) is

$$\mathbf{A}_i \mathbf{M}_1 = \mathbf{B}_i \tag{54}$$

$$\mathbf{B}_i \mathbf{M}_2 = \mathbf{A}_i \tag{55}$$

Leung and Cheung (1991) propose to use the Householder transformation to find $\mathbf{M}_1$ and $\mathbf{M}_2$. If one defines

$$\mathbf{A} = [\mathbf{A}_1^{\mathrm{T}}, \ldots, \mathbf{A}_N^{\mathrm{T}}]_{n \times N}$$

and

$$\mathbf{B} = [\mathbf{B}_1^{\mathrm{T}}, \ldots, \mathbf{B}_N^{\mathrm{T}}]_{p \times N}$$

then the matrix form of (54) and (55) becomes

$$\mathbf{M}_1^{\mathrm{T}} \mathbf{A} = \mathbf{B} \tag{56}$$

$$\mathbf{M}_2^{\mathrm{T}} \mathbf{B} = \mathbf{A} \tag{57}$$

Suppose that all $\mathbf{A}_i'$ and $\mathbf{B}_j'$ are linearly independent. Then rotation matrices $\mathbf{R}_A$ and $\mathbf{R}_B$ with dimensions $n \times n$ and $p \times p$ respectively can be found such that

$$\mathbf{Y}_A = \mathbf{R}_A \mathbf{A} = \begin{bmatrix} \mathbf{Y}_{A0} \\ \cdots \\ \mathbf{0}_{(n-N) \times N} \end{bmatrix}_{n \times N} \tag{58}$$

$$\mathbf{Y}_B = \mathbf{R}_B \mathbf{B} = \begin{bmatrix} \mathbf{Y}_{B0} \\ \cdots \\ \mathbf{0}_{(p-N) \times N} \end{bmatrix}_{p \times N} \tag{59}$$

$\mathbf{Y}_{A0}$ and $\mathbf{Y}_{B0}$ are upper triangular matrices with dimension $N \times N$. $\mathbf{0}_{(n-N) \times N}$ and $\mathbf{0}_{(p-N) \times N}$ are zero matrices with the dimensions specified.

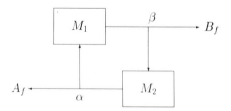

Fig. 5.6. An associative memory structure proposed by Okajima. This two-directional associative scheme uses two different correlation matrices for two different decoding directions. It is different from Kosko's scheme, which uses only one correlation matrix.

Since both $\mathbf{R}_A$ and $\mathbf{R}_B$ are orthonormal matrices, from (58) and (59),

$$\mathbf{A} = \mathbf{R}_A^T \mathbf{Y}_A$$

$$\mathbf{B} = \mathbf{R}_B^T \mathbf{Y}_B$$

Furthermore, $\mathbf{R}_A$ ($\mathbf{R}_B$) can be reduced to dimension $N \times n$ ($N \times p$) by eliminating the last $n - N$ ($p - N$) rows. The new rotation matrices are denoted by $\hat{\mathbf{R}}_A$ and $\hat{\mathbf{R}}_B$, so

$$\mathbf{A} = \hat{\mathbf{R}}_A^T \mathbf{Y}_{A0}$$

$$\mathbf{B} = \hat{\mathbf{R}}_B^T \mathbf{Y}_{B0}$$

Substituting these equations into (56) and (57), one obtains

$$\mathbf{M}_1^T \hat{\mathbf{R}}_A^T \mathbf{Y}_{A0} = \hat{\mathbf{R}}_B^T \mathbf{Y}_{B0} \qquad (60)$$

$$\mathbf{M}_2^T \hat{\mathbf{R}}_B^T \mathbf{Y}_{B0} = \hat{\mathbf{R}}_A^T \mathbf{Y}_{A0} \qquad (61)$$

Since $\hat{\mathbf{R}}_A^T \hat{\mathbf{R}}_A$ and $\hat{\mathbf{R}}_B^T \hat{\mathbf{R}}_B$ are identity matrices, from (60) and (61),

$$\mathbf{M}_1 = (\hat{\mathbf{R}}_B^T \mathbf{Y}_{B0} \mathbf{Y}_{A0}^{-1} \hat{\mathbf{R}}_A)^T$$

$$\mathbf{M}_2 = (\hat{\mathbf{R}}_A^T \mathbf{Y}_{A0} \mathbf{Y}_{B0}^{-1} \hat{\mathbf{R}}_B)^T$$

Furthermore, there are possibly n (p) linearly independent $\mathbf{A}_i$ ($\mathbf{B}_i$), and thus the capacity bound of BAM using these correlation matrices is $\min(n, p)$. For further details, see Leung and Chung (1991).

Hassoun (1989a, b) also proposed to use *generalized-inverse* and Ho–Kashyap encoding methods for this Okajima structure. The details can be found in Chapter 1.

5.5. COMPUTER SIMULATIONS

5.5.1. A Pattern Recognition Example for BAM

Multiple Training

Consider the problem of using a discrete BAM to distinguish among airplanes, tanks, and helicopters from data obtained from satellite observations. The solution begins by training a BAM with patterns which are assumed to have the form shown in Fig. 5.7. Figure 5.8 shows the results of attempts to recall the patterns using Kosko's encoding method. Note that it was not possible to recall the trained pattern correctly. The multiple training method was used with $q_1 = 3$ and $q_2 = q_3 = 4$. Pair 1 (plane) was trained an additional two times, pair 2 (tank) three additional times, and pair 3 (helicopter) three additional times. After this training all three pairs were recalled correctly by using their

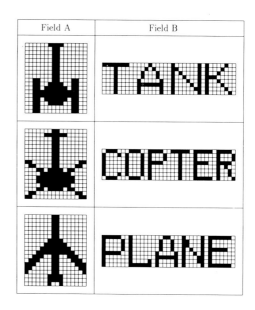

Field A	Field B

Fig. 5.7. The three training pairs for a BAM for the example in Section 5.1. The pictures correspond to field A while the letters correspond to field B. "ON" is represented by a shaded square while "OFF" is represented by an unshaded square. (Adapted from Wang et al., 1990a.)

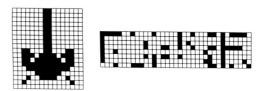

Fig. 5.8. Using Kosko's original encoding/decoding method, and starting from the correct "tank" and "letters" pair in Fig. 5.7, the process converged to the "tank" picture and garbled "letters" as shown. (Adapted from Wang et al., 1990a.)

patterns as the initial conditions. For example, if the pair shown in Fig. 5.9 is used as initial condition, the same pair is retrieved after the BAM decoding process. Figure 5.10 shows an initial condition corrupted with arbitrary amounts of noise. The result of the BAM decoding process for this noisy data after multiple training is shown in Fig. 5.9.

Dummy Augmentation

Figure 5.11 shows the results of using our dummy augmentation method; note the dummy augmentation parts at the bottom of each picture.

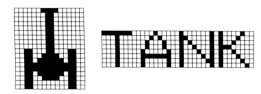

Fig. 5.9. For the "tank" picture, the multiple training method leads to the correct "tank" letters. (Adapted from Wang et al., 1990a.)

Fig. 5.10. A noisy pattern as an initial condition for the "tank" picture and a blank field B. Using multiple training, the noiseless "tank" picture and "tank" letters are recovered, as shown in Fig. 5.9. Using the original Kosko encoding method, the process converged to the incorrect garbled pair shown in Fig. 5.8. (Adapted from Wang et al., 1990a.)

Field A	Field B

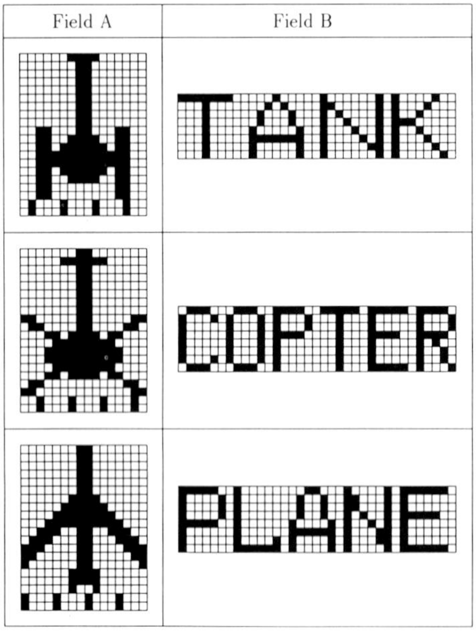

Fig. 5.11. Training pairs for the example in Section 5.1 using dummy augmentation. (Adapted from Wang et al., 1990a.)

The dummy data are expressed in bipolar mode and strictly noise-free. Note that all three pairs of signal data are recalled correctly after training.

It is recognized that it may be difficult to generate dummy elements during the decoding phase. However, identifying the incoming signals may be feasible by sequentially attaching dummy elements and corresponding initial values for field B to the incoming signal for BAM processes.

Figure 5.12 shows that correct identification of a *tank* results from using sequential tests of each dummy augmentation as an initial condition. Similar correct identification of *plane* and *copter* in this example can be achieved using this technique. Use of sequential tests of the dummy augmentation resulted in a correct identification for a noisy tank pattern as shown in Fig. 5.13.

5.5.2. Performance Comparison

A series of simulations provides comparisons between the "multiple training" and original BAM coding strategies.

Capacity

The first simulation is to compare the capacity between an original BAM and a BAM with multiple training. For values of $n = p = 128$, N was varied from 1 to 22. For each N, 100 sets of different data were tested. Each set of data included N training pairs which were provided by a random generator. An event was defined as a "success" if and only if all N pairs could be recalled. Three different strategies were tested. These were the original BAM coding strategy, sequential multiple training (SMT), and linear programming/multiple training (LP/MT). For the sequential multiple training strategy, an upper bound, U, was set for the additional training times where $\sum_{i=1}^{N} (q_i - 1) \leq U$. Values of $U = 100, 200$, and 300 were used. For the linear programming portion, the simplex method was applied to find feasible solutions for minimizing

$$J = \sum_{i=1}^{N} q_i \qquad (62)$$

subject to

$$\sum_{k=1}^{N} q_k \eta_{ik}^{Fj} \geq 1, \quad \text{for } i = 1, \ldots, N, \text{ and } \mathbf{F} = \mathbf{A}, \mathbf{B}$$

$$q_i \geq 0, \quad \text{for } i = 1, \ldots, N \qquad (63)$$

The data for the LP/MT curve were obtained directly by determining if the correlation matrix,

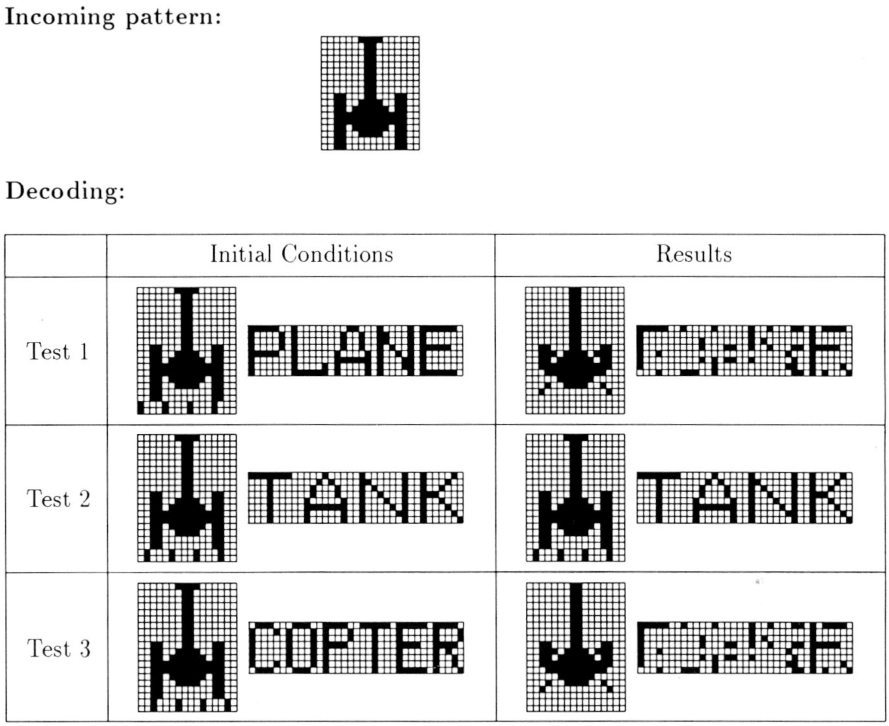

Incoming pattern:

Decoding:

	Initial Conditions	Results
Test 1		
Test 2		
Test 3		

Fig. 5.12. A simulation using dummy augmentation. "Incoming pattern" is the incoming signal for field A using a "tank" picture. The left-hand side of the decoding part shows three different augmentations of the field A with the "letters" in field B as initial conditions. The right-hand side of the decoding part shows that the second augmented field A and field B for the tank recovers the correct pair. (Adapted from Wang et al., 1990a.)

which used $\{q_j | j = 1, \ldots, N\}$ from the output of the simplex method subroutine, could recall all pairs. If the simplex method could not find feasible solutions, then that particular event is identified as a "failure."

Figure 5.14 indicates that the multiple training approach yields better results than the original BAM coding strategy.

Error Correction

One of the advantages of associative memories is that the stored data may be retrieved under certain conditions using inexact initial data. This ability is a means of achieving error correction. In this section simulations are reported which permit comparison of the error correction ability of the original BAM and the LP/MT. $n = p = 8$ was chosen and experiments were conducted for $N = 3, 4,$ and 5. One hundred sets of testing data were generated randomly for each N. Each set of data contained N training pairs. Within these 100

sets of data, only those which could be completely recalled using LP/MT were recorded.

Twenty thousand testing pairs were generated randomly to test the error correction rate after a set of N training pairs had been trained. The error correction rate is defined to be the number of testing pairs which converge correctly to one of the training pairs divided by the total number of testing pairs. The Hamming distance of a testing pair in this simulation is the Hamming distance between the testing pair and the trained pair which it converges to if the testing pair converges to a trained pair. Otherwise, the Hamming distance is defined to be the Hamming distance between the testing pair and the closest training pair. Figure 5.15 depicts the error correction vs. the Hamming distances in our experiments.

Spurious Memories

As we pointed out in Section 5.5.4, spurious memories should be eliminated in associative

Incoming pattern:

Decoding:

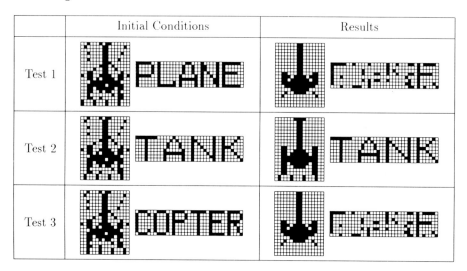

Fig. 5.13. "Incoming pattern" is the noisy "picture" of a "tank" for field A. The left-hand side of the decoding part shows three different augmentations of the noisy field A together with initial conditions for field B. The right-hand side of the decoding part shows that the second augmentation and initial condition yields the correct pair. (Adapted from Wang et al., 1990a.)

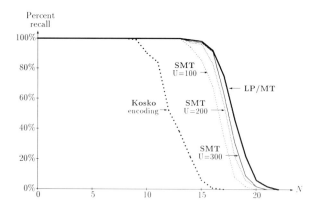

Fig. 5.14. Simulation results for the capacity comparison of Section 5.1. (Adapted from Wang et al., 1991.).

memories. A simulation was performed in which the number of spurious memories in a BAM with $n = p = 8$ was computed for both the original BAM and the LP/MT BAM. One hundred sets of data were generated randomly for each N, and

N was varied from 3 to 5. However, only those sets which could be completely recalled by LP/MT were recorded. The average number of spurious memories (undesired local minima) for each N is shown in Table 5.2.

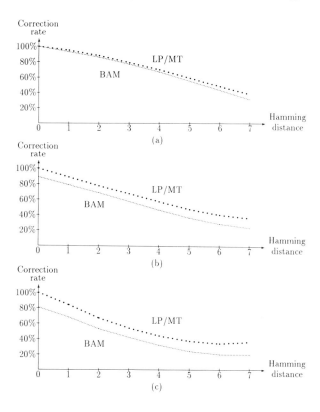

Fig. 5.15. The error correction rate represents the ability for error correction of an associative memory. Both original BAM and LP/MT BAM were used for simulation. The light dotted line is for original BAM and the bold dotted line is for LP/MT. (a), (b), and (c) are for $N = 3$, $N = 4$, and $N = 5$, respectively.

Table 5.2. The average number of spurious memories for the original BAM and the LP/MT BAM. $n = p = 8$ and N was chosen to be 3, 4, and 5

	BAM	LP/MT
$N = 3$	60.3	27.7
$N = 4$	50.7	12.8
$N = 5$	42.1	9.1

5.6. SUMMARY

Hopfield invented a very attractive one-layer autoassociator. Kosko extended that model to be two-layer and two-directional with a smaller correlation matrix. This network is called *bidirectional associative memory* (BAM). BAM has attracted attention because of its simplicity and economy. In particular, it is a one-to-one mapping scheme which provides a backward verification. However, similar to other Hebbian type networks, low capacity is a major disadvantage of BAM. Moreover, the original correlation matrix provided by

Kosko cannot guarantee recall of any training pairs.

Many coding methods have been developed in the past few years to overcome these problems and to increase the capacity. In this chapter, some of these methods have been discussed. Depending on the characteristics of the application, one has a choice of different coding schemes. Since BAM can be implemented in a relatively simple structure, it provides a convenient introduction to neural networks. With deeper understanding of BAM, we may be able to learn more about neural networks, especially memories and data-processing procedures.

REFERENCES

Gass, S. I. (1985). *Linear Programming*, 5th edn. McGraw-Hill, New York.

Haines, K. and Hecht-Nielsen, R. (1988). "A BAM with Increased Information Storage Capacity," in *Proc. IJCNN-88*, Vol. 1, pp. 181–190.

Hassoun, M. H. (1989a). "Dynamic Heteroassociative Neural Memories," *Neural Networks*, **2**, 275–287.

Hassoun, M. H. (1989b). "Adaptive Dynamic Hetero-associative Neural Memories for Pattern Classification," *Optical Pattern Recognition*, H. Liu, ed., *SPIE* **1053**, 75–83.

Hopfield, J. J. (1982). "Neural Networks and Physical Systems with Emergent Collective Computational Abilities," *Proc. Nat. Acad. Sci. U.S.A.*, **79**, 2554–2558.

Hopfield, J. J. (1984). "Neurons with Graded Response Have Collective Computational Properties Like Those of Two State Neurons," *Proc. Nat. Acad. Sci. U.S.A.*, **81**, 3088–3092.

Hopfield, J. J. and Tank, D. W. (1985). "Neural Computation of Decisions in Optimization Problems," *Biological Cybernetics*, **52**, 141–152.

Kosko, B. (1987a). "Constructing an Associative Memory," *Byte*, **12**(10), 137–144.

Kosko, B. (1987b). "Adaptive Bidirectional Associative Memories," *Appl. Opt.*, **26**(23), 4947–4960.

Kosko, B. (1988). "Bidirectional Associative Memories," *IEEE Trans. Systems, Man, and Cybernetics*, **18**(1), 49–60.

Leung, C. S. and Cheung, K. F. (1991). "Householder Encoding for Discrete Bidirectional Associative Memory," *Proc. IJCNN91*, Singapore, pp. 237–241.

McEliece, R. J., Posner, E. C., Rodemich, E. R., and Venkatesh, S. S. (1987). "The Capacity of the Hopfield Associative Memory," *IEEE Trans. Information Theory*, **IT-33**(4), 461–482.

Okajima, K., Tanaka, S., and Fujiwara, S. (1987). "A Heteroassociative Memory Network with Feedback Connection," *Proc. IJCNN 87*, pp. II-711–718.

Simpson, P. K. (1990). "Higher-ordered and Intraconnected Bidirectional Associative Memory," *IEEE Trans. Systems, Man, and Cybernetics*, **20**(3), 637–653.

Srinivasan, V. and Chia, C.-S. (1991). "Improving Bidirectional Associative Memory Performance by Unlearning," *Proc. IJCNN 91*, Singapore, pp. 2472–2477.

Varaiya, P. P. (1972). *Notes on Optimization*. Van Nostrand Reinhold, New York.

Wang, T., Zhuang, X., Xing, X., and Lu, F. (1991). "Optimal Learning Algorithm in Bidirectional Associative Memories," *Proc. IJCNN 91*, Seattle, Vol. 2, pp. 169–172.

Wang, Y.-F., Cruz, Jr., J. B., and Mulligan, Jr., J. H. (1989). "An Enhanced Bidirectional Associative Memory," in *Proc. IJCNN 89*, Vol. 1, pp. 105–110.

Wang, Y.-F., Cruz, Jr., J. B., and Mulligan, Jr., J. H. (1990a). "Two Coding Strategies for Bidirectional Associative Memory," *IEEE Trans. Neural Networks*, **1**(1), 81–92.

Wang, Y.-F., Cruz, Jr., J. B., and Mulligan, Jr., J. H. (1990b). "On Multiple Training for Bidirectional Associative Memory," *IEEE Trans. Neural Networks*, **1**(3), 275–276.

Wang, Y.-F., Cruz, Jr., J. B., and Mulligan, Jr., J. H. (1990c). "Multiple Training for Multiple Patterns on Bidirectional Associative Memory," *Proc. IASTED, Expert Systems and Neural Networks*, pp. 63–65.

Wang, Y.-F., Cruz, Jr., J. B., and Mulligan, Jr., J. H. (1991). "Guaranteed Recall of All Training Pairs for Bidirectional Associative Memory," *IEEE Trans. Neural Networks*, **2**(6), 559–567.

6

High-Density Associative Memories[1]

AMIR DEMBO

6.1. INTRODUCTION

Most of the work on neural associative memories is structure oriented; i.e., given the network architecture, efforts are directed to the analysis of its performance capability and limitations, and to devising algorithms for setting up the free parameters of the networks (typically the synaptic weights), as a function of the desired memories. Such an approach is adopted in Chapters 10 and 11 of this book. A different methodology guides this chapter, where a few essential properties of the associative neural memory are first postulated. Efforts are then geared towards characterizing all networks with these properties, and to the study of other aspects of these networks.

In the setup considered here, the probe is an N-dimensional real-valued vector $\mathbf{v} \in \mathbb{R}^N$. Applying this data as the initial state of a dynamical system yields the state trajectory $\{\mathbf{x}_t\}_{t \geq 0}$ where $\mathbf{x}_0 = \mathbf{v}$. Assuming that this system is asymptotically stable in the sense that for almost any value of $\mathbf{v}$, $\lim_{t \to \infty} \mathbf{x}_t = \mathcal{M}(\mathbf{v})$ exists, we may interpret $\mathcal{M}(\mathbf{v})$ as the memory value associated with $\mathbf{v}$. Hence, the design of associative memories is just the design of the map $\mathcal{M}$. The study of the latter is typically aided by the associated Liapunov function (also called the potential or energy function), $V: \mathbb{R}^N \to \mathbb{R}$. This function has the property that along each trajectory, $V(x_t)$ is nonincreasing in t. Thus the range of $\mathcal{M}$, which is the set of stored memories, is a subset of local minima of the potential function $V(\cdot)$.

To have a concrete example in mind, suppose throughout that $V(\cdot)$ is differentiable and that

$$\dot{\mathbf{x}}_t = \nabla V(\mathbf{x}_t) \qquad t > 0 \qquad \mathbf{x}_0 = \mathbf{v} \qquad (1)$$

where ∇ stands for the gradient operator. This is by no means the only dynamical system with potential function $V(\cdot)$.

In the basic model considered, the location of $K < \infty$ desired memories is specified. Throughout, think of a memory as a charged particle, and of $V(\cdot)$ as the potential associated with a configuration of particles. In general, spread-out charge is allowed, i.e., each desired memory α is specified in terms of a charge density μ_α, and different potentials might be contributed by different memories (this affects the shape of the basins of attraction).

Therefore, the associative network is determined via a map T whose argument is the collection of charge densities $\{\mu_\alpha\}$ and whose output is the potential function $V(\cdot)$.

The following requirements are made regarding this map:

(P1) It is invariant to translations and rotations of coordinates. [This requirement may be omitted, but then the mathematical analysis is more complicated; see Eq. (7).]

(P2) It is linear with respect to adding particles in the sense that the potential of two particles should be the sum of the potentials induced by the individual particles (i.e., interparticle interactions are not allowed; ses however, the discussion in Section 6.4.1).

(P3) Particle locations are the only possible sites of stable memory locations.

To allow for various particle types define a "type space" A, and assume that integration over A is well defined. The specific examples of A that we have in mind are a finite set $A = \{1, 2, \ldots, K\}$ or $A = \mathbb{R}^N$. The memory-building process is $V(\mathbf{x}) = T(\mu(,))$, where $\mu \in M(\mathbb{R}^N \times A)$ is a signed measure over $\mathbb{R}^N \times A$. For example, a single particle of type α located at $\mathbf{x}_0$ yields the potential $V = T(\delta_{(\mathbf{x}_0, \alpha)})$.

(P1) to (P3) now read in the following way.

(i) $T(\mu(\mathbb{R}^N \times A)) = V(\mathbf{x}) \Rightarrow T(\mu((\mathbf{c}\mathbb{R}^N + \eta) \times A)) = V(\mathbf{c}\mathbf{x} + \eta)$, where $\eta \in \mathbb{R}^N$, $\mathbf{c} \in \mathbb{R}^{N \times N}$ is a nonsingular, orthogonal matrix, and $\mathbf{c}\mathbb{R}^N + \eta$ is the $\mathbf{c}$ rotation, η translation of $\mathbb{R}^N$.

[1] This chapter is based on Dembo and Zeitouni (1988).

129

(ii) T is a linear transformation on $M = M(\mathbb{R}^N \times A)$; i.e.,

$$T(\mu(\mathbb{R}^N \times A))$$
$$= \int_{\mathbb{R}^N \times A} f(\mathbf{x}, \mathbf{y}, \boldsymbol{\alpha}) \mu(d\mathbf{y}, d\boldsymbol{\alpha}) \quad (2)$$

where $f(\mathbf{x}, \mathbf{y}, \boldsymbol{\alpha})$ is the kernel of the transformation and $f(\cdot)$ is assumed to be such that (2) makes sense.

(iii) Let $V(\mathbf{x}) = T(\mu(\mathbb{R}^N \times A))$. Then $V(\mathbf{x})$ does not possess minima in any open set U such that $\mu(U, A) = 0$, i.e., in all regions where particles are absent.

In addition, assume throughout the necessary smoothness and growth conditions on $f(\mathbf{x}, \mathbf{y}, \boldsymbol{\alpha})$. Specifically, if $f(\mathbf{x}, \mathbf{y}_0, \boldsymbol{\alpha}_0)$ is finite, then it is twice continuously differentiable with respect to $\mathbf{x}$, with μ-integrable derivatives.

The following structure theorem characterizes all maps T satisfying (i) to (iii) in a universal manner, i.e., with respect to every $\mu \in M$. Note that this does not preclude other maps from satisfying (i) to (iii) for specific choices of μ.

THEOREM 1. $T(\cdot)$ *satisfies (i) to (iii) in a universal manner with respect to every* $\mu \in M$ *if and only if*

$$V(\mathbf{x}) = \int_{\mathbb{R}^N \times A} f_{\boldsymbol{\alpha}}(\|\mathbf{x} - \boldsymbol{\eta}\|^2) \mu(d\boldsymbol{\eta}, d\boldsymbol{\alpha}) \quad \forall \mu \in M$$
$$(3)$$

where

$$\forall \boldsymbol{\alpha} \in A \quad \forall d > 0 \quad f_{\boldsymbol{\alpha}}''(d)d + \frac{N}{2} f_{\boldsymbol{\alpha}}'(d) \leq 0 \quad (4)$$

Remarks. (a) By local minimum (maximum) we mean a strict minimum (maximum). Strict inequality in (4), assures also the nonexistence of nonstrict extrema.

(b) Equality in (4), for all $\boldsymbol{\alpha}$, implies that there are also no strict local maxima of $V(\cdot)$ in any open set U such that $\mu(U, A) = 0$.

This structure theorem is derived in Dembo and Zeitouni (1988) under the most general condition. In typical associative memory applications a finite set of memories is considered. Hence, hereafter let $A = \{1, \dots, K\}$ where $K < \infty$ and $\mu = \sum_{i=1}^{K} \delta_{(\mathbf{u}^{(i)}, i)}$, i.e., there are K memories (particles) to store at positions $\{\mathbf{u}^{(1)}, \dots, \mathbf{u}^{(K)}\}$ in $\mathbb{R}^N$. In this case

$$V(\mathbf{x}) = \sum_{i=1}^{K} f_i(\|\mathbf{x} - \mathbf{u}^{(i)}\|^2) \quad (5)$$

For example, when equality holds in (4), for all $\boldsymbol{\alpha}$, this potential specializes to

$$V(\mathbf{x}) = \sum_{i=1}^{K} \frac{f_i'(1)}{\|\mathbf{x} - \mathbf{u}^{(i)}\|^{(N-2)}} \quad (6)$$

which is exactly the electrostatical potential when $N = 3$.

For any value of $N \geq 3$, and $V(\cdot)$ given by (5), with $f_i(\cdot)$ satisfying (4), we distinguish between three types of memories:

(a) *Attractive memories*, for which

$$\lim_{\mathbf{x} \to \mathbf{u}^{(i)}} V(\mathbf{x}) = -\infty,$$

i.e., they are global minima of the potential function.

(b) *Nonadmissible memories*, for which

$$\lim_{\mathbf{x} \to \mathbf{u}^{(i)}} V(\mathbf{x}) = +\infty,$$

i.e., they are the global maxima of the potential function. An example is the electrostatic potential of a particle with positive charge [with negative charge corresponding to (a) above]. They are of interest if one wishes to avoid specific locations.

(c) *Repulsive memories*, for which $V(\mathbf{u}^{(i)})$ is finite. An example is $f(d) = -d$ ("repulsive spring"), and those forces are weak in the short range but strong in the long range.

In particular, for the electrostatic form of the potential given in (6), $V(\cdot)$ is a harmonic function outside $\{\mathbf{u}^{(i)}\}_{i \in A}$, thus it possesses all its local minima in the attractive memories and all its local maxima in the nonadmissible memories. One can then store two kinds of objects in the same system. While the recall process using (1) will give rise to objects of type (a), the same recall process with $-V(\cdot)$ instead of $V(\cdot)$ will give rise to objects of type (b).

The potential given by (4) and (5) possess the major property one expects from an associative memory. The desired memories are arbitrarily chosen, with their recall being guaranteed and their number and distribution unrestricted. Furthermore, assumption (iii) together with the properties of the potential-type ODEs in (1) (cf. Coddington and Levinson, 1955) guarantees that, except for a set of measure zero of saddle points, every initial probe $\mathbf{x}_0$ will converge to a desired memory $\mathbf{u}^{(i)}$ of type (a).

Assumptions (i) and (ii) make the mathematical analysis tractable. When some are omitted, the class of potentials with property (iii) is enlarged. For example, without (i) we obtain (3) with $f_{\boldsymbol{\alpha}}(\mathbf{x}, \boldsymbol{\eta})$ instead of $f_{\boldsymbol{\alpha}}(\|\mathbf{x} - \boldsymbol{\eta}\|^2)$, and (4) is replaced by

$$\forall \eta \in \mathbb{R}^N \quad \forall \mathbf{x} \in \mathbb{R}^N - \{\eta\} \quad \Delta_{\mathbf{x}} f_{\alpha}(\mathbf{x}, \eta) \leq 0 \quad (7)$$

where $\Delta_{\mathbf{x}}$ stands for the Laplacian operator with respect to $\mathbf{x}$. This corresponds to a nonhomogeneous state space but complicates the mathematical analysis.

Let us return now to study the functions $f_{\alpha}(d)$ that satisfy (4) and to relate them to the memory types (a) to (c) mentioned above. Without loss of generality, assume that $f_{\alpha}(d)$ is not constant so $\exists d_0 > 0$, such that $f'_{\alpha}(d) \neq 0$. Furthermore, by adding a proper constant to $f_{\alpha}(d)$, we can assume that

$$f_{\alpha}(d_0) = -\frac{f'_{\alpha}(d_0) d_0}{[(N/2) - 1]} \quad (8)$$

Then the following is true for all $N \geq 3$ (for proofs, see Dembo and Zeitouni, 1988).

LEMMA 1. (a) *Any solution of* (4) *at most can possess one local maximum (and no local minimum) for* $d \in (0, \infty)$, *and satisfies the following:*

$$f_{\alpha}(d) \geq f_{\alpha}(d_0) \left[\frac{d}{d_0}\right]^{-[(N/2)-1]} \quad \text{for } d \geq d_0 \quad (9)$$

$$f_{\alpha}(d) \leq f_{\alpha}(d_0) \left[\frac{d}{d_0}\right]^{-[(N/2)-1]} \quad \text{for } d \leq d_0 \quad (10)$$

(b) *A memory is attractive if and only if* $\exists d_0 > 0$, *such that* $f'_{\alpha}(d_0) > 0$, *and only in that case* $f_{\alpha}(d)$ *may possess a local maximum in* $(0, \infty)$.

(c) *A memory is nonadmissible if and only if* $\lim_{d \downarrow 0} f_{\alpha}(d) = +\infty$ *and repulsive if and only if* $\lim_{d \downarrow 0} f_{\alpha}(d)$ *is finite, in both cases* $f'_{\alpha}(d) \leq 0$ *in* $(0, \infty)$.

Remark. Observe that (10) implies that attractive and nonadmissible memories usually correspond to strong short-range interactions.

To compare the class of neural networks discussed here with the models of Hopfield (1982), Personnaz et al. (1985), Dembo (1989) and Kanter and Sompolinsky (1987), as well as the information theory bounds on error-correcting codes, we derive the discrete-time finite state-space analogue of the evolution (1).

Consider the state space as the unit hypercube in $\mathbb{R}^N$, to be denoted by H^N. For any potential function $V(\mathbf{x})$, the relaxation algorithm is as follows (in the spirit of Baldi, 1988).

(a) According to some predetermined probability measure which is nowhere zero, pick a point $\mathbf{y} \in H^N$ having a Hamming distance of 1 from the current state $\mathbf{x} \in H^N$.

(b) If $V(\mathbf{y}) < V(\mathbf{x})$, then the new state will be $\mathbf{y}$, otherwise it remains $\mathbf{x}$. In both cases, return to step (a).

It is well known that for any $V(\cdot)$ and $\mathbf{x}_0$, this algorithm converges to a fixed point in H^N (see Baldi, 1988, for example). For any practical memory of this type, $\{\mathbf{u}^{(\alpha)}\}_{\alpha \in A} \subset H^N$, and thus A is a finite set of size $K \leq 2^N$.

6.2. BASINS OF ATTRACTION AND CONVERGENCE RATE

For analyzing the performance as an associative memory of the system (1) with the potential function of (5), assume for simplicity that all memories are attractive with $f_i(\cdot)$ independent of i and $\|\mathbf{u}^{(i)} - \mathbf{u}^{(j)}\| \geq 1$, for every $i \neq j$. We shall investigate the worst case (over all possible memory locations) size of the smallest basin of attraction, which is

$$\varepsilon_{\max} \doteq \min_{\{\mathbf{u}^{(i)}\}, i \in A} \{\max[\rho : \|\mathbf{x}_0 - \mathbf{u}^{(i)}\| \leq \rho$$
$$\text{implies } \lim_{t \to \infty} \mathbf{x}_t = \mathbf{u}^{(i)}]\} \quad (11)$$

It is clear from symmetry arguments that $\varepsilon_{\max} \leq \frac{1}{2}$ (where the outer minimization is over all possible positions of $\{\mathbf{u}^{(i)}\}_{i \in A}$ in $\mathbb{R}^N$). It is also clear that $\varepsilon_{\max}$ is a monotonically nonincreasing function of K. Our aim is to show that a proper choice of $f(\cdot)$ (which is subharmonic) will lead to $\varepsilon_{\max}$ as close to $\frac{1}{2}$ as desired, thus the maximal basin of attraction can be guaranteed. The following lower bound on $\varepsilon_{\max}$ (which is independent of K!) is derived in Dembo and Zeitouni (1988) by bounding the maximal contribution of farther-away particles to forces in the $\varepsilon_{\max}$ ball around $\mathbf{u}^{(i)}$.

LEMMA 2. (a) $\varepsilon_{\max}$ *is larger than any value of* $r < 1$ *for which*

$$f'(r^2) r \geq \int_1^\infty -\frac{d}{dt}\{(t-r) f'[(t-r)^2]\}$$
$$\times (2t+1)^N dt \quad (12)$$

(b) $\varepsilon_{\max} \to 0$ *as* $K \to \infty$ *if and only if the r.h.s.* (*right-hand side*) *of* (12) *diverges for every* $r > 0$.

We shall now restrict our attention to $f(d) = -k(d/d_0)^{-m}$, with $m \geq [(N/2) - 1]$, which satisfies Eq. (4).

The r.h.s. of (12) is finite if and only if $m > (N/2) - \frac{1}{2}$, and then for integer m, (12) yields the bound

$$\tfrac{1}{2} \geq \varepsilon_{\max}(m, N) \geq \{1 + [\tfrac{1}{2}3^{(N+1)}]^{1/(2m+1)}\}^{-1} \tag{13}$$

In particular, for fixed N, $\lim_{m \to \infty} [\varepsilon_{\max}(m, N)] = \tfrac{1}{2}$, and for any $k \geq 1$ fixed,

$$\varepsilon_{\max}\{\tfrac{1}{2}[k(N+1) - 1], N\} \geq (1 + 3^{1/k})^{-1}$$

For example, $\varepsilon_{\max}(N/2, N) \geq \tfrac{1}{4}$.

Remarks. (1) If the $\{u^{(i)}\}_{i \in A}$ are restricted to be contained in a ball of radius $\tilde{\rho}$, then the integral in the r.h.s. of (12) will have upper limit $2\tilde{\rho}$, and the additional term $(2\tilde{\rho} - r)f'[(2\tilde{\rho} - r)^2](4\tilde{\rho} + 1)^N$ would be added there. It is thus finite for every value of m (including the harmonic case $m = (N/2) - 1$), implying that $\varepsilon_{\max}$ is then never zero.

(2) For $N \to \infty$ and fixed k the $(1 + 3^{1/k})^{-1}$ behavior is maintained even if only memories $\{u^{(i)}\}_{i \in A}$ contained in the finite ball $(\tilde{\rho} = 1)$ are considered.

As for the rate of convergence, it is easy to verify that for a large value of m and x_0 far from all the $\{u^{(i)}\}_{i \in A}$, it will take a long time [for the evolution in (1)] before x_t will be near one of the $\{u^{(i)}\}_{i \in A}$. However, the following lemma is proved in Dembo and Zeitouni (1988).

LEMMA 3. *Let Q be any closed set in $\mathbb{R}^N$ whose interior includes the convex hull of $\{u^{(i)}\}_{i \in A} \cup \{\bar{u}\}$, where $\bar{u}$ is an arbitrary point in $\mathbb{R}^N$. Then, adding $g(\|x - \bar{u}\|^2)1_{x \notin Q}$ to $V(x)$, where $g(x)$ is any nondecreasing, differentiable function, will not disturb (iii) nor create additional fixed points to (1), provided all the $f_i(\cdot)$ which compose $V(x)$ are monotonically increasing.*

Therefore, if, for example, $g(d) = d^\beta$ is added (with $\beta > 1$), then the convergence from x_0 at infinity to a point with squared distance d_0 from the points $\{u^{(i)}\}_{i \in A}$ and $\bar{u}$ (with d_0 much larger than the squared distances between these $K + 1$ points) takes the time $T \sim d_0^{-(\beta-1)}/4\beta(\beta - 1)$. Thus, by using β large enough, convergence from infinity to the boundary of Q can take an arbitrarily small time.

Global investigation of the rate of convergence inside the convex hull of $\{u^{(i)}\}_{i \in A}$ is quite cumbersome. Thus, let us again restrict the discussion to the case where $\|u^{(i)} - u^{(j)}\| \geq 1$, $f(d) = -k(d/d_0)^{-m}$ and $m \geq N/2$ is an integer. Furthermore, let x_0 satisfy $\|x_0 - u^{(i)}\| \leq \theta\hat{\varepsilon}(m, N)$, where $\theta < 1$ and $\hat{\varepsilon}(m, N)$ is the lower bound on $\varepsilon_{\max}(m, N)$ given by the r.h.s. of (13).

We have seen already that for every $\theta \leq 1$,

$$\lim_{t \to \infty} x_t = u^{(i)}$$

The following lemma, proved in Dembo and Zeitouni (1988), states that only finite time is needed for convergence to $u^{(i)}$.

LEMMA 4. *Under the above conditions, $x_T = u^{(i)}$, where*

$$T = \left[\frac{\theta\hat{\varepsilon}(m, N)}{\sqrt{d_0}}\right]^{(2m+2)} \left[\frac{d_0}{2mk}\right]$$

$$\times \left[1 - \frac{\theta(1 - \hat{\varepsilon}(m, N))}{(1 - \theta\hat{\varepsilon}(m, N))}^{2m+1}\right]^{-1} \tag{14}$$

Hence, for m large enough, $\sqrt{d_0} = \hat{\varepsilon}(m, N)$ and $k = d_0/2m$, we obtain $\log T \sim 2(m + 1) \log \theta$. Again, by enlarging m while preserving θ fixed, T can be made arbitrarily small.

In summary, the maximal basins of attraction can be guaranteed by enlarging m, i.e., choosing strong, short-range interactions. This will also speed up the convergence within these basins of attraction, which is completed within a small *finite* time. The convergence from infinity to the neighborhood of $\{u^{(i)}\}_{i \in A}$ is governed by adding a long-range field outside a proper set Q, without affecting all those properties.

6.3. EVOLUTION ON THE UNIT HYPERCUBE

Whereas the potential function of the discrete-time algorithm presented in Section 6.1 has no local minima outside $(u^{(i)})_{i \in A}$, this algorithm might posses *fixed points* out of this set. The reason for this phenomenon is the rigidity of the algorithm, which might not allow descent in the gradient direction as the search for lower potential is limited to the nearest neighbors of the current state.

It is shown in Dembo and Zeitouni (1988) that the class of potential functions discussed here is optimal according to the information theory bounds (as $N \to \infty$), and in particular can be used to design error-correcting codes with a positive rate (cf. McEliece, 1977, for definitions).

Let us restrict the discussion to potentials of the form (5) with $f(d) = -d^{-m}$, and $m \geq (N/2) - 1$. Denote the normalized Hamming distance $1/(2N) \sum_{i=1}^{N} |x_i - y_i|$ by $\|x - y\|_H$ and assume that $\forall i \neq j$, $\|u^{(i)} - u^{(j)}\|_H \geq \rho$, with $1/2 \geq \rho > 1/N$, fixed. Therefore, the code words

$\{\mathbf{u}^{(i)}\}_{i=1}^{K}$ can tolerate up to $\frac{1}{2}\rho N$ errors in N coordinates.

THEOREM 2. *For every* $\mathbf{x}_0$ *such that*

$$\|\mathbf{x}_0 - \mathbf{u}^{(i)}\|_H \le \theta\rho, \qquad \tfrac{1}{2} \ge \theta \ge 0$$

and

$$(K-1) \le \left[\frac{1-\theta}{\theta}\right]^m \left[\frac{1 - [1 + (2/N\rho)]^{-m}}{[1 - (2/N\rho)]^{-m} - 1}\right] \tag{15}$$

(a) *The discrete-time algorithm will generate a sequence of states* $\{\mathbf{x}_n\}$ *such that*

$$\|\mathbf{x}_n - \mathbf{u}^{(i)}\|_H < \|\mathbf{x}_{n-1} - \mathbf{u}^{(i)}\|_H$$

whenever $\mathbf{x}_{n-1} \ne \mathbf{x}_n$.

(b) *There are exactly* $N\|\mathbf{x}_0 - \mathbf{u}^{(i)}\|_H$ *distinct states in this sequence, and if each coordinate has positive probability to be chosen as the updated coordinate then* $\mathbf{x}_n$ *converges to* $\mathbf{u}^{(i)}$ *with probability 1 within finite time.*

Consider now $\rho > 0$ and $\theta < \frac{1}{2}$ fixed with $m \ge N\{\log_{2+\varepsilon}[(1-\theta)/\theta]\}^{-1}$ (where $\varepsilon > 0$ is arbitrarily small). Then, if N is large enough, the r.h.s. of (15) is larger than 2^N, thus K is determined only by bounds on the maximal number of points in H^N satisfying $\forall i \ne j$, $\|\mathbf{u}^{(i)} - \mathbf{u}^{(j)}\|_H \ge \rho$, i.e., information theory asymptotic sphere packing bounds for error-correcting codes (cf. McEliece et al., 1977).

In summary, Theorem 2 guarantees that for short-range forces [i.e., $m(N)$ large enough] and large enough dimension, convergence to the nearest code word is obtained independently of the number of code words and their locations, provided that the initial distance between the probe and the nearest code word is smaller than $\frac{1}{2}\min_{i \ne j}\|\mathbf{u}^{(i)} - \mathbf{u}^{(j)}\|_H$.

6.4. DISCUSSION

6.4.1. Possible Generalizations

The associative-memory models presented in Sections 6.2 and 6.3 are capable of both storing information and recalling it.

The discrete-time version of the algorithm can be easily extended to any finite graph whose vertices are embedded in $\mathbb{R}^N$ (cf. Baldi, 1988).

The process of storage and recall of information described in this chapter does not involve any learning or generalization (in the sense of Cooper, 1984). It is also incapable of creating periodic orbits. However, by using charge densities and omitting assumption (i), one can define the class of potentials having a predefined stable attractor set, part of which consists of periodical orbits.

Likewise, learning can be incorporated by modifying the locations of $\{\mathbf{u}^{(i)}\}_{i \in A}$ during the recall operation, either in response to the distribution of the initial states $\mathbf{x}_0$ or to an external teaching procedure. These modifications can be implemented within the evolution (1) by allowing the state $\mathbf{x}_t$ to be represented by a nonnegligible particle which applies *forces* on the given $\{\mathbf{u}^{(i)}\}_{i \in A}$ or by adding interparticles interactions. Generalization, in the sense of Kohonen (1984), i.e., spontaneous clustering, becomes possible once the $\{\mathbf{u}^{(i)}\}_{i \in A}$ particles are allowed to apply forces one on the other and change their locations. Of course, for learning and generalization, goals should be defined rigorously, and then rules to achieve them can be incorporated within this framework.

6.4.2. Comparison with Other Memory Models

For comparison purposes, we deal with three associative memory algorithms on H^N. The first one is the classical Hamming decoder with respect to $\{\mathbf{u}^{(i)}\}_{i=1}^{K} \subset H^N$. It involves the parallel computation of the K correlations $\langle \mathbf{u}^{(i)}, \mathbf{x}_0 \rangle$ (where $\mathbf{x}_0 \in H^N$ as well), followed by a search for the maximal value, implemented in a tree structure. Thus, KN multiplications are needed together with $K \log K$ comparisons of pairs of numbers, and the delay of the algorithm is $(\log K + 1)$ "unit" times (where comparison and multiplication are assumed equivalent throughout).

The second algorithm is the one suggested in Hopfield (1982). It corresponds to $f_i(d) = \frac{1}{2}[N - (N - \frac{1}{2}d)^2]$, which does not satisfy (4) [and some of its extensions do not satisfy assumptions (i) and (ii) at all]. Each iteration involves KN multiplications and N comparisons of pairs of numbers. Note that this algorithm is effectively limited to $K \le N$ as shown in Amit et al. (1985, 1987), Weisbuch and Fogelman-Soulie (1985), McEliece et al. (1987), and Dembo (1989). The time delay of this algorithm is just the number of iterations, which is believed to be almost independent of K.

The last algorithm is the one suggested in Section 6.1 with $f_i(d) = -d^m$. One possible network which implements this algorithm is presented in Dembo and Zeitouni (1988). An implementation for the continuous-time case, which is even simpler, is also hinted at. This implementation is *by no means unique*, and may be not even the simplest one. It consists of $(N + 1)$

artificial neurons which are delay elements and K pointwise nonlinear functions which may be interpreted as delayless, intermediate neurons. There are NK synaptic connections between those two layers of neurons. Since the intermediate neuron corresponds to the memory $\mathbf{u}^{(j)}$, damage to this neuron might completely destroy this memory while not affecting any of the other memories. Thus, it shares some of the properties of "grandmother cell" architectures. The overall complexity of this implementation is determined by the KN multipliers, and its time delay is a small constant, as implied by Theorem 2.

Thus, for $K \leq N$, the new algorithm has the same complexity as the scheme of Hopfield (1982), and the classical Hamming decoder with smaller time delay for the first two algorithms.

This result is true also for $K \sim 2^{h(\rho)N}$, but then the memory scheme of Hopfield (1982), which has long-range forces, should not be used as the maximal number of memories is bounded above by N even for $\theta = 0$ (recall with a perfect probe). Thus, this model has *zero rate* when referred to as an error-correcting code (cf. McEliece, 1977). Even when it converges, the convergence time might grow exponentially with N, unlike the linear time guaranteed by Theorem 2 (when the coordinates are updated in a cyclic manner). Various extensions of this model exist with higher degrees of interconnections permitted (cf. Baldi, 1988; Peretto and Niez, 1986; Dembo et al., 1991), i.e., higher-order polynomial potentials are used. In this case, the new algorithm has complexity which is linear in K, i.e., exponential in N. Hence it is better (in time delay) than the classical Hamming decoder but does not admit the polynomial complexity of some of the special error-correcting codes used in coding theory (cf. McEliece, 1977).

6.5. SUMMARY

The conclusion of this chapter is that a memory can be built such that:

1. Capacity is guaranteed to be, in principle, unlimited without undesired memories.
2. Basins of attraction for each memory are guaranteed to be of the obvious size, that is, half the minimal distance between desired memories.
3. A discrete-time version retains most of the properties of the continuous-time model.

However, it seems that in a typical implementation of this memory, each additional stored memory requires an additional neuron to be added to the system.

REFERENCES

Amit, D. J., Gutfreund, H., and Sompolinsky, H. (1985). "Spin-glass Models of Neural Networks," *Phys. Rev. A*, **32**, 1007–1018.

Amit, D. J., Gutfreund, H., and Sompolinsky, H. (1987). "Information Storage in Neural Networks with Low Levels of Activity," *Phys. Rev. A*, **35**, 2293–2303.

Baldi, P. (1988). "Neural Networks, Orientations of the Hypercube and Algebraic Threshold Functions," *IEEE Trans. Information Theory*, **IT-34**, 523.

Coddington, E. A. and Levinson, N. (1955). *Theory of Ordinary Differential Equations*. McGraw-Hill, New York.

Cooper, L. N. (1984). *J. C. Maxwell, The Sesquicentennial Symposium*, M. S. Berger, ed. North-Holland, Amsterdam.

Dembo, A. (1989). "On the Capacity of Associative Memories with Linear Threshold Functions," *IEEE Trans. Information Theory*, **IT-35**, 709–720.

Dembo, A. and Zeitouni, O. (1988). "General Potential Surfaces and Neural Networks," *Phys. Rev. A*, **37**, 2134–2143.

Dembo, A., Farotimi, O., and Kailath, T. (1991). "High Order Absolutely Stable Neural Networks," *IEEE Trans. Circuits and Systems*, **38**, 57–65.

Hopfield, J. J. (1982). "Neural Networks and Physical Systems with Emergent Collective Computational Abilities," *Proc. Nat. Acad. Sci. U.S.A.*, **79**, 2554–2558.

Kanter, I. and Sompolinsky, H. (1987). "Associative Recall of Memory without Errors," *Phys. Rev. A*, **35**, 380–392.

Kohonen, T. (1984). *Self Organization and Associative Memory*. Springer-Verlag, Berlin.

McEliece, R. J. (1977). *The Theory of Information Theory and Coding*. Addison-Wesley, Reading, MA, Vol. e.

McEliece, R. J., Posner, E. C., Rodemich, E. R., and Venkatesh, S. S. (1987). "The Capacity of the Hopfield Associative Memory," *IEEE Trans. Information Theory*, **IT-33**, 461–483.

Peretto, P. and Niez, J. J. (1986). "Long Term Memory Storage Capacity of Multiconnected Neural Networks," *Biological Cybernetics*, **54**, 53–63.

Personnaz, L., Guyon, I., and Dreyfus, G. (1985). "Information Storage and Retrieval in Spin-Glass like Neural Networks," *J. Phys. (Paris) Rev. Lett.*, **46**, L359–L365.

Weisbuch, G. L. and Fogelman-Soulie, F. (1985). "Scaling Laws for the Attractors of Hopfield Networks," *J. Phys. (Paris) Lett.*, **46**, L623–630.

7
A Normal Form Projection Algorithm for Associative Memory

BILL BAIRD and FRANK EECKMAN

7.1. INTRODUCTION

The learning rules described here are specifically designed to permit the construction of biological models and exploration of engineering or cognitive networks—with analytically guaranteed associative memory function—that employ the recurrent architectures and the type of dynamics found in the brain. Patterns of 40 to 80 Hz oscillation have been observed in the large-scale activity (local field potentials) of olfactory cortex (Freeman, 1975) and visual neocortex (Gray and Singer, 1987), and shown to predict the olfactory (Freeman and Baird, 1987) and visual (Freeman and van Dijk, 1987) pattern recognition responses of a trained animal. Similar observations of 40 Hz oscillation in auditory and motor cortex (in primates), and in the retina and electromyogram (EMG) have been reported. It thus appears that cortical computation in general may occur by dynamical interaction of resonant modes, as has been thought to be the case in the olfactory system. The oscillation can serve as a macroscopic clocking function and entrain or "blind" the relevant microscopic activity of disparate cortical regions into a well-defined phase-coherent collective state or "gestalt." This can override irrelevant microscopic activity and produce coordinated motor output. There is further evidence that though the collective activity is roughly periodic, it is actually chaotic (nonperiodic) when examined in detail (Freeman, 1987).

If this view is correct, then oscillatory and possibly chaotic network modules form the actual cortical substrate of the diverse sensory, motor, and cognitive operations now studied in static networks. It must then be shown how those functions can be done with oscillatory and chaotic dynamics, and what advantages may be gained thereby. Our challenge is to accomplish real tasks that brains can do, using ordinary differential equations, in networks that are as

faithful as possible to the known dynamics and anatomical structure of cortex. It is our expectation that nature makes good use of this dynamical complexity, and our intent is to search here for novel design principles that may underlie the superior performance of biological systems in pattern recognition, robotics, and intelligent behavior. These may then be applied in artificial systems to engineering problems to advance the art of computation.

To this end, we discuss at the close of the chapter a parallel distributed processing architecture that is inspired by the structure and dynamics of cerebral cortex. It is constructed of recurrently interconnected associative memory modules whose theoretical analysis forms the bulk of this chapter. Each module is a network model of a "patch" of cortex that can store oscillatory and chaotic attractors by a Hebb rule. The modules can learn connection weights between themselves which will cause the system to evolve under a clocked "machine cycle" by a sequence of transitions of attractors within the modules, much as a digital computer evolves by transitions of its binary flip-flop states. Thus the architecture employs the principle of "computing with attractors" used by macroscopic systems for reliable computation in the presence of noise. Clocking is done by rhythmic variation of certain *bifurcation parameters* which hold sensory modules clamped at their attractors while motor states change, and then clamp motor states while sensory states are released to take new states based on input from external motor output and internal feedback.

The mathematical foundation for the network modules described in this paper is contained in the *projection theorem*, which details the associative memory capabilities of networks utilizing the *normal form projection algorithm* for storage of periodic attractors. The algorithm was originally designed, using dynamical systems theory, to allow learning and pattern recognition with oscillatory attractors in models of olfactory cortex. Here we concentrate on mathematical analysis and engineering-oriented applications of the algorithm, and briefly discuss biological models at the end. We focus attention on the storage of periodic attractors, since that is the best understood unusual capability of this system. The storage of static and chaotic attractors is discussed as variations on this theme. We hope to give intuitive discussion and geometric perspectives to complement and clarify the formal analysis. Other approaches to oscillatory memory may be found in Grossberg and Elias (1975), Freeman et al. (1988), Wang et al. (1990), Horn and Usher (1991), and Li and Hopfield (1989).

The normal form projection algorithm provides one solution to the problem of storing analog attractors in a recurrent neural network. Associative memory storage of analog patterns and continuous periodic sequences in the same network is analytically guaranteed. For a network with N nodes, the capacity is N static attractors. Periodic attractors which are simple cycles can be stored at a capacity of $N/2$. There are no spurious attractors, and there is a Liapunov function in a special coordinate system which governs the approach of transient states to stored periodic trajectories. For storage of different types of attractors in the same network, there are N units of total capacity in an N-node network. It costs one unit of capacity per static attractor, two per Fourier component of each sequence, and at least three per chaotic attractor.

A key feature of a net constructed by the projection algorithm is that the underlying dynamics is explicitly isomorphic to any of a class of standard, well understood nonlinear dynamical systems—a *normal form* (Gluckenheimer and Holmes, 1983). This system is chosen in advance, independent of both the patterns to be stored and the learning algorithm to be used. This confers control over the network dynamics since it permits the design of important aspects of the dynamics independent of the particular patterns to be stored. Stability, basin geometry, and rates of convergence to attractors can be programmed in the standard dynamical system.

The storage of oscillation amplitude patterns by the projection algorithm begins with the specification of oscillations by coupled ordinary differential equations in a special polar coordinate system, where the equations for the amplitudes of the oscillations are independent of those for the phase rotations. These equations are the normal form for the *Hopf bifurcation*. Amplitude coupling coefficients are chosen to give stable

fixed points on the coordinate axes of the vector field of amplitudes (see Fig. 7.2) where the Liapunov function is defined, frequencies are chosen for the phase equations, and the polar system is then transformed to Cartesian complex conjugate coordinates. The axes of this system of nonlinear ordinary differential equations are then linearly transformed into desired spatial or spatiotemporal patterns by projecting the system into network coordinates (the standard basis) using the desired vectors as columns of the transformation matrix. As we discuss below, this method of network synthesis is roughly the inverse of the usual procedure in *bifurcation theory* for analysis of a given physical system. These operations may all be compactly expressed in network coordinates as a formula or "learning rule" for the coupling weights given the desired patterns and frequencies. Although it is never apparent in network coordinates, since no restrictions are imposed on the patterns that can be stored, these networks are always implicitly the product of a space of phases where frequencies are determined, and an independent space of amplitudes where attractors are programmed.

In the general case, the network resulting from the projection algorithm has third-order synaptic weights or "sigma pi" units (Maxwell et al., 1986), which appear as a four-dimensional matrix T_{ijkl} or *tensor* of coupling in the network equations. In the projection algorithm, the projection of the additional weights T_{ijkl} for the competitive cubic terms $-x_j x_k x_l$ allows us to guarantee exact storage of N static or $N/2$ oscillatory memories, without "spurious attractors." This compares with roughly $1.5N$ static attractor capacity for Hopfield nets. For engineering purposes, there are optical systems which can implement these higher-order networks directly (Psaltis et al., 1988), and fast network simulation architectures that may allow application of these systems to real-world engineering problems in pattern recognition, process control, and robotics.

The *projection network* is an alternative network for implementation of the normal form projection algorithm. The autoassociative case of this network is formally equivalent to the general higher-order network realization which is used later as a biological model. A matrix inversion determines network weights, given prototype patterns to be stored. It has $3N^2$ weights instead of $N^2 + N^4$, and is more useful for engineering applications. All the mathematical results proved for the projection algorithm in the network above carry over to this new architecture, but more general versions can be trained and applied in novel ways.

For biological modeling, where possibly the oscillation amplitude patterns to be stored can be assumed to be sparse and nearly orthogonal, the learning operation for periodic patterns reduces to a kind of periodic or phase-dependent outer product rule that permits local, incremental learning. The system can be truly self-organizing because a synapse of the net can modify itself according to its own activity during external driving by an input pattern to be learned. Standing or traveling waves may be stored to mimic the different patterns of neurophysiological activity observed in the olfactory system. Between units of equal phase, or for static patterns, learning reduces further to the usual Hebb rule.

It is argued that some dimensionality of the higher-order synapses in the mathematical model may be realized in a biological system by synaptic interactions in the dense axodendritic interconnection plexus (the *neuropil*) of local neural populations—given the information immediately available on the primary long-range connections W_{ij}. Theoretical work shows further that only N^2 of the N^4 possible higher-order weights are required in principle to approximate the performance of the projection algorithm.

The method of network synthesis contained in the projection algorithm is roughly the inverse of the usual procedure in bifurcation theory for analysis of a given physical system. A *bifurcation* is an important qualitative change in the dynamical behavior of a system as a parameter is varied—the creation or destruction of attractors, or a change of stability of an attractor to become a repellor. A *phase transition* that a physicist sees at a *critical point* in a physical system may be viewed mathematically as a bifurcation in the stochastic dynamical equations describing that system. Where *dynamics* describes the continuous temporal evolution of a system, *bifurcations* describe discontinuous changes in the dynamical possibilities. We have found that mathematical analysis is simplified, and the power of our learning algorithm increased, when this additional dimension of organization about a *bifurcation point* (critical point) is considered in the design of the network.

This approach describes how associative memory can be programmed and work precisely with continuous dynamics and smooth sigmoids that are weakly nonlinear, when it is operated in the vicinity of such a critical point. This may be thought of as operation in the low-gain limit near the origin, as opposed to the high-gain limit on a hypercube, where analytic results are usually obtained (Hopfield, 1984; Psaltis and Venkatesh,

1989; Sompolinsky and Kantner, 1986; Hemmen, 1987; Herz et al., 1989). The effective decision making nonlinearity in this system is in the higher-order synapses, not the axons. It is nonlinear *coupling*, as opposed to nonlinear activation.

Although developed from a bifurcation theory approach, where the higher-order terms are usually obtained from a truncated Taylor's expansion, the results on the networks discussed in this paper are global and exact since the only nonlinearity used is the cubic terms. The results are not restricted to the neighborhood of any bifurcation point, and for simplicity will be discussed without reference to bifurcation theory. There is, however, a multiple bifurcation at critical values of different bifurcation parameters that can be used to generate the attractors from a single equilibrium at the origin. This kind of parameter is essential to the control of attractor transitions in the subnetworks of the hierarchical system discussed at the end.

7.2. PROJECTION ALGORITHM

7.2.1. Cycles

The oscillatory patterns that can be stored by the projection algorithm may be idealized mathematically as a spatiotemporal pattern or *cycle*, $r\mathbf{x}e^{i\boldsymbol{\theta}}e^{i\omega t}$. Such a cycle is a *periodic attractor* if it is *stable*. We will first supply some concrete imagery about these cycles. A cycle is shown in Fig. 7.1, where the components of the amplitude vector $\mathbf{x}$ are shown laid out in space as compartments of a "spatial pattern" in one dimension. The notions of "relative" (vector: $\mathbf{x}, \boldsymbol{\theta}$) and "global" (scalar: r, w, θ) amplitude and phase are illustrated in the figure. The global amplitude r is just a scaling factor for the pattern $\mathbf{x}$, and the global phase ω in $e^{i\omega t}$ is a periodic scaling that scales $\mathbf{x}$ by a factor between ± 1 at frequency ω as t varies.

The same vector $\mathbf{x}^s$ or "spatial pattern" of relative amplitudes of a set of components can appear as a standing wave, like that seen in the olfactory bulb, if the relative phase θ_i^s of each component is the same, $\theta_{i+1}^s = \theta_i^s$, or as a traveling wave, if the relative phase components of $\boldsymbol{\theta}^s$ form a gradient in space, $\theta_{i+1}^s = (1/\alpha)\theta_i^s$. The traveling wave will "sweep out" the amplitude pattern $\mathbf{x}^s$ in time, but the root-mean-square amplitude measured in an experiment (Freeman, 1975) will be the same $\mathbf{x}^s$, regardless of the phase pattern. At least one network component of the attractor pattern, however, must be out of phase

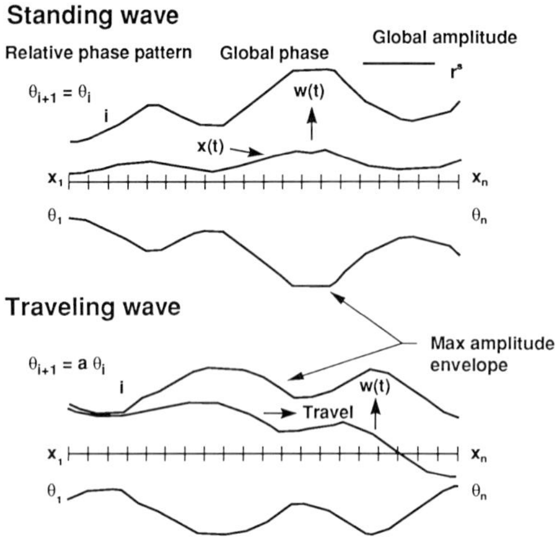

Standing wave

Relative phase pattern   Global phase   Global amplitude

Traveling wave

Fig. 7.1. Standing and traveling wave examples of spatiotemporal patterns or cycles, $r\mathbf{x}e^{i\boldsymbol{\theta}}e^{i\omega t}$, that can be stored as *periodic attractors* by the projection algorithm. Shown are outlines of the histogram of the components of the vector $\mathbf{x}$ at an instant of time $\mathbf{x}(t)$. The maximum amplitude envelope of positive and negative excursions (the "spatial pattern") is outlined in each case as well.

with the rest, since all components cannot pass through zero at the same time, because zero is a fixed point of these equations. For a randomly chosen phase vector, these simple single-frequency cycles can make very complicated-looking spatiotemporal patterns. From the mathematical point of view, the relative phase pattern $\boldsymbol{\theta}$ is a degree of freedom in the kind of patterns that can be stored. Patterns of uniform amplitude $\mathbf{x}$ that differed only in the phase pattern $\boldsymbol{\theta}$ could be stored as well.

Although slight concentric phase gradients that vary in location with each inspiration have been found in the olfactory bulb, we model the bulb's activity as a standing wave. To store the kind of patterns found in the bulb, the amplitude vector $\mathbf{x}$ is assumed to be parsed into equal numbers of excitatory and inhibitory components, with an excitatory/inhibitory pair forming a local oscillator at each point in space. The components of each class have identical phase, but there is a phase difference of 60–90 degrees between the classes. There is also a slight but regular traveling wave of activity in the prepyriform cortex from rostral to caudal, which can easily be modeled here by introducing a slight phase gradient into both the excitatory and inhibitory classes.

7.2.2. Projection Theorem

The central mathematical result of this work is given in the following theorem.

THEOREM 2.1. *Any set* S, $s = 1, 2, \ldots, N/2$, *of cycles:* $r\mathbf{x}^s e^{i\boldsymbol{\theta}^s}e^{i\omega^s t}$ *of linearly independent vectors of relative component amplitudes* $\mathbf{x}^s \in \mathbb{R}^N$ *and phases* $\boldsymbol{\theta}^s \in T^N$ *(N-torus), with frequencies* $\omega^s \in \mathbb{R}$ *and global amplitudes* $r^s \in \mathbb{R}$, *may be established in the vector field of the analog third-order network:*

$$\dot{x}_i = -\tau x_i + \sum_{j=1}^{N} T_{ij}x_j - \sum_{jkl=1}^{N} T_{ijkl}x_j x_k x_l + b_i\delta(t) \tag{1}$$

by the projection operation (learning rule):

$$\left.\begin{array}{l} T_{ij} = \sum_{mn=1}^{N} P_{im} J_{mn} P_{nj}^{-1} \\[2mm] T_{ijkl} = \sum_{mn=1}^{N} P_{im} A_{mn} P_{mj}^{-1} P_{nk}^{-1} P_{nl}^{-1} \end{array}\right\} \tag{2}$$

Here the $N \times N$ *matrix* $\mathbf{P}$ *contains the real and imaginary components* $[\mathbf{x}^s \cos\boldsymbol{\theta}^s, \mathbf{x}^s \sin\boldsymbol{\theta}^s]$ *of the complex eigenvectors* $\mathbf{x}^s e^{i\boldsymbol{\theta}^s}$ *as columns.* $\mathbf{J}$ *is an* $N \times N$ *matrix of complex conjugate eigenvalues in diagonal blocks, and* A_{mn} *is an* $N \times N$ *matrix of* 2×2 *blocks of repeated coefficients* a_{ij} *of the normal form equations (3, 4) below.*

$$\mathbf{J} = \begin{bmatrix} \alpha_1 & -\omega_1 & & & \\ \omega_1 & \alpha_1 & & & \\ & & \alpha_2 & -\omega_2 & \\ & & \omega_2 & \alpha_2 & \\ & & & & \ddots \end{bmatrix}$$

$$\mathbf{A} = \begin{bmatrix} a_{11} & a_{11} & a_{12} & a_{12} & \\ a_{11} & a_{11} & a_{12} & a_{12} & \\ a_{21} & a_{21} & a_{22} & a_{22} & \\ a_{21} & a_{21} & a_{22} & a_{22} & \\ & & & & \ddots \end{bmatrix}$$

The vector field of the dynamics of the global amplitudes r_s *and phases* ϕ_s *is then given by the normal form equations for a multiple Hopf bifurcation:*

$$\dot{r}_s = u_s r_s - r_s \sum_{j=1}^{N/2} a_{sj}r_j^2 \tag{3}$$

$$\dot{\phi}_s = \omega_s + \sum_{j=1}^{N/2} b_{sj}r_j^2 \tag{4}$$

In particular, if $u_s = \alpha_s - \tau > 0$, $a_{sk} > 0$, and $a_{ss}/a_{ks} < 1$ $\forall s$ and k, the cycles $s = 1, 2, \ldots, n/2$ are stable, and have amplitudes $r_s = (u_s/a_{ss})^{1/2}$. Input $b_i\delta(t)$ is a delta function of time that establishes an initial condition.

7.2.3. Proof of the Projection Theorem

The normal form amplitude equations (3) describe the approach of the network to the desired oscillatory patterns because the network equations (1) are equivalent to the normal form equations (3, 4). They can be constructed from them by two simple coordinate transformations. We proceed by showing first that there are always fixed points on the axes of the normal form amplitude equations (3), whose stability is given by the condition stated in the theorem for the coefficients a_{ij} of the nonlinear terms. The network is then constructed by a transformation of the normal form equations first from polar to Cartesian coordinates, and then by a linear transformation or *projection* from these *mode* coordinates into the standard basis $e_1, e_2, \ldots, e_N$, or *network coordinates*. This second transformation constitutes the "learning rule" (2), shown in the theorem, because it transforms the simple fixed points of the amplitude equations (3) into the specific spatiotemporal memory patterns desired for the network.

Amplitude Fixed Points

Because the amplitude equations are independent of the rotation ϕ, the fixed points of the normal form amplitude equations characterize the asymptotic states of the underlying oscillations (see Fig. 7.2). The stability of these cycles is therefore given by the stability of the fixed points of the amplitude equations. On each axis r_s, the other components r_j are zero, by definition, hence

$$\dot{r}_j = r_j\left(u_j - \sum_{k=1}^{N/2} a_{jk}r_k^2\right) = 0$$

for $r_j = 0$, which leaves

$$\dot{r}_s = r_s(u_s - a_{ss}r_s^2)$$

and $\dot{r}_s = 0$, when $r_s^2 = u_s/a_{ss}$.

Thus we have an equilibrium on each axis s, at $r_s = (u_s/a_{ss})1/2$ as claimed. Now the Jacobian $\mathbf{U}$ of the amplitude equations at some fixed point $\hat{r}$ has elements

$$U_{ij} = -2a_{ij}\hat{r}_i\hat{r}_j \quad U_{ii} = u_i - 3a_{ii}\hat{r}_i^2 - \sum_{j=1}^{N/2} a_{ij}\hat{r}_j^2 \quad (5)$$

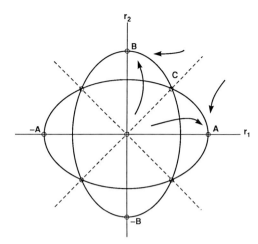

Fig. 7.2. Two-dimensional normal form amplitude vector field with symmetry. There are four static attractors $(A, -A, B, -B)$, or two oscillation amplitude attractors (A, B) which must be positive. C is a saddle point on the dotted *unstable manifold* that forms the *separatrix* which partitions the vector field into the "wedge" shaped basins of attraction. The equilibria occur at intersections of the axes and the ellipsoids which are *nullclines* where $\dot{r}_i = 0$ for each component. By neglecting $\dot{\phi}_2$, the plane can be viewed roughly as a Poincaré section, for $\phi = 0$, that is pierced periodically by the orbit of mode r_1.

For a fixed point $\hat{r}_s$ on axis s, $U_{ij} = 0$, since $\hat{r}_i$ or $\hat{r}_j = 0$, making $\mathbf{U}$ a diagonal matrix whose entries are therefore its eigenvalues. Now $U_{ii} = u_i - a_{is}\hat{r}_s^2$, for $i \neq s$, and $U_{ss} = u_s - 3a_{ss}\hat{r}_s^2$. Since $\hat{r}_s^2 = u_s/a_{ss}$, $U_{ss} = -2u_s$, and $U_{ii} = u_i - a_{is}(u_s/a_{ss})$. This gives $a_{is}/a_{ss} > u_i/u_s$ as the condition for all negative eigenvalues that assures the stability of $\hat{r}_s$. Choice of $a_{ji}/a_{ii} > u_j/u_i$, $\forall i, j$, therefore guarantees stability of all axis fixed points.

Coordinate Transformations

We now construct the neural network from these well-behaved equations by two transformations. The first is polar to Cartesian, (r_s, θ_s) to (v_{2s-1}, v_{2s}). Using $v_{2s-1} = r_s \cos \theta_s$, $v_{2s} = r_s \sin \theta_s$, and differentiating these gives

$$\dot{v}_{2s-1} = \dot{r}_s \cos \theta_s - r_s \sin \theta_s \dot{\theta}_s$$
$$\dot{v}_{2s} = \dot{r}_s \sin \theta_s + r_s \cos \theta_s \dot{\theta}_s$$

by the chain rule. Now substituting $r_s \cos \theta_s = v_{2s-1}$, and $r_s \sin \theta_s = v_{2s}$, gives

$$\dot{v}_{2s-1} = (v_{2s-1}/r_s)\dot{r}_s - v_{2s}\dot{\theta}_s$$
$$\dot{v}_{2s} = (v_{2s}/r_s)\dot{r}_s + v_{2s-1}\dot{\theta}_s$$

Entering the expressions of the normal form for $\dot{r}_s$ and $\dot{\theta}_s$ gives

$$\dot{v}_{2s-1} = (v_{2s-1}/r_s)\left(u_s r_s + r_s \sum_{j=1}^{N/2} a_{sj} r_j^2\right)$$

$$- v_{2s}\left(\omega_s - \sum_{j=1}^{N/2} b_{sj} r_j^2\right)$$

and, since $r_s^2 = v_{2s-1}^2 + v_{2s}^2$,

$$\dot{v}_{2s-1} = u_s v_{2s-1} - \omega_s v_{2s}$$
$$- \sum_{j=1}^{N/2} [v_{2s-1} a_{sj} - v_{2s} b_{sj}](v_{2j-1}^2 + v_{2j}^2)$$

Similarly,

$$\dot{v}_{2s} = u_s v_{2s} + \omega_s v_{2s-1}$$
$$- \sum_{j=1}^{N/2} [v_{2s} a_{sj} + v_{2s-1} b_{sj}](v_{2j-1}^2 + v_{2j}^2)$$

Setting the $b_{sj} = 0$ for simplicity, choosing $u_s = \alpha_s - \tau$ to get a standard network form, and reindexing $i, j = 1. 2, \ldots, N$, we get the Cartesian equivalent of the polar normal form equations.

$$\dot{v}_i = -\tau v_i + \sum_{j=1}^{N} J_{ij} v_j - v_i \sum_{j=1}^{N} A_{ij} v_j^2 \qquad (6)$$

Here $\mathbf{J}$ is a matrix containing 2×2 blocks along the diagonal of the local couplings of the linear terms of each pair of the previous equations v_{2s-1}, v_{2s}, with $-\tau$ separated out of the diagonal terms α, so that $\alpha_s = u_s + \tau$. The matrix $\mathbf{A}$ has 2×2 blocks of identical coefficients a_{sj} of the nonlinear terms from each pair.

$$\mathbf{J} = \begin{bmatrix} \alpha_1 & -\omega_1 & & & \\ \omega_1 & \alpha_1 & & & \\ & & \alpha_2 & -\omega_2 & \\ & & \omega_2 & \alpha_2 & \\ & & & & \ddots \end{bmatrix}$$

$$\mathbf{A} = \begin{bmatrix} a_{11} & a_{11} & a_{12} & a_{12} & \\ a_{11} & a_{11} & a_{12} & a_{12} & \\ a_{21} & a_{21} & a_{22} & a_{22} & \\ a_{21} & a_{21} & a_{22} & a_{22} & \\ & & & & \ddots \end{bmatrix}$$

Learning Transformation—Linear Term

The second transformation is linear. $\mathbf{J}$ can be viewed as the canonical (diagonalized) form of a real matrix with complex conjugate eigenvalues, where the conjugate pairs appear in blocks along the diagonal as shown ($\mathbf{J} - \tau \mathbf{I}$ is the Jacobian of this system). The Cartesian normal form equations describe the interaction of these linearly uncoupled complex *modes* due to the coupling of the nonlinear terms. We can interpret the normal form equations as network equations in eigenvector (=mode) or "memory" coordinates. In other words, we can think of them as having been produced from some network by a diagonalizing coordinate transformation $\mathbf{P}$, so that each axis pair v_{2s-1}, v_{2s} is a component of the complex eigenvector basis, and $\mathbf{J}$ displays the complex conjugate eigenvalues of some coupling matrix $\mathbf{T}$ in network coordinates. In this case, for the linear term, we know that

$$\mathbf{J} = \mathbf{P}^{-1}\mathbf{T}\mathbf{P}$$

Then it is clear that

$$\mathbf{T} = \mathbf{P}\mathbf{J}\mathbf{P}^{-1}$$

The matrix $\mathbf{P}$ is usually constructed of columns that are the eigenvectors calculated from the matrix $\mathbf{T}$ to be diagonalized. For the present purpose, we work backwards (Lenihan, 1987; Psaltis and Venkatesh, 1989), choosing eigenvectors for $\mathbf{P}$ and eigenvalues for $\mathbf{J}$, then constructing $\mathbf{T}$ by the formula above. We choose as columns the real and imaginary vectors $[\mathbf{x}^s \cos \theta^s, \mathbf{x}^s \sin \theta^s]$ of the cycles $\mathbf{x}^s e^{i\theta^s}$ of the set S to be learned. Any linearly independent set of complex eigenvectors in $\mathbf{P}$ can be chosen to transform the system and become the patterns "stored" in the vector field of network dynamics. This learning process might also be thought of as giving a vector representation to nodes in idealized coordinates by projecting out of that space onto the vectors of the standard basis— hence the name "projection algorithm." If we write the matrix expression for $\mathbf{T}$ above in component form, we recover the expression given in the theorem for T_{ij},

$$T_{ij} = \sum_{mn}^{N} P_{im} J_{mn} P_{nj}^{-1}$$

Nonlinear Term Projection

The nonlinear terms are transformed as well, but the expression cannot be easily written in matrix form. Using the component form of the transformation,

$$x_i = \sum_{j}^{N} P_{ij} v_j \qquad \dot{x}_i = \sum_{j}^{N} P_{ij} \dot{v}_j \qquad v_j = \sum_{k}^{N} P_{jk}^{-1} x_k$$

and substituting in the Cartesian normal form,

$$\dot{x}_i = \sum_j^N P_{ij}\dot{v}_j$$

$$= \sum_j^N P_{ij}\left[-\tau v_j + \sum_{k=1}^N J_{jk}v_k - v_j\sum_{k=1}^N A_{jk}v_k^2\right]$$

Now, using $v_j = \sum_k^N P_{jk}^{-1}x_k$, we get

$$\dot{x}_i = (-\tau + 1)\sum_j^N P_{ij}\left(\sum_k^N P_{jk}^{-1}x_k\right)$$

$$+ \sum_j^N P_{ij}\sum_k^N J_{jk}\left(\sum_l^N P_{kl}^{-1}x_l\right)$$

$$- \sum_j^N P_{ij}\left(\sum_k^N P_{jk}^{-1}x_k\right)$$

$$\times \sum_l^N A_{jl}\left(\sum_m^N P_{lm}^{-1}x_m\right)\left(\sum_n^N P_{ln}^{-1}x_n\right)$$

Rearranging the orders of summation gives

$$\dot{x}_i = (-\tau + 1)\sum_k^N\left(\sum_j^N P_{ij}P_{jk}^{-1}\right)x_k$$

$$+ \sum_l^N\left(\sum_k^N\sum_j^N P_{ij}J_{jk}P_{kl}^{-1}\right)x_l$$

$$- \sum_n^N\sum_m^N\sum_k^N\left(\sum_l^N\sum_j^N P_{ij}P_{jk}^{-1}A_{jl}P_{lm}^{-1}P_{ln}^{-1}\right)x_k x_m x_n$$

Finally, performing the bracketed summations and relabeling indices gives us the network of the theorem,

$$\dot{x}_i = -\tau x_i + \sum_j^N T_{ij}x_j - \sum_{jkl}^N T_{ijkl}x_j x_k x_l$$

with the expression for the tensor of the nonlinear term,

$$T_{ijkl} = \sum_{mn}^N P_{im}A_{mn}P_{mj}^{-1}P_{nk}^{-1}P_{nl}^{-1} \qquad \square$$

7.2.4. Numerical Example

Everything mathematically required to implement the network is given in the statement of the theorem. One can pick desired amplitude patterns, frequencies, $(\mathbf{x}^s, \boldsymbol{\theta}^s, \omega^s, \alpha^s)$, and normal form coefficients $0 < a_{ss}/a_{ks} < 1$. Then form matrices $\mathbf{P}$, $\mathbf{J}$, and $\mathbf{A}$. Numerically calculate $\mathbf{P}^{-1}$ and execute the specified projection operation (2) to find the weights T_{ij} and T_{ijkl} that store the desired patterns as periodic attractors. Inputs $b_i\delta(t)$ which establish nearby initial conditions will "retrieve" these patterns as the cycle of activity into which the network settles. The only restriction on the patterns to be stored is that the

columns of $\mathbf{P}$ must be linearly independent so that $\mathbf{P}$ may be inverted. This means, for example, that the relative phases θ_i of any pattern cannot all be identical.

This procedure may be done incrementally by replacing previously unused (randomly chosen) $\mathbf{P}$ matrix columns with new desired patterns, setting the corresponding $\mathbf{J}$ and $\mathbf{A}$ matrix entries (previously zero) to desired values, and reexecuting the projection rule (2). Any pre-existing set of stable patterns or sequences may also be modified by simply choosing new $\mathbf{P}$ matrix columns corresponding to those sequences to be modified and projecting again for the new weights.

We give here a simple numerical example of the algorithm. This is a four-dimensional system—four units or neural populations ($N = 4$)—that stores two oscillatory attractors of the same frequency $\omega_1 = \omega_2 = 250$ rad/sec $= 40$ Hz.

$$\mathbf{P} = \frac{1}{\sqrt{5}}\begin{bmatrix} 1 & 0 & 2 & 0 \\ 2 & 0 & -1 & 0 \\ 0 & 1 & 0 & 2 \\ 0 & 2 & 0 & -1 \end{bmatrix}$$

$$\mathbf{J} = \begin{bmatrix} 2 & -250 & 0 & 0 \\ 250 & 2 & 0 & 0 \\ 0 & 0 & 2 & -250 \\ 0 & 0 & 250 & 2 \end{bmatrix}$$

$$\mathbf{A} = \begin{bmatrix} 1 & 1 & 10 & 10 \\ 1 & 1 & 10 & 10 \\ 10 & 10 & 1 & 1 \\ 10 & 10 & 1 & 1 \end{bmatrix}$$

The patterns in the matrix $\mathbf{P}$ to be stored as attractors are chosen for numerical simplicity and to satisfy the constraints of the Hebb rule to be introduced in the next section and the biological model introduced much later. With these constraints, the projection rule coincides with the Hebb rule, and will generate a network of biological form. This illustrates how precisely networks may be programmed by the projection learning rule. The patterns are chosen to be orthonormal and each is divided into two sets of components (excitatory and inhibitory) of constant phase within the set. The first two columns contain the real and imaginary parts $[\mathbf{x}^s\cos\boldsymbol{\theta}^s, \mathbf{x}^s\sin\boldsymbol{\theta}^s]$ of the complex eigenvectors $|\mathbf{x}^s|e^{i\boldsymbol{\theta}^s}$ which determine the first pattern. We have chosen $\theta_1 = \theta_2 = 0$ degrees for the first two components, which represent excitatory neurons in the biological model. Since the cosine is 1 and the sine is 0, we see just the chosen amplitudes, $x_1 = 1$ and $x_2 = 2$, in the first column, with the second column zero. Now we choose $\theta_3 = \theta_4 = 90$ degrees phase difference from θ_1, θ_2, but with the sample amplitudes, to represent the activity of the inhibitory neurons in the biological model. Here we get the nonzero column switched because the cosine is now zero, and the sine is 1. For the second attractor pattern to be specified by the next two columns, we use the same phase structure, but choose a different amplitude pattern where the larger amplitude value occurs now in the first component $x_1 = 2$ and $x_2 = -1$. The minus sign makes these two patterns orthogonal now, because the sum of the product of the corresponding components (inner product) is zero. The absolute value (magnitude) of these numbers determines the amplitude of the oscillation, so the sign does not affect the amplitude specification, but reverses the phase of this component by $180°$. The norm of these column vectors is the same, $\sqrt{(1^2 + 2^2)} = \sqrt{5}$, so we can divide the whole matrix by this value to get normalized vectors.

The chosen frequencies, $\omega_1 = \omega_2 = 40\text{ Hz} = 40(2\pi) = 250$ rad/sec, are placed in the $\mathbf{J}$ matrix, along with real parts $\alpha_1 = \alpha_2 = 2$ of the complex conjugate eigenvalues. We will choose $\tau = 1$ in the network equations, so that $u = \alpha - \tau = 2 - 1 = 1$ in the normal form amplitude equations. Since we have chosen $a_{ss} = 1$, we will get oscillations with identical global amplitude multipliers $r_s = \sqrt{(u_s/a_{ss})} = \sqrt{(1/1)} = 1$, as shown earlier. We will show later that our convergence rates near these periodic attractors will be $e^{-2u_s t} = e^{-2t}$ along the s axis direction, and $e^{(u_i - a_{is}(u_s/a_{ss}))t} = e^{(1 - 10(1/1))t} = e^{-9t}$ in the direction of the other axes.

Since the $\mathbf{P}$ matrix is orthonormal, $\mathbf{P}^{-1} = \mathbf{P}^{T}$, and need not be calculated numerically. Now applying the matrix multiplications of the projection rule for attractors meeting these particular specifications to find the weights T_{ij} will yield the matrix $\mathbf{T}$ shown below:

$$\mathbf{P}^{-1} = = \mathbf{P}^{T} = \frac{1}{\sqrt{5}}\begin{bmatrix} 1 & 2 & 0 & 0 \\ 0 & 0 & 1 & 2 \\ 2 & -1 & 0 & 0 \\ 0 & 0 & 2 & -1 \end{bmatrix}$$

$$\mathbf{T} = \begin{bmatrix} 2 & 0 & -250 & 0 \\ 0 & 2 & 0 & -250 \\ 250 & 0 & 2 & 0 \\ 0 & 250 & 0 & 2 \end{bmatrix}$$

This is nearly the coupling structure (19) of the biological model discussed in Section 7.6, which consists of identical oscillators of excitatory and inhibitory neural populations coupled only by excitatory connections. The excitatory coupling is diagonal in this case because the eigenvalues α_s are equal, but full cross-coupling of oscillators is still given by the higher-order terms. Here also there are diagonal entries in the lower right 2×2 block because for numerical simplicity (orthogonality) we used exactly 90 degree excitatory/inhibitory phase difference in the patterns to be stored. As explained in Baird (1990b), we must use nonorthogonal real and imaginary columns ($\mathbf{x}^s \cos \boldsymbol{\theta}^s, \mathbf{x}^s \sin \boldsymbol{\theta}^s]$ differing slightly from this (for example, try $\theta_3 = \theta_4 = 85$ degrees, $\theta_1 = \theta_2 = 0$ degrees), and calculate $\mathbf{P}^{-1}$ numerically to get exactly the biological model. The biological model is thus generated as a special case of the projection learning rule for the particular kinds of attractors found in cortex. In the general case, the amplitude and phase patterns and frequencies of attractors can be chosen arbitrarily—up to the requirement of linear independence of the columns of the $\mathbf{P}$ matrix.

With these values chosen and calculated, we can execute the matrix multiplications specifying the projection learning rule (2) to find the weights T_{ijkl}. Simulate the network equations (1) with initial conditions close to each stored pattern on different runs to observe convergence to the two attractors with their different relative component amplitudes. Observe that oscillation amplitudes shrink and the attractors disappear in a multiple Hopf bifurcation as τ exceeds 2. Setting frequencies to zero in $\mathbf{J}$ and projecting again will give a network with two static attractors with the same basins of attraction as the oscillatory attractors.

Discussion

In addition to the linear terms T_{ij}, the network has a specific cubic nonlinearity programmed by the T_{ijkl} weights, instead of the usual sigmoid nonlinearity $T_{ij}g(x_j)$ intimately tied to the T_{ij} terms. Similar terms would appear if the symmetric sigmoid functions of such a network were Taylor-expanded up to cubic terms (quadratic terms are killed by the symmetry). These competitive (negative) cubic terms constitute a highly programmable nonlinearity that is independent of the linear terms. Normal form theory shows that these cubics are the essential nonlinear terms required to store oscillations, because of the (odd) phase shift symmetry required in the vector field.

They serve to create multiple periodic attractors by causing the oscillatory modes of the linear term to compete, much as the sigmoidal non-linearity does for static modes in a network with static attractors like the analog Hopfield network. The nonlinearity here is in the *coupling* instead of the activation functions of the network. Intuitively, these cubic terms may be thought of as sculpting the maxima of a "saturation" (energy) landscape into which the linear modes with positive eigenvalues expand, and positioning them to lie in the directions specified by the eigenvectors to make them stable. A Liapunov function for this landscape will be constructed later for the amplitude equations (3) of the polar coordinate form of this network.

The normal form equations for $\dot{r}_s$ and $\dot{\phi}_s$ determine how r_s and ϕ_a for pattern s evolve in time in interaction with all the other patterns of the set S. This could be thought of as the process of phase locking of the pattern that finally emerges. The unusual power of this algorithm lies in the ability to precisely specify these nonlinear interactions.

The terms b_{ij} in the phase equations ϕ of the general Hopf normal form (4) represent a nonlinear dependence of the frequency of a mode on its amplitude r^2 and those of others of nonzero amplitude during transient relaxation to a single mode. They can be used to model oscillations with "hard" or "soft" elasticity (springs) where the frequency increases or decreases, respectively, with greater amplitude of oscillation. For the sake of simplicity in this chapter we set all b_{ij} to zero.

There is a multiple Hopf bifurcation of codimension $N/2$ at $\tau = \alpha = 2$, where the real parts of all the complex eigenvalues of the Jacobian are zero. All of the stored attractors will arise from a single equilibrium at the origin as τ decreases further. Since there are no approximations here, however, the theorem is not restricted to the neighborhood of this bifurcation, and can be discussed without further reference to bifurcation theory.

The general difficulty of ensuring stability of limit sets which are trajectories is due to the difficulty of finding a simple "measure" to characterize them. When we start with this special polar coordinate description of periodic orbits, with amplitude equations (3) independent of the phase equations (4), then the amplitude equations give as a "handle" on these trajectories. Stable fixed points of the amplitude equations correspond to stable limit cycles in the network vector field. The usual tools of analysis and synthesis of nonlinear systems with fixed points, such as Liapunov functions, can be applied to

the amplitude equations to control the under-lying oscillations. Although it will never be apparent in network coordinates, because no restriction is imposed on the patterns that can be stored, these systems may always be decomposed in normal form coordinates into the Cartesian product of a space of amplitude variables independent of a space of phase variables.

Later we prove that the attractors on the amplitude axes are the *only* stable fixed points in this system—there are no "spurious" attractors. We show that further restrictions on the **A** matrix produce a very well-behaved vector field with identical scale- and rotation-invariant basins that are bounded by hyperplanes. When $a_{ii} = c$, and $a_{ij} = d$ for all i, j, the amplitude component equations are identical, and the normal form vector field is symmetric under permutation of the axes (the components). For the two-component vector field in the plane shown earlier in Fig. 7.2, it is obvious that this symmetry forces the diagonal separatrices through the interior saddle points to be straight lines, since only a line evenly dividing the quadrant is consistent with this symmetry under reflection of the axes. These basins are identical two-dimensional "wedges" of infinite extent, and are thus naturally scale invariant. They define "categories" or classes of directions of vectors. The hyperplane boundaries are linearly transformed by the learning algorithm, and so remain hyperplanes in the network vector field. Whenever there are symmetries chosen for the normal form equations, then the network equations inherit these symmetries as well, and the network attractors are interrelated by a group of trans-formations. These are given from the learning process even though there are no restrictions on the patterns being learned.

7.2.5. Learning Rule Extensions

This is the core of the mathematical story, and there are many ways in which it may be extended. For biological modeling, where the patterns to be stored may be assumed to be orthogonal as discussed previously, the linear projection operation described above becomes a generalized periodic outer product rule that reduces to the usual Hebb-like rule for the storage of fixed points. There are mechanisms in neural layers that can normalize inputs (Freeman, 1975), and we assume these here. When the columns of **P** are orthonormal, it follows that $\mathbf{P}^{-1} = \mathbf{P}^{T}$, and the formula above for the linear network coupling becomes $\mathbf{T} = \mathbf{PJP}^{T}$. To see how an outer product rule arises, suppose first that we wanted to store static attractors. We fill the columns of the $\mathbf{P}$ matrix with real orthonormal eigenvectors $\mathbf{x}^s$ to be learned, and for each one choose real eigenvalues α_s to be placed in the corresponding column on the diagonal of the matrix $\mathbf{J}$. ($\mathbf{T}$ is the identity of course if the eigenvalues are all identical, but full cross-couplings arise for the case of complex eigenvalues.) Noting that $J_{mn} = \alpha_n \delta_{mn}$, and $P_{nj}^{\mathrm{T}} = P_{jn}$,

$$T_{ij} = \sum_{m=1}^{N}\sum_{n=1}^{N} P_{im} J_{mn} P_{nj}^{\mathrm{T}} = \sum_{n=1}^{N}\sum_{m=1}^{N} \alpha_n \delta_{mn} P_{im} P_{jn}$$

$$= \sum_{n=1}^{N} \alpha_n P_{in} P_{jn} = \sum_{s=1}^{N} \alpha^s x_i^s x_j^s$$

Here δ_{ij} is the Kronecker delta: $\delta_{ij} = 1$, for $i = j$ and zero otherwise. Writing the matrix multiplication in component form, we get

$$T_{ij} = \sum_{n=1}^{N} \alpha_n P_{in} P_{jn}$$

This is a summation over the column entries of $\mathbf{P}$, which we know by our construction to be the patterns we want to learn. Thus the $\mathbf{T}$ matrix can be written in the familiar form of the Hebb rule $T_{ij} = \sum_s T_{ij}^s = \sum_s \alpha^s x_i^s x_j^s$, where the eigenvalue α^s is revealed to be the "learning rate" or relative frequency of pattern s in the summation of patterns.

In a similar fashion to the derivation above, the learning rule for the higher-order nonlinear terms for static attractors becomes a multiple outer product rule when the matrix $\tilde{\mathbf{A}}$ of the normal form amplitude equations is chosen to have a simple form that will be shown later to provide a vector field whose only attractors are the desired ones. Here we distinguish $\tilde{\mathbf{A}}$ from the previously displayed matrix $\mathbf{A}$ for the Cartesian normal form (6), which was shown in the proof of the projection theorem to be formed by doubling the entries of the matrix $\tilde{\mathbf{A}}$ of coupling coefficients a_{ij} in the normal form amplitude equations (3). For static attractors, the amplitude equations alone are the normal form, since there are no phase equations. The $\tilde{\mathbf{A}}$ matrix is chosen to have uniform entries $a_{ij} = c$ for all its off-diagonal entries, and uniform entries $a_{ij} = c - d$ for the diagonal entries. Thus $\tilde{\mathbf{A}} = x\mathbf{R} - d\mathbf{I}$, where $\mathbf{R}$ is a matrix of uniform entries $R_{ij} = 1$ for all i, j, and $\mathbf{I}$ is the identity matrix.

Choice of $c > d > 0$ then satisfies the condition for stable-axes fixed points derived earlier. Now the expression for the higher-order terms, when $\mathbf{P}^{-1} = \mathbf{P}^{\mathrm{T}}$ becomes

$$T_{ijkl} = \sum_{m=1}^{N}\sum_{n=1}^{N} P_{im} \tilde{A}_{mn} P_{mj}^{\mathrm{T}} P_{nk}^{\mathrm{T}} P_{nl}^{\mathrm{T}}$$

$$= \sum_{m=1}^{N}\sum_{n=1}^{N} P_{im} [cR_{mn} - d\delta_{mn}] P_{mj}^{\mathrm{T}} P_{nk}^{\mathrm{T}} P_{nl}^{\mathrm{T}}$$

$$= c \sum_{m=1}^{N} P_{im} P_{mj}^{\mathrm{T}} \left[\sum_{n=1}^{N} P_{nk}^{\mathrm{T}} P_{nl}^{\mathrm{T}} \right]$$

$$- d \sum_{m=1}^{N}\sum_{n=1}^{N} P_{im} \delta_{mn} P_{mj}^{\mathrm{T}} P_{nk}^{\mathrm{T}} P_{nl}^{\mathrm{T}}$$

(using $P_{nj}^{\mathrm{T}} = P_{jn}$, and substituting patterns)

$$T_{ijkl} = c\delta_{ij}\delta_{kl} - d \sum_{n=1}^{N} P_{in} P_{jn} P_{kn} P_{ln}$$

$$= c\delta_{ij}\delta_{kl} - d \sum_{s=1}^{N} x_i^s x_j^s x_k^s x_l^s$$

For the case of complex eigenvectors and eigenvalues to be learned to store oscillating patterns, the derivation is basically the same, but is notationally complicated by the fact that the patterns to be learned come in pairs—each eigenvector is a real and imaginary pair of columns in the $\mathbf{P}$ matrix, and each corresponding eigenvalue in the matrix $\mathbf{J}$ is a complex conjugate pair in a diagonal block, as shown previously.

$$T_{ij} = \sum_{m=1}^{N}\sum_{n=1}^{N} P_{im} J_{mn} P_{nj}^{\mathrm{T}}$$

$$= \sum_{n=1}^{N} \left[\sum_{m=1}^{N} P_{im} J_{mn} \right] P_{nj}^{\mathrm{T}} = \sum_{k=1}^{N} \tilde{P}_{ik} P_{kj}^{\mathrm{T}}$$

$$= \sum_{s=1}^{N/2} [\tilde{P}_{i,2s-1} P_{2s-1,j}^{\mathrm{T}} + \tilde{P}_{i,2s} P_{2s,j}^{\mathrm{T}}]$$

Now the summation of patterns s is over pairs of columns and rows. The entries from the matrix $\tilde{\mathbf{P}}$ are

$$\tilde{P}_{i,2s-1} = (\alpha^s x_i^s \cos\theta_i^s + \omega^s x_i^s \sin\theta_i^s)$$

$$\tilde{P}_{i,2s} = (-\omega^s x_i^s \cos\theta_i^s + \alpha^s x_i^s \sin\theta_i^s)$$

after the matrix multiplication by the complex eigenvalue block. Those from $\mathbf{P}^{\mathrm{T}}$ are $P_{2s-1,j}^{\mathrm{T}} = x_j^s \cos\theta_j^s$, and $P_{2s,j}^{\mathrm{T}} = x_j^s \sin\theta_j^s$, the jth real and imaginary components of pattern s. Thus

$$T_{ij} = \sum_{s=1}^{N/2} T_{ij}^s$$

$$= \sum_{s=1}^{N/2} (\alpha^s x_i^s \cos\theta_i^s + \omega^s x_i^s \sin\theta_i^s) x_j^s \cos\theta_j^s$$

$$+ (-\omega^s x_i^s \cos\theta_i^s + \alpha^s x_i^s \sin\theta_i^s) x_j^s \sin\theta_j^s$$

$$= \sum_{s=1}^{N/2} x_i^s x_j^s [(\alpha^s \cos\theta_i^s + \omega^s \sin\theta_i^s) \cos\theta_j^s$$

$$+ (-\omega^s \cos\theta_i^s + \alpha^s \sin\theta_i^s) \sin\theta_j^s]$$

Then,

$$T_{ij} = \sum_{s=1}^{N/2} x_i^s x_j^s [\alpha^s \cos(\theta_i^s - \theta_j^s) + \omega^s \sin(\theta_i^s - \theta_j^s)]$$

(7)

Thus for orthogonal cycles or fixed points, the projection operation can be expressed as a local, incremental learning procedure (Baird, 1989). The network can be truly self-organizing, because a synapse of the net can modify itself according to the pre- and postsynaptic activities it experiences during external driving by an input pattern to be learned.

The learning rule above may be thought of as a "periodic" outer product rule, or a "phase-dependent" Hebb rule, and there is evidence for it in the hippocampus. Between units of equal phase, or when $\theta_i^s = \theta_j^s = 0$ for a nonoscillating pattern, then this reduces to the usual Hebb rule. For $\theta_i^s - \theta_j^s$ from 0 to 90 degrees, the rule implies a positive weight change, as is found in long-term potentiation (LTP) in the hippocampus (Stanton and Sejnowski, 1989), with the frequency-dependent term $\omega^s \sin(\theta_i^s - \theta_j^s)$ growing to a maximum at 90, while the cosine declines to zero. For $(\theta_i^s - \theta_j^s) > 90$ degrees, however, the synaptic change can be negative, depending on ω^s, since the cosine term is negative. It must be maximally negative for phase differences of 180 to 270 degrees between pre- and postsynaptic depolarizations, since both sinusoidal terms are negative there. This kind of long-term depression, or anti-Hebbian learning, was found in hippocampus when the phase differences between applied presynaptic stimulation and postsynaptic depolarizations was in this range—i.e., when stimulation is paired with postsynaptic hyperpolarization (Stanton and Sejnowski, 1989).

Again, the oscillation learning rule for the higher-order nonlinear terms becomes a multiple outer product rule when the matrix $\mathbf{A}$, shown in the projection theorem, is chosen to have a simple form. Recall again that the $\mathbf{A}$ matrix for the Cartesian normal form (6) was shown in the proof to be formed by doubling the entries a_{ij} of the matrix $\tilde{\mathbf{A}}$ of coupling coefficients in the normal form amplitude equations (3). It is chosen here to have uniform entries $A_{ij} = c$ for all its off-diagonal 2×2 blocks, and uniform entries $A_{ij} = c - d$ for the diagonal blocks. Thus $\mathbf{A} = c\mathbf{R} - d\mathbf{S}$, where $\mathbf{R}$ is a matrix of uniform entries $R_{ij} = 1$ for all i, j, and $\mathbf{S}$ is a matrix of nonzero 2×2 diagonal blocks of uniform entries $S_{ij} = 1$.

$$\mathbf{R} = \begin{bmatrix} 1 & 1 & 1 & \\ 1 & 1 & 1 & \\ 1 & 1 & 1 & \\ & & & \ddots \end{bmatrix} \quad \mathbf{S} = \begin{bmatrix} 1 & 1 & & \\ 1 & 1 & & \\ & & & \ddots \end{bmatrix}$$

Choice of $c > d > 0$ then satisfies the condition for stable-axes fixed points derived earlier.

Now the expression for the higher-order terms, when $\mathbf{P}^{-1} = \mathbf{P}^{\mathrm{T}}$, becomes (where δ_{ij} is the Kronecker delta)

$$T_{ijkl} = \sum_{m=1}^{N} \sum_{n=1}^{N} P_{im} A_{mn} P_{mj}^{\mathrm{T}} P_{nk}^{\mathrm{T}} P_{nl}^{\mathrm{T}}$$

$$= \sum_{m=1}^{N} \sum_{n=1}^{N} P_{im} [cR_{mn} - dS_{mn}] P_{mj}^{\mathrm{T}} P_{nk}^{\mathrm{T}} P_{nl}^{\mathrm{T}}$$

$$= c \sum_{m=1}^{N} P_{im} P_{mj}^{\mathrm{T}} \left[\sum_{n=1}^{N} P_{nk}^{\mathrm{T}} P_{nl}^{\mathrm{T}} \right]$$

$$- d \sum_{m=1}^{N} \sum_{n=1}^{N} P_{im} S_{mn} P_{mj}^{\mathrm{T}} P_{nk}^{\mathrm{T}} P_{nl}^{\mathrm{T}}$$

(using $P_{nk}^{\mathrm{T}} = P_{kn}$)

$$= c\delta_{ij}\delta_{kl} - d \sum_{s=1}^{N/2}$$

$$\times [P_{i,2s-1} S_{2s-1,2s-1} P_{2s-1,j}^{\mathrm{T}} P_{2s-1,k}^{\mathrm{T}} P_{2s-1,l}^{\mathrm{T}}$$

$$+ P_{i,2s} S_{2s,2s-1} P_{2s,j}^{\mathrm{T}} P_{2s-1,k}^{\mathrm{T}} P_{2s-1,l}^{\mathrm{T}}$$

$$+ P_{i,2s-1} S_{2s-1,2s} P_{2s-1,j}^{\mathrm{T}} P_{2s,k}^{\mathrm{T}} P_{2s,l}^{\mathrm{T}}$$

$$+ P_{i,2s} S_{2s,2s} P_{2s,j}^{\mathrm{T}} P_{2s,k}^{\mathrm{T}} P_{2s,l}^{\mathrm{T}}]$$

(using $S_{ij} = 1$)

$$= c\delta_{ij}\delta_{kl} - d \sum_{s=1}^{N/2}$$

$$\times [P_{i,2s-1} P_{2s-1,j}^{\mathrm{T}} P_{2s-1,k}^{\mathrm{T}} P_{2s-1,l}^{\mathrm{T}}$$

$$+ P_{i,2s} P_{2s,j}^{\mathrm{T}} P_{2s-1,k}^{\mathrm{T}} P_{2s-1,l}^{\mathrm{T}}$$

$$+ P_{i,2s-1} P_{2s-1,j}^{\mathrm{T}} P_{2s,k}^{\mathrm{T}} P_{2s,l}^{\mathrm{T}}$$

$$+ P_{i,2s} P_{2s,j}^{\mathrm{T}} P_{2s,k}^{\mathrm{T}} P_{2s,l}^{\mathrm{T}}]$$

(using $P_{ij}^{\mathrm{T}} = P_{ji}$)

$$= c\delta_{ij}\delta_{kl} - d \sum_{s=1}^{N/2}$$

$$\times [P_{i,2s-1} P_{j,2s-1} P_{k,2s-1} P_{l,2s-1}$$

$$+ P_{i,2s} P_{j,2s} P_{k,2s-1} P_{l,2s-1}$$

$$+ P_{i,2s-1} P_{j,2s-1} P_{k,2s} P_{l,2s}$$

$$+ P_{i,2s} P_{j,2s} P_{k,2s} P_{l,2s}]$$

(substitute patterns)

$$T_{ijkl} = c\delta_{ij}\delta_{kl} - d \sum_{s=1}^{N/2} x_i^s x_j^s x_k^s x_l^s$$

$$\times [\cos \theta_i^s \cos \theta_j^s \cos \theta_k^s \cos \theta_l^s$$
$$+ \sin \theta_i^s \sin \theta_j^s \cos \theta_k^s \cos \theta_l^s$$
$$+ \cos \theta_i^s \cos \theta_j^s \sin \theta_k^s \sin \theta_l^s$$
$$+ \sin \theta_i^s \sin \theta_j^s \sin \theta_k^s \sin\theta_l^s]$$

This reduces to the multiple outer product derived above,

$$T_{ijkl} = c\delta_{ij}\delta_{kl} - d \sum_{s=1}^{N/2} x_i^s x_j^s x_k^s x_l^s \qquad (8)$$

for a nonoscillating system, when the phase θ is zero for all patterns. As we have seen, $c > d$ guarantees stability of all stored patterns (Baird, 1990c).

Because the numerical example given earlier used orthonormal attractor patterns, the Hebbian rules for those patterns can be used here to find weights that store them directly in this network, without any matrix inversions or multiplications. To implement that example with this rule, $c = 10$, $d = 9$, $\tau = 1$, $\alpha = 2$, and the pattern phases are 0 and 90 degrees, with amplitudes and frequencies as described.

7.2.6. Other Normal Forms

Given the projection algorithm for constructing networks with vector fields that are isomorphic to (given by a linear transformation of) the vector field of a normal form or any system ordinary differential equations with uncoupled linear terms and polynomial nonlinearities, a great many possibilities arise. One application of the projection algorithm is to construct networks for the exploration of the computational capabilities of more exotic vector fields involving, for example, quasiperiodic or chaotic flows. A neural network can probably be synthesized to realize any nonlinear behavior that is known in the catalog of well-analyzed normal forms. Later we give an example of a network that stores multiple Lorenz attractors, and we have constructed other networks for Roessler attractors, and the "double scroll" attractor of Chua (Chua et al., 1986).

The competitive Lokta–Volterra equations are an example of a different normal form that could be used for the projection algorithm for the storage of static patterns. These equations correspond to the normal form equations for a multiple transcritical bifurcation, and are of the form of our amplitude equations, but with

quadratic instead of cubic terms:

$$\dot{r}_s = u_s r_s - r_s \sum_{j=1}^{N} a_{sj} r_j$$

Projection to store desired patterns then gives a network architecture with only third-order correlations that does not have the additional attractors ($-x$ is an attractor if x is) given by odd symmetry of the Hopf normal form, when it is used to store static attractors.

$$\dot{x}_i = -\tau x_i + \sum_{j=1}^{N} T_{ij}x_j - \sum_{jk=1}^{N} T_{ijk}x_j x_k$$

where

$$T_{ijk} = \sum_{mn=1}^{N} P_{im} A_{mn} P_{mj}^{-1} P_{nk}^{-1}$$

7.2.7. Networks with Sigmoids

The normal form theorem (Guckenheimer and Holmes, 1983) asserts that there is a large class of ordinary, partial, and integrodifferential equations which give local vector fields that are topologically equivalent to that of any normal form under consideration. This implies that there are many other possible network architectures to be discovered that can realize the performance of that produced by the projection algorithm. In fact, all of the above results hold as well for networks with sigmoids, provided their coupling is such that they have a Taylor's expansion which is equal to the above networks up to third order (Baird, 1989). For example, the network

$$\dot{x}_i = -\tau x_i + \sum_{j=1}^{N} T_{ij}g(x_j)$$

$$- \sum_{jkl=1}^{N} T_{ijkl}g(x_j)g(x_k)g(x_l) \qquad (9)$$

is a valid projection of the Hopf normal form if

$$T_{ijkl} = \sum_{mn=1}^{N} P_{im} A_{mn}^* P_{mj}^{-1} P_{nk}^{-1} P_{nl}^{-1}$$

where

$$A_{mn}^* = A_{mn} - (g'''(0)J_{mn}/6g'(0)^3)$$

and $g(x)$ has odd symmetry $g(-x) = -g(x)$. The odd symmetry assures that there are no even-order terms in the expansion about the origin (i.e., the inflexion point, where $g''(0) = g''''(0) = 0$), and the correction A_{mn}^* to A_{mn} removes, in normal form coordinates, the cubic terms that arise from the expansion of the sigmoids of the $\sum_{j=1}^{N} T_{ij}g(x_j)$ term in the network equations. The results now hold for bifurcation parameter values in the neighborhood of the bifurcation point for which the truncated expansion is accurate. The expected performance of this system has been verified in simulations, and in practice the valid region is large.

7.3. PROJECTION NETWORK

The *projection network* is an alternative network for implementation of the normal form projection algorithm (Baird and Eeckman, 1991). The autoassociative case of this network is formally equivalent to the higher-order network realization (2) shown above, which is used later as a biological model. It has $3N^2$ weights instead of $N^2 + N^4$, and is more useful for engineering applications and for simulations of the biological model. All the mathematical results proved for the projection algorithm in the network above carry over to this new architecture, but more general versions can be trained and applied in novel ways.

In the projection net for autoassociation, the algebraic projection operation into and out of memory coordinates is done explicitly by a set of weights in two feed-forward linear networks characterized by weight matrices $\mathbf{P}^{-1}$ and $\mathbf{P}$. These two maps map inputs into and out of the nodes of the recurrent dynamical network in memory coordinates which is sandwiched between the two maps. This kind of network, with explicit input and output projection maps that are inverses, may be considered an "unfolded" version of the purely recurrent networks described above.

This network is shown in Fig. 7.3. Input pattern vectors $\mathbf{x}'$ are applied as pulses which project onto each vector of weights (row of the $\mathbf{P}^{-1}$ matrix) on the input to each unit i of the dynamic network to establish an activation level v_i which determines the initial condition for the relaxation dynamics of this network. The recurrent weight matrix $\mathbf{A}$ of the dynamic network can be chosen so that the unit or predefined subspace of units which receives the largest projection of the input will converge to some state of activity, static or dynamic, while all other units are suppressed to zero activity.

The evolution of the activity in these memory coordinates appears in the original network coordinates at the output terminals as a spatio-temporal pattern which may be fully distributed across all nodes. Here the state vector of the dynamic network has been transformed by the $\mathbf{P}$ matrix back into the coordinates in which the input was first applied. At the attractor $\mathbf{v}^*$ in

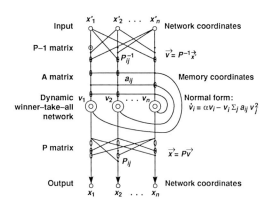

Fig. 7.3. Projection network—$3N^2$ weights. The $\mathbf{A}$ matrix determines a k-winner-take-all-net—programs attractors, basins of attraction, and rates of convergence. The columns of $\mathbf{P}$ contain the output patterns associated to these attractors. The rows of $\mathbf{P}^{-1}$ determine category centroids.

memory coordinates, only a linear combination of the columns of the $\mathbf{P}$ weight matrix multiplied by the winning nonzero modes of the dynamic net constitute the network representation of the output of the system. Thus the attractor retrieved in memory coordinates reconstructs its learned distributed representation $\mathbf{x}^*$ through the corresponding columns of the output matrix $\mathbf{P}$, e.g. $\mathbf{P}^{-1}\mathbf{x}' = \mathbf{v}, \mathbf{v} \to \mathbf{v}^*, \mathbf{P}\mathbf{v}^* = \mathbf{x}^*$.

For the special case of content-addressable memory or autoassociation, which we have been describing here, the actual patterns to be learned form the columns of the output weight matrix $\mathbf{P}$, and the input matrix is its inverse $\mathbf{P}^{-1}$. These are the networks that can be "folded" into higher-order recurrent networks, as shown above. For orthonormal patterns, the inverse is the transpose of this output matrix of memories, $\mathbf{P}^{-1} = \mathbf{P}^{\mathrm{T}}$, and no computation of $\mathbf{P}^{-1}$ is required to store or change memories—just plug the desired patterns into appropriate rows and columns of $\mathbf{P}$ and $\mathbf{P}^{\mathrm{T}}$.

In the autoassociative network, the input space, output space, and normal form state space are each of dimension N. The input and output linear maps require N^2 weights each, while the normal form coefficients determine another N^2 weights. Thus the net needs only $3N^2$ weights, instead of the $N^2 + N^4$ weights required by the folded recurrent network (2). The $2N^2$ input and output weights could be stored off-chip in a conventional memory, and the dynamic form network, with its fixed weights, could conceivably be implemented in analog VLSI (Cruz and Chua, 1991) for fast relaxation.

7.3.1. Learning Extensions for the Projection Network

More generally, for a heteroassociative net (i.e., a net designed to perform a map from input space to possibly different output space) the linear input and output maps need not be inverses, and may be noninvertible. They may be found by any linear map learning technique such as Widrow–Hoff or by finding pseudoinverses (see Chapter 1 for a detailed discussion of optimal linear associative mappings).

Learning of all desired memories may be instantaneous, when they are known in advance, or may evolve by many possible incremental methods, supervised or unsupervised. The standard competitive learning algorithm, where the input weight vector attached to the winning memory node is moved toward the input pattern, can be employed. We can also decrease the tendency to choose the most frequently selected node by adjusting parameters in the normal form equations, to realize the more effective frequency-selective competitive learning algorithm (Ahalt et al., 1990). Supervised algorithms like bootstrap Widrow–Hoff may be implemented as well, where a desired output category is known. The weight vector of the winning normal form node is updated by the competitive rule if it is the right category for that input, but moved away from the input vector if it is not the desired category, and the weight vector of the desired node is moved toward the input.

Thus the input map can be optimized for clustering and classification by these algorithms, as the weight vectors (row vectors of the input matrix) approach the centroids of the clusters in the input environment. The output weight matrix may then be constructed with any desired output pattern vectors in appropriate columns to place the attractors corresponding to these categories anywhere in the state space in network coordinates that is required to achieve a desired heteroassociation.

If either the input or the output matrix is learned, and the other chosen to be its inverse, then these competitive nets can be folded into oscillating biological versions, to see what the competitive learning algorithms correspond to there. Now either the rows of the input matrix may be optimized for recognition, or the columns of the output matrix may be chosen to place attractors, but not both. We are searching for a learning rule in the biological network which we can prove will accomplish competitive learning, using the unfolded form of the network. Thus the work on engineering applications feeds

back on the understanding of the biological system.

7.3.2. Programming the Normal Form Network

The power of the projection algorithm to program these systems lies in the freedom to chose a well-understood normal form for the dynamics, independent of the patterns to be learned. Here we give an intuitive discussion of vector field programming of the normal form to preview the formal proof that follows in Section 7.4.5. The Hopf normal form, shown here again in Cartesian coordinates,

$$\dot{v}_i = -\tau v_j + \sum_{j=1}^{N} J_{ij} v_j - v_i \sum_{j=1}^{N} A_{ij} v_j^2$$

is especially easy to work with for programming periodic attractors, but handles fixed points as well, when the frequencies ω in $\mathbf{J}$ are set to zero. $\mathbf{J}$ is a matrix with real eigenvalues for determining static attractors, or complex conjugate eigenvalue pairs in blocks along the diagonal for periodic attractors.

Since the A_{ij} and the v_j^2 are always positive, and the summation $\sum_{j=1}^{N} A_{ij} v_j^2$ is multiplied by $-v_i$, the cubic terms are always "competitive"— that is, of a sign that is opposite to the present sign of v_i. The real parts of the eigenvalues of the Jacobian $\mathbf{J} - \tau\mathbf{I}$ are set to be positive, and cause initial states to move away from the origin until the competitive cubic terms dominate at some distance, and cause the flow to be inward from all points beyond. The off-diagonal cubic terms cause competition between directions of flow within a spherical middle region and thus create multiple attractors and basins. The larger the eigenvalues in $\mathbf{J}$ and off-diagonal weights in $\mathbf{A}$, the faster the convergence to attractors in this region.

It is easy to choose blocks of coupling along the diagonal of the $\mathbf{A}$ matrix to produce different kinds of attractors, static, periodic, or chaotic, in different coordinate subspaces of the network. The sizes of the subspaces can be programmed by the sizes of the blocks. The basin of attraction of an attractor determined within a subspace is guaranteed to contain the subspace (Baird, 1990c). Thus basins can be programmed, and "spurious" attractors can be ruled out when all subspaces have been included in a programmed block.

This can be accomplished simply by choosing the $\mathbf{A}$ matrix entries outside the blocks on the diagonal (which determine coupling of variables within a subspace) to be greater (causing more competition) than those within the blocks. The principle is that this makes the subspaces defined by the blocks compete exhaustively, since inter-subspace competition is greater than subspace self-damping. Within the middle region, the flow is forced to converge laterally to enter the subspaces programmed by the blocks.

A simple example is a matrix of the form,

$$\mathbf{A} = \begin{bmatrix} d & & & & & \\ & d & & & (g) & \\ & & \begin{bmatrix} d & c \\ c & d \end{bmatrix} & & & \\ & & & \begin{bmatrix} d & d & c & c \\ d & d & c & c \\ c & c & d & d \\ c & c & d & d \end{bmatrix} & \\ & (g) & & & & \ddots \end{bmatrix}$$

where $0 < c < d < g$. There is a static attractor on each axis (in each one-dimensional subspace) corresponding to the first two entries on the diagonal, by the argument above. In the first two-dimensional subspace block there is a single fixed point in the interior of the subspace on the main diagonal, because the off-diagonal entries within the block are symmetric and less negative than those on the diagonal. The components do not compete, but rather combine. Nevertheless, the flow from outside is into the subspace, because the entries outside the subspace are more negative than those within it.

The last subspace contains entries appropriate to guarantee the stability of a periodic attractor with two frequencies (Fourier components) chosen in the **J** matrix. The doubling of the entries is because these components come in complex conjugate pairs (in the **J** matrix blocks) which get identical **A** matrix coupling. Again, these pairs are combined by the lesser off-diagonal coupling within the block to form a single limit-cycle attractor. A large subspace can store a complicated continuous periodic spatiotemporal sequence with many component frequencies.

7.4. VECTOR FIELD PROGRAMMING

The key to the power of the projection algorithm is the ability to specify the global vector field by projecting the nonlinear terms of the normal form. The resulting network is a superposition of a purely linear part and a highly specified cubic nonlinearity. The network vector field may thus

be programmed independently of the patterns assigned to equilibria by the linear term. The linear terms of the networks constructed above program the basis eigenvectors of the normal form, as we have seen above, and thereby determine the location of all limit sets (equilibria, cycles, sequences) in the network state space. The a_{ij} of the nonlinear terms then program the ambient flow about these limit sets—stability, basin geometry, rates of convergence. In this section we will see analytically how restrictions on the matrix **A** give first a strict Liapunov function for the amplitude equations, and then a very well-behaved vector field with identical scale- and rotation-invariant basins that are bounded by hyperplanes.

7.4.1. Liapunov Function

Looking again at the Jacobian matrix of the amplitude equations at any fixed point $\hat{r}$, with entries

$$U_{ij} = -2a_{ij}\hat{r}_i\hat{r}_j \qquad U_{ii} = u_i - 3a_{ii}\hat{r}_i^2 - \sum_{j \neq i}^{N/2} a_{ij}\hat{r}_j^2$$

we see that it is clearly symmetric $U_{ij} = U_{ji}$ for symmetric coupling, $a_{ij} = a_{ji}$. From this symmetry condition, it follows directly that the vector field is "irrotational" (curl = 0), and it is therefore the gradient of some potential function $V(\mathbf{r})$ (Hirsch and Smale, 1974).

If we write the amplitude equations (including an input term s_i),

$$\dot{r}_i = u_i r_i - r_i \sum_{j=i}^{N/2} a_{ij} r_j^2 + s_i$$

then taking anti-derivatives of the negative of the right-hand terms yields the ith term of a scalar potential $V(\mathbf{r})$,

$$V_i(\mathbf{r}) = -\tfrac{1}{2}u_i r_i^2 + \tfrac{1}{2}r_i^2 \sum_{j=i}^{N/2} a_{ij} r_j^2 - r_i s_i$$

so

$$V(\mathbf{r}) = -\frac{1}{2}\sum_{i=1}^{N/2} u_i r_i^2 + \frac{1}{2}\sum_{i=1}^{N/2}\sum_{j=i}^{N/2} a_{ij} r_i^2 r_j^2 - \sum_{i=1}^{N/2} r_i s_i$$

The reverse of this calculation clearly shows that $\dot{\mathbf{r}} = -\nabla V(\mathbf{r})$. Now $\dot{V}(\mathbf{r}) = DV(\mathbf{r})\dot{\mathbf{r}} = -\langle \dot{\mathbf{r}}, \dot{\mathbf{r}} \rangle$, by the chain rule, showing that $V(\mathbf{r})$ is a strict Liapunov function for the amplitude equations. However, $\dot{\mathbf{r}} = -\nabla V(\mathbf{r})$ is a much stronger condition than that given by a strict Liapunov function, and such a vector field is called a *gradient vector field*. The potential $V(\mathbf{r})$ is mathematically dual to the vector field and provides an alternative description containing the same information. All trajectories are orthogonal to the level surfaces of the potential function $V(\mathbf{r})$, and must converge to fixed points. The Jacobian matrix at every isolated equilibrium is symmetric with real eigenvalues, and spirals, limit cycles, or chaos in the flow are strictly ruled out (Hirsch and Smale, 1974). This guaranteed convergence to fixed points of the amplitude equations, of course, translates to convergence to periodic attractors for the network equations.

If static attractors are being stored, then there are no phase equations, and the amplitude equations alone are the normal form for the network equations. Static attractors can be mixed in with periodic ones simply by setting $\omega^s = 0$ for some s. The Cartesian normal form then contains two static modes in place of the usual complex mode. They are made stable by choosing the self-coupling within the diagonal $\mathbf{A}$ matrix block less than the coupling between the modes. Real eigenvalues may be chosen for these in the $\mathbf{J}$ matrix, and two corresponding desired static patterns placed in the projection matrix $\mathbf{P}$.

7.4.2. Rates of Convergence

The parameters u_i in the amplitude equations are growth rates for the ith axis or mode r_i, since, in the absence of nonlinear terms, $r_i(t) = r_i(0)\, e^{u_i t}$. Choice of large growth rates and strong nonlinear competition $a_{ij} \gg a_{ii}$ in the amplitude equations has been shown in simulations to produce fast convergence of an initial input state to the attractor in that basin. This is consistent with an intuitive picture of nonlinear mode competition. The input state is in the basin of a mode or axis attractor, and moves toward it under the "force" of the linear growth term and the nonlinear damping terms.

The eigenvalues of the axis attractors, which we have previously calculated to get conditions for their stability, give a quantitative estimate of this convergence rate. For mode r_s there is one eigenvalue $-2u_s$, and $N-1$ eigenvalues $u_i - a_{is}(u_s/a_{ss})$, where the subscript i denotes the other modes r_i. We therefore have exponential convergence near an axis attractor s at the rates $e^{-2u_s t}$ along the s axis, and $e^{(u_i - a_{is}(u_s/a_{ss}))t}$ along the other axes. Thus large u_s and $a_{is}/a_{ss} \gg u_i/u_s$ program fast convergence.

7.4.3. Interior Fixed Points

With further restrictions on the coefficients of the amplitude equations, $a_{ij} = c$, $a_{ii} = d$, for all i, j, where $c > d$, to maintain the stability of axis fixed

points, and with a uniform eigenvalue spectrum u, $u_i = u$, the entire vector field becomes symmetric under permutations of the axes, since the component equations are now identical. Looking again at the condition for an equilibrium in the amplitude equations,

$$\dot{r}_j = r_j\left(u_j - \sum_{k=1}^{N/2} a_{jk} r_k^2\right) = 0$$

it is clear that it is satisfied in every subspace that is defined by some of the components r_j being zero, wherever the remaining nonzero r_k satisfy $u_j - \sum_{k=1}^{N/2} a_{jk} r_k^2 = 0$, for all j, k of the subspace. In vector form, $\mathbf{u} = \mathbf{A}\mathbf{r}^2$, where $\mathbf{r}^2$ is a vector of elements r_i^2, for all i in the subspace. Thus, all subspace fixed points are at $\mathbf{r}^2 = \mathbf{A}^{-1}\mathbf{u}$, hence $r_i = \pm(\sum_{j=1}^{N/2} A_{ij}^{-1} u_j)^{1/2}$, exhibiting the Z_2 symmetry of the system under flips of the axes.

Geometrically, each fixed point can be viewed as an intersection of nullclines. The nullclines (locus of points for which $\dot{r}_s = 0$), for each variable r_s, where $r_s \neq 0$, are given by the states in r space where $u_s = \sum_{j=1}^{N/2} a_{sj} r_j^2$. These are ellipsoids symmetric about the origin under flips of each axis (see Fig. 7.2). Thus equilibria not on an axis are given by the points of common intersection of all subsets of nullclines in the subspace of nonzero variables corresponding to that subset. In general, depending on the choice of a_{ij} there may be few of these fixed points or there may be at most one in each orthant (because of the Z_2 symmetry) of each subspace, which in general are given by a transversal intersection of the nullclines and are therefore structurally stable. The stability of the interior fixed points is again specified by the a_{ij}, which determine the eigenvalues of the Jacobian, and therefore control the direction of the flow in the state space.

It has been shown for the competitive Lokta–Volterra equations that stability of axis fixed points is sufficient to guarantee that all interior fixed points are saddles (M. Zeeman, personal communication). This kind of theorem probably holds for our amplitude equations as well and will be investigated in the future. For the present, for the $\mathbf{A}$ matrix above, we prove later that all the interior fixed points are saddle points, hence the only attractors in the state space are the fixed points on the axes.

7.4.4. Fourier Synthesis of Sequences

The stability of fixed points on the axes of the normal form amplitude equations (3, 4) can be transferred to an "interior" (nonaxis) fixed point in a subspace of any dimension $m < N/2$ by violating the axis stability condition stated in the projection theorem, and increasing the size of the diagonal $\mathbf{A}$ matrix entries a_{ii} to something greater than the off-diagonal entries a_{ij} in that subspace. The temporal pattern in the network corresponding to this "mixed mode" is then a linear combination of the m axis patterns at their different frequencies,

$$\mathbf{x}(t) = \sum_{s=1}^{m} r^s \mathbf{x}^s e^{i\boldsymbol{\theta}^s} e^{i\omega^s t}$$

or

$$x_j^s(t) = \sum_{s=1}^{m} r^s x_j^s e^{i\theta_j^s} e^{i\omega^s t}$$

for spatial component j of this spatiotemporal pattern. This is a Fourier series description of a periodic function (discrete spectrum), where $r^s x_j^s e^{i\theta_j^s}$ is the magnitude ($r^s x_j$) and phase ($e^{i\theta_j^s}$) of the complex Fourier coefficient F_j^s for frequency ω^s. Given some continuous temporal pattern $x_j(t)$ of period T to be stored for component j, we can find the real $\Re$ and imaginary $\Im$ parts of this coefficient F_j^s, as

$$\Re F_j^s = \frac{1}{T} \int_0^T x_j(t) \cos w^s t \, dt$$

$$\Im F_j^s = \frac{1}{T} \int_0^T x_j(t) \sin w^s t \, dt$$

hence

$$x_j(t) = \sum_{s=1}^{m} F_j^s e^{iw^s t}$$

where $w^s = k2\pi/T$, for some k. For an arbitrary $x_j(t)$, the sum must in principle be infinite, but in practice the finite sum given by a discrete Fourier transform will approximate most signals of interest. If the desired sequence is given as a set of discrete steps to be learned, these may be treated as if they were discrete samples in the time domain of a continuous sequence, and the discrete Fourier transform could be applied directly to find the Fourier components required for network storage.

We construct N-dimensional complex vectors $\mathbf{F}^s$ of the real and imaginary components of all the Fourier coefficients F_j^s for a given frequency ω^s. These go into the corresponding columns of the $\mathbf{P}$ matrix of the projection algorithm as the eigenvectors for that frequency. Network couplings are then found by the projection operation (2) as before. To make efficient use of the network storage capacity, we must reuse as many of the same frequencies ω^s as possible in the series representation for each spatial component j. We choose only as many of the principal

frequencies (with the largest coefficients) as required to approximate a multiple-frequency pattern to some desired level of tolerance. The temporal "complexity" of a spatiotemporal pattern here is the minimum number m of spectral components required to reconstruct it. This is equivalent to the network storage capacity required to store and resynthesize it, which is the dimension m of the subspace of the amplitude equations (the number of subspace axes of different frequencies) that must be dedicated to it.

The discrete Fourier transform of a set of samples of a sequence in space and time can be input directly to the $\mathbf{P}$ matrix as a set of complex columns corresponding to the frequencies in $\mathbf{J}$ and the subspace programmed in $\mathbf{A}$. $N/2$ total DFT samples of N-dimensional time-varying spatial vectors may be placed in the $\mathbf{P}$ matrix, and parsed by the $\mathbf{A}$ matrix into $m < N/2$ separate sequences as desired, with separate basins of attraction guaranteed (Baird, 1990c). For a symmetric $\mathbf{A}$ matrix, as we have seen above, there is a Liapunov function, in the amplitude equations of the polar coordinate version of the normal form (3), which governs the approach of initial states to stored amplitude trajectories.

Enforcing Phase Convergence

Although at this point we have proved convergence to the desired mixed mode amplitudes, there is inherently no attraction to any particular relative phase relation in the phase equations ϕ for these modes. Since the modes have different integer multiples of the lowest frequency, we need for all cycles to have a particular fixed phase relationship at their appropriate multiple of the lowest frequency. This is equivalent, for example, to having all component cycles begin at the same phase, as is often the convention for Fourier synthesis. This kind of requirement constitutes a special initial condition for the system. There is neutral stability in the phase equations as they stand, so any starting set of relative phases will persist. Any starting pattern of initial conditions that is close to a stored trajectory will establish initial mode phases that are correct, and the remainder of the stored spatiotemporal trajectory will be generated properly. What we have proved thus far is the existence of invariant attracting manifolds (tori) in network coordinates on which all starting relative mixed-mode phases are free to vary. The desired trajectory lies on this manifold, but it is not attracting.

In the projection network, we can easily enforce convergence to the desired phase synchrony between the modes with multiple frequencies. This corresponds to producing convergence to a desired sequence trajectory from any initial condition in its basin of attraction. This is done by using the polar form of the normal form equations in mode coordinates (6) as the dynamic network and adding a term to the phase equations $\dot{\phi}_s$ of the form $-\alpha \sin(\phi_s - n\phi_0)$. Here ϕ_0 is the phase of the lowest frequency mode of the set γ that combines to form a particular multifrequency attractor, and n is the multiple of that frequency ω_0^γ. The normal form, for the set $s \in \gamma$ with these new phase equations and the $b_{ij} = 0$, is now

$$\dot{r}_s = u_s r_s - r_s \sum_{j=1}^{N/2} a_{sj} r_j^2 \tag{10}$$

$$\dot{\phi}_s = n\omega_0^\gamma - \alpha \sin(\phi_s - n\phi_0) \tag{11}$$

It is easy to show convergence of each phase ϕ_s of the set γ contributing to the sequence. Forming a differential equation for the phase difference ψ between $\dot{\phi}_s$ and $n\dot{\phi}_0$, we get

$$\dot{\psi} = \dot{\phi}_s - n\dot{\phi}_0 = n\omega_0^\gamma - \alpha \sin(\phi_s - n\phi_0) - n\omega_0^\gamma$$

which, substituting $\psi = \phi_s - n\phi_0$, leads to

$$\dot{\psi} = -\alpha \sin \psi$$

This has fixed points at $\psi = 0$ and $\psi = \pi$ where $\dot{\psi} = -\alpha \sin \psi = 0$. The Jacobian is $-\alpha \cos \psi$, which, evaluated at π, gives the eigenvalue α and evaluated at 0 gives $-\alpha$. This shows π to be an unstable fixed point, and phase difference $\psi = 0$ to be an attractor with exponential convergence rate α, since $\psi(t) = e^{-\alpha t}$ in the neighborhood.

These phase equations cannot be transformed back into equations in Cartesian and network coordinates without introducing higher-order terms of order n for each frequency multiple n. In the projection network, however, the ψ and $\dot{r}$ equations of the polar normal form (3) may be simulated directly, and the state (r, θ) of this system transformed into Cartesian coordinates (u, v) by $u = r \cos \phi$ and $v = r \sin \phi$. These activities may then be transformed into network coordinates in the projection network as before to reconstruct stored spatiotemporal patterns. Input may of course be mapped from Cartesian into the polar normal form states by the conversion $r = \sqrt{(u^2 + v^2)}$ and $\phi = \arctan(v/u)$.

The rate of output of a sequence may be varied simply by scaling all its component frequencies simultaneously. Since there are no hidden units thus far in this system, uniqueness of solutions for these ordinary differential equations requires that no stored trajectories can cross. Furthermore, no two normalized columns of the P matrix can be the same, because it must be invertible.

The same frequencies may be used in different subspaces to store different sequences, but a given spatial vector of Fourier coefficients may not be reused. All this may be subverted at the cost of adding extra state space dimensions (hidden nodes) in the network, which are not to be used for output. The entries in the P matrix for these may be randomly chosen to break the degeneracy of any columns that are identical over the output variables in the P matrix. Stored trajectories may then cross in the projection subspace of the output units as often as desired, since they are still distinct in the true state space that includes the hidden units.

7.4.5. Guaranteed Basins of Attraction—No Spurious Attractors

Here we consider specific vector field programming to establish separate basins of attraction for different subspace dynamics. This can be done simply by choosing the A matrix entries outside the blocks on the diagonal (which determine coupling of variables within a subspace) to be greater than those within the blocks. The principal is the same as that we have used for making axis fixed points stable (Baird, 1989), but here we make the flow enter a whole subspace by choosing intersubspace competition to be greater than subspace self-damping.

As an example of this kind of vector field programming, we consider normal form couplings for the amplitude equations (3) in the $N/2 \times N/2$ matrix $\tilde{A}$. The $N \times N$ matrix A of the Cartesian coordinate normal form (6) for oscillatory systems displayed and discussed thus far was easily derived in the proof of the projection theorem by a doubling of the entries in $\tilde{A}$. The results here apply directly to networks with static attractors and to the amplitudes of systems with periodic attractors. Consider an $\tilde{A}$ matrix the following form,

$$\tilde{A} = \begin{bmatrix} \begin{bmatrix} d & c \\ c & d \end{bmatrix} & & (g) & \\ & \begin{bmatrix} d & (c) \\ & d & \\ & & d \\ (c) & & d \end{bmatrix} & \\ (g) & & & \ddots \end{bmatrix}$$

where $0 < c < d < g$. Any number of blocks of different dimensions m are on the diagonal, as described above, and ( ) indicates the uniform entries in a region.

Looking at the condition for an equilibrium in the amplitude equations (3),

$$\dot{r}_i = r_i\left(u - \sum_{j=1}^{N/2} a_{ij}r_j^2\right) = 0$$

where $u = 1 - \tau$, it is clear that it is satisfied in every subspace that is defined by some of the components r_j being zero, wherever the remaining nonzero r_i satisfy $u - \sum_{k=1}^{N/2} a_{ij}r_j^2 = 0$ for all j, k of the subspace. Furthermore, all subspaces are invariant under the flow since there can be no components of the flow in any direction of the zeroed components not in the subspace (since $r_j = 0$ implies $dr_j/dt = 0$). Once a component goes to zero during the evolution of the system, it never comes back; or in other words, once in a subspace (the origin, an axis, a hyperplane of axes), you never get out. A small amount of noise in a real system, however, will prevent getting hung up in any subspace that does not contain an attractor.

THEOREM 4.1. *Within any subspace of dimension $m < N/2$ of the amplitude equations (3) (when $u = 1 - \tau > 0$), defined by a diagonal block of the $\tilde{A}$ matrix containing identical diagonal entries d and off-diagonal entries c within the block, and where all entries g outside the diagonal blocks are identical, with $0 < c < d < g$, the only stable fixed point is the interior (nonaxis) fixed point on the main diagonal of the subspace at $r_i^2 = \pm u/(mc - c + d)$.*

PROOF. To show this, it suffices to observe the expression for the Jacobian matrix $\mathbf{U}$ at all fixed points within the subspace under the restrictions given. In any subspace of dimension m, there is a fixed point $\hat{r}$ where

$$u = a_{ii}\hat{r}_i^2 + \sum_{j\neq i}^{N/2} a_{ij}\hat{r}_j^2$$

for all nonzero variables i, j of the subspace. Since $a_{ii} = d$ for all i, and $a_{ij} = c$ within the subspace, the equations are identical, $u = d\hat{r}^2 + (m-1)c\hat{r}^2$. Thus the components of the fixed point $\hat{r}$ are identical, $\hat{r}_i^2 = \pm u/(mc - c + d)$, and therefore lie on the main diagonal within the subspace. Defining $d = c - e$ to get simpler expressions, we have $\hat{r}_i^2 = \pm u/(mc - e)$. Now the entries of the Jacobian $\mathbf{U}$ are,

$$U_{ij} = -2a_{ij}\hat{r}_i\hat{r}_j = -2c\hat{r}_i^2 = -\frac{2cu}{(mc - e)}$$

for $\hat{r}_i$ and $\hat{r}_j \neq 0$, and 0 otherwise. Similarly, for components i where $\hat{r}_i \neq 0$,

$$U_{ii} = u - 3a_{ii}\hat{r}_i^2 \sum_{j\neq i}^{N/2} a_{ij}\hat{r}_j^2$$

$$= u - 3(c - e)\hat{r}_i^2 - (m - 1)c\hat{r}_i^2$$

$$= u - \frac{3(c - e)u}{(mc - e)} - \frac{(m - 1)cu}{(mc - e)}$$

$$= -\frac{u2(c - e)}{(mc - e)}$$

Now for components i where $\hat{r}_i = 0$, $a_{ij} = g$, so $\tilde{U}_{ii} = u - \sum_{j=1}^{N/2} g\hat{r}_j^2$, and since m of the $\hat{r}_j^2 = u/(mc - e)$, and the rest are zero,

$$\tilde{U}_{ij} = u - \frac{mgu}{(mc - e)} = u\left[1 - \frac{mc}{(mc - e)}\right]$$

$$= u\left[1 - \frac{mg}{(mc - c + d)}\right]$$

Thus the nonzero off-diagonal entries U_{ij} of the Jacobian at any interior fixed point are identical, as are the diagonal entries U_{ii} for the components where the fixed point is nonzero, and the components $\tilde{U}_{ii}$, where the fixed point is zero. By rearranging the location of row and column numbers (which does not change the eigenvalues of the matrix), the Jacobian for the fixed point in any subspace of dimension m may be written as an $m \times m$ block $\mathbf{D}$ on the diagonal, $D = U_{ij}\mathbf{B} + (U_{ii} - U_{ij})\mathbf{I}$, or $\mathbf{D} = b\mathbf{B} + (a - b)\mathbf{I}$. Here $\mathbf{B}$ is a matrix of all 1's, and $\mathbf{I}$ is the identity matrix. This block is followed by constant diagonal entries $\tilde{U}_{ii} = f$ for the $\hat{r}_i = 0$ components, up to the full rank n of the Jacobian.

$$\mathbf{U} = \begin{bmatrix} a & b & b & & & \\ b & a & b & & 0 & \\ b & b & a & & & \\ & & & f & & \\ & 0 & & & \ddots & \\ & & & & & f \end{bmatrix}$$

Since $\mathbf{U}$ is the direct sum of the eigenspaces of the block $\mathbf{D}$ on the diagonal, and the diagonal entries f, its eigenvalues are those corresponding to those eigenspaces. Since $\mathbf{U}$ is in diagonal form for the f entries, they directly constitute $N/2 - m$ of the eigenvalues of $\mathbf{U}$. The expression for these eigenvalues f contributed by the zero variables not in the subspace is negative, in spite of $c < d$, because we have $g > d$. Thus the flow is into the fixed point from out of the ambient state space beyond the subspace.

The eigenvalues of the m directions within the subspace are given by the matrix $\mathbf{D}$, and all are negative, since $c < d$. The eigenvalues of $\mathbf{D} = b\mathbf{B} + a\mathbf{I}$ are $a - b$ with multiplicity $m - 1$, and $bm + a - b$ with multiplicity 1, since the rank of $\mathbf{B}$ is 1 (all its columns are identical), and its eigenvalues are m with multiplicity 1, and 0 with multiplicity $m - 1$. Now

$$a - b = U_{ii} - U_{ij}$$

$$= -\frac{u2(c - e)}{(mc - e)} + \frac{2cu}{(mc - e)} = \frac{2ue}{(mc - e)}$$

which is negative, since $0 < c < d$ implies that $e = c - d$ must be negative. Finally,

$$bm + a - b = U_{ij}m + a - b$$

$$= -\frac{2cum}{(mc - e)} + \frac{2ue}{(mc - e)} = -2u$$

which is always negative, since u is positive by our assumption.

Since all eigenvalues of this interior fixed point are negative, and all main diagonal interior fixed points defined by the diagonal blocks of the $\tilde{\mathbf{A}}$ matrix in this system have the same expressions for their eigenvalues, we have proved that all are stable and must thus have their own basin of attraction. We can show further that this basin must extend at least to the boundaries of the subspace, since the subspace is invariant under the flow, and we will now show that all other equilibria within the subspace have some positive eigenvalues, and hence are saddle points.

For any subspace of dimension $p < m$ of variables within any diagonal block, the expression for the fixed point is identical to that above, with p replacing m, $\hat{r}_i^2 = \pm u/(pc - e)$. We have also the same expressions for the Jacobian entries corresponding to the nonzero variables defining this subsubspace—the $\mathbf{D}$ matrix above (now of dimension $p \times p$). The $\tilde{U}_{ii} = f$ entries for all variables outside the full m-dimensional subspace are also as above, with p replacing m, $\tilde{U}_{ii} = u[1 - pg/(pc - c + d)]$. All these entries imply negative eigenvalues, as we have just shown. However, the remaining entries $\tilde{U}_{ii} = f$ for the variables outside the subspace, but still within the m-dimensional subspace have $a_{ij} = c$ instead of g, and this leads to positive eigenvalues. $\tilde{U}_{ii} = u - \sum_{j=1}^{N/2} a_{ij}\hat{r}_j^2$, where $a_{ij} = c$, and $\hat{r}_j^2 = u/(pc - e)$, within the subsubspace (and $\hat{r}_j^2 = 0$ otherwise), so

$$\tilde{U}_{ii} = u - pc\hat{r}^2 = u\left[1 - \frac{pc}{(pc - e)}\right]$$

$$= u\left[1 - \frac{pc}{(pc - c + d)}\right]$$

This is positive, since $pc < (pc - c + d)$. Thus the desired fixed point on the main diagonal is the *only* stable fixed point (attractor) in the subspace defined by any diagonal block. Since all fixed points are in some programmed subspace, we have proved, in addition, that there are no spurious attractors in the network.   □

Since the $\tilde{\mathbf{A}}$ matrix is symmetric, the amplitude equations define a gradient vector field, as shown in Baird (1989, 1992b), and the Liapunov function constructed there characterizes the approach of the system trajectories. Furthermore, should all these diagonal blocks be of the same dimension, then the amplitude equations are equivariant under subspace permutations, and symmetry arguments, like those discussed for axis fixed points in Baird (1989), guarantee that the basin boundaries in the amplitude state space are hyperplanes that define scale-invariant "wedges" as basins of attraction.

7.4.6. Basin Geometry

The *stable manifold* of a saddle point consists of all points in the state space on trajectories that asymptotically approach that point (Guckenheimer and Holmes, 1983). At a saddle point, the stable manifold is tangent to the eigenspaces (directions) of the state space, which correspond to the negative eigenvalues of the saddle point. A stable manifold of dimension one less than the dimension of the state space can form part of a *separatrix* that divides the state space into separate compartments or basins. This is because the flow diverges away from the stable manifold of a saddle point, as water does from the top of a ridge. When the $\tilde{\mathbf{A}}$ matrix is of the form discussed in the previous theorem, we have just proved that the number of negative eigenvalues of an interior saddle point is $N - m + 1$, where N is the dimension of the state space, and m is the dimension of the subspace containing the saddle point. Only the saddle points in the two-dimensional subspaces defined by pairs of axes have stable manifolds of dimension $N - 1$ that can form parts of separatrices. Because the flow everywhere converges to equilibria, the basin boundaries are made up of pieces of the stable manifolds of these equilibria.

Furthermore, with this $\tilde{\mathbf{A}}$ matrix, the normal form component equations are identical, and the normal form vector field is symmetric under permutation of the axes (the components). For the two-mode vector field in the plane shown in Fig. 7.2, it is obvious that this symmetry forces the diagonal separatrices through the interior saddle points to be straight lines, since only a line evenly dividing the quadrant is consistent with this symmetry under reflection of the axes. These basins are identical two-dimensional "wedges" of infinite extent, and are thus naturally scale-invariant. They define classes of directions of vectors.

The picture of basins of attraction as wedges is also clear in three dimensions. The basins of the axes attractors are bounded by planes that are linear subspaces that pass through the origin and extend from the diagonals of the planes defined by pairs of the axes (i, j, k) to intersect on the main diagonals of points where all components are equal. These basins are therefore identical three-dimensional wedges. In N dimensions we should get "hyperwedges" whose boundaries are the unstable manifolds of the saddle points in all the i, j planes as shown in Fig. 7.2, which are $N - 1$ dimensional linear subspaces or hyperplanes.

Symmetry Group of Equivariant Vector Fields

To formalize this argument we need a more precise definition of the "symmetry" of a vector field (Golubitsky and Schaefer, 1985). If the coupling of a system is such that the network equations are *equivariant*, i.e., can be shown to commute with the action a *compact Lie group* Γ, that is, for $dx/dt = F(x)$, if $F(\gamma x) = \gamma F(x)$ for all γ in Γ, then the vector field inherits the isotropy subgroup lattice structure associated with the action of that group. The compact Lie groups can be identified with the subgroups of $O(n)$, a subgroup of the general linear group of N-dimensional invertible matrices.

If $x(t)$ is a solution, including in particular a fixed point x_0, then $\gamma x(t)$ is also a solution—either equal to $x(t)$, in which case γ characterizes a symmetry invariance of that solution, or different from $x(t)$, in which case an *orbit* of other solutions can be found from all group elements whose actions are not in the isotropy subgroup of elements which leave $x(t)$ fixed. Any network using symmetric sigmoids like $\tanh(x_i)$ or $\sin^{-1}(x_i)$, where $g(-x_i) = -g(x_i)$, for example, has such a symmetry group—namely, Z_2. For every fixed point stored in such a net, there is an orbit of spurious (unintended) group-related attractors. For the networks with this Z_2 symmetry, for example, $-x$ is also an attractor if x is an attractor, as seen in the vector field of Fig. 7.2 above.

Symmetric Normal Form Example

For a concrete example, consider the system of four identical normal form amplitude equations that results from the choice of $\tilde{A}$ matrix, considered in earlier sections, which guarantees that only the axis fixed points are stable. This system is characterized by the tetrahedral symmetry group of even permutations of four objects. The equivariance or symmetry of the dynamical equations is seen as the fact that you can swap the location of elements in the net by a permutation $x \rightarrow \gamma x$, and the differential equation for the activity of that element is unchanged, $F(\gamma x) = \gamma F(x)$. The particular structure of this symmetry group now governs the kind of fixed point solutions $\hat{x}$ (i.e., patterns of activity at an axis attractor or interior saddle point) that are possible. There are 12 transformations in the tetrahedral group—8 different rotations or permutations that leave the activity at one node fixed, 3 different flips that swap the activity at two separate pairs of nodes in the net, and the identity permutation that leaves everything the same.

This group structure characterizes the number of possible fixed point patterns and subspaces of each symmetry type in the normal form state space. A solution such as the trivial solution $x = 0$, which is a uniform spatial pattern of activity, is completely symmetric or invariant under the action of the full group of transformations. It has an isotropy subgroup of all 12 elements. If an attractor is stored that has uniform activity at all but one node (an axis attractor) then this solution has the symmetry of one of the isotropy subgroups with the next largest number of elements—two rotations leave the pattern unchanged.

For interior saddle points in the two-dimensional subspaces, with two activities at one amplitude value and two at another, there is a single flip for which the pattern remains invariant. Finally, for any three or more differing amplitudes of activity across the four nodes, the symmetry is fully broken and only the identity transformation leaves it unperturbed. All 11 of the other permutations of such a solution now give an orbit of 11 other patterns which must also be solutions. Thus as the spatial symmetry of solutions is successively broken, the number of group-related solutions on the orbit with that symmetry goes up. The fixed points in each subspace of a given dimension in our normal form state space are on the orbit of an isotropy subgroup. There are four positive axis attractors, generated by the four permutations γ_i

that move the single nonzero element in the pattern.

When there are symmetries like this chosen for the normal form equations, then the network vector field given by the learning algorithm is equivariant as well, and the network attractors and basins are interrelated by a group of transformations. These are given from the learning process (the projection) even though there are no restrictions on the patterns being learned, and none of the attractors is spurious. This is because any symmetry group operation, with matrix representation $\mathbf{B}$ (a permutation matrix in our case) of a symmetry group Γ which characterizes the vector field of the normal form equations, becomes $\mathbf{PBP}^{-1}$ in the network coordinate system after projection. Thus the symmetries conferred on the network vector field vary with the patterns to be learned in the $\mathbf{P}$ matrix, and may be very complicated, but they are characterized exactly in this expression by our knowledge of the patterns learned and the symmetries of the normal form.

7.4.7. Subspace Basin Boundaries

The vector field for even arbitrary nonlinear $F(x)$ has invariant linear *fixed point subspaces* given by sums of the isotropy subgroups of the action of elements of Γ. To see this, let Σ be a subgroup of Γ. The fixed point subspace $\mathrm{Fix}(\Sigma) = \{\mathbf{x} \in V : \gamma \mathbf{x} = \mathbf{x}, \forall \gamma \in \Sigma\}$, where V is the state space of the Γ-equivariant vector field F. By the definition of $\mathrm{Fix}(\Sigma)$, $F(\mathbf{x}) = F(\gamma \mathbf{x})$, and by equivariance, $F(\gamma \mathbf{x}) = \gamma F(\mathbf{x})$, hence $F(\mathbf{x}) = \gamma F(\mathbf{x})$. In the two-dimensional equivariant vector field above, the diagonals are one-dimensional fixed point subspaces of the action of the permutation group of two elements (flip and identity), which are therefore invariant under the flow. The saddle point C is in this subspace. If we can show that the eigenspace of the negative eigenvalues of $\mathbf{D}f$ at C is in this subspace, then it follows that the stable manifold of C is restricted to this subspace, since it is tangent to this eigenspace at C and invariant under the flow. From the Jacobian $\mathbf{D}f$ at interior fixed points shown earlier, the eigenvector corresponding to the negative eigenvalue at C is $(1, 1)$, which is indeed in the diagonal fixed point subspace.

In higher dimensions, there are further eigenvectors spanning the subspace of negative eigenvalues given by the diagonal $\tilde{U}_{ii}$ entries of $\mathbf{D}f$ at C, which are orthogonal, since $\mathbf{D}f$ is symmetric. In three dimensions, the unstable eigenspace of C is two-dimensional, and is spanned by $(1, 1, 0)$ and $(0, 0, 1)$, which is the plane through the

diagonal containing C, perpendicular to the i, plane. This is again a fixed point subspace invariant under the flow, and under the i, j permutations, which are a subgroup of the permutations on i, j, k. The planes from the saddles in each face intersect at the main diagonal from the origin and constitute the boundaries of the basins of the axis attractors. Knowing the invariant subspaces and the location and eigenvalues and eigenvectors of all fixed points, we can extend this line of argument to prove that the basin boundaries in arbitrary dimensions are linear subspaces (hyperplanes). The general form of the stable subspace extending from the saddle point in any plane i, j (following the form of the Jacobian after component rearrangement) will be the subspace spanned by $(1, 1, 0, \ldots, 0)$, $(0, 0, 1, 0, \ldots, 0)$, $(0, 0, 0, 1, 0, \ldots, 0)$, $\ldots$, $(0, 0, \ldots, 1)$, where all eigenvectors are invariant under i, j reflection of the first two components. The subspace is therefore invariant under this subgroup Σ, and $\equiv \mathrm{Fix}(\Sigma)$. Thus the stable manifold $\in \mathrm{Fix}(\Sigma)$ and is of the same dimension, hence it is a linear subspace, hence it is a hyperplane.

7.4.8. Basin Geometry in Network Coordinates

For networks with *static* attractors, the basins of the normal form are linearly transformed under the projection learning procedure and so remain hyperwedges. The subspace basin boundaries of these wedges are positioned (rotated) by the learning transformation to realize clustering of inputs into categories and retrieval by nearest Euclidean distance. This is an analog network which may be restricted to categorize like a binary network of linear threshold logic units, and still produce a specified *analog* output.

For networks with periodic attractors, the basin geometry is more difficult to visualize. The state space of the amplitude equations is simpler to start with, since only the positive orthant is included—because amplitudes of oscillation must be positive. This is a very concrete difference in the operation of the oscillatory system from the static systems of bistable elements. Here the Z_2 symmetric sigmoid and cubic normal form does not result in attractors which are the negative of each attractor intentionally stored as it did in the figure above. Instead, there is the additional rotational symmetry ϕ, which transforms the two-dimensional wedge basins of amplitude equations in the plane into three-dimensional cones in the four-dimensional normal form state space that includes the phase equations $d\phi_s/dt = \omega_s$. The learning transformation then deforms this picture further, but may still leave scale- and rotation-invariant (conelike) basin boundaries made up of "sheaves" of straight rays. The basins of periodic attractors are invariant under oscillation and so may be said to recognize or categorize oscillatory patterns, as the pre-pyriform cortex must recognize the oscillatory output of the bulb.

All of the above group theory holds (in our system with its independent phase equations) for periodic solutions $\mathbf{x}(t)$, by the addition of the circle group of phase shifts S_1 to the definition of a symmetry group, to give $\Gamma \otimes S_1$ (Golubitsky and Schaefer, 1985). There is not only a spatial, but now a temporal component to the symmetry of the limit set, which must be invariant under both γ and phase shift ϕ to be a member of a *twisted* isotropy subgroup of the full group $\Gamma \otimes S_1$. There are now orbits of new limit cycles from both spatial transformations γ and phase shifts ϕ. Multiple Hopf bifurcations with symmetry are handled within this framework. Again, after projection into network coordinates, very complicated well-characterized spatiotemporal symmetries result. These tools may help give insight into the process of sequence recognition by analysis of the symmetry invariances of basins of attraction of mixed-mode attractors.

7.4.9. Chaotic Attractors

Chaotic attractors may be created in the Hopf normal form, with sigmoid nonlinearities added to the right-hand side, $v_i \to g(v_i) = \tanh(v_i)$.

$$\dot{v}_i = -\tau v_i + \sum_{j=1}^{N} J_{ij} g(v_j) - g(v_i) \sum_{j=1}^{N} A_{ij} g(v_j)^2 \tag{12}$$

The sigmoids yield a spectrum of higher-order terms that break the phase shift symmetry of the system. Two oscillatory pairs of nodes like those programmed in the 4×4 block above can then be programmed to interact chaotically. In our simulations, for example, if we set the upper block of d entries to -1, and the lower to 1, and replace the upper c entries with 4.0, and the lower with -0.4, we get a chaotic attractor of dimension less than 4, but greater than 3.

This is "weak" or "phase-coherent" chaos that is still nearly periodic. It is created by the broken symmetry, when a homoclinic tangle occurs to break up an invariant 3-torus in the flow (Guckenheimer and Holmes, 1983). This is the Ruelle–Takens route to chaos and has been observed in Taylor–Couette flow when both cylinders are rotated. Experiments of Freeman have suggested that chaotic attractors of the above dimension occur in the olfactory (Freeman, 1987). These might most naturally be created by the interaction of oscillatory modes.

We have demonstrated in simulations that sets of Lorentz equations in three-dimensional subspace blocks can be used in a projection network as well. Multiple Lorenz attractors have been created simply by adding off-diagonal normal form competitive terms to couple sets of the three Lorenz equations

$$\dot{v}_1 = s(-v_1 + v_2) - v_1 \sum_{j=1}^{N} A_{1j} v_j^2 \tag{13}$$

$$\dot{v}_2 = rv_1 - v_2 - v_1 v_3 - v_2 \sum_{j=1}^{N} A_{2j} v_j^2 \tag{14}$$

$$\dot{v}_3 = -bv_3 + v_1 v_2 - v_3 \sum_{j=1}^{N} A_{3j} v_j^2 \tag{15}$$

$$\vdots \tag{16}$$

The $\mathbf{A}$ matrix here has constant off-diagonal coupling, and three-dimensional blocks of zeros along the diagonal to remove the self-damping terms from the subspaces containing the Lorenz equations. The famous "butterfly" attractor appears for $s = 10.0$, $r = 28.0$, $b = 2.666\,666\,66$.

This network has the fascinating property that when the competitive coupling coefficients are reduced beyond a critical threshold, the system begins to "percolate" and jumps at intermittent times to different subspaces, as though "day-dreaming" at random from memory to memory. It is, however, a completely deterministic system, and may therefore have in fact one large chaotic attractor with many "wings" that are visited intermittently, just as the two wings of the Lorenz butterfly are visited at seemingly random times. Unlike static or oscillatory attractors, the orbits of a Lorenz attractor may approach zero at intermittent times. This releases the competitive suppression of the other attractors and allows another to grow in activity and suppress the rest until it also passes near enough to zero to allow another transition. The orbits of all the Lorenz attractors may thus connect with each other through a region of the state space in the neighborhood of the origin. Varying the degree of competition seems to vary the size of this region, and therefore the frequency of transitions between attractors. We expect that such behavior may be useful as a search process during reinforcement learning, and will compare it to transitions between attractors induced by additive noise.

In the brain, as in fluid mechanics, the statistical (thermal) molecular synaptic events which are often appealed to as a source of noise are far too microscopic to be an effective perturbation of the ensemble activity of neurons at the network level. Chaos in the macroscopic dynamics of neural networks may therefore be essential to generate the kind of random behavioral output (like "babbling" in early speech learning) that is necessary for trial-and-error reinforcement learning. This "fuzzy search" function is one of the most obvious and popular uses of chaos, since it is already employed everywhere in the form of the chaotic one-dimensional maps that generate random numbers for computer programs of all kinds.

These equations may be projected into network coordinates to give network equations,

$$\dot{x}_i = -\tau x_i + \sum_{j=1}^{N} T_{ij}x_j + \sum_{jk=1}^{N} T_{ijk}x_j x_k$$
$$- \sum_{jkl=1}^{N} T_{ijkl}x_j x_k x_l \qquad (17)$$

where

$$\mathbf{T} = \mathbf{PJP}^{-1} \quad T_{ijk} = \sum_{mno=1}^{N} P_{im}B_{mno}P_{nj}^{-1}P_{ok}^{-1} \left.\vphantom{\sum_{mn=1}^{N}}\right\}$$
$$T_{ijkl} = \sum_{mn=1}^{N} P_{im}A_{mn}P_{mj}^{-1}P_{nk}^{-1}P_{nl}^{-1} \qquad (18)$$

Here the matrix $\mathbf{A}$ is as described above, but the matrix $\mathbf{J}$ now contains the linear coupling terms of the Lorenz equations in three-dimensional blocks along the diagonal, and zeros everywhere else. The additional quadratic coupling terms of the Lorenz equations are given as $+1$ and -1 near the diagonal in a new tensor B_{ijk}, with zeros everywhere else, and these are projected like the other terms to yield the quadratic terms with coupling tensor T_{ijk} in the network equations. The diagonal blocks $\mathbf{J}^B$ in $\mathbf{J}$ have the form

$$\mathbf{J}^B = \begin{bmatrix} -s & s & 0 \\ r & -1 & 0 \\ 0 & 0 & -b \end{bmatrix}$$

and the nonzero elements of B_{ijk} may be generated as

$$B_{i+1,i,i+2} = -1 \qquad B_{i+2,i,i+1} = 1$$

where i goes from 1 to N by increments of 3.

We lack at present a rigorous theoretical justification for why multiple chaotic attractors may so easily be stored, but the intuitive picture given in the previous section suggests that any kind of dynamics may be implemented in competing subspaces by the addition of these competition terms. We have well-behaved simulations containing multiple static, oscillatory, and chaotic attractors of three different kinds—Ruelle–Takens, Lorenz, and Roessler—in different competing subspaces of the same network (Baird et al., 1991). This is done by coupling sets of the required equations by competitive terms as shown above for the Lorenz equations.

In this network, the projection network, or its folded biological version, the chaotic attractors have a basin of attraction in the N-dimensional state that constitutes a category, just like any other attractor in the system. They are, however, "fuzzy" attractors, and there may be computational advantages to the basins of attraction (categories) produced by chaotic attractors, or to the effects their outputs have as fuzzy inputs to other network modules. In the projected or folded network coordinates, the particular N-dimensional spatiotemporal patterns learned for the four components of the chaotically paired oscillatory modes, or the three components of the Lorenz system, may be considered a coordinate-specific "encoding" of the chaotic attractor, which may constitute a recognizable input to another network, if it falls within some learned basin of attraction. While the details of the trajectory of a chaotic attractor in any physical continuous dynamical system are lost in the noise, there is still a particular structure to the attractor which is a recognizable "signature." This allows communication and "recognition" of chaotic attractors.

7.5. DIGIT RECOGNITION

Handwriting exemplifies the kind of individual biological variability in motor output that plagues most artificial pattern recognition systems, whose capabilities are so easily exceeded by real neural networks. The performance of our system on this problem may thus predict its performance on more difficult visual and speech recognition problems.

In cortex it appears that dynamic representations (oscillations, chaos) are used even for recognition of static visual inputs (Freeman and van Dijk, 1987; Gray and Singer, 1987; Eckhorn et al., 1988). We have therefore approached this real-time digit recognition problem with the normal form projection algorithm in order to produce dynamic representations in continuous-time recurrent networks like those found in cortex (Baird, 1986, 1990a, b, 1992a). We suspect that there are computational advantages to such representations which we seek to demonstrate.

Handwritten characters have a natural scale- and translation-invariant analog representation in terms of a sequence of angles that parametrize the pencil trajectory. We remove the writing velocity variation from the raw set of (x, y) samples of a digit trajectory by interpolating a new set of N (x, y) points equally spaced along the curve. A vector of the sines and cosines of the angles from the horizontal made by the line segment emanating from each point then becomes the preprocessed representation of the digit to be learned or input to the network for recognition. We have thus far constructed an on-line system which anyone may train and submit input to by mouse or digitizing pad and observe the performance of the system for themselves in immediate comparison to their own internal recognition response. The performance of different networks with static, periodic, and two types of chaotic attractors—Lorenz and Roessler—may be tested.

Learning in this system can be as fast as recognition. We have seen that the projection algorithm supplies a formula that allows one-shot learning of prototypes and immediately establishes a basin of attraction that determines the generalization response of the network to future inputs. We are interested determining the differences in the basins of attraction established by the different types of attractors.

We have only qualitative results to report at this point. Using a projection network architecture where the input vector is of dimension $N = 40$ or 80, with only this one-shot learning of prototypes (by placing them in the $\mathbf{P}$ matrices), and very small databases for a single writer at a time, all attractors seem to allow roughly 95% correct recognition responses. Use of 80 samples of the digit, and use of four different prototypes as weight vectors (in the $\mathbf{P}^{-1}$ matrix) for the normal form nodes defining attractors in four-dimensional subspaces improves recognition performance. Supervised or unsupervised competitive learning of weight vectors should bring considerable improvement.

When the $\mathbf{A}$ matrix has identical off-diagonal entries, then, for static and oscillatory attractors in subspaces of any dimension, the classification rule appears to be that an input vector converges to the attractor in the *subspace* that is nearest in Euclidean distance. We prove this in Baird (1989) for static attractors in one-dimensional subspaces. The basin boundaries in network coordinates are given by the points in the network state space that have equal magnitudes (norms) of projection on any combination of any number of the subspaces containing attractors. Unexpected properties have been found in the systems utilizing chaotic attractors as discussed above. It is clear that chaotic attractors have different basin boundaries than static or oscillatory attractors, but we have yet to characterize them in detail.

7.6. MINIMAL BIOLOGICAL NETWORK MODEL

We have determined a biologically "minimal" model that is intended to assume the least anatomically justified coupling sufficient to allow function as an oscillatory associative memory. The network is meant only as a cartoon of the real biology that is designed to reveal the general mathematical principles and mechanisms by which the actual system might function.

Long-range excitatory to excitatory connections are well known as "associational" connections in olfactory cortex (Haberly and Bower, 1989) and cortico–cortico connections in neocortex. Since our units are neural populations, we can expect that some density of full cross-couplings exists in the system (Haberly and Bower, 1989), and our weights are taken to be the average synaptic strengths of these connections. Local inhibitory "interneurons" are a ubiquitous feature of the anatomy of cortex (Gray et al., 1989; Engel et al., 1990). It is unlikely that they make long-range connections (> 1 min) by themselves. These connections, and even the debated interconnections between them, are therefore left out of a minimal coupling model. The resulting network is a fair cartoon of the well-studied circuitry of olfactory (pyriform) cortex. Since almost all cortex has this type of structure in the brains of amphibia and reptiles, our supernetwork of these submodules has the potential to become a reasonable caricature of the full cortical architecture in these animals. Although the neocortex of mammals is more complicated, we expect the model to provide useful suggestions about the principles of oscillatory computation there as well.

For an N-dimensional system, this minimal coupling structure is described mathematically by the matrix

$$\mathbf{T} = \begin{bmatrix} \mathbf{W} & -h\mathbf{I} \\ g\mathbf{I} & 0 \end{bmatrix} \quad (19)$$

$\mathbf{W}$ is the $N/2 \times N/2$ matrix of excitatory interconnections, and $g\mathbf{I}$ and $h\mathbf{I}$ are $N/2 \times N/2$ identity matrices multiplied by the positive

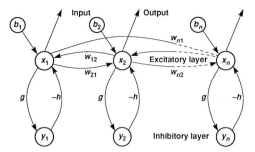

Fig. 7.4. Biological subnetwork of excitatory cell populations x_i, inhibitory cell populations y_i, inputs b_i, adaptive excitatory to excitatory connections W_{ij}, and constant local inhibitory feedback connections g and $-h$.

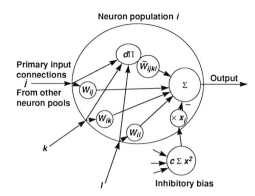

Fig. 7.5. Neural population subnetwork acting as a sigma–pi unit. It uses secondary higher-order synaptic weights $\tilde{W}_{ijkl}$ on products of the activities of incoming primary connections W_{ij}, and receives a global inhibitory bias.

scalars g and h. These give the strength of coupling around local inhibitory feedback loops. A state vector is composed of local average cell voltages for $N/2$ excitatory neuron populations $\mathbf{x}$ and $N/2$ inhibitory neuron populations $\mathbf{y}$. Intuitively, since the inhibitory units receive no direct input and give no direct output, they act as hidden units that create oscillation for the amplitude patterns stored in the excitatory cross-connections $\mathbf{W}$. This may perhaps be viewed as a specific structure addition to a recurrent analog higher-order network architecture to convert its static attractors to periodic attractors. Here the symmetric sigmoid functions of such a network are Taylor-expanded up to cubic terms with third-order weights (quadratic terms are killed by the symmetry). Network equations with the first-order coupling (19) shown in Fig. 7.4, plus these higher-order excitatory synapses, are shown below in component form:

$$\dot{x}_i = -\tau x_i - hy_i + \sum_{j=1}^{N/2} W_{ij} x_j$$

$$- \sum_{jkl=1}^{N/2} W_{ijkl} x_j x_k x_l + b_i \qquad (20)$$

$$\dot{y}_i = -\tau y_i + gx_i \qquad (21)$$

(Baird, 1989, 1990c). We use this network directly as our biological model. From a physiological point of view, (20) may be considered a model of a biological network which is operating in the linear region of the known *axonal* sigmoid nonlinearities, and contains instead the higher-order *synaptic* nonlinearities.

Adding the higher-order weights corresponds, in connectionist language, to increasing the complexity of the neural population nodes to become "higher-order" or "sigma–pi" units.

Clusters of synapses within a population unit can *locally* compute products of the activities on incoming primary connections W_{ij}, during higher-order Hebbian learning, to establish a weight W_{ijkl} (see Fig. 7.5). The secondary higher-order synapses are then used in addition to the synapses W_{ij}, during operation of the overall network, to weight the effect of triple products of inputs in the output summation of the population.

Using only the long-range excitatory *connections* W_{ij} available, some number of the higher-order synaptic weights W_{ijkl} could be realized locally within a neural population in the axo-dendritic interconnection plexus known as "neuropil" (Baird, 1990b). Only $(N/2)^2$ of these $(N/2)^4$ possible higher-order weights are required in principle to approximate the performance of the projection algorithm (Baird, 1992b). The size of our cortical patches is limited by this number, and is itself motivation for modularity.

7.6.1. Hebbian Learning

Only the higher-order weights W_{ijkl} between excitatory populations shown in the minimal model (20) are required for pattern storage that closely approximates that of the full projection (2). The minimal network coupling for $\mathbf{T}$ results from the projection operation (2) when a specific biological form is chosen, in the columns s of $\mathbf{P}$, for the patterns to be stored. This complex form for $\mathbf{P}^s$ and the corresponding asymptotic solutions $\mathbf{X}^s(t)$ established are shown below in (23). The numerical example for the projection learning rule shown earlier gives an example of these patterns, and was chosen specifically to demonstrate this result. The oscillatory chaotic attractors discussed previously may easily be stored when the **A** matrix has the form required for them and sigmoids are used in the network equations. This is because, as we have seen, the Taylor expansion of the system, transformed into normal form coordinates, can be equivalent to that for the normal form equations with sigmoids (13).

As we have shown earlier, for orthonormal static patterns $\mathbf{x}^s$, the projection operation for the **W** matrix reduces to an outer product, or "Hebb" rule, and the projection for the higher-order weights becomes a *multiple* outer product rule:

$$
\left.
\begin{aligned}
W_{ij} &= \sum_{s=1}^{N/2} \alpha^s x_i^s x_j^s \\
W_{ijkl} &= c\delta_{ij}\delta_{kl} - d \sum_{s=1}^{N/2} x_i^s x_j^s x_k^s x_l^s
\end{aligned}
\right\}
\tag{22}
$$

The first rule, with $\alpha^s > \tau$, is analytically guaranteed to establish desired patterns $\mathbf{x}^s$ as eigenvectors of the **W** matrix with corresponding eigenvalues α^s (Baird, 1990b). Here again, δ_{ij} is the Kronecker delta, $\delta_{ij} = 1$, for $i = j$ and zero otherwise.

In the minimal net, these real eigenvalues and eigenvectors (modes) learned for **W** are converted by the network structure into complex conjugate eigenvalues and eigenvectors (modes) for the complete coupling matrix **T** which includes the local inhibitory feedback. These complex modes are standing-wave oscillations whose amplitudes are the magnitudes of the eigenvectors. The frequency w^s of each stored oscillation s is a specified function of the eigenvalue α^s corresponding to its amplitude pattern $\mathbf{x}^s$ and the strength of the fixed feedback connections g and h within the local oscillators,

$$
\lambda_{1,2}^s = \tfrac{1}{2}[\alpha_s \pm \sqrt{(\alpha_s^2 - 4hg)}]
$$

and $\omega^s = \tfrac{1}{2}[\sqrt{(4hg - \alpha_s^2)}]$ (Baird, 1990b).

The second rule, with $c > d$, gives higher-order weights W_{ijkl} for the cubic terms in (20) which ensure, by the projection theorem, that these complex modes of the coupling matrix **T** become the only attractors in the vector field of the full nonlinear system. The forms of the eigenvectors of **T** and the corresponding asymptotic solutions $\mathbf{X}^s(t)$ established by this learning rule are

$$
\mathbf{P} = \begin{bmatrix} |\mathbf{x}^s| e^{i\boldsymbol{\theta}_x^s} \\ \sqrt{(g/h)}|\mathbf{x}^s| e^{i\boldsymbol{\theta}_y^s} \end{bmatrix} \Rightarrow
$$

$$
\mathbf{X}^s(t) = \begin{bmatrix} |\mathbf{x}^s| e^{i\boldsymbol{\theta}_x^s + i\omega^s t} \\ \sqrt{(g/h)}|\mathbf{x}^s| e^{i\boldsymbol{\theta}_y^s + i\omega^s t} \end{bmatrix}
\tag{23}
$$

Because of the restricted coupling, the only oscillations possible in this network have zero phase lag across space, i.e., are standing waves. The phase θ_x, θ_y is constant over the components of each kind of neural population x and y, and differs only between them. This is basically what is observed in the olfactory bulb (primary olfactory cortex) and prepyriform cortex (Baird, 1990b). The phase of inhibitory components θ_y always lags the phase of the excitatory components θ_x by approximately 90 degrees. In Baird (1990b) we show that this is a natural feature of the minimal network structure, since the phase lag for attractors s approaches -90 degrees asymptotically for all values of $hg \gg \alpha_s$ (or frequencies $\omega \gg \alpha_s$). This condition is satisfied in normal operation of the network, since attractor frequencies are typically orders of magnitude larger numbers than the positive eigenvalues of the **W** matrix, as is demonstrated in the numerical example below.

When the Hebbian learning rule (22) is used, the higher-order weights W_{ijkl} of the network model (20) can be decomposed so that (20) becomes

$$
\dot{x}_i = -\tau x_i - h y_i + \sum_{j=1}^{N/2} W_{ij} x_j
$$

$$
+ d \sum_{jkl=1}^{N/2} \tilde{W}_{ijkl} x_j x_k x_l - c x_i \sum_{j=1}^{N/2} x_j^2 + b_i \tag{24}
$$

$$
\dot{y}_i = -\tau y_i + g x_i \tag{25}
$$

where $\tilde{W}_{ijkl} = \sum_{s=1}^{N/2} x_i^s x_j^s x_k^s x_l^s$ comes from the multiple outer product in (22), and $-c x_i \sum_{j=1}^{N/2} x_j^2$ comes from the $c\delta_{ij}\delta_{jk}$ term in (22). As we discussed earlier, for the Hebb rule derived from the general projection algorithm, this single-weight negative term $-c x_i \sum_{j=1}^{N/2} x_j^2$ constitutes a shunting inhibitory bias which depends on the global excitatory activity of the network. "Shunting" here means multiplied by the current average cell voltage x_i of population i.

This is an input which is identical for all excitatory neural populations and could be calculated by a single node of the network which receives input from all excitatory populations as shown in Fig. 7.5. Such a node might correspond to one of the nuclei which lie below the prepyriform cortex. These send and receive the diffuse projections required to and from prepyriform cortex. The constants c and d of Eq. (24) give the magnitude of the inhibitory bias c and the average higher-order weight d. These constants are derived from entries in the normal form matrix **A**, and, as we have seen, $c > d$ guarantees stability of all stored patterns. The greater the bias c relative to d, the greater the "competition" between stored patterns, the more robust is the stability of the present attractor, and vice versa. This is the mechanism employed below in the biological sensory–motor architecture for central control of attractor transitions within modules.

Numerical Example

As we have noted before, the attractors of the numerical example given earlier were chosen so that they could also be stored by this network using these Hebbian rules. Here, the biological network structure we begin with determines the frequencies and phase patterns of all attractors. We *start* now with $g = h = 250$ for the **T** matrix, which we found by projection before. We know this will give frequencies of

$$\omega^s = \tfrac{1}{2}[\sqrt{(4hg - \alpha_s^2)}] = \tfrac{1}{2}[\sqrt{(4hg - 4)}]$$

$$= \tfrac{1}{2}[2\sqrt{(hg - 1)}] \cong \sqrt{h^2} = 250/2\pi \cong 40 \text{ Hz}$$

This Hebb rule (unlike the previous more general one) is applied directly to the amplitudes of the *excitatory* components *only* to get the matrix **W** and the higher-order weights $\tilde{W}_{ijkl}$ above. As for the previous Hebb rule, $\alpha = 2$, $c = 10$, and $d = 9$. The orthonormal excitatory amplitude patterns from the example are

$$\mathbf{x}_1 = \frac{1}{\sqrt{5}}\begin{bmatrix} 1 \\ 2 \end{bmatrix} \quad \text{and} \quad \mathbf{x}_2 = \frac{1}{\sqrt{5}}\begin{bmatrix} 2 \\ -1 \end{bmatrix}$$

Calculating **W** by hand,

$$W_{11} = 2(1/\sqrt{5})(1/\sqrt{5}) + 2(2/\sqrt{5})(2/\sqrt{5})$$
$$= 2/5 + 8/5 = 2$$

$$W_{22} = 2(2/\sqrt{5})(2/\sqrt{5}) + 2(-1/\sqrt{5})(-1/\sqrt{5})$$
$$= 8/5 + 2/5 = 2$$

$$W_{12} = W_{21}$$
$$= 2(1/\sqrt{5})(2/\sqrt{5}) + 2(2/\sqrt{5})(-1/\sqrt{5})$$
$$= 2/5 - 2/5 = 0$$

Thus

$$\mathbf{W} = \begin{bmatrix} 2 & 0 \\ 0 & 2 \end{bmatrix}$$

hence

$$\mathbf{T} = \begin{bmatrix} 2 & 0 & -250 & 0 \\ 0 & 2 & 0 & -250 \\ 250 & 0 & 0 & 0 \\ 0 & 250 & 0 & 0 \end{bmatrix}$$

as was found before by projection. The **W** matrix is diagonal in this case only because the eigenvalues α_s are equal, but full cross-coupling

of oscillators is still given by the higher-order terms. Similarly we can calculate $\tilde{W}_{ijkl}$ for these amplitude vectors,

$$\tilde{W}_{1111} = (1/\sqrt{5})(1/\sqrt{5})(1/\sqrt{5})(1/\sqrt{5})$$
$$+ (2/\sqrt{5})(2/\sqrt{5})(2/\sqrt{5})(2/\sqrt{5})$$
$$= 1/25 + 16/25 = 17/25 = 0.68$$

and so forth, to give the tensor components which can be displayed as four 2×2 matrices,

$$\tilde{W}_{11kl} = \begin{bmatrix} 0.68 & -0.24 \\ -0.24 & 0.32 \end{bmatrix}$$

$$\tilde{W}_{12kl} = \begin{bmatrix} -0.24 & 0.32 \\ 0.32 & 0.24 \end{bmatrix}$$

$$\tilde{W}_{21kl} = \begin{bmatrix} -0.24 & 0.32 \\ 0.32 & 0.24 \end{bmatrix}$$

$$\tilde{W}_{22kl} = \begin{bmatrix} 0.32 & 0.24 \\ 0.24 & 0.68 \end{bmatrix}$$

We get some negative weights here, and a phase reversal component in the oscillation of the second attractor, because we used a negative amplitude pattern value $x_2^2 = -1$ to get orthogonal patterns in this small example. In a larger more biologically realistic system where synapses respond to the mean amplitude of 3 to 5 cycles of 50 Hz activity over the 100-msec period of sensitivity for LTP, sparse patterns with all positive amplitudes would be used to get a biologically correct $\tilde{W}_{ijkl}$ with all positive weights, and biologically correct standing wave attractors. Now simulate the equations (24) above (with $\tau = 1$) to see the same relative amplitude attractors as in the previous example, even though only half of the higher-order terms are being used here (Eq. (22) for W_{ijkl} must be used to compare to the numerical values of the previous W_{ijkl}).

7.6.2. Architecture of Subnetwork Modules

Given the sensitivity of neurons to the location and arrival times of dendritic input (Jack et al., 1983), the successive volleys of pulses (the "wave packet" of Freeman) that are generated by the collective oscillation of a neural net may be ideal for reliable long-range transmission of the collective activity of one cortical area to another. On the hypothesis that oscillatory network modules like those described above form the actual substrate of the diverse sensory, motor, and cognitive operations found in various cortical areas, we are investigating a parallel distributed processing architecture that is designed to model, however simplistically, the architecture of cerebral cortex (Baird and Eeckman, 1992).

Because we can work with this class of mathematically well-understood associative memory networks, we can take a constructive approach to building a cortical architecture, using the networks as modules, in the same way that digital computers may be designed from well-behaved continuous analog flip-flop circuits. The operation of the architecture can be precisely specified, even though the outcome of a particular computation set up by the programmed connections may be unknown. Even though it is constructed from a system of continuous nonlinear ordinary differential equations, the system can operate as a discrete-time symbol processing architecture, like a cellular automaton, but with analog input and oscillatory or chaotic subsymbolic representations.

The architecture is such that the larger system is itself a special case of the type of network of the submodules, and can be analyzed with the same tools used to design the subnetwork modules. The modules can learn connection weights between themselves which will cause the system to evolve under a clocked "machine cycle" by a sequence of transitions of attractors within the modules, much as a digital computer evolves by transitions of its binary flip-flop states. Thus the architecture employs the principle of "computing with attractors" used by macroscopic systems for reliable computation in the presence of noise. Clocking is done by rhythmic variation of certain bifurcation parameters which hold sensory modules clamped at their attractors while motor states change, and then clamp motor states while sensory states are released to take new states based on input from external motor output and internal feedback.

Supervised learning by recurrent back-propagation or unsupervised reinforcement learning can then be used to train the connections between modules. When the inputs from one module to the next are given as impulses that establish initial conditions, the behavior of a module is exactly predicted by the projection theorem. Each module is described in normal form or mode coordinates as the k-winner-take-all network discussed above, where the winning set of units may have static, periodic, or chaotic dynamics. By choosing modules to have only two attractors, networks can be built which are similar to networks using binary units. There can be fully recurrent connections between modules. The entire supernetwork of connected modules, however, is itself a polynomial network that can be projected into standard network coordinates like those for the biological model. The attractors

within the modules may then be distributed patterns like those described for the biological model and observed experimentally in the olfactory system (Freeman and Baird, 1987). The system is still equivalent to the architecture of modules in normal form, however, and it may easily be designed, simulated, and theoretically evaluated in these coordinates.

The identity of the individual patches is preserved by the transformation between network and normal form coordinates. The connections *within* patches in network coordinates determine the particular distributed patterns that correspond to nodes in the normal form coordinates are transformed from those learned to effect transitions between winning nodes in normal form coordinates; they now effect transitions between the corresponding distributed patterns in the patches. Thus a network can be designed or analysed as a "spreading activation" style network where "concepts" correspond to single nodes, but be transformed to operate with the more fault-tolerant and biologically plausible distributed representations.

The "machine cycle" of the system is implemented by varying a "competition" bifurcation parameter in the sensory and motor patches, to alternately clamp one set of attractors, while the other is allowed to make any attractor transitions prescribed by the present sensory states. Motor states feed back to sensory states to cause transitions, so that internal activity can be self-sustaining, as in any recurrent network. After training, this internal activity can also be guided by a sequence of sensory inputs to produce a learned sequence of motor outputs. It can behave, in other words, like a finite-state automaton, with oscillatory or chaotic states. It can therefore be applied to problems like system identification and control, and grammatical inference and language recognition problems. We are exploring a version of the architecture that is equivalent to the "simple" recurrent "Elman" networks that are currently being applied to these problems (Elman, 1991); Pollack, 1990; Giles et al., 1992). The capabilities of this system are being investigated by application to the problem of word recognition from character sequences.

At low values of the clocked bifurcation parameter there is no internal "competition" between attractors within a module. The learned internal dynamics is effectively turned off, and the state vector is determined entirely by a clamped input vector. If there is only a single attractor within a module then that is a scalar multiple of whatever input vector is applied. The previous attracting state of a module is completely erased at this point. As the competition increases again, this attractor is forced to a new position in state space determined by the nearest learned attractor. The other learned attractors reappear by saddle–node bifurcations at a distance from this attractor and create other basins of attraction.

The greater the clock parameter, the greater the competition between attractors and the faster the convergence to a chosen attractor. At high values of competition the module is "clamped" at its current attractor, and the effect of the changing input it receives by feedback from the modules undergoing transition is negligible. The feedback between sensory and motor modules is therefore effectively cut. The supersystem can thus be viewed as operating in discrete time by transitions between a finite set of states. This kind of clocking and "buffering" of some states while other states relax is essential to the reliable operation of digital architectures. In our simulations, if we clock all modules to transition at once, the programmed sequences lose stability, and we get transitions to unprogrammed fixed points and simple limit cycles for the whole system. This is a primary justification for the use of the "machine cycle."

7.6.3. Simulations

Simulations show robust storage of oscillatory and chaotic attractor transition sequences in a system with a sinusoidal clock and continuous intermodule driving. We have shown that the dynamic attractors—oscillatory or chaotic— within the different modules of this architecture must rapidly synchronize or "bind" to effectively communicate information and produce reliable transitions (Baird and Eeckman, 1992). In these simulations, we synchronized Lorenz attractors for successful operation in the architecture using techniques of coupling developed by Chua (Tang et al., 1983; Endo and Chua, 1991) for secure communication by a modulated chaotic carrier wave.

Because communication between modules in the architecture is by continuous time-varying analog vectors, the process is more one of signal detection and pattern recognition by the modules of their inputs than it is "message passing" in the manner of a more conventional parallel architecture like the connection machine. This is why the demonstrated performance of the modules in handwritten character recognition is significant, and why we expect there are important possibilities in the architecture for the kinds of chaotic signal processing studied by Chua (Endo and Chua, 1991).

An important element of intracortical communication in the brain, and between modules in this architecture, is the ability of a module to detect and respond to the proper input signal from a particular module when inputs from other modules which are irrelevant to the present computation are contributing cross-talk and noise. We believe that synchronization is one important aspect of how the brain solves this coding problem. Attractors in these modules may be frequency coded during learning if biologically plausible excitatory-to-inhibitory connections are added in the lower left quadrant of the T matrix. Such attractors will synchronize only with the appropriate active attractors in other modules that have a similar resonant frequency. The same hardware (or "wetware") connection matrix can thus subserve many different networks of interaction between modules at the same time without cross-talk problems. Chaotic attractors may be "chaos coded," since only chaotic attractors with similar parameters will synchronize.

We have shown in these modules a superior stability of oscillatory attractors over otherwise identical modules with static attractors (frequencies set to zero) in the presence of additive Gaussian noise perturbations with the $1/f$ spectral character of the noise found experimentally in the brain (Freeman and Baird, 1987). This may help explain why the brain uses dynamic attractors. The oscillatory attractor acts like a bandpass filter and is effectively immune to the many slower macroscopic bias perturbations in the theta–beta range (3–25 Hz) below its 40–80 Hz passband, and the more microscopic perturbations of single neuron spikes in the 100–1000 Hz range. In an environment with this spectrum of perturbation, modules with static attractors cannot operate reliably. This suggests that the hierarchical organization of periodic activity in the brain is a form of large-scale frequency coding.

Exciting recent work of Brown, Chua, and Popp (unpublished) demonstrates that the vector fields of chaotic systems are robust to noise perturbation in a way that allows them to become detectors of signals buried in noise up to 45 dB—far below what can be detected by any present conventional signal detection technique. The Brown and Chua result indicates that chaotic attractors may be even more robust to noise perturbation than periodic attractors and lend special signal processing capabilities to the mechanisms of intercortical communication.

Many variations of the basic system configuration and operation, and various training algorithms, such as reinforcement learning with adaptive critics, are being explored. The ability to operate as a finite automaton with oscillatory/chaotic "states" is an important benchmark for this architecture, but only a subset of its capabilities. At low competition, the suprasystem reverts to one large continuous dynamical system. We expect that this kind of variation of the operational regime, especially with chaotic attractors inside the modules, though unreliable for habitual behaviors, may nonetheless be very useful in other areas such as the search process of reinforcement learning.

7.7. SUMMARY

The *normal form projection algorithm* with various extensions and architectures for associative memory storage of analog patterns, continuous sequences, and chaotic attractors in the same network has been described and mathematically analysed. A matrix inversion and a learning rule determines network weights, given prototype patterns to be stored. There are N units of capacity in an N-node *projection network* with $3N^2$ weights. It costs one unit per static attractor, two per Fourier component of each sequence, and at least three per chaotic attractor. There are no spurious attractors, and there is a Liapunov function in a special coordinate system which governs the approach of transient states to stored trajectories. Unsupervised or supervised incremental learning algorithms for pattern classification, such as competitive learning or bootstrap Widrow–Hoff, can easily be implemented in this network. Network behavior was illustrated by a discussion of its application to the problem of real-time handwritten digit recognition.

The architecture can be "folded" into a recurrent network with higher-order weights that can be used as a model of cortex that stores oscillatory and chaotic attractors by a Hebb rule. Hierarchical sensory–motor networks may further be constructed of interconnected "cortical patches" of these network modules. Computation then occurs as a sequence of transitions of attractors in these associative memory subnetworks—much as a digital computer evolves by transitions of state in its binary flip-flop elements. The architecture thus employs the principle of "computing with attractors" used by macroscopic systems for reliable computation in the presence of noise.

ACKNOWLEDGMENTS

This work was supported by AFOSR-87-0317 and a grant from LLNL. It is a pleasure to acknowledge the invaluable assistance of Morris Hirsch, Walter Freeman, Todd Troyer, and Jim Vacarro.

REFERENCES

Ahalt, C., Krishnamurthy, A., Chen, P., and Melton, D. (1990). "Competitive Learning Algorithms for Vector Quantization," *Neural Networks*, **3**, 277–290.

Baird, B. (1986). "Nonlinear Dynamics of Pattern Formation and Pattern Recognition in the Rabbit Olfactory Bulb," in *Evolution, Games, and Learning: Models for Adaptation in Machines and Nature*, D. Farmer, A. Lapedes, N. Packard, and B. Wendroff, eds. North-Holland, Amsterdam, pp. 150–176. (Also in *Physica D*, Vol. 22.)

Baird, B. (1989). "A Bifurcation Theory Approach to Vector Field Programming for Periodic Attractors," in *Proc. Int. Joint Conference on Neural Networks*, Washington, D.C., Vol. 1, pp. 381–388.

Baird, B. (1990a). "Associative Memory in a Simple Model of Oscillating Cortex," in *Advances in Neural Information Processing Systems 2*, D. S. Touretsky, ed. Morgan Kaufman, San Mateo, CA, pp. 68–75.

Baird, B. (1990b). "Bifurcation and Learning in Network Models of Oscillating Cortex," in *Emergent Computation*, S. Forest, ed. North-Holland, Amsterdam, pp. 365–384. (Also in *Physica D*, Vol. 42.)

Baird, B. (1990c). "A Learning Rule for Cam Storage of Continuous Periodic Sequences," in *Proc. Int Joint Conference on Neural Networks*, San Diego, Vol. 3, pp. 493–498.

Baird, B. (1992a). "Learning with Synaptic Nonlinearities in a Coupled Oscillator Model of Olfactory Cortex," in *Analysis and Modeling of Neural Systems*, F. Eeckman, ed. Kluwer, Norwell, MA, pp. 319–327.

Baird, B. (1992b). *Bifurcation Theory Approach to the Analysis and Synthesis of Neural Networks for Engineering and Biological Modeling*. Research Notes in Neural Computing. Springer-Verlag, New York, in press.

Baird, B. and Eeckman, F. H. (1991). "Cam Storage of Analog Patterns and Continuous Sequences with $3n^2$ Weights," in *Advances in Neural Information Processing Systems 3*, D. S. Touretsky, ed. Morgan Kaufman, San Mateo, CA, pp. 192–198.

Baird, B. and Eeckman, F. H. (1992). "A Hierarchical Sensory–motor Architecture of Oscillating Cortical Area Subnetworks," in *Analysis and Modeling of Neural Systems II*, F. H. Eeckman, ed. Kluwer, Norwell, MA, pp. 196–204.

Baird, B., Freeman, W., Eeckman, F., and Yao, Y. (1991). "Applications of Chaotic Neurodynamics in Pattern Recognition," in *SPIE Proc.*, **1469**, 12–23.

Chua, L. O., Motomasa, K., and Matsumoto, T. (1986). "The Double Scroll Family," *IEEE Trans. Circuits and Systems*, **CAS-33**(11), 1073–1118.

Cruz, J. and Chua, L. (1991). 'A CNN Chip for Connected Component Detection,' *IEEE Trans. Circuits and Systems*, **38**(7), 812–817.

Eckhorn, R., Bauer, R., Jordan, W., Brosch, M., Kruse, W., Munk, M., and Reitboeck, H. (1988). "Coherent Oscillations: A Mechanism of Feature Linking in the Visual Cortex?" *Biological Cybernetics*, **60**, 121.

Elman, J. (1991). "Distributed Representations, Simple Recurrent Networks and Grammatical Structure," *Machine Learning*, **7**(2/3), 91.

Endo, T. and Chua, L. O. (1991). "Synchronization and Chaos in Phase-locked Loops," *IEEE Trans. Circuits and Systems*, **38**(12), 620–626.

Engel, A. K., Konig, P., Gray, C., and Singer, W. (1990). "Synchronization of Oscillatory Responses: A Mechanism for Stimulus-dependent Assembly Formation in Cat Visual Cortex," in *Parallel Processing in Neural Systems and Computers*, R. Eckmiller, ed. Elsevier, Amsterdam, pp. 105–108.

Freeman, W. (1975). *Mass Action in the Nervous System*. Academic Press, New York.

Freeman, W. (1987). "Simulation of Chaotic EEG Patterns with a Dynamic Model of the Olfactory System," *Biological Cybernetics*, **56**, 139.

Freeman, W. and Baird, B. (1987). "Relation of Olfactory EEG to Behavior: Spatial Analysis," *Behavioral Neurosci.*, **101**, 393–408.

Freeman, W. J. and van Dijk, B. W. (1987). "Spatial Patterns of Visual Cortical EEG During Conditioned Reflex in a Rhesus Monkey," *Brain Res.*, **422**, 267.

Freeman, W. J., Yao, Y., and Burke, B. (1988). "Central Pattern Generating and Recognizing in Olfactory Bulb: A Correlation Learning Rule," *Neural Networks*, **1**(4), 277.

Giles, C., Miller, C. B., Chen, D., Chen, H., Sun, G., and Lee, Y. (1992). "Learning and Extracting Finite State Automata with Second Order Recurrent Neural Networks," *Neural Computation*, in press.

Golubitsky, M. and Schaefer, D. (1985). *Singularities and Groups in Bifurcation Theory*, Vol. I. Springer-Verlag, New York.

Gray, C. M. and Singer, W. (1987). "Stimulus Dependent Neuronal Oscillations in the Cat Visual Cortex Area 17," *Neuroscience* [Suppl.], **22**, 1301P.

Gray, C., Konig, P., Engel, A., and Singer, W. (1989). "Oscillatory Responses in Cat Visual Cortex Exhibit Intercolumnar Synchronization Which Reflects Global Stimulus Properties," *Nature* (*London*), **338**, 334–337.

Grossberg, S. and Elias, S. (1975). "Pattern Formation, Contract Control, and Oscillations in the Short Term Memory of Shunting On-center Off Surround Networks," *Biological Cybernetics*, **20**, 69.

Guckenheimer, J. and Holmes, D. (1983). *Nonlinear Oscillations, Dynamical Systems, and Bifurcations of Vector Fields*. Springer-Verlag, New York.

Haberly, L. B. and Bower, J. M. (1989). "Olfactory Cortex: Model Circuit for Study of Associative Memory?" *Trends in Neurosci.*, **12**(7), 258.

Hemmen, J. v. (1987). "Nonlinear Neural Networks Near Saturation," *Phys. Rev. A*, **36**, 1959.

Herz, A., Sulzer, B., Kuhn, R., and van Hemmen, J. L. (1989). "Hebbian Learning Reconsidered: Representation of Static and Dynamic Objects in Associative Neural Nets," *Biological Cybernetics*, **60**, 457.

Hirsch, M. W. and Smale, S. (1974). *Differential Equations, Dynamical Systems, and Linear Algebra*. Academic Press, New York.

Hopfield, J. (1984). "Neurons with Graded Response Have Collective Computational Properties Like Those of Two State Neurons," *Proc. Nat. Acad. Sci. U.S.A.*, **81**, 3088.

Horn, D. and Usher, M. (1991). "Parallel Activation of Memories in an Oscillatory Neural Network," *Neural Computation*, **3**(1), 31–43.

Jack, J., Noble, D., and Tsien, R. (1983). *Electric Current Flow in Excitable Cells*. Clarendon Press, Oxford.

Lenihan, S. (1987). *Spontaneous Oscillations in Neural Networks*. Technical report, U.S. Army Yuma Proving Ground.

Li, Z. and Hopfield, J. (1989). "Modeling the Olfactory Bulb and Its Neural Oscillatory Processings," *Biological Cybernetics*, **61**, 379.

Maxwell, T., Giles, C., Lee, Y., and Chen, H. (1986). "Nonlinear Dynamics of Artificial Neural Systems," in *Neural Networks for Computing—AIP Conf. Proc. 151*. AIP, New York, p. 299.

Pollack, J. (1990). *The Induction of Dynamical Recognizers*. Tech. Report 90-JP-Automata, Dept. of Computer and Information Science, Ohio State University.

Psaltis, D. and Venkatesh, S. S. (1989). "Neural Associative Memory," in *Evolution, Learning, and Cognition*, Y. Lee, ed. World Scientific, New York.

Psaltis, D., Park, C. H., and Hong, J. (1988). "Higher Order Associative Memories," *Neural Networks*, **1**(2), 149.

Sompolinsky, H. and Kantner, L. (1986). "Temporal Association in Asymmetric Neural Networks," *Phys. Rev. Lett.*, **57**, 2861.

Stanton, P. K. and Sejnowski, T. (1989). "Storing Covariance by the Associative Long-term Potentiation and Depression of Synaptic Strengths in the Hippocampus," in *Advances in Neural Information Processing Systems, I*, D. Touretzky, ed. Morgan Kaufmann, San Mateo, CA, p. 402.

Tang, Y., Mees, A., and Chua, L. (1983). "Synchronization and Chaos," *IEEE Trans. Circuits and Systems*, **CAS-30**(9), 620–626.

Wang, D., Buhmann, J., and von der Malsburg, C. (1990). "Pattern Segmentation in Associative Memory," *Neural Computation*, **2**(1), 94–106.

PART III

ANALYSIS OF MEMORY DYNAMICS AND CAPACITY

8

Statistical Neurodynamics of Various Types of Associative Nets

SHUN-ICHI AMARI AND HIRO-FUMI YANAI

8.1. INTRODUCTION

A couple of decades have been passed since the correlation type associative memory was proposed (Anderson, 1972; Kohonen, 1972; Nakano, 1972). Its statistical analysis was given by Uesaka and Ozeki (1972), and dynamical properties of recalling processes were analyzed by Amari (1972). Amari (1972) studied not only the stability of memorized patterns as equilibria, but also the stability of recall of memorized pattern sequences by using a net with asymmetric connections.

It was Hopfield (1982) who opened a new direction of research by introducing the concept of the energy function (Liapunov function) in associative memory nets of symmetric connections. This was pointed out in analogy with the statistical physics of spin glass, which has invited a large group of physicists to this field. He also remarked the capacity m of the correlation type associative net, which is the maximum number of patterns memorized in a net with n neurons, and suggested by computer simulation that the capacity is about

$$m = 0.15n$$

when n is large if a small percentage of recollection error bits is allowed. This capacity is called the relative capacity.

A lot of research has appeared since then. The absolute capacity, on the other hand, denotes the maximum number of memorized patterns which can be recalled without any error. This exact recall required that the memorized patterns be equilibria of the network dynamics. The absolute capacity is proved asymptotically to be

$$m = \frac{n}{4 \log n}$$

by McEliece et al. (1987) by the combinatorial and information-theoretic approach (refer also to

Chapters 1 and 12). An epoch-making work was the theoretical calculation of the relative capacity (Amit et al., 1985a, b)

$$m = 0.14n$$

by using the replica method, which is not yet justified but is believed to give a good approximate value. Gardner (1986), employed a statistical-mechanical technique to calculate the distribution of equilibrium states in the associative net. It was shown how many and where in the state space equilibria exist relative to the memorized patterns. It was also shown that the correlation associative memory has a large number of spurious memories. However, the application of the statistical-mechanical method is mostly limited to the properties of equilibrium states.

The dynamical process of recall, in particular the transient process, is much more interesting than identifying equilibria but is difficult to analyze. Amari and Maginu (1988) applied the method of statistical neurodynamics (Amari, 1974; Amari et al., 1977) to the analysis of the transient dynamical behavior of memory recall by introducing new macroscopic variables, and obtained the relative capacity $m = 0.16n$. Although this theory is not exact but approximate, it is known that the dynamical equations are in very good agreement with simulation results when recall is successful, but not so good when recall fails (Nishimori and Ozeki, 1992). See also Zagrebnov and Chvyrov (1989) and Patrick and Zagrebnov (1991) for dynamic treatments.

From these works, it is now widely recognized that the simple prototype of the autocorrelation associative memory shows very complex dynamical behaviors. There are various modifications of the simple prototype correlation associative memory model. One direction is to introduce a cascaded or cyclic structure in the model. The bidirectional associative memory (BAM) is a simple cyclic structure model (Kosko, 1986) (also, see Chapter 5 in this volume). Meir and Domany (1987) introduced a cascaded model, and gave exact analysis of the recall processes (see also Domany et al., 1991). Amari (1988) showed a unified statistical-neurodynamical approach to various architectures of associative nets.

Another direction is to introduce a restriction on connectivity. Shinomoto (1987) showed that an associative memory model having two types of neurons, namely, excitatory and inhibitory neurons, has an interesting property when recall fails. Yanai et al. (1991) introduced a structural restriction on possible synaptic connections, and showed that the memory capacity per connection increases with the sparsity of connections. It was pointed out by Willshaw et al. (1969) and Palm (1980) that the capacity of an associative memory model increases when patterns to be memorized have sparsity in excitations, that is, when the number of excited elements in a pattern is extremely small compared to that of quiescent elements. Memory capacity increases drastically as the sparsity increases. A scaling law of the effect of sparsity was studied in detail by Meunier et al. (1991) (see also Amari, 1989). It was shown by Amari et al. (1977) that one can obtain the exact macroscopic dynamical equation of state transition for a randomly connected network for the case of extremely sparse connections. A similar result was obtained by Derrida et al. (1987) for autocorrelation associative nets.

In yet other directions of research, Kohonen and Ruohonen (1973) introduced the generalized inverse matrix to replace the correlation matrix (see Chapter 1 for details), and Amari (1977) analyzed its dynamical property. Gardner (1988) showed that the memory capacity becomes $m = 2n$ when an optimal connection matrix is introduced. All of these results show that the absolute capacity can be increased to be proportional to n, instead of $n/\log n$. However, none of these models is free of spurious memories. Recently, Morita et al. (1990) introduced a model consisting of neurons with nonmonotonic activation and showed that the absolute capacity increases to $m = 0.4n$, and that the model is almost free from spurious memories. In his model, the state exhibits a chaotic behavior when recall fails. Yoshizawa et al. (see Chapter 13 in this volume) recently gave a statistical-neurodynamical analysis for such a model.

The statistical-neurodynamical method is useful and important in analyzing the dynamic behavior of associative memory models. The present chapter is an introduction to the statistical-neurodynamical method, which is different from the earlier physicists' statistical-neurodynamical analysis approach. Here we present a unified approach to the analysis of various architectures of associative memories such as the simple cross-correlation associative memory model, cascaded model, cyclic model, BAM, autocorrelation model, and associative sequence generator model.

8.2. ASSOCIATIVE MEMORY MODELS

This section is an introduction to the various architectures of associative memory models which have been proposed so far. The behavior of these models will be analyzed by a unified mathematical method called the statistical neurodynamical method (note that this is not the same as the spin-glass statistical-mechanical method).

8.2.1. Cross-Associative Nets

A cross-associative net is a one-layer neural network which associates input patterns with corresponding output patterns. Let us denote by $(\mathbf{s}^{(1)}, \mathbf{t}^{(1)}), (\mathbf{s}^{(2)}, \mathbf{t}^{(2)}), \dots, (\mathbf{s}^{(m)}, \mathbf{t}^{(m)})$ m pairs of input and output memory patterns to be memorized. When $\mathbf{s}^{(\mu)}$ is input to the net, its output is expected to be $\mathbf{t}^{(\mu)}$, $\mu = 1, 2, \dots, m$. Here, $\mathbf{s}^{(\mu)}$ and $\mathbf{t}^{(\mu)}$ are vectors whose components $s_i^{(\mu)}$ and $t_i^{(\mu)}$ take on the binary values 1 and -1. The Greek letters μ, λ, ν, etc., are used to denote pattern labels and the Roman indices i, j, k, etc., are reserved for denoting components of vectors.

Now we describe the behavior of a cross-associative net. It is a one-layer net (Fig. 8.1), receiving a vector input $\mathbf{x}$ and emitting a vector output. Let w_{ij} be the synaptic connection weight from the jth input to the ith output neuron. The input–output behavior is then described by

$$y_i = f\left(\sum_{j=1}^{n} w_{ij} x_j \right) \tag{1}$$

where

$$f(u) = \begin{cases} 1 & u \geq 0 \\ -1 & u < 0 \end{cases} \tag{2}$$

Here, the ith component of the vectors $\mathbf{x}$ and $\mathbf{y}$ is denoted by x_i and y_i, respectively, and they take on the values 1 and -1; 1 for the firing state and -1 for the resting state. If one prefers a vector notation, we can rewrite the above equation as

$$\mathbf{y} = f(\mathbf{W}\mathbf{x}) \tag{3}$$

where $\mathbf{W} = (w_{ij})$ is the connection matrix.

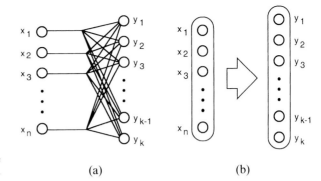

Fig. 8.1. Architecture of a cross-associative net (a) and its simplified representation (b).

(a)   (b)

The correlation type associative memory uses the following connection matrix $\mathbf{W} = (w_{ij})$,

$$w_{ij} = \frac{1}{n} \sum_{\mu=1}^{m} t_i^{(\mu)} s_j^{(\mu)} \qquad (4)$$

Here w_{ij} is the correlation of the excitation of the ith and jth neurons of the input–output pairs to be memorized. In vector notation, we have

$$\mathbf{W} = \frac{1}{n} \sum_{\mu=1}^{m} \mathbf{t}^{(\mu)} \mathbf{s}^{(\mu)\prime} \qquad (5)$$

where $\mathbf{s}$ and $\mathbf{t}$ are column vectors, $'$ denotes transposition, and n is the vectors' dimension. It is not necessary to assume that the input and output vectors have the same dimension but it is assumed here only to simplify notation.

The present simple model can be used as a prototype for other models described later.

8.2.2. Cascaded Associative Nets

When we connect a number of one-layer cross-associative nets, as shown in Fig. 8.2, we have a cascaded associative net. There are L layers connected in cascade. Neurons are connected via synapses from layer to layer. Synaptic connections exist only between neighboring

layers and connections are one-way. Although, in general, the number of neurons in each layer can be different. Here, for simplicity, we use an equal number of neurons for every layer. The results which follow can easily be extended to the case where the number of neurons varies from layer to layer.

Let us consider m sequences $S^1, S^2, \ldots, S^m$ of patterns of length $L + 1$:

$$S^1: \mathbf{s}^{(1)}(1) \to \mathbf{s}^{(1)}(2) \to \cdots \mathbf{s}^{(1)}(L + 1)$$
$$S^2: \mathbf{s}^{(2)}(1) \to \mathbf{s}^{(2)}(2) \to \cdots \mathbf{s}^{(2)}(L + 1)$$
$$\vdots$$
$$S^m: \mathbf{s}^{(m)}(1) \to \mathbf{s}^{(m)}(2) \to \cdots \mathbf{s}^{(m)}(L + 1)$$

The cascaded associative net is expected to have the following behavior: when the head pattern $\mathbf{s}^{(\mu)}(1)$ of sequence S^μ is presented, the second layer outputs $\mathbf{s}^{(\mu)}(2)$, the third layer outputs $\mathbf{s}^{(\mu)}(3)$, and so on, and the last layer outputs the tail $\mathbf{s}^{(\mu)}(L + 1)$, recalling the whole sequence S^μ for any μ ($\mu = 1, \ldots, m$).

Here, correlation type connections are employed. The synaptic connections from the lth layer to $(l + 1)$st ($l = 1, 2, \ldots, L$) are given by

$$w_{ij}(l) = \frac{1}{n} \sum_{\mu=1}^{m} s_i^{(\mu)}(l + 1) s_j^{(\mu)}(l) \quad (i \neq j) \qquad (6)$$

We also impose a subsidiary condition that

$$w_{ii}(l) = 0$$

for the sake of simplicity. However, our analysis is still valid without this condition.

When an arbitrary pattern $\mathbf{x}(1)$ is presented, the net outputs $\mathbf{x}(2), \mathbf{x}(3), \ldots, \mathbf{x}(L + 1)$ from the corresponding layers. The state of the ith neuron in the lth layer is determined by the following rule:

$$x_i(l + 1) = f\left(\sum_{j=1}^{n} w_{ij}(l) x_j(l) \right) \quad l = 1, 2, \ldots, L$$

$$(7)$$

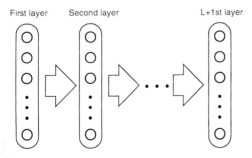

First layer   Second layer   L+1st layer

Fig. 8.2. A cascaded associative net.

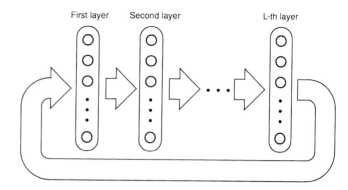

Fig. 8.3. A cyclic associative net.

The cascaded associative net was proposed by Meir and Domany (1987) and a rigorous mathematical analysis was given. In this chapter, we will give a different analysis which can also be used for other models.

8.2.3. Cyclic Associative Nets and BAM

A cyclic associative net is obtained by connecting the outputs of the Lth layer of a cascaded associative net to the inputs of the first layer (Fig. 8.3). This is a cyclically connected net.

Let us consider m cyclic sequences C^μ ($\mu = 1, \ldots, m$) of length L,

$$C^\mu : \mathbf{s}^\mu(1) \to \mathbf{s}^\mu(2) \to \cdots \to \mathbf{s}^\mu(L) \to \mathbf{s}^\mu(1) \to \cdots$$

The cyclic net is expected to reproduce any of the cycles when $\mathbf{s}^\mu(l)$ or a noisy version of it is presented at the input of the lth layer. The connections from the lth layer to the $(l + 1)$st layer are determined by (6), and those from the Lth layer to the first layer are given by

$$w_{ij}(L) = \frac{1}{n} \sum_{\mu=1}^{m} s_i^{(\mu)}(1) s_j^{(\mu)}(L)$$

$$w_{ii}(L) = 0$$

When the number L of layers is sufficiently large, the dynamical property of the cyclic net is considered to be identical to the cascaded net which has been rigorously investigated by Meir and Domany (1987). In the following analyses, we can see that the dynamics is approximately described by the same equation if L is greater than or equal to 3 (Amari, 1988). If L is equal to 1 or 2, however, the dynamics is governed by different equations. A cyclic net with $L = 2$ is called a BAM (bidirectional associative memory) (Kosko, 1988) and the net with $L = 1$ is an autoassociative net and will be considered next.

Since we will not deal with BAMs in detail (the reader is referred to Chapter 5 in this volume), we just note that cyclic-associative nets with $L \geq 3$ can store memories more efficiently than BAMs, and BAMs are more efficient than autoassociative nets. This property can also be understood qualitatively by the result on sparsely connected autoassociative nets, where the efficiency of memory storage increases as the sparsity of connections increases.

8.2.4. Autoassociative Nets

An autoassociative net is a recurrently connected net in which m patterns $\mathbf{s}^{(1)}, \ldots, \mathbf{s}^{(m)}$ are memorized in the form of its stable equilibrium states. Memorized patterns are recalled by the state transition dynamics of the net. This net could also be regarded as a cyclic-associative net with $L = 1$ (see also the next section). In this net, neurons are connected with other neurons via recurrent synapses. Thus, the synaptic weights are defined as follows,

$$w_{ij} = \frac{1}{n} \sum_{\mu=1}^{m} s_i^{(\mu)} s_j^{(\mu)} (i \neq j) \quad w_{ii} = 0 \quad (8)$$

Let $\mathbf{x}(t)$ be the state of the net at time t, $x_i(t)$ being its ith component. The dynamics of the net are given by

$$x_i(t + 1) = f\left(\sum_{j=1}^{n} w_{ij} x_j(t) \right) \quad (9)$$

where t stands for discrete times ($= 0, 1, 2, \ldots$). A state $\mathbf{s}$ is said to be an equilibrium when

$$s_i = f\left(\sum_{j=1}^{n} w_{ij} s_j \right)$$

is satisfied. Dynamical proprties of this net were

analyzed as early as 1972 by Amari (1972). Later, Hopfield (1982) pointed out its interesting dynamical aspects (e.g., the existence of a potential function and its capacity) in analogy with magnetic phenomena in theoretical physics, and then many research papers appeared including the work of Amit et al. (1985a, b), Gardner et al. (1987), Gardner (1988) and Amari and Maginu (1988).

8.2.5. Associative Sequence Generator

Let us consider a sequence of patterns

$$\mathbf{s}^{(1)} \to \mathbf{s}^{(2)} \to \mathbf{s}^{(\mu)} \to \cdots \to \mathbf{s}^{(m+1)}$$

It is a cycle when $\mathbf{s}^{(1)} = \mathbf{s}^{(m+1)}$. A net with recurrent connections (like the autoassociative net) is called a sequence-generating associative net or a sequence generator when its synaptic weights are defined by

$$w_{ij} = \frac{1}{n} \sum_{\mu=1}^{m} s_i^{(\mu+1)} s_j^{(\mu)} \quad (i \neq j) \quad w_{ii} = 0 \quad (10)$$

The dynamics of this net are defined by Eq. (9). In general, it is possible to memorize a number of sequences and/or cycles, and the numbers of patterns included in these sequences and/or cycles can be different. Even in the general cases, dynamical properties of the net remain unchanged within the level of approximation employed here if the lengths of the cycles are greater than or equal to 3 (see Amari, 1972; Fukushima, 1973). Therefore, we confine ourselves to the study of a net with a single stored sequence for the sake of simplicity.

8.3. A UNIFIED TREATMENT OF VARIOUS TYPES OF ASSOCIATIVE NETS

Thus far, we have described various architectures of associative nets. In order to make a unified theoretical analysis and to understand the nature of such associative nets, we have to discuss a unified method of approach. We will then be able to discuss advantages and disadvantages of certain architectures of nets. To this end, we try to find a framework covering all of these architectures.

The most elementary prototype is the cross-associative net, where the connection matrix w_{ij} is not symmetric, as is shown in (4). We will show the statistical analysis of the behavior of this network in the next section.

The second prototype is the cascaded associative net which is a concatenation of cross-associative nets. However, its property is not merely a concatenation of those of cross-associative nets, but a new type of mutual statistical interactions arises because randomly assigned output patterns $\mathbf{s}^{\mu}(l)$ of the lth layer are used as the input patterns to the $(l+1)$st layer. Therefore, the connection matrices $\mathbf{W}(l) = [w_{ij}(l)]$ and $\mathbf{W}(l+1) = [w_{ij}(l+1)]$ of the lth and $(l+1)$st layers share common factors as is seen in (6), and so that they are not independent. Therefore, when analyzing the behavior of such nets, these correlations should be taken into account.

The thid prototype is the autoassociative net, in which the output of the net is fed back to its input. This implies that the memorized input patterns and the output patterns are identical, so that the connection matrix $\mathbf{W}$ is symmetric as is seen in (8). This gives rise to complicated correlations when analyzing the dynamics of the net. On the other hand, the sequence-generating associative net has the same architecture of recurrent connections as that of the autoassociative net, but with different input and output patterns to be memorized. Therefore, the connection matrix $\mathbf{W}$ is not symmetric as is shown in (10), and its analysis is easier.

The cyclic-associative nets, including the BAM, can be considered as a special architecture of the autoassociative net. A cyclic-associative net is a version of the autoassociative net in which a number of synapses are disconnected, that is, w_{ij} is put equal to 0 when the synapse from the jth neuron is disconnected.

Let us consider a cyclic-associative net of L layers, each layer consisting of n neurons, so that there are Ln neurons in all. This can be regarded as a version of a recurrently connected autoassociative net where a memorized pattern is constructed from a set of L memory patterns for the cyclic net. Combining L patterns in series, we define

$$\tilde{\mathbf{s}}^{(1)} = \mathbf{s}^{(1)}(1) \oplus \mathbf{s}^{(1)}(2) \oplus \cdots \oplus \mathbf{s}^{(1)}(L)$$
$$\tilde{\mathbf{s}}^{(2)} = \mathbf{s}^{(2)}(1) \oplus \mathbf{s}^{(2)}(2) \oplus \cdots \oplus \mathbf{s}^{(2)}(L)$$
$$\vdots$$
$$\tilde{\mathbf{s}}^{(m)} = \mathbf{s}^{(m)}(1) \oplus \mathbf{s}^{(m)}(2) \oplus \cdots \oplus \mathbf{s}^{(m)}(L)$$

and from these patterns the enlarged total synaptic weight matrix is constructed as

$$\tilde{w}_{ij} = \frac{1}{n} c_{ij} \sum_{\mu=1}^{m} \tilde{s}_i^{(\mu)} \tilde{s}_j^{(\mu)} \quad \tilde{w}_{ii} = 0 \quad (11)$$

where $c_{ij} = 1$ if $(l-1)n + 1 \leq i \leq ln$ and

$$(l-2)n + 1 \leq j \leq (l-1)n \quad (l = 2, 3, \ldots, L)$$

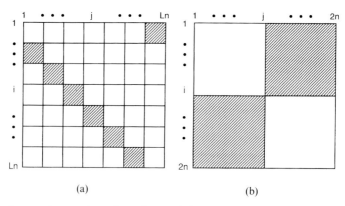

Fig. 8.4. Structure of connection matrices of a cyclic associative net (a) and a BAM (b).

or $1 \leq i \leq n$ and $(L-1)n + 1 \leq j \leq Ln$ hold, and $c_{ij} = 0$ otherwise. If we represent the distribution of c_{ij} in matrix form, we have Fig. 8.4a, where the (i, j)-element of the matrix represents the value of c_{ij} and the hatched regions are for $c_{ij} = 1$. The dynamics of the net is defined by

$$x_i(t + 1) = f\left(\sum_{j=1}^{Ln} \tilde{w}_{ij} x_j(t) \right) \tag{12}$$

Dynamical properties of this autoassociative net are equivalent to those of the corresponding cyclic-associative net. However, since $c_{ij} \neq c_{ji}$, the total connection matrix $\tilde{\mathbf{W}} = [\tilde{w}_{ij}]$ is not symmetric. Hence, this reduces correlations in dynamics.

Similarly, the dynamics of a BAM is equivalent to that of the autoassociative net with $2n$ neurons and with a c_{ij}-matrix shown in Fig. 8.4b. In this way, a variety of classes of associative nets can be regarded as special cases of the autoassociative net. Hence we may compare the capabilities of various associative nets based on a common standard (Yanai et al., 1991).

8.4. BEHAVIOR OF CROSS-ASSOCIATIVE NETS

A cross-associative net memorizes m input–output pairs $(\mathbf{s}^{(\mu)}, \mathbf{t}^{(\mu)})$, $\mu = 1, \ldots, m$, such that $\mathbf{t}^{(\mu)}$ is the output from the net when $\mathbf{s}^{(\mu)}$ is presented at the input. It is desirable that when a pattern $\mathbf{x}$ close to $\mathbf{s}^{(\mu)}$ is presented, the net outputs $\mathbf{y}$ which is also very close to $\mathbf{t}^{(\mu)}$. In order to evaluate this property of recall, we introduce a distance in the set of patterns.

The distance between two signal patterns $\mathbf{x}$ and $\mathbf{y}$ is defined by

$$d(\mathbf{x}, \mathbf{y}) = \frac{1}{2n} \sum_{j=1}^{n} |x_j - y_j|$$

where x_j and y_j are components of $\mathbf{x}$ and $\mathbf{y}$, respectively, taking on the binary values 1 and -1. Therefore, the distance d denotes the number of components in which $\mathbf{x}$ differs from $\mathbf{y}$ divided by n. It is called the relative Hamming distance, satisfying $0 \leq d \leq 1$.

Let $\mathbf{x}$ be an input which is close to one of the memorized inputs, say $\mathbf{s}^{(\mu)}$. Let

$$d = d(\mathbf{x}, \mathbf{s}^{(\mu)})$$

be its distance from $\mathbf{s}^{(\mu)}$. Let

$$\mathbf{y} = f(\mathbf{W}\mathbf{x})$$

be the output corresponding to $\mathbf{x}$. We evaluate the association performance of the net by the distance

$$d' = d(\mathbf{y}, \mathbf{t}^{(\mu)})$$

of the output from the desired $\mathbf{t}^{(\mu)}$ as a function of d. If

$$d' \ll d$$

holds, we may say that the recall of the memorized pattern is excellent since the output $\mathbf{y}$ is very close to $\mathbf{t}^{(\mu)}$ compared to the distance of the input $\mathbf{x}$ from $\mathbf{s}^{(\mu)}$. In order to analyze the performance, we use a statistical method by assuming that each component $s_i^{(\mu)}$ or $t_i^{(\mu)}$ of all the $\mathbf{s}^{(\mu)}$ and $\mathbf{t}^{(\mu)}$ are randomly and independently generated, taking values 1 and -1 with probability 0.5 each.

Let $\mathbf{x}$ be an input pattern whose distance from the νth input pattern $\mathbf{s}^{(\nu)}$ is d. It is required that the corresponding output pattern $\mathbf{y}$ is as close to the vth output pattern $\mathbf{t}^{(v)}$ as possible. Also, let d' be the distance between $\mathbf{y}$ and $\mathbf{t}^{(v)}$. These distances are given by

$$
\left.
\begin{aligned}
d &= \frac{1}{2n} \sum_{j=1}^{n} |x_j - s_j^{(v)}| \\
d' &= \frac{1}{2n} \sum_{i=1}^{n} |y_i - t_i^{(v)}|
\end{aligned}
\right\} \tag{13}
$$

By using Eq. (4), the weighted sum u_i of the stimuli which the ith output neuron receives is

$$
\begin{aligned}
u_i &= \sum_{j=1}^{n} w_i^j x_j = \frac{1}{n} \sum_{\mu=1}^{m} \sum_{j=1}^{n} t_i^{(\mu)} s_j^{(\mu)} x_j \\
&= \frac{1}{n} \sum_{\mu=1}^{m} t_i^{(\mu)} (\mathbf{s}^{(\mu)} \cdot \mathbf{x})
\end{aligned} \tag{14}
$$

where the dot means the scalar product. The output y_i is 1 when $u_i \geq 0$ and -1 otherwise.

Since the distance between $\mathbf{x}$ and $\mathbf{s}^{(v)}$ is d, it is easy to show that their inner product is

$$
\frac{1}{n} \mathbf{s}^{(v)} \cdot \mathbf{x} = 1 - 2d \tag{15}
$$

This is the term in u_i contributing to the emission of the desired output $t_i^{(v)}$, and all the other terms are regarded as noises disturbing recollection. So we write u_i as

$$
\left.
\begin{aligned}
u_i &= t_i^{(v)} (1 - 2d) + N_i \\
N_i &= \frac{1}{n} \sum_{\mu \neq v} \sum_{j=1}^{n} t_i^{(\mu)} s_j^{(\mu)} x_j
\end{aligned}
\right\} \tag{16}
$$

We now evaluate the noise term N_i, by taking into account the statistical property of $t_i^{(\mu)}$ and $s_i^{(\mu)}$ being randomly and independently chosen subject to the same distribution. The central limit theorem guarantees the following lemma.

LEMMA 1. *The noise or interference term N_i is distributed subject to the normal distribution with the average and variance given, respectively, by*

$$
E[N_i] = 0 \qquad V[N_i] = r
$$

where $r = m/n$ is called the loading ratio (a measure of loading of memories to the net).

PROOF. Since $t_i^{(\mu)}$, $s_i^{(\mu)}$ are independent and their expectations are 0, for any fixed $\mathbf{x}$,

$$
E[N_i] = \frac{1}{n} \sum \sum E[t_i^{(\mu)}] E[s_j^{(\mu)}] E[x_j] = 0
$$

As for the variance, we remark that N_i is the sum of $n(m-1)$ independent random variable

$t_i^{(\mu)} s_j^{(\mu)} x_j$ divided by n. Since the variance of each variable is 1, we have

$$
V[N_i] = \frac{n(m-1)}{n^2} \cong \frac{m}{n} = r \qquad \square
$$

This can be proved by calculating directly the following terms,

$$
\begin{aligned}
V[N_i] &= E[N_i^2] - E[N_i]^2 \\
&= E\left[(1/n^2) \sum_j \sum_{j'} \sum_{\mu \neq v} \sum_{\mu' \neq v} t_i^{(\mu)} s_j^{(\mu)} x_j t_i^{(\mu')} s_{j'}^{(\mu')} x_{j'} \right]
\end{aligned}
$$

Now N_i is a random variable subject to a normal distribution with mean 0 and variance r, so we normalize it as

$$
N_i = \sqrt{r}\, \varepsilon_i
$$

with ε_i being the unit normal distribution (mean 0 and variance 1) subject to $N(0, 1)$. Then, according to Eq. (1), the ith output is written as

$$
y_i = f(t_i^{(v)} (1 - 2d) + \sqrt{r}\, \varepsilon_i)
$$

The distance of the output $\mathbf{y}$ from the desired $\mathbf{t}^{(v)}$ is the number of components of $\mathbf{y}$ different from those of $\mathbf{t}^{(v)}$, or

$$
d' = \frac{1}{n} \sum_{i=1}^{n} \tau_i
$$

where $\tau_i = 1$ when $y_i \neq t_i^{(v)}$ and is 0 otherwise. The event $y_i \neq t_i^{(v)}$ or $y_i t_i^{(v)} < 0$ occurs when

$$
1 - 2d + \sqrt{r}\, t_i^{(v)} \varepsilon_i < 0
$$

Since $t_i^{(v)}$ and ε_i are independent, this probability is given by

$$
\begin{aligned}
p_e &= \Pr\left\{ \varepsilon_i < -\frac{1 - 2d}{\sqrt{r}} \right\} \\
&= \int_{-\infty}^{-(1-2d)/\sqrt{r}} \frac{1}{\sqrt{2\pi}} \exp\left(-\frac{t^2}{2} \right) dt \\
&= \phi\left(\frac{1 - 2d}{\sqrt{r}} \right)
\end{aligned} \tag{17}
$$

where

$$
\phi(u) = \int_{-\infty}^{-u} \frac{1}{\sqrt{2\pi}} \left(-\frac{t^2}{2} \right) dt \tag{18}
$$

is the error integral. Since d' is the arithmetic average of independent random variables τ_i, by the law of large numbers when n is large, we have the transform relation

$$
d' = \phi\left(\frac{1 - 2d}{\sqrt{r}} \right) \tag{19}
$$

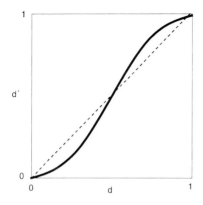

Fig. 8.5. Transfer of distance of a cross-associative net for $r = 0.2$.

This relation is shown in Fig. 8.5 for the case $r = 0.2$. It should be remarked that, even when $d = 0$, $d' = \phi(1/\sqrt{r})$ is not zero. Therefore, although the error rate is small when $r < 0.3$, the association is imperfect. This analysis was first performed by Uesaka and Ozeki (1972). Amari (1977) and Kinzel (1985) used this relation to evaluate the recalling dynamics of the auto-associative nets. For this parameter, we can see that the distance is reduced by the associative net for most of $d < 0.5$, whereas the distance is enlarged for values of d close to zero. By analyzing Eq. (19) we can see that the smaller the loading ratio r, the better the noise-reduction properties.

8.5. RECALL PROPERTIES OF CASCADED ASSOCIATIVE NETS, CYCLIC ASSOCIATIVE NETS AND ASSOCIATIVE SEQUENCE GENERATORS

When an input pattern $\mathbf{x}$ is in a neighborhood of the first component $\mathbf{s}^{(v)}(1)$, it is expected that a cascaded net recalls the whole sequence S^v. Although a cascaded associative net is just a concatenation of cross-associative nets, the recalling property of the former is not a simple concatenation property of the latter. That implies that when recalling a memorized pattern sequence S^v, the distance at the lth layer d_l does not follow the recurrence relation

$$d_{l+1} = \phi\left(\frac{1 - 2d_l}{\sqrt{r}}\right) \qquad (20)$$

if $l \geq 2$ in spite of the fact that Eq. (20) is valid for the first one-layer transition. This is because the output $\mathbf{x}(l)$ of neurons at $l \geq 2$ are correlated with the connections $w_{ij}(l)$. This arises from the

fact that $\mathbf{W}(l - 1)$ and $\mathbf{W}(l)$ share common patterns $\mathbf{s}^{(\mu)}(l)$. Therefore, when calculating the variance of N_i, we need to take this correlation into account.

Instead of Eq. (20), it will soon be shown that the transition law of the distance for a cascaded-associative net is described by an equation of the form

$$d_{l+1} = \phi\left(\frac{1 - 2d_l}{\sigma_l}\right) \qquad (21)$$

where σ_l is determined recursively. That is, the variance of the interference (noise) term N_i is not fixed but changes in the course of transition. We shall explain in the rest of this section how the variance is calculated by the method of statistical neurodynamics.

Now we treat the case of recalling the vth sequence S^v so that the output $\mathbf{x}(l)$ of the lth layer is in a neighborhood of $\mathbf{s}^{(v)}(l)$. From Eqs. (6) and (7) we can see that

$$x_i(l + 1) = f\left(\sum_{j=1}^{n} w_{ij}(l)x_j(l)\right)$$

$$= f\left(s_i^{(v)}(l + 1)(1 - 2d_l)\right.$$

$$\left. + \frac{1}{n} \sum_{\mu \neq v} \sum_{j \neq i} s_i^{(\mu)}(l + 1)s_j^{(\mu)}(l)x_j(l)\right)$$

$$= s_i^{(v)}(l + 1)f(1 - 2d_l + N_i^{(v)}(l)) \qquad (22)$$

where

$$N_i^{(v)}(l) = \frac{1}{n} s_i^{(v)}(l + 1)$$

$$\times \sum_{\mu \neq v} \sum_{j \neq i} s_i^{(\mu)}(l + 1)s_j^{(\mu)}(l)x_j(l) \qquad (23)$$

is the interference term subject to a normal distribution. Lemma 2 shows this property, and the proof is given in the Appendix.

LEMMA 2. *The expectation and the variance of the noise term $N_i(l)$ is given by*

$$E[N_i^{(v)}(l)] = 0$$

$$V[N_i^{(v)}(l)] = \sigma_i^2 = r + 4\gamma\left(\frac{1 - 2d_{l-1}}{\sigma_{l-1}}\right)^2$$

with

$$\sigma_1 = \sqrt{r}$$

$$\gamma(u) = \frac{1}{\sqrt{2\pi}}\exp\left(-\frac{u^2}{2}\right)$$

From this, we easily have the theorem describing the behavior of a concatenated associative net.

THEOREM 1. *The dynamic property of a concatenated associative net is given by*

$$
\left.\begin{aligned}
d_{l+1} &= \phi\left(\frac{1 - 2d_l}{\sigma_l}\right) \\
\sigma_{l+1}^2 &= r + 4\gamma\left(\frac{1 - 2d_l}{\sigma_l}\right)^2 \\
\sigma_1 &= \sqrt{r}
\end{aligned}\right\} \quad (24)
$$

This coincides with the corresponding equation in Meir and Domany (1987), which was derived by a different method.

Now let us briefly describe the characteristic behavior of the change in the distance d_l in the process of recalling the sequence S^ν. For a fixed range $0 \leq r \leq r_{\text{crit}}$, there is a threshold value $d_{\text{th}}(r)$ such that for an initial distance d_1 when $d_1 < d_{\text{th}}(r)$, d_l monotonically decreases to $d_\infty(r)$ which is close to zero, and if $d_1 > d_{\text{th}}(r)$, d_l tends to $1/2$, failing recollection. The threshold $d_{\text{th}}(r)$ is a monotone decreasing function of r, and $d_\infty(r)$ is a monotone increasing function of r; they coincide at $r \cong 0.27$ with $d_{\text{th}}(r) = d_\infty(r) \cong 0.08$. This implies that

$$r_{\text{crit}} = 0.27$$

implying that the cascaded-associative net can store about $0.27n$ pattern sequences. We say that the capacity of this net is 0.27. The same line of calculations hold for the sequence generators and cyclic-associative nets, provided that any cycle (if it exists) has a period not shorter than 3 and that long-range correlations can be neglected. This is confirmed by computer simulation. This means that the dynamics of cyclic-associative nets and sequence generators are essentially the same. For sequence generators, the loading ratio r is defined by the whole number of sequence patterns divided by the number of neurons, and the layer number l is replaced by a discrete time t.

However, the behavior of BAM which has period-2 cycles is a little different. The autoassociative net which has fixed points or period 1 cycles is much more difficult to analyze, as we see in the next section.

8.6. RECALL DYNAMICS OF AUTOASSOCIATIVE NETS

An autoassociative net stores m patterns $\mathbf{s}^{(1)}, \ldots, \mathbf{s}^{(m)}$ in its synapses in a distributed and superimposed way. It recalls any one of its memory patterns through the dynamics of updating its state. It is thought to be a prototype which includes a variety of nets as its special cases, as mentioned previously.

Let $\mathbf{x}(0)$ be the initial state in a neighborhood of $\mathbf{s}^{(\nu)}$ and let d_0 be the distance between them. Similarly, let d_t be the distance between $\mathbf{s}^{(\nu)}$ to be recalled and the state $\mathbf{x}(t)$ of the network at time t, which is calculated recursively by

$$\mathbf{x}(t + 1) = f(\mathbf{W}\mathbf{x}(t)) \quad (25)$$

This can again be rewritten as

$$
\left.\begin{aligned}
x_i(t + 1) &= f[u_i(t)] \\
u_i(t) &= s_i^{(\nu)}(1 - 2d_t) + N_i^{(\nu)}(t) \\
N_i^{(\nu)} &= \frac{1}{n}s_i^{(\nu)} \sum_{\mu \neq \nu} \sum_{j \neq i} s_i^{(\mu)}s_j^{(\mu)}x_j(t)
\end{aligned}\right\} \quad (26)
$$

When we evaluate the probability distribution of $N_i^{(\nu)}(t)$, a problem arises. Since $x_j(t)$ is determined recursively from (25), it also includes $s_i^{(\mu)}$ and $s_j^{(\mu)}$. Therefore, we cannot neglect the correlation between the coefficients $s_i^{(\mu)}$ or $s_j^{(\mu)}$ and the same ones nonlinearly included in $\mathbf{x}(t)$. Moreover, $\mathbf{x}(t)$ is determined from $\mathbf{x}(t - 1)$, which also includes $s^{(\mu)}$'s. This implies that such correlations depend on the whole history of $\mathbf{x}(t - T)$, $T = 0, 1, 2, \ldots$.

When $t = 0$, $\mathbf{x}(0)$ is the initial value independent of $\mathbf{s}^{(\nu)}$'s. Therefore, $N_i^{(\nu)}(0)$ can be calculated as before and is subject to $N(0, \sigma_0^2)$, and

$$d_1 = \phi\left(\frac{1 - d_0}{\sigma_0}\right) \quad (27)$$

holds. To calculate the distance at $t = 2$, complex correlations should be taken into account, but only two-step correlations are sufficient, because $\mathbf{x}(0)$ is again independent of $\mathbf{s}^{(\mu)}$. Therefore, by taking the two-step correlations into account, we have the exact form of d_2. In this case, the noise term $N_i^{(\nu)}$ is proved to be asymptotically normally distributed, and we can calculate its mean and variance exactly.

However, we cannot continue this argument further. This is because $N_i^{(\nu)}(t)$ depends on all the past history of $\mathbf{x}(0), \ldots, \mathbf{x}(t)$ so that we cannot have a simple dynamical state equation. We therefore make the following assumptions.

1. The cross-talk noise $N_i^{(\nu)}(t)$ is subject to the normal distribution $N(0, \sigma_t^2)$.
2. The variance σ_t^2 of $N_i^{(\nu)}$ is calculated from (26) by taking the direct correlations up to two steps between $s_i^{(\mu)}$ and those in $x_j(t)$, and all the other past effects are renormalized in the noise terms of variance σ_{t-1}^2.

From these assumptions, we have the macroscopic dynamical equations governing the transient recall dynamics,

$$
\left.
\begin{aligned}
d_{t+1} &= \phi\left(\frac{1 - 2d_t}{\sigma_t}\right) \\
\sigma_{t+1}^2 &= r + 4\gamma\left(\frac{1 - 2d_t}{\sigma_t}\right)^2 \\
&\quad + \frac{4r(1 - 2d_t)(1 - 2d_{t+1})}{\sigma_t}\gamma\left(\frac{1 - 2d_t}{\sigma_t}\right)
\end{aligned}
\right\}
$$

$$(28)$$

where $\sigma_0 = \sqrt{r}$. The derivation of this equation is along the same line as for cascaded associative nets though it is more complicated, resulting in an additional term in the variance. The equation is correct for $t = 0$. We can calculate the correct equation for $t = 1$, but we need to include a bias term,

$$
b_1 = \frac{2r}{\sigma_0}\gamma\left(\frac{1 - 2d_0}{\sigma_0}\right)
$$

However, this term is neglected in the dynamical equation (28).

According to Eq. (28), the dynamical properties of the autoassociative net is qualitatively the same as the cascaded associative net. The differences are in numerical values: For the autoassociative net, the capacity is about 0.16, where $d_\infty \cong 0.05$. For a given value of the loading ratio, the autoassociative net has smaller basins of attraction and larger d_∞ (carrying more errors after recall).

Since the dynamical equations are not exact but approximate, it is surprising that the equations give good agreement with computer simulated results. The agreement is almost perfect when the recall process is successful (Nishimori and Ozeki, 1992). Therefore, the equations can explain various interesting dynamical properties of recall such as capacity, the value of threshold for recall, the transient overshooting phenomenon when recall fails, etc. However, this is not the case when the net fails to recall the nearest memory pattern. That is, according to the equation, the distance should converge to $1/2$ as t tends to infinity, whereas the distance remains between 0 and $1/2$ in simulations. This is surely because of the long-range correlations between successive states of the net that are not considered in deriving Eq. (28). In cascaded associative nets, there is no such long-range correlation, hence the theory is exact.

8.7. STRUCTURALLY RESTRICTED CONNECTIONS AND DIFFERENT CODING SCHEMES

8.7.1. Autoassociative Nets with Randomly Disconnected Synapses

When we want to analyze more realistic systems, we have to take various restrictions such as nonuniformities of connections into account. Here, we show results on the dynamics of autoassociative nets with sparse connections and nonuniform connections (Yanai et al., 1991). In this section we try to gain insight on the effects of the network's structure.

To represent the structural connectivity of the net, we have introduced c_{ij} (see Eq. (11)). Here we discuss the uniformly sparse case where $c_{ij} = 1$ with probability c and $c_{ij} = 0$ with probability $1 - c$, implying that two neurons are not connected with probability $1 - c$. Since we are treating a net with a large number of neurons (hence synapses), the value c ($0 < c \leq 1$) represents the actual connection rate. The dynamical equation for the transition of the distance can be calculated in a similar way to the fully connected autoassociative net. Using the same approximation scheme as we did to derive Eq. (28), we have the following dynamical equation for this net:

$$
\left.
\begin{aligned}
d_{t+1} &= \phi\left(\frac{1 - 2d_t}{\sigma_t}\right) \\
\sigma_{t+1}^2 &= \rho + 4\left(1 - \frac{1 - c}{\sigma_t^2}\rho\right)\gamma^2 \\
&\quad + \frac{4c\rho(1 - 2d_t)(1 - 2d_{t+1})}{\sigma_t}\gamma
\end{aligned}
\right\}
$$

$$(29)$$

with $\sigma_0 = \sqrt{\rho}$, where $\rho \equiv m/(cn) = r/c$ and $\gamma = \gamma((1 - 2d_t)/\sigma_t)$. Here ρ is introduced to show the loading ratio which represents the number of memorized patterns per synapse. By putting $t = 0$ in Eq. (29), we can see that the initial one-step transform of distance is characterized only by ρ. From the equation we can see that, if the normalized loading ratio is fixed, the smaller the connection rate c, the more easily the net can retrieve its memories. In other words, the net can make more effective use of its synapses under a smaller connection rate.

Next we would like to draw the reader's attention to the effect of the symmetry in synaptic disconnections. What happens if we impose the symmetry $c_{ij} = c_{ji}$ on the disconnection of synapses? If we use the same approximation as used to derive Eq. (29), no change appears in the dynamics. We would like to know if this is true in the exact sense.

In the physicists' approach, symmetry of synaptic weights ($w_{ij} = w_{ji}$) is often assumed in order to use the methods established in thermodynamics. Hence if those methods are to be used, synaptic disconnection should also be symmetric ($c_{ij} = c_{ji}$) (cf. Sompolinsky, 1986), whereas in our statistical-neurodynamical method we do not have to make the assumption of symmetry. Taking advantage of our method, we shall show the effect of symmetry of synaptic disconnections on the dynamics of the net through the exact solutions for the initial two time steps. We can derive the exact solution for the distance at $t = 2$ as

$$d_2 = (1 - d_1)\phi\left(\frac{1 - 2d_1 + b_1}{\sigma_1}\right)$$
$$+ d_1\phi\left(\frac{1 - 2d_1 - b_1}{\sigma_1}\right) \quad (30)$$

where d_1 and σ_1 are determined by Eq. (29) irrespective of the symmetry, and

$$b_1 = \begin{cases} \dfrac{2\rho}{\sigma_0}\gamma\left(\dfrac{1 - 2d_0}{\sigma_0}\right) & \text{symmetric} \\[3mm] \dfrac{2\rho}{\sigma_0}\gamma\left(\dfrac{1 - 2d_0}{\sigma_0}\right)c & \text{asymmetric} \end{cases} \quad (31)$$

If b_1 is positive, it tends to make the net remain at the current state, and since b_1 is larger in the symmetric case, the net with symmetric synapses has a tendency to stabilize the memories. Although the effects of b_1 are not simply restricted to that, we can conclude that the symmetry of synaptic disconnections affects the net's dynamics. Moreover, from Eq. (31) we can see that if c is small, then Eq. (29) is a fairly good approximation to the exact dynamics of the asymmetric net. This is because if c (hence b_1) is small enough, then Eq. (30) almost coincides with Eq. (29).

Finally we would like to show how nonuniformity of neurons affects the performance of the autoassociative net with disconnected synapses. Let the nonuniformity be introduced into the net as a noise in the threshold. That is, the dynamics of the net is

$$x_i(t + 1) = f\left(\sum_{j=1}^{n} w_{ij}x_j(t) - h_i\right)$$

where $h_i = \sigma_h \varepsilon_i$ (ε_i is subject to a unit normal distribution). The main conclusion is that there is a unimodal relation between the connection rate c and the normalized capacity. That is, if the

noise level σ_h is given, there exists an optimal value of c that makes the most efficient use of synapses. For example, when $\sigma_h = 0.05$, the optimal connection rate is $c \approx 0.25$.

8.7.2. Correlation of Successive Patterns in Cyclic Associative Nets

Let us study the effects of the correlation of successive memory patterns in cyclic-associative nets or cascaded associative nets. In Section 8.5 we derived the exact dynamical equations for uncorrelated memory patterns randomly generated with equal probability for 1 and -1 (see Eq. (24)). Here we generate every memory pattern in the first layer as before, and the memory patterns in the consecutive layers are generated by the rule

$$q = 1 - 2\,\Pr\{s_i^{(\mu)}(l + 1) \neq s_i^{(\mu)}(l)\}$$

where $q(-1 \leq q \leq 1)$ is a correlation, μ is a memory number ($\mu = 1, 2, \ldots, m$), i is a neuron number ($i = 1, 2, \ldots, n$), and l is a layer number ($l = 1, 2, \ldots, L - 1$). This means that patterns are assumed to be correlated only within the same sequences and between the corresponding neurons of the successive layers. The numbers of 1s and -1s are, again, almost equal in all of the memory patterns. When $q = 0$, there are no correlations. When $q > 0$, neighboring memories in a sequence are similar (positively correlated) to each other; in particular when $q = 1$, the model is equivalent to autoassociative nets, and so on. We have to note that in the present situation the resultant equation does not apply to sequence generators that have recurrent connections.

By assuming zero expectation for the interference term, and assuming enough layers, we can derive approximated solution for the dynamics of the net as

$$d_{l+1} = \phi\left(\frac{1 - 2d_l}{\sigma_l}\right)$$
$$\sigma_{l+1}^2 = r + 4\gamma\left(\frac{1 - 2d_l}{\sigma_l}\right)^2$$
$$+ \frac{4|q|r(1 - 2d_l)(1 - 2d_{l+1})}{\sigma_l}\gamma\left(\frac{1 - 2d_l}{\sigma_l}\right)$$
(32)

with $\sigma_l = \sqrt{r}$, where initial conditions are applied to the first layer. When $q = 0$, this equation is exact and coincides with Eq. (24); and when $|q| = 1$ this equation coincides with Eq. (28), which is the equation for autoassociative nets.

From this equation, we can show that the net can memorize about $0.27n$ patterns when $q = 0$, and about $0.16n$ when $|q| = 1$; the number decreases monotonically with increasing $|q|$.

8.7.3. Sparsity of Coding in Autoassociative Nets

A simple but very interesting coding scheme is sparse coding, which uses sparsely coded patterns as memories. A sparsely coded pattern contains few 1s, i.e., the activity of a pattern is very low. If the number of 1s in a pattern is on the same order as -1s, associative nets can store $O(n/\log n)$ patterns as stable states, which is the absolute capacity. If the fraction of 1s is of $O((\log n)/n)$, the net can store $O((n/\log n)^2)$ patterns (Willshaw et al., 1969; Palm, 1980; Amari, 1989; Meunier et al., 1991; Gardner, 1988). By using the sparse coding scheme, not only a larger number of memory patterns but also larger total information can be stored in a net. In this section we compare results for a wide range of sparsity of coding which have been obtained by the method of statistical neurodynamics (Amari, 1989; Meunier et al., 1991).

In the previous sections we have dealt with patterns with 1s and -1s, and regarded 1s as the firing state of neurons and -1s as the resting state. In the following we shall use 0s instead of -1s to represent the resting state, and call the fraction of 1s the activity. There exists a transformation between these two models by choosing appropriate threshold values for the neurons. But even if we use a fixed threshold value for every neuron, those results differ only by a factor. For example, in the case of the covariance rule (defined below), the maximum

number of memory patterns that can be stored stably is

$$
m_{\max} =
\begin{cases}
\dfrac{1}{24}\left(\dfrac{n}{\log n}\right)^2 & \{1, 0\}\ \text{model} \\[3ex]
\dfrac{1}{12}\left(\dfrac{n}{\log n}\right)^2 & \{1, -1\}\ \text{model}
\end{cases}
$$

when the activity is $(\log n)/n$.

We shall derive the memory capacity of the autoassociative net for any sparsity of coding. In doing so, we use two learning rules, the covariance rule and the Hebb rule:

$$
w_{ij} =
\begin{cases}
\dfrac{1}{n}\displaystyle\sum_{\mu=1}^{m}(s_i^{(\mu)} - a)(s_j^{(\mu)} - a) & \text{covariance} \\[3ex]
\dfrac{1}{n}\displaystyle\sum_{\mu=1}^{m} s_i^{(\mu)} s_j^{(\mu)} & \text{Hebb}
\end{cases}
\quad (i \neq j)
$$

$$
w_{ii} = 0
$$

$$(33)$$

where $s_i^{(\mu)} = 0$ or 1, and a is the activity of memory patterns. The dynamics are given by

$$
x_i(t + 1) = g\left(\sum_{j=1}^{n} w_{ij} x_j(t) - h\right) \quad (34)
$$

where $t = 0, 1, 2, \ldots$ is a discrete time, h is a threshold, and

$$
g(u) =
\begin{cases}
1 & (u \geq 0) \\
0 & (u < 0)
\end{cases}
\quad (35)
$$

Let the activity of memory patterns be characterized by

$$
a = cn^{-e} \quad (c > 0, 0 \leq e \leq 1) \quad (36)
$$

Then the number of active neurons is $A = an = cn^{1-e}$.

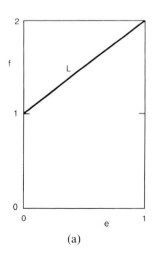

(a)

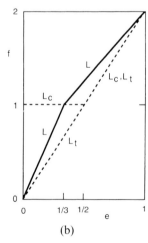

(b)

Fig. 8.6. Phase diagrams of sparsely coded associative nets for the covariance rule (a) and for the Hebb rule (b).

To show the relation between m_{max} and a, let us show the phase diagrams in the (e, f) plane, where e is a dominant exponent for the activity of memory patterns as stated above, $a \sim n^{-e}$, and f is defined as a dominant exponent for the number of memory patterns, $m \sim n^f$. In Fig. 8.6, transition lines are shown for the covariance rule (a) and the Hebb rule (b) in the case where n tends to infinity. The transition lines are

- L is the line below which memory patterns are all stable by the appropriate choice of the threshold. Thus L determines the capacity.
- L_t is the line below which the optimal threshold relative to the signal strength is finite and above which it is infinite.
- L_c is the line below which the interference terms of two distinct neurons are independent (correlation coefficient equals to zero), and above which they are correlated with correlation coefficient being unity.

The diagram for the covariance rule has a single phase transition line L; the optimal threshold is always finite and the correlation coefficient is zero everywhere. In the case of the Hebb rule, the diagram is more complicated. Deterioration of memories by overloading occurs in a different way depending on the exponent e when we deal with the Hebb rule.

Finally in this section let us show results on the basins of attractions of memory patterns, or the error-correction property of the net. Let d be the relative Hamming distance ($0 \le d \le 1$) as used above and let the activity of the initial state be ka ($0 < k \le 1$). For the covariance rule, the memory pattern is perfectly recalled by one-step dynamics if

- (e, f) is below L and $d < ka$

and for the Hebb rule, if either of the following is satisfied

- (e, f) is below L_t and $d < ka$, or
- (e, f) is between L and L_t, $k = 1$ and $d < a$.

The derivation of the above relation is not simple. The reader may refer to Meunier et al. (1991) for such derivation.

8.8. SUMMARY

We have analyzed the dynamics of various types of associative nets in a unified way through the method of statistical neurodynamics. In addition, we have presented example calculations explaining the method. Contrary to other methods presented in the literature, our results are based on a set of clearly stated hypotheses and should help deepen the understanding of the dynamics of associative nets.

APPENDIX: Proof of Lemma 2

The expection of $N_i(l)$ is easily calculated as

$$E[s_i^{(\mu)}(l+1)s_j^{(\mu)}(l)x_j(l)] = E[s_i^{(\mu)}(l+1)] \cdot E[s_j^{(\mu)}(l)x_j(l)]$$
$$= 0$$

This is because $s_i^{(\mu)}(l+1)$ is not included in $x_j(l)$ and is independent of $s_j^{(\mu)}(l)x_j(l)$. So we can see that the expectation of the interference term $N_i(l)$ is equal to zero.

The variance of the interference term is therefore equal to the expectation of the square $\{N_i(l)\}^2$ of the term. However, $N_i(l)$ is not a sum of independent variables because $x_j(l)$ is correlated with $s_j^{(\mu)}(l)$, so that calculations are complicated. From

$$V[N_i(l)] = E[\{N_i(l)\}^2]$$
$$= \frac{1}{n^2} E\left[\left\{\sum_{\mu \neq v}\sum_{j \neq i} s_i^{(\mu)}(l+1)s_j^{(\mu)}(l)x_j(l)\right\} \right.$$
$$\left. \times \left\{\sum_{\mu' \neq v}\sum_{j' \neq i} s_i^{(\mu')}(l+1)s_j^{(\mu')}(l)x_{j'}(l)\right\}\right]$$

we put

$$V[N_i(l)] = E\left[\left(\frac{1}{n^2}\right)\sum_{\mu \neq v}\sum_{\mu' \neq v}\sum_{j \neq i}\sum_{j' \neq i} S(j, j'; \mu, \mu')\right]$$

where

$$S = S(j, j'; \mu, \mu')$$
$$= s_i^{(\mu)}(l+1)s_j^{(\mu)}(l)x_j(l)s_i^{(\mu')}(l+1)s_j^{(\mu')}(l)x_{j'}(l)$$

We calculate the expectation of $S(j, j'; \mu, \mu')$. Since the value is different depending on whether $j = j'$ and/or $\mu = \mu'$, we consider the four cases separately.

1. $j \neq j'$, $\mu \neq \mu'$. There are $(n-1)(m-1)$ such terms. Since

$$S(j, j; \mu, \mu) = 1$$
$$E[S] = 1$$

2. $j \neq j'$, $\mu \neq \mu'$. There are $(n-1)(m-1)(m-2)$ such terms. Since $\{x_j(l)\}^2 = 1$,

$$E[S(j, j; \mu, \mu')]$$
$$= E[s_i^{(\mu)}(l+1)s_j^{(\mu)}(l)s_i^{(\mu')}(l+1)s_j^{(\mu')}(l)]$$
$$= E[s_i^{(\mu)}(l+1)]E[s_j^{(\mu)}(l)]E[s_i^{(\mu')}(l+1)]E[s_j^{(\mu')}(l)]$$
$$= 0$$

3. $j \neq j'$, $\mu = \mu'$. There are $(n-1)(n-2)(m-1)$ such terms. When we check the term

$$S(j, j'; \mu, \mu) = s_j^{(\mu)}(l)x_j(l)s_{j'}^{(\mu)}(l)x_{j'}(l)$$

carefully, we find that $x_j(l)$ and $x_{j'}(l)$ include, respectively, $s_j^{(\mu)}(l)$ and $s_{j'}^{(\mu)}(l)$. They are surely correlated, so that we single out these in the expression of $x_j(l)$ and $x_{j'}(l)$,

$$x_j(l) = f(s_j^{(\nu)}(l)(1 - 2d_{l-1}) + s_j^{(\mu)}(l)R + Q)$$
$$x_{l'}(l) = f(s_{j'}^{(\nu)}(l)(1 - 2d_{l-1}) + s_{j'}^{(\mu)}(l)R + Q')$$

where

$$R = \frac{1}{n}\sum_{k \neq j, j'} s_k^{(\mu)}(l-1)x_k(l-1)$$

$$Q = \frac{1}{n}\sum_{\alpha \neq \nu} \sum_{k \neq j} s_j^{(\alpha)}(l)s_k^{(\alpha)}(l-1)x_k(l-1)$$
$$+ \frac{1}{n} s_j^{(\mu)}(l)s_j^{(\mu)}(l-1)x_{j'}(l-1)$$

$$Q' = \frac{1}{n}\sum_{\alpha \neq \nu, \mu} \sum_{k \neq j} s_{j'}^{(\alpha)}(l)s_k^{(\alpha)}(l-1)x_k(l-1)$$
$$+ \frac{1}{n} s_{j'}^{(\mu)}(l)s_{j'}^{(\mu)}(l-1)x_j(l-1)$$

Variables R, Q, and Q' are subject to normal distributions with mean zero and variances,

$$V[Q] \cong V[Q'] \cong \frac{m-2}{m-1}\sigma_{l-1}^2 \cong \sigma_{l-1}^2$$

$$V[R] \cong \frac{1}{m}\sigma_{l-1}^2$$

because Q and Q' are the major parts of $N_j(l-1)$ and $N_{j'}(l-1)$ subject to the normal distribution as is assumed in the previous layer $l-1$. Moreover, the correlation between $x_j(l)$ and $x_{j'}(l)$ originates from the common term R.

Since $s_j^{(\mu)}(l)$ assumes values 1 or -1 equally likely, we first take the expectation of $s_j^{(\mu)}(l)x_j(l)$ with respect to $s_j^{(\mu)}(l)$ only and hold $s_j^{(\nu)}(l)$ and R fixed. Mathematically speaking, this is the conditional expectation:

$$E[s_j^{(\mu)}(l)x_j(l)|s_j^{(\nu)}(l), R]$$
$$= \frac{1}{2}[f(s_j^{(\nu)}(l)(1 - 2d_{l-1}) + R + Q)$$
$$- f(s_j^{(\nu)}(l)(1 - 2d_{l-1}) - R + Q)]$$

We then take the expectation with respect to Q, yielding

$$\frac{1}{2}s_j^{(\nu)}(l)E[f(1 - 2d_{l-1} + s_j^{(\nu)}(l)R + s_j^{(\nu)}(l)Q)$$
$$- f(1 - 2d_{l-1} - s_j^{(\nu)}(l)R + s_j^{(\nu)}(l)Q)]$$
$$= \frac{1}{2}s_j^{(\nu)}(l)\left[\phi\left(\frac{1 - 2d_{l-1} + s_j^{(\nu)}(l)R}{\sigma_{l-1}}\right)\right.$$
$$\left. - \phi\left(\frac{1 - 2d_{l-1} - s_j^{(\nu)}(l)R}{\sigma_{l-1}}\right)\right]$$

Here we have used the fact that the distribution of Q is the same as that of $-Q$. Then by noting that $s_j^{(\nu)}(l)$ also takes 1 or -1 with equal probability, since R is small, we expand the above result with respect to R, so that the conditional expectation is

$$E[s_j^{(\mu)}(l)x_j(l)|R]$$
$$= \phi\left(\frac{1 - 2d_{l-1} + R}{\sigma_{l-1}}\right) - \phi\left(\frac{1 - 2d_{l-1} - R}{\sigma_{l-1}}\right)$$
$$= 2\phi'\left(\frac{1 - 2d_{l-1}}{\sigma_{l-1}}\right)\frac{R}{\sigma_{l-1}} + O(R^3)$$

where $\phi'(u) \equiv d\phi(u)/du = -\gamma(u)$ and R is still kept as a random variable.

By the same procedure we get the same result for the term $E[s_j^{\prime(\mu)}x_j'(l)|R]$. Since we can show that R, Q, and Q' are mutually uncorrelated, by taking the expectation with respect to the common R, we finally have

$$E[S(j, j'; \mu, \mu)] = E[E[s_j^{(\mu)}(l)x_j(l)|R] \cdot E[s_{j'}^{(\mu)}(l)x_{j'}(l)|R]]$$
$$= 4\left\{\phi'\left(\frac{1 - 2d_{l-1}}{\sigma_{l-1}}\right)\frac{1}{\sigma_{l-1}}\right\}^2 E[R^2]$$
$$= \frac{4}{m}\gamma\left(\frac{1 - 2d_{l-1}}{\sigma_{l-1}}\right)^2$$

4. $j \neq j'$ $\mu \neq \mu'$. There are $(n-1)(n-2)(m-1)(m-2)$ such terms. It is not difficult to show

$$E[S(j, j'; \mu, \mu')] = 0$$

in this case. Collecting the results from (1) through (4) together, we have

$$\sigma_l^2 \cong \frac{1}{n^2}(n-1)(m-1)$$
$$+ \frac{1}{n^2}(n-1)(n-2)(m-1)\frac{4}{m}\gamma\left(\frac{1 - 2d_{l-1}}{\sigma'_{-1}}\right)^2$$

and this proves Eq. (24).

REFERENCES

Amari, S. (1972). "Learning Patterns and Pattern Sequence by Self-organization Nets of Threshold Elements," *IEEE Trans. Computers*, **C-21**, 1197–1206.

Amari, S. (1974). "A Method of Statistical Neurodynamics," *Kybernetik*, **14**, 201–215.

Amari, S. (1977). "Neural Theory of Association and Concept-formation," *Biological Cybernetics*, **26**, 175–185.

Amari, S. (1988). "Associative Memory and its Statistical-Neurodynamical Analysis," *Neural and Synergetic Computers*, H. Haken, ed. Springer Series in Synergetics, No. 42, Springer-Verlag, New York.

Amari, S. (1989). "Characteristics of Sparsely Encoded Associative Memories," *Neural Networks*, **2**, 451–457.

Amari, S. and Maginu, K. (1988). "Statistical Neuro-dynamics of Associative Memory," *Neural Networks*, **1**, 63–73.

Amari, S., Yoshida, K., and Kanatani, K. (1977). "A Mathematical Foundation for Statistical Neuro-dynamics," *SIAM J. Appl. Math.*, **33**, 95–126.

Amit, D. J., Gutfreund, H., and Sompolinsky, H. (1985a). "Storing Infinite Number of Patterns in a Spin-glass Model of Neural Networks," *Phys. Rev. Lett.*, **55**, 1530–1533.

Amit, D. J., Gutfreund, H., and Sompolinsky, H. (1985b). "Spin-glass Model of Neural Networks," *Phys. Rev.*, **A32**, 1007–1018.

Anderson, J. A. (1972). "A Simple Neural Network Generating Interactive Memory," *Math. Biosci.*, **14**, 197–220.

Derrida, B., Gardner, E., and Zippelius, A. (1987). "An Exactly Solvable Asymmetric Neural Network Model," *Europhys. Lett.*, **4**, 167–173.

Domany, E., Kinzel, W., and Meir, R. (1989). "Layered Neural Networks," *J. Phys. A: Math. Gen.*, **22**, 2081–2102.

Fukushima, K. (1973). "A Model of Associative Memory in the Brain," *Kybernetik*, **12**, 58–63.

Gardner, E. (1986). "Structure of Metastable States in the Hopfield Model," *J. Phys. A: Math. Gen.*, **19**, L1047–1052.

Gardner, E. (1988). "The Space of Interactions in Neural Network Models," *J. Phys. A: Math. Gen.*, **21**, 257–270.

Gardner, E., Derrida, B., and Mottishaw, P. (1987). "Zero Temperature Parallel Dynamics for Infinite Range Spin-glasses and Neural Networks," *J. Physique*, **48**, 741–755.

Hopfield, J. J. (1982). "Neural Networks and Physical Systems with Emergent Collective Computational Abilities," *Proc. Nat. Acad. Sci. U.S.A.*, **79**, 2445–2458.

Kinzel, W. (1985). "Learning and Pattern Recognition in Spin Glass Models," *Z. Angew. Phys.*, **B60**, 205–213.

Kohonen, T. (1972). "Correlation Associative Memories," *IEEE Trans. Computers*, **C-21**, 353–359.

Kohonen, T. and Ruohonen, M. (1973). "Representation of Associated Data by Matrix Operators," *IEEE Trans. Computers*, **C-22**(7), 701–702.

Kosko, B. (1988). "Bidirectional Associative Memories," *IEEE Trans. Systems, Man, and Cybernetics*, **SMC-18**, 49–60.

McEliece, R. J., Posner, E. C., Rodemich, E. R., and Venkatesh, S. S. (1987). "The Capacity of the Hopfield Associative Memory," *IEEE Trans. Info. Theory*, **IT-33**, 461–482.

Meir, R. and Domany, E. (1987). "Exact Solution of a Layered Neural Network Model," *Phys. Rev. Lett.*, **59**, 359–362.

Meunier, C., Yanai, H-F., and Amari, S. (1991). "Sparsely Coded Associative Memories: Capacity and Dynamical Properties," *Network*, **2**, 469–487.

Morita, M., Yoshizawa, S., and Nakano, K. (1990). "Analysis and Improvement of the Dynamics of Autocorrelation Associative Memory," *Trans. Institute Electronics, Information and Communication Engineers*, **J73-D-III**(2), 232–242.

Nakano, K. (1972). "Associatron—a Model of Associative Memory," *IEEE Trans. Systems, Man, and Cybernetics*, **SMC-2**, 381–388.

Nishimori, H. and Ozeki, T. (1992). "Retrieval Dynamics of Associative Memory of the Hopfield Type," to appear.

Palm, G. (1980). "On Associative Memory," *Biological Cybernetics*, **36**, 19–31.

Patrick, A. E. and Zagrebnov, V. (1991). "A Probabilistic Approach to Parallel Dynamics for the Little–Hopfield Model," *J. Phys. A: Math. Gen.*, **24**, 3413–3426.

Shinomoto, S. (1987). "A Cognitive and Associative Memory," *Biological Cybernetics*, **57**, 197–206.

Sompolinsky, H. (1986). "Neural Networks with Nonlinear Synapses and a Static Noise," *Phys. Rev. A*, **34**, 2571–2574.

Uesaka, Y. and Ozeki, K. (1972). "Some Properties of Associative Memory" (in Japanese), *J. Inst. Electr. Comm. Eng. Jap.*, **55-D**, 323–330.

Willshaw, D. J., Buneman, O. P., and Longuett-Higgins, H. C. (1969). "Nonholographic Associative Memory," *Nature*, **222**, 960–962.

Yanai, H-F., Sawada, Y., and Yoshizawa, S. (1991). "Dynamics of an Auto-associative Neural Network Model with Arbitrary Connectivity and Noise in the Threshold," *Network*, **2**, 295–314.

Zagrebnov, V. A. and Chvyrov, A. S. (1989). "The Little–Hopfield Model: Recurrence Relatins for Retrieval-pattern Errors," *Sov. Phys.—JETP*, **68**, 153–157.

9

Convergence Analysis of Associative Memories

JÁNOS KOMLÓS and RAMAMOHAN PATURI

9.1. INTRODUCTION

In this chapter we consider the convergence analysis of associative memories. We present a detailed rigorous mathematical analysis of the fully connected associative memory model of Hopfield (1982). In the following, we first make some preliminary remarks about systems exhibiting associative memory properties. We then give a precise description of the Hopfield model of associative memory and formulate the mathematical problems associated with it.

Assume that a dynamical system has a large number of stable states with a substantial domain of attraction around them. That is, the system started at *any* state in the domain of attraction would converge to the stable state. We can then regard such a system as an associative memory. In this framework, stored items are represented by stable states, nearby states represent partial information given a suitable metric. The process of retrieving full information from partial information corresponds to a state in the domain of attraction converging to the stable state. One can think of associative memory as correcting errors in a noisy input.

Many times, full information is not obtainable. Often, we can relax the requirement that a stored item corresponds to a stable state. We merely require selected states to have large domains of attraction around them such that if we start anywhere in the domain we will eventually get within a small distance from the stored item (*residual error* in recall). What is important is that we have a significant amount of error correction.

Another desirable feature of such a system is a *learning* mechanism, by which the system adapts itself to remember new items. With this general picture in mind, we now look at the specific details of the Hopfield model.

9.1.1. Model

The model consists of a system of $n \geq 1$ *fully* interconnected neurons or linear threshold elements where each interconnection is symmetric and has a certain weight. Each neuron in the system can be in one of two states ± 1. The state of the entire system can be represented by an n-dimensional vector $\mathbf{x}$, where the ith component x_i of $\mathbf{x}$ is the state of the ith neuron. The weight of each interconnection is given by real numbers w_{ij} with $w_{ij} = w_{ji}$. The weights are conveniently represented as a symmetric matrix $\mathbf{W}$, with zeros in the diagonal.

Each neuron updates its state based on whether a linear form of the current states of the other neurons, computed with the weights of the interconnections, is above or below its threshold value. We will assume in this paper that all thresholds are zero. Hence, with the system in state $\mathbf{x}$, neuron i resets its state to $\text{sgn}(\sum_{j \neq i} w_{ij} x_j)$ where the function sgn is defined as

$$\text{sgn}(x) = \begin{cases} +1 & \text{if } x \geq 0 \\ -1 & \text{otherwise} \end{cases}$$

9.1.2. System Dynamics

We consider two modes of dynamic operation of the system. In the *synchronous* mode, at every time step, every neuron updates its state simultaneously. Thus, after one synchronous step, the state $\mathbf{x}$ is reset to state $\text{sgn}(\mathbf{Wx})$. In the *asynchronous* mode, at each time step, at most one neuron resets its state, with each neuron eventually getting its turn.

In either case, we say that a state $\mathbf{y}$ *reaches* a state $\mathbf{x}$ if the system, when started in state $\mathbf{y}$, will be in state $\mathbf{x}$ after a succession of transitions.

A state $\mathbf{v}$ is called *stable* if no transition out of it is possible. More precisely, for each i, $v_i = \text{sgn}(\sum_{j=1}^{n} w_{ij} v_j)$. Note that the notion of stability does not depend on the mode (synchronous or asynchronous) of operation.

Hopfield described the (asynchronous) dynamics of the system using an energy surface. The energy of the system is given by the negative of the quadratic form associated with the weight matrix. More precisely, the *energy*

$\mathscr{E}(\mathbf{x})$ of the system in state $\mathbf{x}$ is given by $-\frac{1}{2}\mathbf{x}^{\mathrm{T}}\mathbf{W}\mathbf{x} = -\frac{1}{2}\sum_{i,j} w_{ij}x_i x_j$. This concept is important if we note that the energy of the system decreases at every asynchronous move. Indeed, we have, for any two vectors $\mathbf{x}$ and $\mathbf{v}$,

$$\mathscr{E}(\mathbf{x}) - \mathscr{E}(\mathbf{v}) = -2 \sum_{\substack{i \in D \\ j \notin D}} w_{ij}v_i v_j$$

where D is the set of indices where $\mathbf{x}$ and $\mathbf{v}$ differ.

If $\mathbf{x}$ differs from $\mathbf{v}$ only in the ith coordinate then

$$\mathscr{E}(\mathbf{x}) - \mathscr{E}(\mathbf{v}) = 2v_i \sum_{j \neq i} w_{ij}v_j$$

This shows that an asynchronous step in which the state $\mathbf{v}$ changes to state $\mathbf{x}$ (where the states $\mathbf{x}$ and $\mathbf{v}$ differ in one position) does not increase the energy of the system, thus guaranteeing that a stable state is reached eventually in an asynchronous mode of operation. (Note that the definition $\mathrm{sgn}(0) = +1$ guarantees that we cannot get into a cycle.) Such a convergence is not guaranteed in the case of synchronous operation.

We can also see that stable states are local minima of the *energy landscape*. Indeed, a state $\mathbf{v}$ is a local minimum for the energy function if $\sum_{j=1}^{n} w_{ij}v_i v_j \geq 0$ for all i. Clearly, stable states satisfy this condition and hence are local minima.

9.1.3. Hebb's Learning Rule

Hopfield used the following Hebb type rule (Hebb, 1949) to select the weights of the interconnections. To store a single vector in the system, we require that each interconnection remember the correlation of the states of the two neurons it interconnects. More precisely, we set the weights $w_{ij} = v_i v_j$, $i \neq j$, and $w_{ii} = 0$ to remember a single vector $\mathbf{v} = (v_1, v_2, \ldots, v_n)$. With this choice of weights, the system has a stable point at state $\mathbf{v}$.

Moreover, for this choice of weights, when the system is started at a state $\mathbf{x}$ within a distance $n/2$ from the stored vector $\mathbf{v}$, it reaches state $\mathbf{v}$ in one synchronous step. Thus, $\mathrm{sgn}(\mathbf{W}\mathbf{x})$ maps every vector $\mathbf{x}$ within distance $n/2$ from $\mathbf{v}$ into $\mathbf{v}$. In the asynchronous mode of operation, the state of the system converges monotonically to the stable state $\mathbf{v}$. (As customary, we measure the distance between two ± 1 vectors or states by their Hamming distance: the number of components in which they differ.) It is this *attracting* nature of the system that gives it an *error-correcting* capability.

To store many vectors $\mathbf{v}^1, \ldots, \mathbf{v}^m$, we simply take the sum of the corresponding weights. Thus,

our matrix $\mathbf{W}$ of weights is defined by

$$w_{ij} = \sum_{t=1}^{m} v_i^t v_j^t \quad (i \neq j), \text{ and } w_{ii} = 0$$

The hope is that if the stored vectors are sufficiently different, such a linear addition of weights would not cause much interference in the error-correcting behavior of the system. Note that the system does not remember the individual vectors $\mathbf{v}^t$, but only the weights w_{ij} which basically represent the correlation among the vectors. If a new vector $\mathbf{v}$ is to be stored, the system "learns" on-line by adding $v_i v_j$ to the existing weight w_{ij}. We call each such stored vector a *fundamental memory*. (This storage receipe is equivalent to the correlation recording technique discussed in Chapter 1.)

This completes the description of the Hopfield model, its mode of operation, and the learning rule. We now consider the question of its convergence analysis.

9.1.4. Error-Correction Behavior

If we wish the system to "remember" the fundamental memories, we expect each of them to be stable and to attract all the vectors within a distance ρn for some constant $\rho > 0$. Or more generally, we consider the system to be error-correcting, if *every* vector within a distance ρn from a fundamental memory eventually ends up within a distance of εn for some $\varepsilon < \rho$. We call this εn residual error. We would also be interested in the rate with which the errors will be corrected, or the *rate of convergence* of the system.

When we store a single vector, we have already seen that the fundamental memory is a stable state of the system which attracts all vectors within a distance $n/2$ in one synchronous step. When we have a number of fundamental memories, the retrieval of a memory will be disturbed by the *noise* created by the other fundamental memories. Yet, we hope that this noise is not overwhelming when the number of fundamental memories is not too large. *Hence, the main question is to determine the amount of error correction and the rate of convergence as a function of the numbers m of fundamental memories and n.*

For a given associative memory system (specified by the matrix $\mathbf{W}$), we define that a state $\mathbf{x}$ has (ρ, ε) error-correcting behavior if every vector within a distance ρn from $\mathbf{x}$ eventually ends up within a distance of εn from $\mathbf{x}$, and every vector within a distance εn from $\mathbf{x}$ will forever stay within this distance. We say that a state $\mathbf{x}$ has a *domain of attraction* of radius ρn if the system started at *any* state **y** at distance not more than ρn from **x** eventually reaches the state **x**. In other words, if the state **x** has a domain of attraction of radius ρn, then **x** has $(\rho, 0)$ error-correcting behavior.

Ideally we would like to store *any* set of m fundamental memories and obtain good error-correcting behavior for each of the fundamental memories. But this requirement is somewhat self-contradictory. For we cannot expect to store vectors that are too close to each other. (Closeness is not the only potential problem. If we require stability of the stored vectors, then, even for $m = 4$, there exist m pairwise distant vectors that cannot be stored as fundamental memories, at least not by using the above storage method of Hopfield.) Hence, a reasonable minimal requirement is that we would like to store almost all sets of m vectors. Therefore, we will take a set of m *random vectors* as our set of fundamental memories, and expect the system to remember them with *probability near* 1. (This randomness is often achieved by coding the input first.) In probabilistic terms, the main question can be restated as follows:

Given m, determine ρ and ε such that with probability near 1 all fundamental memories in a set of m randomly chosen fundamental memories have the (ρ, ε) error-correcting behavior.

In addition, we are also interested in *extraneous* stable states or memories that exhibit error-correcting properties similar to those of the fundamental memories. We would like to estimate the number of such memories and describe their error-correcting behavior.

It should also be mentioned that the error-correcting behavior of the system might be different in different modes (synchronous or asynchronous) of operating the system.

9.1.5. Worst-case and Random Errors

So far, we are requiring that each state within distance ρn from a fundamental memory end up within a distance of εn. Such a requirement guarantees that even in the *worst case* we make a significant error correction. In fact, under this definition, a fundamental memory with $(\rho, 0)$ error-correcting behavior will have a *domain of attraction* of radius at least ρn.

Sometimes, we can relax this requirement and only ask that, with probability near 1, a *randomly chosen* vector within ρn distance from the fundamental memory will eventually end up within a

distance of εn. In most applications, correcting such random errors may be satisfactory.

Yet, it is interesting to find out if stable fundamental memories can attract *all* the vectors within a distance of ρn for some positive constant ρ. In other words, we are interested in establishing a domain of attraction of radius ρn around each fundamental memory. For this, one cannot rely on simulations, since simulations (due to the prohibitively large number of error patterns) can only reveal the behavior of the system in the presence of random errors.

When restricted to random starting states, the system will exhibit a quantitatively different behavior. In particular, a one-step synchronous convergence will now be possible, whereas, in the case of worst-case errors, one synchronous step cannot even correct $\sqrt{n}$ errors. Moreover, one can have a domain of attraction of radius *near* $1/2$ in the case of random errors, as shown by McEliece et al. (1987). However, for the worst-case errors, $\rho > 1/8$ is already impossible even when m is as small as 3, as shown by Montgomery and Vijaya Kumar (1986). Thus, one can only hope for a gradual convergence and a domain of attraction of smaller radius in the case of worst-case errors. In this chapter, we devote our attention to analyzing the more general case of worst-case errors.

9.1.6. Outline of the Chapter

In the following sections we develop the techniques to answer some of the questions posed in this section. In Section 9.2, we present a brief description of some of the rigorous analyses of the Hopfield associative memory model. In the rest of the chapter, we present the convergence analysis of the authors (1988). In Section 9.3, we consider the convergence analysis in the synchronous case. We present the main lemma used in analyzing the synchronous as well as the asynchronous convergence behavior. This lemma measures the progress made in each synchronous step of the system. Using this lemma, we derive the results regarding the error-correcting behavior in the synchronous case. In addition, we use this lemma to conclude that there cannot be stable states in certain annuli around each of the fundamental memories, which in turn will be useful in establishing the asynchronous convergence. In Section 9.4, we prove an energy-barrier lemma and deduce the convergence results in the asynchronous mode. In Section 9.5, we consider the extraneous memories in the system and provide an analysis of the convergence behavior of these memories. We also prove that an exponential number of such memories exists. In Section 9.6, we present the summary of the chapter. We also include an appendix where we derive the large-deviation theorems used in the chapter.

9.2. SUMMARY OF EARLIER WORK

Several researchers have described and predicted the features of the Hopfield model using various techniques. Chapters 8 by Amari and Yamai, 11 by Hui et al., and 13 by Yoshizawa et al. in this book give an account of other approaches to the analysis of associative memory models. In this chapter we focus primarily on the results concerning the rigorous convergence analysis of the Hopfield associative memory model.

A number of authors contributed to the growing literature on the convergence analysis of associative memories. The following gives an account of some of the basic results in this regard.

Basic questions about the absolute stability of the global pattern formation in dynamical systems have been studied by Grossberg (1982), and Cohen and Grossberg (1983) using Liapunov functions.

For the fully connected Hopfield associative memory, McEliece, Posner, Rodemich, and Venkatesh (1987) determined the maximum number of stable fundamental memories and the convergence properties in the presence of *random* errors.

- If $m < n(1 - 2\rho)^2/(4 \log n)$, then (with probability near 1) *all* fundamental memories will be stable. Also, for any fundamental memory, the system can correct *most* patterns of less than ρn errors in one synchronous step.
- If

$$(1 - 2\rho)^2 n/(4 \log n) < m < (1 - 2\rho)^2 n/(2 \log n),$$

then still *most* fundamental memories will be stable with the above-described capability of correcting most patterns of errors.

When m is larger than $cn/\log n$, in particular, when $m = \alpha n$, the fundamental memories are not retrievable exactly, but one still may find stable states in their vicinity. This is suggested by the "energy landscape" results of Newman (1988). In particular, Newman proves that

- For all fundamental memories, all the vectors that are exactly at a distance of ρn from the fundamental memory have energy in excess of at least μn^2 above the energy level of the fundamental vector.

Thus, when starting from a fundamental memory, the system cannot wander away too far.

Komlós and Paturi (1988) addressed the question of worst-case errors and proved the following results.

There are absolute constants α_s, α_a, ρ_s, $\rho_a < \rho_b$ such that the following properties hold with probability near 1 for a random choice of m fundamental memories.

- In the synchronous case, if $m \leq \alpha_s n$, and if the system is started *anywhere* within a distance of $\rho_s n$ from a fundamental memory $\mathbf{v}$, then in about $\log(n/m)$ synchronous steps it will end up within a distance $ne^{-n/(4m)}$ from $\mathbf{v}$ and stay within that distance forever. In particular, when $m < n/(4 \log n)$, the system will *converge* to $\mathbf{v}$ in O(log log n) synchronous steps.
- In the asynchronous case, if $m \leq \alpha_a n$, and if the system is started *anywhere* within a distance of $\rho_a n$ from a fundamental memory $\mathbf{v}$, then it will *converge* to a stable state within a distance of $ne^{-n/(4m)}$ from $\mathbf{v}$. In particular, when $m < n/(4 \log n)$, the system will converge to $\mathbf{v}$.
- For any fundamental memory $\mathbf{v}$, the maximum energy of any state *within* a distance of $\rho_a n$ from $\mathbf{v}$ is less than the minimum energy of any state *at* a distance of $\rho_b n$ from $\mathbf{v}$, and there are no stable states in the annuli defined by the radii $\rho_b n$ and $ne^{-n/(4m)}$ centred at $\mathbf{v}$.

Although in this chapter we only consider the fully connected system of neurons, models in which neurons are less densely connected are also interesting since such models might be more realistic and feasible for implementation. A detailed analysis of the effect of connectivity on the emergence of associative memory can be found in Komlós and Paturi (in press).

The convergence results presented in this chapter for the fully interconnected Hopfield model are adopted from the paper of Komlós and Paturi (1988).

Notations. In the rest of this chapter, we use the following notation.

For vectors $\mathbf{x}$, x_i will denote the ith component of $\mathbf{x}$.

If $\mathbf{x}$ and $\mathbf{y}$ are vectors of the same dimension, the Hamming distance $d(\mathbf{x}, \mathbf{y})$ between $\mathbf{x}$ and $\mathbf{y}$ is defined as the number of components in which they differ.

The scalar product $(\mathbf{x}, \mathbf{y})$ of two vectors is defined as $(\mathbf{x}, \mathbf{y}) = \sum_i x_i y_i$.

The norm of a vector $\mathbf{x}$ is defined as $\|\mathbf{x}\| = (\sum_i x_i^2)^{1/2}$.

We write $[n] = \{1, 2, \ldots, n\}$.

$P(A)$ refers to the probability of the event A.
E stands for expected value.
For nonnegative $\rho \leq 1$, we define the entropy function

$$h(\rho) = -\rho \log \rho - (1 - \rho) \log(1 - \rho)$$

For nonnegative ρ and ρ' such that $\rho + \rho' \leq 1$, we define the entropy function

$$h(\rho, \rho') = -\rho \log \rho - \rho' \log \rho'$$
$$- (1 - \rho - \rho') \log(1 - \rho - \rho')$$

where log stands for natural logarithm.
$c_1, c_2, \ldots$ are small positive absolute constants.

We let $\mathbf{v}^1, \mathbf{v}^2, \ldots, \mathbf{v}^m$ denote the randomly selected fundamental memories. The learning rule gives us the weights

$$w_{ij} = \sum_{t=1}^{m} v_i^t v_j^t \qquad i \neq j$$

$i, j = 1, 2, \ldots, n$ and $w_{ii} = 0$

We define Q_i and L_i to be

$$Q_i = Q_i(\mathbf{x}) = x_i L_i(\mathbf{x}) = \sum_{\substack{j=1 \\ j \neq i}}^{n} w_{ij} x_i x_j$$

$$= \sum_{j \neq i} \left(\sum_{t=1}^{m} v_i^t v_j^t \right) x_i x_j$$

We define the m-dimensional vectors

$$\mathbf{u}^i = (v_i^1, \ldots, v_i^m)$$

for $i = 1, 2, \ldots, n$ (reading the matrix of the vectors $\mathbf{v}$ column-wise). We now have

$$Q_i(\mathbf{x}) = \sum_{j \neq i} (x_i \mathbf{u}^i, x_j \mathbf{u}^j) = \left(x_i \mathbf{u}^i, \sum_{j \neq i} x_j \mathbf{u}^j \right)$$

where $(\mathbf{u}^i, \mathbf{u}^j)$ denotes the scalar product of the vectors $\mathbf{u}^i$ and $\mathbf{u}^j$.

9.3. SYNCHRONOUS CONVERGENCE

An important step in establishing our convergence results in the synchronous case is to characterize the *error-correction dynamics* of the system. Let $\mathbf{x}$ be a vector at a distance of $\rho n \leq \rho_s n$ from some fundamental memory. In one synchronous step, the system, started at state $\mathbf{x}$, will move to a state $\mathbf{x}'$, which is at a distance of $\rho' n$ from that fundamental memory. Our goal is to find the relationship between ρ and ρ'; we describe it in the Main Lemma. This relationship will completely determine the behavior of the synchronous algorithm.

9.3.1. Main Lemma

In the following, we present a lemma (Main Lemma) that gives a quantitative picture of the error-correcting behavior in one synchronous step. This picture is basically as follows. Write $\alpha = m/n$. There are constants α_s and ρ_s such that for $\alpha \leq \alpha_s$ the following holds with probability near 1. (Remember that this probability refers to the random selection of the m fundamental memories. After this selection, the system works in a deterministic fashion; neither the transition rule nor the selection of the initial state is random. We will also use the expression "for almost all choices of the m fundamental memories".) If the system is started at any state at a distance ρn ($\rho \leq \rho_s$) from a fundamental memory (ρn "errors"), then it will correct most of the ρn errors in one synchronous step, and be at a much smaller distance $\rho' n$ from the fundamental memory. This ρ' is about ρ^3 if $\rho > \alpha$, and about $\alpha \rho^2$ if $\rho < \alpha$.

A repeated application of the synchronous operation will result in a double exponential shrinkage of error, until there are only about $ne^{-1/(4\alpha)}$ errors left. After that, the system will never depart farther than this distance. (In particular, there are no errors left when $\alpha < 1/(4 \log n)$, i.e., we have synchronous convergence.) It also turns out that the *convergence in the synchronous case is monotone*.

The Main Lemma implies that the energy function has no local minima in the annuli defined by the radii $\rho_s n$ and $ne^{-1/(4\alpha)}$ around any fundamental memory. This will help us establish asynchronous convergence.

In all the following statements, probability $1 - o(1)$ denotes probability approaching 1 as $n \to \infty$.

MAIN LEMMA (One-Step Error Correction). *There is an α_s, and for every $\alpha \leq \alpha_s$ there are two numbers $\varepsilon(\alpha) < \lambda(\alpha)$ with the following properties.*

1. $\lambda(\alpha)$ *is increasing to a constant ρ_o as α tends to 0.*
2. *As α tends to 0, $\varepsilon(\alpha)$ is decreasing as $e^{-1/(4\alpha)}$.*
3. *The following holds with probability $1 - o(1)$: For all $\rho \in [\varepsilon(\alpha), \lambda(\alpha)]$ and for all fundamental memories $\mathbf{x}$, if $\mathbf{y}$ is such that $d(\mathbf{y}, \mathbf{x}) = \rho n$, then $x_i L_i(\mathbf{y}) > 0$ for all but at most $f(\rho)n$ of the indices i where $f(\rho)$ can be chosen as*

$$f(\rho) = \max\{e^{-1/(4\alpha)}, c_1 \rho h(\rho)(\alpha + h(\rho))\}$$

Remark. We can now define $\rho_s = \lambda(\alpha_s)$.

PROOF OF MAIN LEMMA. For a given fundamental vector $\mathbf{x}$, we will compute the probability that there exists a $\mathbf{y}$ such that

$d(\mathbf{y}, \mathbf{x}) = \rho n$ and that there are more than $f(\rho)n$ indices i such that $x_i L_i(\mathbf{y}) \le 0$. Let A be this event. Let K be the set of indices in which $\mathbf{x}$ and $\mathbf{y}$ differ and let $|K| = \rho n$. Let $\rho' = f(\rho)$. Clearly, we have

$$P(A) \le P(\exists K, |K| = \rho n, I, |I| = \rho' n$$

$$\text{such that } \forall i \in I, x_i L_i(\mathbf{y}) \le 0)$$

Note that $\forall i \in I, x_i L_i(\mathbf{y}) \le 0$ implies

$$\sum_{i \in I} x_i L_i(\mathbf{y}) \le 0$$

But

$$\sum_{i \in I} x_i L_i(\mathbf{y}) = \sum_{i \in I} x_i L_i(\mathbf{x}) - \sum_{i \in I} x_i(L_i(\mathbf{x}) - L_i(\mathbf{y}))$$

$$= \sum_{i \in I} Q_i - 2 \sum_{i \in I} \left(x_i \mathbf{u}^i, \sum_{k \in K - i} x_k \mathbf{u}^k \right)$$

$$= \sum_{i \in I} Q_i - 2 \left(\sum_{i \in I} x_i \mathbf{u}^i, \sum_{k \in K} x_k \mathbf{u}^k \right) + 2|I \cap K|$$

For the sake of convenience, let the fundamental vector $\mathbf{x} = \mathbf{v}^1$. Hence, the first components of all the vectors $x_i \mathbf{u}^i$ are 1. Let $\tilde{\mathbf{u}}^i$ denote the $m - 1$-dimensional vector obtained from $x_i \mathbf{u}^i$ after removing the first component. Clearly, $\tilde{\mathbf{u}}^i$ are independent and uniformly distributed ± 1 random vectors of dimension $m - 1$ (for simplicity, we will not change m to $m - 1$ in the formulas).

We now have

$$\sum_{i \in I} Q_i = \left(\sum_{i \in I} x_i \mathbf{u}^i, \sum_{j \in [n]} x_j \mathbf{u}^j \right) - m|I|$$

$$= \left(\sum_{i \in I} x_i \mathbf{u}^i, \sum_{j \in [n] - I} x_j \mathbf{u}^j \right)$$

$$+ \left(\sum_{i \in I} x_i \mathbf{u}^i \right)^2 - m|I|$$

$$= (n - m)|I| + \left(\sum_{i \in I} \tilde{\mathbf{u}}^i, \sum_{j \in [n] - I} \tilde{\mathbf{u}}^j \right)$$

$$+ \left(\sum_{i \in I} \tilde{\mathbf{u}}^i \right)^2$$

$$= (n - m)|I| + T_1 + T_2$$

Thus,

$$\sum_{i \in I} x_i L_i(\mathbf{y}) = (n - m)|I| + T_1 + T_2 - 2T_3$$

$$+ 2|I \cap K|$$

where

$$T_1 = \left(\sum_{i \in I} \tilde{\mathbf{u}}^i, \sum_{j \in [n] - I} \tilde{\mathbf{u}}^j \right) \quad T_2 = \left(\sum_{i \in I} \tilde{\mathbf{u}}^i \right)^2$$

$$T_3 = \left(\sum_{i \in I} x_i \mathbf{u}^i, \sum_{k \in K} x_k \mathbf{u}^k \right)$$

We now estimate from below each of the above terms individually. From Corollary 4 in the Appendix, we get

$$T_1 > -(z_1 + \sqrt{2z_1})m\sqrt{[I|(n - |I|)]}$$

with probability $1 - e^{-\Delta}$, where

$$z_1 = h(\rho')/\alpha + \Delta/m$$

(We used the fact that the inner product has a symmetric distribution.)

T_2 is obviously nonnegative.

To estimate T_3, we use Corollary 5 in the Appendix.

$$T_3 < c_5[\sqrt{[\rho'\rho h(\rho)(\alpha + h(\rho)]} + \rho'(\alpha + h(\rho'))]n^2$$

Hence, if

$$(1 - \alpha)\rho' - \alpha(z_1 + \sqrt{2z_1})\sqrt{[\rho'(1 - \rho')]}$$

$$- 2c_5[\sqrt{[\rho'\rho h(\rho)(\alpha + h(\rho))]}$$

$$+ \rho'(\alpha + h(\rho'))] > 0$$

then the probability $P(\exists$ a fundamental memory $\mathbf{x}$ such that the event A holds$)$ is upper bounded by $me^{-\Delta}$. This probability is small if $\Delta - \log m$ is large.

It is not hard to see that

$$\rho' = f(\rho) = \max\{e^{-1/(4\alpha)}, c_1\rho h(\rho)(\alpha + h(\rho))\}$$

will satisfy the last inequality. (The only case that needs careful analysis is when $\rho' = 1/n$.)

We define the function $\varepsilon(\alpha) = e^{-1/(4\alpha)}$. It is easy to see that there are positive α_s and ρ_s such that $f(\rho) < \rho$ for $\varepsilon(\alpha) < \rho \le \rho_s$, $\alpha \le \alpha_s$. Choose a decreasing $\lambda(\alpha)$, $\rho_s \le \lambda(\alpha) < 1/2$ such that $f(\rho) < \rho$ for $\varepsilon(\alpha) < \rho \le \lambda(\alpha)$, $\alpha \le \alpha_s$.

Refinements. A little more careful analysis of the above equation (separating the case $n\rho' \le k$ from $n\rho' > k$) shows the following more detailed picture. When

$$m \le \frac{k + 1}{k + 2} \frac{n}{2 \log n}$$

the number of errors left is at most k.

At the other end, when $h(\rho) > \alpha$, one gets a better bound by estimating T_3 with the product of the norms of the two factors in the scalar product. We get

$$T_3 < c\sqrt{[\rho\rho'(\alpha + h(\rho))(\alpha + h(\rho'))]}n^2$$

that leads to the following more detailed error correction

$$\rho' < e^{-c/(\rho h(\rho))}$$

as long as $h(\rho') > \alpha$ holds. After this incredibly fast error correction, the following slower (but still double exponential) function takes over:

$$\rho' = \max\{e^{-1/(4\alpha)}, c\alpha\rho h(\rho)\} \qquad \square$$

9.3.2. Convergence Results

In the following, we establish the existence of a synchronous domain of attraction of positive radius ρ_o around each fundamental memory when $m < n/(4 \log n)$. More generally, we show that for $\alpha \le \alpha_s n$, if the system is started within a distance of $\rho_s n$ from a fundamental memory, then, in about $\log(n/m)$ synchronous steps, it will end up within a distance $ne^{-n/(4m)}$ from the fundamental memory, that is, it will eventually get within a distance $ne^{-n/(4m)}$ of the fundamental memory and remains within that distance. This result is obtained by a repeated application of the Main Lemma.

THEOREM 1 (Error-Correction Theorem). *The following holds with probability* $1 - o(1)$. *For all* $\alpha < \alpha_s$ *and for all fundamental memories* **x**, *if the system is started at a vector* **y** *such that* $d(\mathbf{y}, \mathbf{x}) \le \lambda(\alpha)n$, *then, in* $O(\min\{\log(n/m), \log\log n\})$ *synchronous steps, it will end up within a distance of* $\varepsilon(\alpha)n$ *from* **x** *and stay within this distance.*

PROOF OF THEOREM 1. The theorem easily follows from the Main Lemma and the definitions of $\lambda(\alpha)$ and $\varepsilon(\alpha)$. $\qquad \square$

In particular, we establish a domain of attraction when $m < n/(4 \log n)$. The radius of this domain of attraction is ρ_o. Our calculations show that $0.024 < \rho_o < 1/8$. It would be interesting to find the maximum value for ρ_o.

THEOREM 2 (Synchronous Domain of Attraction). *The following holds with probability* $1 - o(1)$. *If* $m < n/(4 \log n)$, **x** *is any fundamental memory, and* **y** *is such that* $d(\mathbf{y}, \mathbf{x}) \le \rho_o n$, *then the system started in state* **y** *will converge to* **x** *within* $O(\log \log n)$ *synchronous steps.*

Remark. In fact, this $O(\log \log n)$ convergence can be considered very fast. We already indicated that one-step synchronous convergence is not possible even with $O(\sqrt{(n/\alpha)})$ arbitrary errors. The idea behind this observation is the following: One-step convergence would mean, e.g., $x_i L_i(\mathbf{y}) \ge 0$ for all **y** close to **x**. By changing the jth bit, we change the quantity $x_i L_i(\mathbf{y})$ with $2w_{ij}x_i y_j$, which is of the order $\sqrt{m}$. Since $x_i L_i(\mathbf{x}) = O(n)$ by changing appropriate $c\sqrt{(n/\alpha)}$ bits of **x**, we can make $x_i L_i(\mathbf{y}) < 0$.

PROOF OF THEOREM 2. The Main Lemma shows that when $\alpha < 1/(4 \log n)$, $\varepsilon(\alpha)n < 1$, i.e., there are no errors left. The radius of attraction is ρ_o. $\qquad \square$

As a result of the error-correction behavior in the annulus defined by $\varepsilon(\alpha)n$ and $\lambda(\alpha)n$, we can conclude that there are no stable states in this region.

THEOREM 3. *The following holds with probability* $1 - o(1)$ *for all* $\alpha \le \alpha_s$: *There are no stable states in the annuli defined by the radii* $\varepsilon(\alpha)n$ *and* $\lambda(\alpha)n$ *around the fundamental memories.*

Remark. Actually, we get that there are not even local minima of the energy function in the annulus defined by $\lambda(\alpha)n$ and $\varepsilon(\alpha)n$.

9.4. ASYNCHRONOUS CONVERGENCE

Convergence results in the asynchronous case require the existence of high energy barriers around the fundamental memories. They will ensure that there is no escape from a fundamental memory to a state far away. These barriers, together with the result that there are no stable states (Theorem 3) unless one gets within $\varepsilon(\alpha)n$ distance from a fundamental memory, ensure eventual convergence to within $\varepsilon(\alpha)n$ distance from the fundamental memories. Here, we use the fact that asynchronous convergence is guaranteed in the Hopfield model as mentioned in Section 9.1.2. First, we present the results related to the high energy barriers.

9.4.1. Energy Barriers

The following theorem establishes upper and lower bounds on the energy of any state in the vicinity of a fundamental memory.

THEOREM 4 (Energy Levels). *The following holds with probability* $1 - o(1)$. *Let* $0 < \rho \le 1/2$ *and* $m = \alpha n$.
If **x** *is a fundamental memory and* **y** *is such that* $d(\mathbf{y}, \mathbf{x}) = \rho n$, *then*

$$|E(\mathbf{y}) - E(\mathbf{x}) - 2\rho(1 - \rho)n^2| \le \delta n^2$$
where
$$\delta = \delta(\alpha, \rho) < 2\left((h(\rho) + \frac{\Delta}{n} + \sqrt{[2\alpha(h(\rho) + \Delta/n)]} \right)\sqrt{[\rho(1 - \rho)]}$$

where Δ *is such that* $\Delta - \log m$ *tends to infinity.*

PROOF OF THEOREM 4. For the sake of convenience, assume that the fundamental memory $\mathbf{x} = \mathbf{v}^1$. Let $\mathbf{y}$ be such that $d(\mathbf{y}, \mathbf{x}) = \rho n$ and let K be the set of coordinates in which $\mathbf{y}$ and $\mathbf{x}$ differ.

It is easy to see that

$$E(\mathbf{y}) - E(\mathbf{x}) = 2\left(\sum_{k \in K} x_k \mathbf{u}^k, \sum_{k \in K} x_k \mathbf{u}^k \right)$$

Since $\mathbf{x} = \mathbf{v}^1$, we get

$$E(\mathbf{y}) - E(\mathbf{x}) = 2\rho(1 - \rho)n^2 + 2\left(\sum_{k \in K} \tilde{\mathbf{u}}^k, \sum_{k \in K}^N \tilde{\mathbf{u}}^k \right)$$

where $\tilde{\mathbf{u}}^i$ is obtained from $x_i \mathbf{u}^i$ by removing the first component. Note that the $\tilde{\mathbf{u}}^i$ are independent and uniformly distributed ± 1 random vectors of dimension $m - 1$.

From Corollary 4 in the Appendix we get

$$|E(\mathbf{y}) - E(\mathbf{x}) - 2\rho(1 - \rho)n^2|$$
$$< 2(z + \sqrt{2z})m\sqrt{|K|(n - |K|)]}$$

with probability $1 - me^{-\Delta}$, where

$$z = h(\rho)/\alpha + \Delta/m$$

Choose again a Δ much beyond $\log m$. (The exact equation is

$$|E(\mathbf{y}) - E(\mathbf{x}) - 2\rho(1 - \rho)n^2|$$
$$< 2\alpha\delta\sqrt{[\rho(1 - \rho)]}n^2$$

with probability $< e^{-(\alpha/2)p_2(\delta)n}$, which is equivalent to Newman's equation.) We say that b_2 is a *barrier* for b_1 if, for all fundamental memories $\mathbf{x}$,

$$\max\{E(\mathbf{y}) : d(\mathbf{y}, \mathbf{x}) \le b_1 n\}$$
$$< \min\{E(\mathbf{y}) : d(\mathbf{y}, \mathbf{x}) = b_2 n\} \qquad \square$$

Theorem 4 gives the following energy barrier result, which is a generalization of Newman's (1988) result.

THEOREM 5. *There exists a threshold* α_{th} *and two positive constants* $b_1 < b_2$ *such that the following holds with probability* $1 - o(1)$ *for all* $m \le \alpha_{th} n$: b_2 *is a barrier for* b_1. *In fact, any pair* b_1 *and* b_2 *are good for which*

$$2b_1(1 - b_1) + \delta(\alpha, b_1) < 2b_2(1 - b_2) - \delta(\alpha, b_2)$$

where $\delta(\alpha, \rho)$ *is as given in Theorem 4.*

9.4.2. Convergence Results

To establish convergence in the asynchronous case, we will select $\alpha_a < \alpha_s$ such that $\lambda(\alpha_a)$ is a barrier for $\varepsilon(\alpha_a)$. We write $\rho_1 = \varepsilon(\alpha_a)$ and $\rho_2 = \lambda(\alpha_a)$. (More precisely, since $\lambda(\alpha_a)$ may be too large, one should choose $\rho_2 = \min\{b_2, \lambda(\alpha_a)\}$, and assume that ρ_2 is a barrier for ρ_1.)

This implies, in the asynchronous mode, that if $\alpha < \alpha_a$ then any state $\mathbf{y}$ within a distance $\rho_1 n$ from a fundamental memory $\mathbf{x}$ will converge to a state within a distance of $\varepsilon(\alpha)n$ from $\mathbf{x}$. Indeed, by Theorem 3, there are no stable states in the annulus defined by $\lambda(\alpha)n$ and $\varepsilon(\alpha)n$, and $\lambda(\alpha) > \lambda(\alpha_a) \ge \rho_2$. But the system was started within $\rho_1 n$ of the fundamental memory, and, since ρ_2 is a barrier for ρ_1, it cannot escape from the neighborhood with radius $\rho_2 n$. Thus, we get the following result.

THEOREM 6 (Asynchronous Convergence Theorem). *The following holds with probability* $1 - o(1)$ *for all* $m \le \alpha_a n$.

For all fundamental memories $\mathbf{x}$, *if* $\mathbf{y}$ *is such that* $d(\mathbf{y}, \mathbf{x}) \le \rho_1 n$, *then the vector* $\mathbf{y}$ *will converge to a vector within a distance of* $\varepsilon(\alpha)n$ *from* $\mathbf{x}$.

In particular, we get an asynchronous domain of attraction when $m < n/(4 \log n)$.

THEOREM 7. *The following holds with probability* $1 - o(1)$ *for all* $m < n/(4 \log n)$. *For all fundamental memories* $\mathbf{x}$, *if* $\mathbf{y}$ *is such that* $d(\mathbf{y}, \mathbf{x}) \le \rho_1 n$, *then the vector* $\mathbf{y}$ *will converge to* $\mathbf{x}$.

9.5. EXTRANEOUS MEMORIES

In this section we will establish the existence of an exponential number of stable states, and extend some of our previous results to these stable states.

9.5.1. Stability

Note that our convergence proofs were based on the fact that the gradient at the fundamental memories was large. More precisely, all Q_i are large when $m < cn/\log n$, and still most Q_i are large when $m = cn$. We observe that this property is sufficient to extend our proofs. Hence, we introduce the notion of β-stable vectors.

Given $0 < \beta < 1$, a vector $\mathbf{x}$ is β-*stable* if $Q_i(\mathbf{x}) \ge \beta n$ for all i.

Note that a β-stable vector is not only stable but also a deep energy minimum. In fact, β-stable vectors have a large domain of attraction. Furthermore, all the fundamental vectors are β-stable when $m < cn/\log n$.

When $m = \alpha n$ for some constant α, we cannot establish the existence of β-stable vectors. A weaker notion of stability (valid for all fundamental memories) will be introduced, and used to derive some error-correcting properties.

For $0 < \beta < 1$ and $0 \le \varepsilon < 1$, we define that a vector is (β, ε)-*stable* if for all but at most εn indices i, $Q_i(\mathbf{x}) \ge \beta n$.

Even though (β, ε)-stable vectors are not stable themselves, we will later see that *there are stable states in their close vicinity.*

THEOREM 8. *Let* $0 < \beta < 1$. *There exists* $c_o = c_o(\beta)$ *such that, with probability* $1 - o(1)$, *if* $m < c_o n/\log n$, *then all fundamental memories are* β-*stable.* (c_o *can be made arbitrary close to* $\frac{1}{4}$ *by choosing a small enough* β.)

THEOREM 9. *Let* $0 < \beta < 1$ *and* $\varepsilon > 0$. *There exists* $\alpha_o = \alpha_o(\beta, \varepsilon)$ *such that, with probability* $1 - o(1)$, *if* $m \le \alpha_o n$, *then, all fundamental memories are* (β, ε)-*stable.*

In addition to the fundamental memories, there are an exponential number of other stable vectors. Some are there accidentally (true extraneous memories), but c^m of them are there for a reason. The whole model is based on a linear method, so one is not surprised to see that it remembers not only individual vectors, but the whole subspace generated by them. (This has been observed by several researchers before.)

Given vectors $\mathbf{v}^1, \mathbf{v}^2, \ldots$, we define

$$S(\mathbf{v}^1, \mathbf{v}^2, \ldots) = \left\{ \text{sign}\left(\sum_i d_i \mathbf{v}^i \right): d_i = \pm 1 \right\}$$

THEOREM 10. *For every positive* ε, *there is an* $\alpha^* = \alpha^*(\varepsilon)$ *such that, with probability* $1 - o(1)$, *if* $m \le \alpha^* n$, *then more than half of the* 2^m "*linear combinations*" *of the fundamental memories (i.e., elements of* $S(\mathbf{v}^1, \ldots, \mathbf{v}^m)$) *are* $(0.5, \varepsilon)$-*stable.*

For the proof of this theorem, we need the following simple lemma.

LEMMA 1. *Let* $Y_{i,j}$ *be independent and uniformly distributed* ± 1 *random variables. Then,*

$$P\left(\frac{1}{N} \sum_{i=1}^{N} \left| \sum_{j=1}^{m} Y_{i,j} \right| < d\sqrt{m} - 1 \right) < (4\sqrt{d})^N$$

PROOF. Let $Y_i = |\sum_{j=1}^{m} Y_{i,j}|$, and write $D = d\sqrt{m} - 1$. There exists more than $N/2$ i such that $Y_i < 2D$. Thus, the probability in question is bounded by

$$2^N [P(Y_i < 2D)]^{N/2} < 2^N \left[(4D + 1) \binom{m}{m/2} 2^{-m} \right]^{N/2}$$

$$< (4\sqrt{d})^N \qquad \square$$

PROOF OF THEOREM 10. We show first that for any specific choice of the coefficients d_i, the vector $\mathbf{x} = \text{sign}(\sum_{i=1}^{m} d_i \mathbf{v}^i)$ is $(0.5, \varepsilon)$-stable with probability $1 - o(1)$.

Given any specific d_i, we can replace the vectors $\mathbf{v}^i$ with the vectors $d_i \mathbf{v}^i$ since the scalar products $(\mathbf{u}^j, \mathbf{u}^k)$ are invariant under the sign changes of the vectors $\mathbf{v}^i$. Hence, without loss of generality, we take $d_i = 1$ for all i.

Let $\mathbf{x} = \text{sign}(\sum_{i=1}^{m} \mathbf{v}^i)$. Write

$$\tilde{\mathbf{u}}^i = x_i \mathbf{u}^i = \text{sign}\left(\sum_{j=1}^{m} \mathbf{u}_j^i \right) \mathbf{u}^i$$

$\tilde{\mathbf{u}}^i$ is obtained by flipping over the components of $\mathbf{u}^i$ if $\mathbf{u}^i$ has more -1's than $+1$'s. Otherwise, $\tilde{\mathbf{u}}^i$ equals $\mathbf{u}^i$.

Let I denote the set of indices i such that $(x_i \mathbf{u}^i, \sum_{j \ne i} x_j \mathbf{u}^j) < 0.5n$, and let us write $\varepsilon = |I|/n$. Then, we have

$$\left(\sum_{i \in I} x_i \mathbf{u}^i, \sum_{j=1}^{n} \tilde{\mathbf{u}}^j \right) - m|I| < 0.5n|I|$$

The vectors $\tilde{\mathbf{u}}^i$ will be written as $\mathbf{u}^i + 2\mathbf{z}^i$ where $\mathbf{z}^i$ is a $(0, 1)$-vector defined as follows.

Let $d = (\mathbf{1}, \mathbf{u}^i)$ where $\mathbf{1}$ is an m-dimensional vector all of whose components are equal to 1. If $d \ge 0$, then $\mathbf{z}^i$ is a 0 vector. Otherwise, we will randomly select $|d|$ of the indices j where $\mathbf{u}^i$ is -1, and set $\mathbf{z}^i$ to 1 for these j and 0 elsewhere. It is clear that $\tilde{\mathbf{u}}^i$ and $\mathbf{u}^i + 2\mathbf{z}^i$ have the same distribution. Furthermore, the probability p that a component of $\mathbf{z}^i$ is 1, is approximately $c/\sqrt{m}$.

We now have

$$\left(\sum_{i \in I} x_i \mathbf{u}^i, \sum_{j=1}^{n} \tilde{\mathbf{u}}^j \right) = \left(\sum_{i \in I} x_i \mathbf{u}^i, \sum_{j=1}^{n} (\mathbf{u}^i + 2\mathbf{z}^i) \right)$$

We estimate the two terms

$$S_1 = \left(\sum_{i \in I} x_i \mathbf{u}^i, \sum_{j=1}^{n} \mathbf{u}^i \right)$$

$$S_2 = \left(\sum_{i \in I} x_i \mathbf{u}^i, \sum_{j=1}^{n} \mathbf{z}^i \right)$$

separately. We estimate S_2 in the following way:

$$S_2 = \left(\sum_{i \in I} x_i \mathbf{u}^i, E \sum_{j=1}^{n} \mathbf{z}^j \right)$$

$$+ \left(\sum_{i \in I} x_i \mathbf{u}^i, \sum_{j=1}^{n} (\mathbf{z}^j - E\mathbf{z}^j) \right)$$

$$= S_{21} + S_{22}$$

Since $x_i \mathbf{u}^i$ is the flipped-over version of $\mathbf{u}^i$, it follows that

$$S_{21} = np \sum_{i \in I} (x_i \mathbf{u}^i, \mathbf{1}) = np \sum_{i \in I} \left| \sum_{j=1}^{m} \mathbf{u}^i_j \right|$$

It follows from Lemma 1 that, with probability $1 - o(1)$, for all $|I|$, $S_{21} > c\varepsilon^2 n|I| = c\varepsilon^3 n^2$. We also have that

$$|S_{22}| < \left\| \sum_{i \in I} x_i \mathbf{u}^i \right\| \left\| \sum_{j=1}^{n} (\mathbf{z}^j - E\mathbf{z}^j) \right\|$$

$$E \left\| \sum_{j=1}^{n} (\mathbf{z}^j - E\mathbf{z}^j) \right\|^2 = \sum_{k=1}^{m} E \left[\sum_{j=1}^{n} (z^j_k - Ez^j_k) \right]^2$$

$$= \sum_{k=1}^{m} \sum_{j=1}^{n} E(z^j_k - Ez^j_k)^2$$

$$\leq \sum_{k=1}^{m} \sum_{j=1}^{n} E(z^j_k)^2 = mnp$$

Thus, since $p \to 0$, $\|\sum_{j=1}^{n} (\mathbf{z}^j - E\mathbf{z}^j)\|^2 = o(n^2)$ with probability $1 - o(1)$. From this and from Corollary 3 in the Appendix, we get that $S_{22} = o(n^2)$ with probability $1 - o(1)$.

To estimate S_1, we use the inequality

$$|S_1| \leq \left\| \sum_{i \in I} x_i \mathbf{u}^i \right\| \left\| \sum_{j=1}^{n} \mathbf{u}^j \right\|$$

Again from Corollary 3, we get that with probability $1 - o(1)$, for all $\mathbf{x}$ and for all $|I|$, $|S_1| \leq c'n\sqrt{mn} = c'\sqrt{\alpha}n^2$.

Comparing S_1 and S_{21}, we get that as long as $\alpha < c''\varepsilon^6$, the number of indices i for which $Q_i(\mathbf{x}) < 0.5n$ is at most εn.

Hence, with probability $1 - o(1)$, $\mathbf{x}$ is a $(0.5, \varepsilon)$-stable vector. It follows that, with probability $1 - o(1)$, most of the 2^m "linear combinations" of the fundamental vectors are $(0.5, \varepsilon)$-stable, which proves the theorem. $\square$

Remark. In fact, the proof shows that the above set of linear combinations can be extended from ± 1 linear combinations to all linear combinations, i.e., to the set $\{\text{sign}(\sum_i d_i \mathbf{v}^i)\}$ with real coefficients d_i. This would improve the lower bound c^m to $e^{c \min\{m^2, n\}}$.

9.5.2. Convergence

In this subsection we show that the stability introduced above guarantees convergence properties similar to those of the fundamental memories.

THEOREM 11. *The Main Lemma, and Theorems 1–7 remain valid if we replace fundamental memories by β-stable vectors, but the numerical*

quantities involved change as follows: The function $f(\rho)$ in the Main Lemma is replaced by

$$f(\rho) = c_2(\beta)\rho h(\rho)(\alpha + h(\rho)) \quad (i.e. \ \varepsilon(\alpha) \ becomes \ 0)$$

(Or, as we remarked after the proof of the Main Lemma, one can take for $f(\rho)$ the smaller of the two quantities $\max\{\alpha, e^{-c/(\rho h(\rho))}\}$ and $c(\beta)\rho h(\rho)(\alpha + h(\rho))$.)

Thus, Theorems 2 and 7 (synchronous and asynchronous convergence) hold even if $m = \alpha n$ with a constant α. The condition $m = O(n/\log n)$ is not necessary any more since we assumed that $\mathbf{x}$ is β-stable.

All constants involved in these theorems will depend on β.

The bound in Theorem 4 changes to

$$|E(\mathbf{y}) - E(\mathbf{x})| \leq c\sqrt{[\rho(\alpha + h(\rho)]n^2}$$

Indeed, the only term that changes in the proof of the Main Lemma is T_1, but this is now assumed to be greater than $\beta n|I|$. The other theorems are corollaries.

The proof of the analog of Theorem 4 follows the same pattern. We start with the identity

$$E(\mathbf{y}) - E(\mathbf{x}) = 2 \left(\sum_{k \in K} x_k \mathbf{u}^k, \sum_{k \in \bar{K}} x_k \mathbf{u}^k \right)$$

and then estimate the scalar product by the lengths of the vectors using Corollary 3 in the Appendix.

The following theorem follows from the above and the definition of (β, ε)-stable vectors.

THEOREM 12. *The Main Lemma, and Theorems 1, 3, 4, 5, and 6 remain valid if we replace fundamental memories by (β, ε)-stable vectors, $\varepsilon(\alpha)$ by ε, and $f(\rho)$ by $f(\rho) + \varepsilon$.*

COROLLARY 1. *If $\mathbf{x}$ is a (β, ε)-stable vector, then there is a stable state within a distance of εn from $\mathbf{x}$.*

COROLLARY 2. *There are constants $\alpha_{\exp} > 0$ and $c > 1$ such that, with probability $1 - o(1)$, if $m \leq \alpha_{\exp} n$, then the number of stable states is more than c^m.*

Indeed, it is easy to see that for a fixed $\varepsilon < 1/4$, the probability that two of the vectors in $S(\mathbf{v}^1, \mathbf{v}^2, \ldots)$ are at a distance $2\varepsilon n$ or less, is exponentially small (in n). Starting with the 2^{m-1} vectors mentioned in Theorem 10, and using a greedy algorithm, one can select c^m of these vectors such that any two of them are at a distance larger than $2\varepsilon n$. For each of them, select a stable vector within a distance εn.

Remark. If one is satisfied with an exponential convergence (time log n) instead of double exponential convergence (time log log n), then the following much simpler proof could be given for the convergence to β-stable vectors. It is based on a lemma that establishes a Lipschitz type property for the gradient of the energy function.

LEMMA 2 (Gradient Lemma). *Let* $0 < \beta < 1$. *Then there exist positive* $\alpha_o(\beta)$, $\rho_o(\beta)$ *such that the following holds with probability* $1 - o(1)$ *for all* $\alpha \leq \alpha_o(\beta)$, $\rho \leq \rho_o(\beta)$.
For all $\mathbf{x}$ *and* $\mathbf{y}$, *if* $d(\mathbf{y}, \mathbf{x}) \leq \rho n$, *then the number of i such that*

$$|L_i(\mathbf{x}) - L_i(\mathbf{y})| \geq \beta n$$

is at most $\rho' = c(\beta)\rho(\alpha + h(\rho)) < \rho/2$.

PROOF. Indeed, if K denotes the set of indices where $\mathbf{x}$ and $\mathbf{y}$ differ, then

$$L_i(\mathbf{x}) - L_i(\mathbf{y}) = 2\left(\mathbf{u}^i, \sum_{k \in K - i} x_k \mathbf{u}^k \right)$$

Thus, if I stands for the set where

$$L_i(\mathbf{x}) - L_i(\mathbf{y}) \geq \beta n$$

then, by Corollary 3 in the Appendix,

$$|I|\beta n \leq \sum_{i \in I} (L_i(\mathbf{x}) - L_i(\mathbf{y}))$$

$$= 2\left(\sum_{i \in I} \mathbf{u}^i, \sum_{k \in K} x_k \mathbf{u}^k \right) - 2m|I \cap K|$$

$$\leq 2 \left\| \sum_{i \in I} \mathbf{u}^i \right\| \left\| \sum_{k \in K} x_k \mathbf{u}^k \right\|$$

$$= 2\sqrt{[c_2 \rho(\alpha + h(\rho))]}\sqrt{[c_2 \rho'(\alpha + h(\rho'))]}n^2$$

$$< \rho'\beta n^2$$

(a contradiction) if

$$\rho' > (c/\beta^2)\rho(\alpha + h(\rho)) < \rho/2$$

assuming ρ and α are small in terms of β. Thus, if $\mathbf{x}$ is β-stable, then it has the error-correcting property of the Main Lemma with $\rho' < c(\beta)\rho(\alpha + h(\rho))$. If $\mathbf{x}$ is (β, ε)-stable, then it also has the error correcting property, but with ρ' above replaced by $\rho' + \varepsilon$. $\qquad \Box$

9.6. SUMMARY

This chapter presents rigorous mathematical proofs for some observed convergence phenomena in an associative memory model

introduced by Hopfield (based on Hebbian rules) for storing a number of random n-bit patterns.

We prove that Hopfield's associative memory model is capable of correcting a linear number of arbitrary errors, thus the existence of a large domain of attraction. More precisely, we prove

- When m, the number of patterns stored, is less than $n/(4 \log n)$, the fundamental memories have a domain of attraction of radius at least ρn with $\rho = 0.024$, and both the synchronous and the asynchronous algorithms converge very fast.
- When $m = \alpha n$ (with α small), *all* patterns within a distance ρn from a fundamental memory end up within a distance εn from the fundamental memory, where ε is about $e^{-1/(4\alpha)}$.

We also extend the description of the "energy landscape," and prove the existence of an exponential number of stable states (extraneous memories) with convergence properties similar to those of the fundamental memories.

REFERENCES

Cohen, M. A. and Grossberg, S. (1983). "Absolute Stability of Global Pattern Formation and Paral lel Memory Storage by Competitive Neural Networks," *IEEE Trans. Systems, Man, and Cybernetics,* **13**, 815–826.

Grossberg, S. (1982). *Studies of Mind and Brain.* Reidel, Boston.

Hebb, D. O. (1949). *The Organization of Behavior.* Wiley, New York.

Hopfield, J. J. (1982). "Neural Networks and Physical Systems with Emergent Collective Computational Abilities," *Proc. Nat. Acad. Sci. U.S.A.,* **79**, 2554–2558.

Komlós, J. and Paturi, R. (1988). "Convergence Results in an Associative Memory Model," *Neural Networks,* **1**, 239–250.

Komlós, J. and Paturi, R. (in press). "Effect of Connectivity in an Associative Memory," *J. Computer and System Sci.*

McEliece, R. J., Posner, E. C., Rodemich, E. R., and Venkatesh, S. S. (1987). "The Capacity of the Hopfield Associative Memory," *IEEE Trans. Information Theory,* **33**, 461–482.

Montgomery, B. L. and Vijaya Kumar, B. V. K. (1986). "Evaluation of the Use of the Hopfield Neural Network Model as a Nearest-neighbor Algorithm," *Appl. Opt.,* **25**, 3759–3766.

Newman, C. M. (1988). "Memory Capacity in Neural Network Models: Rigorous Lower Bounds," *Neural Networks,* **1**, 223–238.

APPENDIX: Large Deviation Theorems

In the previous sections we used estimates for the scalar products of sums of random vectors. In the following, we derive these estimates. Just as in Newman's (1988) paper, we use large deviation theorems to bound the probabilities in question.

Let X be a random variable. The function

$$R_X(t) = Ee^{tX}$$

is called the moment-generating function of X. The most important properties of $R_X(t)$ are listed here.

Fact 1. *If X and Y are independent, then*

$$R_{X+Y} = R_X(t)R_Y(t)$$

In particular, if $X_1, X_2, \ldots, X_n$ are independent and identically distributed (i.i.d.) random variables and $S_n = \sum_{i=1}^{n} X_i$, then

$$R_{S_n}(t) = [R_{X_1}(t)]^n$$

Fact 2. (Chernoff's bound): *For $c > \mu = EX$,*

$$P(S_n > cn) \leq [\inf_{t \geq 0} e^{-ct} R(t)]^n$$

where $R(t) = Ee^{tX_1}$.

This simply follows from the following inequality, known as Markov's inequality: If Y is a nonnegative random variable, then for any $y > 0$, $P(Y \geq y) \leq EY/y$.

Let $X_1, X_2, \ldots, X_n$ be independent ± 1 random variables, with $P(X_i = 1) = P(X_i = -1) = \frac{1}{2}$. As before, $S_n = \sum_{i=1}^{n} X_i$, and, for a set $I \subseteq [n]$, $S_I = \sum_{i \in I} X_i$. (Recall that $[n] = \{1, 2, \ldots, n\}$.)

Fact 3. $Ee^{tS_n/\sqrt{n}} \leq e^{t^2/2}$, $-\infty < t < +\infty$.

Fact 4. $Ee^{tS_n^2/n} \leq 1/\sqrt{1 - 2t}$, $0 \leq t < 1/2$.

Fact 5. $Ee^{tS_I S_J/\sqrt{|I||J|}} \leq 1/\sqrt{1 - t^2}$, $-1 < t < 1$, *where I and J are disjoint sets.*

In the last three inequalities, we used the following observations: if all coefficients in the Taylor series expansion around 0 of a function $f(x)$ are nonnegative, then

$$Ef\left(\frac{S_n}{\sqrt{n}}\right) \leq Ef(\eta)$$

where η is standard normal. This follows from the well-known inequalities

$$E\frac{S_n^{2k}}{n^k} \leq E\eta^{2k}$$

(We assume absolute convergence of the Taylor series to integrable functions with respect to the above measures.)

In the following, we give estimates of the scalar product of sums of random vectors. Let $\mathbf{u}^1, \mathbf{u}^2, \ldots$, be independent and uniformly distributed ± 1 random vectors of dimension m.

The first lemma estimates the norm of a sum of random vectors.

LEMMA 3. *Let r be an integer.*

$$P\left[\left\|\sum_{j=1}^{r} \mathbf{u}^j\right\|^2 \geq (1 + \delta)rm\right] \leq e^{-(m/2)p_1(\delta)}$$

where $p_1(\delta)$ is defined as

$$p_1(\delta) = \delta - \log(1 + \delta)$$

The function p_1 has the property $p_1(z + \sqrt{2z}) > z$ for $z > 0$.

COROLLARY 3.

$$\Pr\left[\left\|\sum_{j=1}^{r} \mathbf{u}^j\right\|^2 \geq (1 + 2z + 2\sqrt{z})rm\right] < e^{-zm}$$

Thus, with probability $1 - e^{-\Delta}$, for all I, $|I| = \rho n$,

$$\left\|\sum_{i \in I} \mathbf{u}^i\right\|^2 < (1 + 2z + 2\sqrt{z})m|I|$$

where $z = h(\rho)/\alpha + \Delta/m$.

Consequently, with probability $1 - o(1)$, for all $\mathbf{x}$ and I, $|I| = \rho n$, $0 < \rho \leq 1/2$,

$$\left\|\sum_{i \in I} x_i \mathbf{u}^i\right\|^2 < c_2\rho(\alpha + h(\rho))n^2$$

In particular, with probability $1 - o(1)$, for all $\mathbf{x}$ and I,

$$\left\|\sum_{i \in I} x_i \mathbf{u}^i\right\|^2 < c_3 n^2$$

PROOF. Let A be the event $\|\sum_{j=1}^{r} \mathbf{u}^j\|^2 \geq (1 + \delta)rm$. A implies that, for all $t > 0$,

$$\exp\left[\frac{t}{r}\left\|\sum_{j=1}^{r} \mathbf{u}^j\right\|^2\right] \geq \exp[t(1 + \delta)m]$$

By using Markov's inequality, we get that

$$P(A) \leq \exp[-t(1 + \delta)m] E \exp\left[\frac{t}{r}\left\|\sum_{j=1}^{r} \mathbf{u}^j\right\|^2\right]$$

Note that in the sums

$$\left(\sum_{j=1}^{r} \mathbf{u}^j\right)_i = u_i^1 + u_i^2 + \cdots + u_i^r$$

the terms are independent and uniformly distributed ± 1 random variables, and different sums are independent of each other. Hence we get that

$$P(A)$$
$$\leq \exp[-t(1 + \delta)m]\left(E \exp\left[\frac{t}{r}(u_1^1 + \cdots + u_1^r)^2\right]\right)^m$$

By using the moment-generating function inequality Fact 4, we get that

$$P(A) \le e^{-t(1+\delta)m} \frac{1}{(\sqrt{1-2t})^m} = e^{-(m/2)[2t(1+\delta)+\log(1-2t)]}$$

The exponent in the above expression achieves its minimum in the range $[0, \frac{1}{2})$ when $t = \delta/(2(1+\delta))$. Hence, the lemma follows. (The property of the function p_1 mentioned in the lemma is standard calculus.)

The first part of the corollary follows from the lemma and the fact that the number of sets I to consider is

$$\binom{n}{\rho n} < e^{h(\rho)n}$$

(This factor

$$\binom{n}{\rho n}$$

makes the difference between random errors and worst-case errors.)

In the second part, we have to multiply by an additional factor $2^{\rho n}$ for the choice of x_i.

Next, the following lemma estimates the scalar product of two vector-sums.

LEMMA 4. *Let r and s be integers. Then,*

$$P\left(\left(\sum_{i=1}^{r} \mathbf{u}^i, \sum_{j=r+1}^{r+s} \mathbf{u}^j\right) \ge \delta m\sqrt{rs}\right) \le e^{-(m/2)p_2(\delta)}$$

where $p_2(\delta)$ is given by

$$p_2(\delta) = [\sqrt{(1+4\delta^2)} - 1] + \log\left(\frac{\sqrt{1+4\delta^2} - 1}{2\delta^2}\right)$$

The function p_2 has the property $p_2(z/2 + \sqrt{z}) > z$ for $z > 0$.

COROLLARY 4.

$$P\left(\left(\sum_{i=1}^{r} \mathbf{u}^i, \sum_{j=r+1}^{r+s} \mathbf{u}^j\right) \ge (z + \sqrt{2z})m\sqrt{rs}\right) < e^{-zm}$$

Thus, with probability $1 - e^{-\Delta}$, for all pairs of disjoint sets I, J, $|I| = \rho'n$, $|J| = \rho''n$,

$$\left(\sum_{i\in I} \mathbf{u}^i, \sum_{j\in J} \mathbf{u}^j\right) < (z + \sqrt{2z})m\sqrt{|I||J|}$$

where $z = h(\rho', \rho'')/\alpha + \Delta/m$.

Consequently, with probability $1 - o(1)$, for all $\mathbf{x}$ and for all pairs of disjoint sets I, J, $|I| = \rho'n$, $|J| = \rho''n$, $0 < \rho', \rho'' \le 1/2$,

$$\left|\left(\sum_{i\in I} \mathbf{u}^i, \sum_{j\in J} \mathbf{u}^j\right)\right| < c_4\sqrt{[\rho'\rho''h(\rho', \rho'')(\alpha + h(\rho', \rho''))]}\,n^2$$

PROOF. Let A denote the event $(\sum_{i=1}^{r} \mathbf{u}^i, \sum_{j=r+1}^{r+s} \mathbf{u}^j) \ge \delta m\sqrt{rs}$. A implies that, for all $t > 0$,

$$\exp\left[\frac{t}{\sqrt{rs}}\left(\sum_{i=1}^{r} \mathbf{u}^i, \sum_{j=r+1}^{r+s} \mathbf{u}^j\right)\right] \ge e^{t\delta m}$$

By Markov's inequality we have

$$P(A) \le e^{-t\delta m} E \exp\left[\frac{t}{\sqrt{rs}}\left(\sum_{i=1}^{r} \mathbf{u}^i, \sum_{j=r+1}^{r+s} \mathbf{u}^j\right)\right]$$

$$= e^{-t\delta m}\left(E \exp\left[\frac{t}{\sqrt{rs}}\left(\sum_{i=1}^{r} \mathbf{u}^i, \sum_{j=r+1}^{r+s} \mathbf{u}^j\right)\right]\right)^m$$

By using Fact 5, we get

$$P(A) \le e^{-t\delta m} \frac{1}{(\sqrt{1-t^2})^m} = e^{-(m/2)[2t\delta + \log(1-t^2)]}$$

The exponent in the above expression will be minimal in the range $[0, 1)$ when $t = (\sqrt{1 + 4\delta^2} - 1)/(2\delta)$. Hence, the lemma follows.

Combining Corollary 3 and Corollary 4, we get

COROLLARY 5. *With probability $1 - o(1)$, for all $\mathbf{x}$ and for all pairs of (not necessarily disjoint) sets I, J, $|I| = \rho'n$, $|J| = \rho''n$, $0 < \rho' \le \rho'' \le 1/2$,*

$$\left|\left(\sum_{i\in I} \mathbf{u}^i, \sum_{j\in J} \mathbf{u}^j\right)\right|$$
$$< c_5\{\sqrt{[\rho'\rho''h(\rho'')(\alpha + h(\rho''))]} + \rho'(\alpha + h(\rho'))\}\,n^2$$

10

Nonlinear Dynamics of Analog Associative Memories

FREDERICK R. WAUGH, CHARLES M. MARCUS, and ROBERT M. WESTERVELT

10.1. INTRODUCTION

Since Hopfield (1982) revived interest in the connection between neural networks and collective phenomena in complex systems (Caianiello, 1961; Grossberg, 1967, 1968; Kohonen, 1974; Little, 1974), the tools of statistical physics have been applied to neural network associative memories with notable success [for reviews see Amit (1989), Hertz et al. (1991); for collections of recent research see Garrido (1990), Domany et al. (1991)]. Some of the most remarkable achievements, such as the phase diagram for Hebb-rule associative memories (Amit et al., 1985a, b, 1987) and the exploration of the space of network interactions (Gardner, 1988), rely on techniques originally developed to study the models of disordered magnetic systems known as spin glasses [for reviews see Binder and Young (1986), Mezard et al. (1987), Fischer and Hertz (1991)]. As a result, much of the associative memory literature is devoted to the behavior of networks of two-state neurons with stochastic dynamics, which are closely related to Ising spin systems (McCulloch and Pitts, 1943; Cragg and Temperley, 1954).

Despite these successes, hardware implementations of theoretical models based on spin glass ideas remain few. An important reason is that several aspects of these models are poorly suited for electronic realization. For example, most models require serial updating of neurons to prevent oscillation, while speed considerations dictate that updating be done in parallel (Goles-Chacc et al., 1985; Goles and Vichniac, 1986; Fontanari and Köberle, 1987, 1988a); and stochasticity places costly demands on circuit size, since each neuron must be equipped with a local noise generator (Alspector et al., 1989, 1991). The lack of implementations of these models is particularly striking because it is widely thought that neural networks must move off general-purpose computers and onto specialized

hardware if they are to achieve useful, real-time applicability.

Rather than trying to force theoretical models onto hardware, an alternate approach is to ask what features are readily implementable in hardware and to design neural networks accordingly. In fact, neural network implementations in conventional analog or VLSI circuitry are numerous (Graf et al., 1986; Holler et al., 1989; Maher et al., 1989; Mead, 1989; Andreou et al., 1991; Fisher et al., 1991; Murray, 1991). Usually, these networks consist of *analog* neurons—characterized by a nonlinear input–output transfer function with finite slope or gain (see Fig. 10.1)—

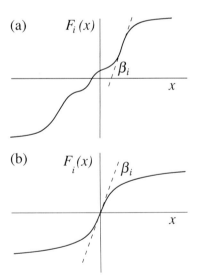

Fig. 10.1. (a) A nonlinear neuron transfer function admissible to the dynamical analysis of Section 10.3. The function is monotonically increasing and rises slower than linearly when its argument is large in magnitude. The maximum slope or gain β_i is also indicated. (b) A sigmoidal or s-shaped nonlinear transfer function that meets the less general conditions assumed for the associative memories in this chapter.

updated *deterministically*, either in continuous time or in discrete time with *parallel* updating. This chapter explores the dynamics of these networks when configured as associative memories. As we will see, analog neurons can be updated in parallel without oscillation, an important advantage over two-state neurons for real-time processing. Moreover, the finite gain of analog neurons can be used much like temperature in stochastic systems to improve associative memory performance by eliminating spurious fixed-point attractors, as discussed below.

The chapter is organized as follows. In Section 10.2 we give a mathematical description of analog neural networks with symmetric interconnections that operate in either continuous or discrete time. In Section 10.3 we construct Liapunov functions to analyze the attractors of these networks. We show that, with continuous-time updating, analog networks converge only to fixed points for all values of neuron gain (Cohen and Grossberg, 1983; Hopfield, 1984); but with discrete-time updating, they may converge also to oscillatory modes unless the gain is reduced below a critical value (Golden, 1986; Marcus and Westervelt, 1989b; Fogelman-Soulie et al., 1989). Phase diagrams for analog associative memories are introduced in Section 10.4 (Marcus et al., 1990; Kühn, 1990; Kühn et al., 1991). These diagrams describe network dynamics as a function of neuron gain and the ratio of stored patterns to neurons. In Section 5 we show how reducing neuron gain can eliminate spurious minima from the energy landscape of an analog associative memory, allowing improved recall of stored patterns (Waugh et al., 1990, 1991; Fukai and Shiino, 1990). The technique of varying analog gain to improve system of performance, known as deterministic annealing, has been used in solving a variety of difficult optimization problems (Hopfield and Tank, 1985, 1986; Koch et al., 1986; Blake and Zisserman, 1987; Durbin and Willshaw, 1987). Section 10.5 describes multiple time-step updating, another method of improving network performance for discrete-time networks (Marcus and Westervelt, 1990; Herz, 1991; Herz et al., 1991). We summarize and identify some key unsolved problems in Section 10.7.

10.2. ANALOG NEURAL NETWORKS

The neural networks we will consider in this chapter are defined by either the continuous-time system,

$$\frac{dx_i(t)}{dt} = -x_i(t) + F_i\left[\sum_{j=1}^{N} T_{ij}x_j(t) + I_i\right]$$

$$i = 1, \ldots, N \quad (1)$$

or the discrete-time system

$$x_i(t+1) = F_i\left[\sum_{j=1}^{N} T_{ij}x_j(t) + I_i\right]$$

$$i = 1, \ldots, N \quad (2)$$

with parallel updating. The real-valued quantity $x_i(t)$ denotes the output of neuron i at time t. Each neuron is characterized by a nonlinear input–output transfer function $F_i(x)$ relating output to input. The input to neuron i, which appears in square brackets in Eqs. (1) and (2), consists of two terms. The first term is a weighted sum of the other neuron outputs, with the weights given by a symmetric, real-valued interconnection matrix T_{ij}. The second term is a time-independent bias I_i, which may represent data from the external world. These two terms are summed and passed through the transfer function to determine the neuron output. For reasons that will be apparent in the next section, we require that the transfer functions be continuous and monotonically increasing and that they rise slower than linearly when their arguments are large in magnitude. These requirements are quite general and are satisfied by virtually all functions of interest for associative memory applications. An important characteristic of a transfer function $F_i(x)$ is its maximum slope or gain, which we will denote by β_i. We will often use the sigmoidal, or s-shaped, function $F_i(x) = \tanh(\beta x)$, which has gain β. Examples of admissible transfer functions are shown in Fig. 10.1.

Equations (1) and (2) are based on the circuit equations of the electronic neural network depicted schematically in Fig. 10.2. When operated in continuous time, the circuit obeys the equations (Hopfield, 1984)

$$C_i\frac{du_i(t')}{dt'} = -\frac{u_i(t')}{R_i} + \sum_j T_{ij}f_j[u_j(t')] + I_i \quad (3)$$

When neurons are updated in discrete time at a frequency $1/\tau$, the circuit obeys the equations

$$\frac{u_i(t'+\tau)}{R_i} = \sum_j T_{ij}f_j[u_j(t')] + I_i \quad (4)$$

In Eqs. (3) and (4), $u_i(t')$ is the input voltage of neuron i at time t', the matrix element T_{ij} is the conductance between neurons i and j, I_i is an externally applied current, and R_i is the total

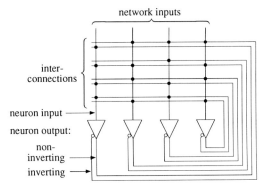

network inputs

inter-
connections

neuron input

neuron output:

non-
inverting

inverting

Fig. 10.2. Schematic of a four-neuron electronic network showing input currents, interconnection conductances, and neurons with noninverting and inverting outputs. The interconnection conductances, represented by the filled circles, are all nonnegative; the noninverting outputs are used for excitatory connections, the inverting outputs for inhibitory ones. The circuit may be operated in continuous time, or in discrete time using an external clock. If operated in continuous time, the circuit obeys Eq. (3), with stray capacitances determining the quantities C_i. If clocked externally, the circuit obeys Eq. (4).

resistance at the input of neuron i:

$$R_i = \sum_j |T_{ij}|^{-1} \qquad (5)$$

In Eq. (3), C_i is the capacitance at the input of neuron i. Sums over the index j in Eqs. (3) to (5) and sums over i and j in the rest of this chapter range from 1 to N. Equation (4) is equivalent to Eq. (2), as can be seen by defining new variables $x_i(t)$ representing the neuron outputs, making the change of variables

$$\frac{u_i(t)}{R_i} = \sum_j T_{ij}x_j(t) + I_i \quad f_i(x) = F_i\left(\frac{x}{R_i}\right) \qquad (6)$$

and measuring time in units of τ. Similarly, Eq. (3) is equivalent to Eq. (1) provided that all neurons have the same time constant R_iC_i, as can be seen by making the transformation (6) and rescaling time according to $t' = R_iC_it$.

10.3. STABILITY RESULTS FOR ANALOG NEURAL NETWORKS

We want to use the analog neural networks described in the previous section to build associative memories that carry out computations by relaxing to fixed-point attractors. However, dynamical systems with parallel updating often have stability problems that lead to sustained

oscillations (Goles-Chacc et al., 1985; Goles and Vichniac, 1986; Fontanari and Köberle, 1987, 1988a). Hence, it is important to understand whether such oscillations can occur and how to eliminate them. In this section, we analyze the dynamics of analog neural networks with arbitrary transfer functions (subject to the restrictions mentioned above) and arbitrary symmetric interconnections. We will see that the continuous-time system (1) has only fixed-point attractors. The discrete-time system (2), on the other hand, can have period-2 limit cycles as well as fixed points. However, these oscillatory modes are eliminated, leaving only fixed-point attractors, when the neuron gains satisfy a simple criterion determined by the eigenvalue spectrum of the interconnection matrix (Golden, 1986; Marcus and Westervelt, 1989b). We will derive this criterion below.

We first consider the continuous-time system (1) and show that it has only fixed-point attractors. Consider the function

$$L(t) = -\frac{1}{2}\sum_{i,j} T_{ij}x_i(t)x_j(t) - \sum_i I_ix_i(t)$$

$$+ \sum_i G_i(x_i(t)) \qquad (7)$$

where $G_i(x)$ is the integral of the transfer function inverse:

$$G_i(x) = \int_0^x F_i^{-1}(z)\,dz \qquad (8)$$

The function $L(t)$ and similar functions have been used by a number of authors to study the dynamics of analog networks (Cohen and Grossberg, 1983; Hopfield, 1984; Golden, 1986; Marcus and Westervelt, 1989b; Fogelman-Soulie et al., 1989). We now prove that $L(t)$ is a Liapunov function by showing (i) that the derivative of $L(t)$ with respect to time is always less than or equal to zero, and (ii) that $L(t)$ is bounded below. Using the symmetry of the interconnection matrix, the time derivative is

$$\frac{dL(t)}{dt} = \sum_i \frac{dx_i(t)}{dt}\left[-\sum_j T_{ij}x_j(t) - I_i + G_i'(x_i(t))\right] \qquad (9)$$

where $G_i'(x) = F_i^{-1}(x)$ is the derivative of $G_i(x)$. The first two terms in square brackets can be rewritten using Eq. (1), with the result

$$\frac{dL(t)}{dt} = \sum_i \frac{dx_i(t)}{dt}\left[-F_i^{-1}\left(x_i(t) + \frac{dx_i(t)}{dt}\right)\right.$$

$$\left. + F_i^{-1}(x_i(t))\right] \qquad (10)$$

As long as each transfer function increases monotonically, the quantity in square brackets in Eq. (10) always has the opposite sign of $dx_i(t)/dt$. Thus

$$\frac{dL(t)}{dt} \le 0 \qquad (11)$$

with equality holding only when $dx_i(t)/dt = 0$ for all i. Furthermore, because $F_i(x)$ rises slower than linearly when its argument is large in magnitude, the third sum in Eq. (7) eventually rises faster than quadratically, dominating the other two sums and causing $L(t)$ to be bounded below. The result (11) and the boundedness of $L(t)$ imply that $L(t)$ is a Liapunov function of the continuous-time network: the network seeks out the local minima of $L(t)$. The continuous-time network therefore has only fixed-point attractors.

The discrete-time network, on the other hand, may have both fixed points and period-2 limit cycles. To see this, we introduce another function (Marcus and Westervelt, 1989b)

$$E(t) = -\sum_{i,j} T_{ij} x_i(t) x_j(t-1)$$

$$- \sum_i I_i [x_i(t) + x_i(t-1)]$$

$$+ \sum_i [G_i(x_i(t)) + G_i(x_i(t-1))] \qquad (12)$$

with $G_i(x)$ defined as in Eq. (8). Just as for the continuous-time network, we prove that $E(t)$ is a Liapunov function by showing that the change in $E(t)$ between *discrete* time steps,

$$\Delta E(t) = E(t+1) - E(t)$$

$$= -\sum_{i,j} T_{ij} x_i(t+1) x_j(t) + \sum_{i,j} T_{ij} x_i(t) x_j(t-1)$$

$$- \sum_i I_i [x_i(t+1) - x_i(t-1)]$$

$$+ \sum_i [G_i(x_i(t+1)) - G_i(x_i(t-1))] \qquad (13)$$

is a nonincreasing function of time and is bounded below. Using Eq. (2) and the symmetry of the interconnection matrix, Eq. (13) can be written as

$$\Delta E(t) = -\sum_i F_i^{-1}[x_i(t+1)] \Delta_2 x_i(t)$$

$$+ \sum_i [G_i(x_i(t+1)) - G_i(x_i(t-1))] \qquad (14)$$

where $\Delta_2 x_i(t) = x_i(t+1) - x_i(t-1)$ is the change in x_i between two time steps. Because the transfer function increases monotonically, the following inequality holds (see Fig. 10.3):

$$G_i(x_i(t+1)) - G_i(x_i(t-1))$$

$$\le G_i'(x_i(t+1)) \Delta_2 x_i(t) \qquad (15)$$

Equality occurs in (15) when $\Delta_2 x_i(t) = 0$ for all i, which implies that the network has reached either a fixed-point or a period-2 attractor. Combining Eqs. (14) and (15) and using the result $G_i'(x) = F_i^{-1}(x)$ leads to

$$\Delta E(t) \le 0 \qquad (16)$$

where again equality holds only when $\Delta_2 x_i(t) = 0$. This result, together with the fact that $E(t)$ (for the same reasons as $L(t)$) is bounded below, implies that $E(t)$ is a Liapunov function for the discrete-time network and that all attractors are either fixed points or period-2 limit cycles.

The period-2 limit cycles, however, do not appear at all values of neuron gain. They are eliminated, leaving only fixed-point attractors, where the gains are reduced sufficiently. To see this, we again consider the function $L(t)$ of Eq. (7), only this time applied to the discrete-time network. The function is a Liapunov function when the neuron gains satisfy the condition

$$\frac{1}{\beta_i} > -\lambda_{min} \qquad (17)$$

where β_i is the gain of the transfer function F_i, and λ_{min} is the smallest eigenvalue of the interconnection matrix T_{ij}. If the interconnection matrix has no negative eigenvalues, then (17) is satisfied for any value of β_i, since $\beta_i > 0$; but if there are one or more negative eigenvalues, then λ_{min} is the most negative of them and (17) places an upper limit on the gain.

To prove that $L(t)$ is a Liapunov function when (17) is satisfied, we consider the change in $L(t)$ between two time steps:

$$\Delta L(t) = L(t+1) - L(t)$$

$$= -\frac{1}{2} \sum_{i,j} T_{ij} x_i(t+1) x_j(t+1)$$

$$+ \frac{1}{2} \sum_j T_{ij} x_i(t) x_j(t)$$

$$- \sum_i I_i [x_i(t+1) - x_i(t)]$$

$$+ \sum_i [G_i(x_i(t+1)) - G_i(x_i(t))] \qquad (18)$$

Using the update equation and the symmetry of

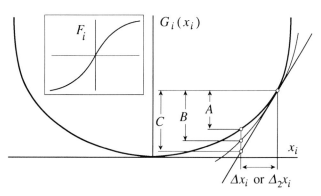

Fig. 10.3. Graphical representation of the inequalities (15) and (20). For the sigmoidal transfer function $F_i(x)$ with gain β_i depicted in the inset, the function $G_i(x_i)$ is concave up with minimum curvature β_i^{-1}. A line and a parabola with curvature β_i^{-1} are tangent to $G_i(x_i)$ at the point $[x_i(t+1), G_i(x_i(t+1))]$. The inequality (15) follows from the fact that $C \leq A$, while (20) follows from the fact that $C \leq B$. (Adapted from Marcus and Westervelt, 1989b, with permission.)

the interconnection matrix, this becomes

$$\Delta L(t) = -\frac{1}{2}\sum_{i,j} T_{ij}\, \Delta x_i(t)\, \Delta x_j(t)$$

$$- \sum_i F_i^{-1}[x_i(t+1)]\, \Delta x_i(t)$$

$$+ \sum_i [G_i(x_i(t+1)) - G_i(x_i(t))] \quad (19)$$

where $\Delta x_i(t) = x_i(t+1) - x_i(t)$ is the change in x_i between consecutive time steps. We now construct an inequality relating the last term in Eq. (19) to $\Delta x_i(t)$. This inequality is similar to Eq. (15) but includes a term quadratic in $\Delta x_i(t)$ (see Fig. 10.3):

$$G_i(x_i(t+1)) - G_i(x_i(t-1))$$

$$\leq G_i'(x_i(t+1))\, \Delta x_i(t) - \frac{1}{2\beta_i}[\Delta x_i(t)]^2 \quad (20)$$

The quantity β_i is the inverse of the minimum curvature of $G_i(x)$, which is just the gain of $F_i(x)$. Combining Eqs. (19) and (20) and using the result $G_i' = F_i^{-1}(x)$ leads to

$$\Delta L(t) \leq -\frac{1}{2}\sum_{i,j}\left(T_{ij} + \delta_{ij}\frac{1}{\beta_i}\right)\Delta x_i(t)\, \Delta x_j(t) \quad (21)$$

where $\delta_{ij} = 1$ for $i = j$ and $\delta_{ij} = 0$ otherwise. If the matrix $(T_{ij} + \delta_{ij}\beta_i^{-1})$ is positive definite, then

$$\Delta L(t) \leq 0 \quad (22)$$

Equality holds in (22) only when $\Delta x_i(t) = 0$, which implies that the network has reached a fixed-point attractor. The requirement that the matrix $(T_{ij} + \delta_{ij}\beta_i^{-1})$ be positive definite is just the inequality (17). Thus Eq. (22) and the boundedness of $L(t)$ together mean that, when the neuron gains all satisfy the inequality (17), $L(t)$ is a Liapunov function for the discrete-time network and the only attractors are fixed points.

We have seen in this section that symmetrically connected analog networks with continuous-time updating have only fixed-point attractors, while those with discrete-time updating have period-2 oscillations in addition to fixed points unless the neuron gains are reduced to satisfy the criterion (17). These results are a necessary first step in designing associative memories in which patterns of neuron activity are stored as attractors of the network dynamics. In the associative memories we will treat in the next section, patterns are stored as fixed points, and the oscillatory modes that appear at high neuron gain in discrete-time networks are an unwanted side-effect. It is possible, however, to put limit cycles to use, for example for storing pattern sequences (Kleinfeld, 1986; Sompolinsky and Kanter, 1986; Dehaene et al., 1987; Guyon et al., 1988; Mori et al., 1989; Herz, 1991) or for signaling that a network does not recognize a particular input (Fontanari and Köberle, 1988b). We will consider pattern sequences briefly in Section 10.6.

10.4. PHASE DIAGRAMS OF ANALOG ASSOCIATIVE MEMORIES

While the stability analysis of the last section provides information about the types of attractors that can exist in analog networks, building an associative memory involves actively programming a set of chosen patterns of neuron activity to be attractors. A whole new set of questions arises regarding the storage and retrieval of these patterns. How many patterns can be stored as fixed points? How does neuron gain affect storage? What other kinds of fixed points are there besides those corresponding to stored patterns? How does the stability criterion (17) affect the ability of the discrete-time network to store and retrieve patterns? We will address these questions below by combining the Liapunov function approach of the previous section with the statistical mechanics of disordered systems.

The analog associative memories we consider in this section are less general than Eqs. (1) and (2). They consist of N neurons with identical sigmoidal transfer functions $F_i(x) = F(x)$ that achieve their maximum slope β at the origin. The external bias terms I_i are zero for all i. The interconnection matrices are constructed using the Hebb rule (Hebb, 1949) for storing p patterns of neuron activity. The patterns, denoted ξ_i^μ for $\mu = 1, \ldots, p$, are assumed to take on the values ± 1 randomly and without bias. The storage rule is

$$T_{ij} = \frac{1}{N} \sum_{\mu=1}^{p} \xi_i^\mu \xi_j^\mu \quad T_{ii} = 0 \tag{23}$$

The Hebb rule is perhaps the simplest of many storage methods; others have been proposed for their increased storage capacity, ease of implementation, or biological plausibility. The pseudo-inverse rule allows the storage of more patterns per neuron than the Hebb rule and is especially useful for correlated patterns (Personnaz et al., 1985, 1986a, b; Diederich and Opper, 1987; Kanter and Sompolinsky, 1987). Other storage rules attempt to capture observed behavior in the cortex by separating the functions of excitatory and inhibitory interconnections: excitatory interconnections implement associative memory while inhibitory ones keep the level of neuron activity low (Amit and Treves, 1989; Buhmann, 1989; Treves and Amit, 1989; Kohring, 1990; Golomb et al., 1990). In addition to these explicit storage rules, iterative procedures have been developed for pattern storage that generate binary ($T_{ij} = \pm 1/N$) interconnections (Köhler et al., 1990) or interconnections that are optimal in the sense of making attractor basins as large and deep as possible (Gardner, 1988). Because this chapter emphasizes analog network dynamics after learning has taken place rather than the learning process itself, we will focus on the rule (23).

Figure 10.4 shows two analytical phase diagrams for analog associative memories with Hebb-rule interactions, one for continuous-time and one for discrete-time updating. These diagrams, valid in the limit of large N, indicate the types of attractors as a function of the neuron gain β and the ratio $\alpha = p/N$ of patterns to neurons. The diagram for continuous-time networks (Fig. 10.4a) contains three regions labeled recall, spin glass, and origin. In the recall region, the networks function reliably as associative memories, with a set of $2p$ fixed points each

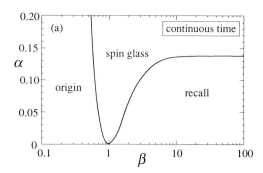

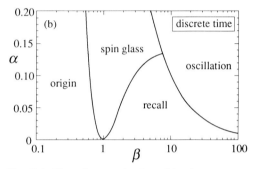

Fig. 10.4. Phase diagrams for the Hebb-rule associative memory with neuron transfer function $F(x) = \tanh(\beta x)$ for (a) continuous-time and (b) discrete-time updating. (Adapted from Marcus et al., 1990, with permission.)

having a large overlap $m_\mu = N^{-1} \sum_i \xi_i^\mu x_i$ with a stored pattern or its inverse. In addition, the recall region contains other fixed points, known as spurious states, that have negligible overlaps with the patterns; their existence degrades associative memory performance. Crossing from the recall region into the spin glass region destroys the pattern recall capability. In the spin glass region, the recall states are no longer fixed points; the only attractors are spurious states. Finally, in the origin region, the network has a single attractor with $x_i = 0$ for all i.

The phase diagram for discrete-time associative memories (Fig. 10.4b) also contains recall, spin glass, and origin regions, with the same definitions as for continuous-time networks. In addition, Fig. 10.4b has a fourth region marked oscillation, in which the stability criterion (17) is violated. Recall and spurious states may still exist in this region, but the network can also become trapped in period-2 limit cycles. Except for the oscillation region, the two phase diagrams of Fig. 10.4 are identical, since continuous-time and discrete-time networks have the same Liapunov function as long as the stability criterion (17) is satisfied.

The boundary separating the recall and spin glass regions, also known as the storage capacity $\alpha_c(\beta)$, was computed by Kühn et al. (1991; see also Kühn, 1990; Marcus et al., 1990; Shiino and Fukai, 1990). These authors treat the Liapunov function $L(t)$ as an energy and, by introducing an auxiliary temperature, construct a partition function and derive a free energy from it using standard techniques of statistical mechanics. The free energy provides information about metastable states separated by energy barriers scaling with system size N. When the auxiliary temperature is set equal to zero, these metastable states are the fixed points of the dynamical system for which $L(t)$ is a Liapunov function. We stress that the networks are completely deterministic; the auxiliary temperature is simply a mathematical device that allows the use of statistical mechanics to learn about the fixed points. In fact, the resulting storage capacity for analog networks differs slightly from the corresponding curve for stochastic networks of two-state neurons (Amit et al., 1985b, 1987).

The averaged free energy per neuron is

$$f = -\frac{1}{\tilde{\beta}N} \langle \ln Z \rangle_\xi$$

$$= -\frac{1}{\tilde{\beta}N} \left\langle \ln \int \prod_i [d\rho(x_i)] \exp(-\tilde{\beta}L) \right\rangle_\xi \quad (24)$$

In Eq. (24), $\tilde{\beta}$ is the inverse of the auxiliary temperature, Z is the partition function, the angled brackets $\langle \cdots \rangle_\xi$ denote an average over all possible realizations of the stored patterns, and L is the Liapunov function appearing as Eq. (7). The measure $d\rho(x_i)$ is chosen to be uniform on the range of the transfer function; in the limit $\tilde{\beta} \to 0$ the interval over which $d\rho(x_i)$ is nonzero, and not its functional form, is all that matters (Kühn, 1990; Kühn et al., 1991).

The free energy is calculated using the replica method (Sherrington and Kirkpatrick, 1975). When the network has a nonzero overlap m_v with a single pattern v, f is determined in the limit of large N by the following saddle-point equations:

$$m_v = \langle\langle \xi^v \hat{x} \rangle\rangle \quad (25)$$

$$q = \langle\langle \hat{x}^2 \rangle\rangle \quad (26)$$

$$C = \sqrt{\frac{1}{\alpha r}} \langle\langle z\hat{x} \rangle\rangle \quad (27)$$

In Eqs. (25) to (27), $\hat{x}$ is determined implicitly by

$$\hat{x} = F[m_v \xi^v + \alpha(\tilde{r} - 1)\hat{x} + \sqrt{\alpha r}\, z] \quad (28)$$

The double angle brackets $\langle\langle \cdots \rangle\rangle$ indicate an average over both the pattern ξ^v, which takes on the values ± 1 with equal probability, and the continuous variable z using a Gaussian distribution:

$$\langle\langle \cdots \rangle\rangle \to \left\langle \int \frac{dz}{\sqrt{2\pi}} \exp(-\tfrac{1}{2}z^2)(\cdots) \right\rangle_{\xi^v} \quad (29)$$

The quantities r and $\tilde{r}$ are

$$r = \frac{q}{(1-C)^2} \qquad \tilde{r} = \frac{1}{1-C} \quad (30)$$

For a given value of β, the storage capacity $\alpha_c(\beta)$ is given by that value of α for which the solution of Eqs. (25) to (27) with $m_v \sim 1$ exists. This solution is stable for smaller values of α and disappears for larger ones. In Fig. 10.4, the storage capacity curves are for the transfer function $F(x) = \tanh(\beta x)$.

The boundary between the spin glass and origin regions may be calculated by looking for solutions of the saddle-point equations (25) to (27) with m_v equal to zero and with small but nonzero q. An equivalent but easier way is to perform local stability analysis about the fixed point $x_i = 0$ that defines the origin region. The linearized equations are

$$\frac{dx_i(t)}{dt} = -x_i(t) + \beta \sum_j T_{ij} x_j(t) \quad (31)$$

for the continuous-time network and

$$x_i(t + 1) = \beta \sum_j T_{ij} x_j(t) \quad (32)$$

for the discrete-time network. For the continuous-time case, the fixed point at the origin disappears when at least one eigenvalue of the matrix $(-\delta_{ij} + \beta T_{ij})$ has a positive real part; for the discrete-time case, it appears when at least one eigenvalue of βT_{ij} leaves the unit circle. The extrema eigenvalues of the interconnection matrix for large N and $\alpha < 1$ are $\lambda_{min} = -\alpha$, $\lambda_{max} = 1 + 2\sqrt{\alpha}$ (Geman, 1980; Crisanti and Sompolinsky, 1987; Le Cun et al., 1991) leading to the same boundary curve for both updating schemes:

$$\beta = \frac{1}{1 + 2\sqrt{\alpha}} \quad (33)$$

Again, this curve differs slightly from the corresponding one for networks of two-state neurons at finite temperature (Amit et al., 1985b, 1987).

The boundary of the oscillation region for discrete-time networks is determined by the stability condition (17). With the result $\lambda_{min} = -\alpha$, the boundary is

$$\beta = \frac{1}{\alpha} \tag{34}$$

The recall–spin glass boundary terminates at the oscillation boundary in Fig. 10.4b because $L(t)$ is not a Liapunov function in the oscillation region. For continuous-time updating, on the other hand, $L(t)$ is a Liapunov function for all values of neuron gain.

Numerical tests of the phase diagram for discrete-time associative memories have been performed; typical results are shown in Fig. 10.5 (Marcus et al., 1990). A set of 20 interconnection matrices with $N = 100$ was generated on a computer using random, unbiased patterns $\xi_i^\mu = \pm 1$, and 50 initial conditions were tested for each matrix. The data, which represent a horizontal

slice through the phase diagram for $\alpha = 0.05, 0.1$, and 0.2, show the fraction of random initial conditions $x_i(0) = \pm 1$ that flow to each of the four possible types of attractors—the origin, a recall state, a spurious state, or an oscillatory state. The results agree well with the theoretical phase diagrams; the lack of sharpness in the appearance of recall, spin glass, and oscillatory states as β increases is the result of finite-size effects.

10.5. SPURIOUS STATES AND DETERMINISTIC ANNEALING

In the previous section, we saw how the critical storage capacity and, for discrete-time updating, the existence of oscillatory attractors can limit

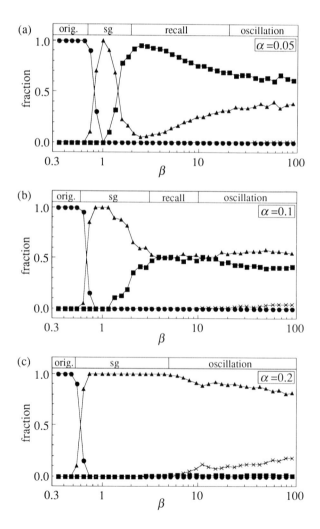

Fig. 10.5. Numerical data for discrete-time Hebb-rule associative memories showing, as a function of neuron gain β, the fraction of randomly chosen initial states that lead to the four types of attractors: the origin (circle), a recall state (square), a spurious state (triangle), or an oscillatory state (cross). For a given value of β, the four data points represent a total of 1000 initial states from 20 matrices constructed from random, unbiased memory patterns with $N = 100$. The three panels are for $\alpha N = 5$, 10, and 20 patterns, and the strip along the top indicates the regions of the phase diagram, Fig. 10.4b, for that value of α. (Adapted from Marcus et al., 1990, with permission.)

the useful operating region of analog associative memories. However, the recall capability can vary considerably even within the recall region. This effect is especially visible in Fig. 10.5a: the fraction of initial states flowing to a recall state increases substantially as neuron gain decreases. The trend is observed for other values of α and for other associative memory storage rules (Marcus et al., 1990). In addition, a number of authors have reported improved performance as gain is reduced in analog computers that solve complex optimization problems such as the traveling salesman problem (Hopfield and Tank, 1985, 1986) and the detection of edges in visual images (Koch et al., 1986; Blake and Zisserman, 1987).

The idea behind the improved performance as gain decreases is that the Liapunov function $L(t)$ becomes smoother, so that shallow local minima are eliminated. This effect is shown schematically in Fig. 10.6. Since the stored patterns tend to lie in wide, deep basins, essentially all of the local minima eliminated are spurious states. The phenomenon is reminiscent of what happens as temperature increases in simulated annealing (Kirkpatrick et al., 1983); in fact, the loose analogy between analog gain and inverse temperature has led to the phrase *deterministic annealing* to describe the analog case. Simulated annealing and determinstic annealing, however, are quite different, both conceptually and in practice. Simulated annealing employs a stochastic mechanism that allows a system to take occasional uphill steps on a rough landscape, while deterministic annealing smoothes the landscape itself. As mentioned above, stochasticity is also considerably more cumbersome to implement than

adjustable analog gain in an electronic circuit (Alspector et al., 1989, 1991).

In this section we investigate the relationship between neuron gain and recall capability both analytically and numerically. We will see that the number of local minima in the Liapunov function of a typical analog associative memory increases exponentially with the number of neurons N as $\exp(Ng)$. The main point of this section is to show how the scaling exponent $g(\alpha, \beta)$ can be calculated as a function of the storage ratio α and the neuron gain β (Waugh et al., 1990, 1991), and how the resulting form for g implies that the number of local minima decreases dramatically as the neuron gain is reduced. For analog associative memories, the elimination of fixed points as gain decreases translates into substantially improved recall performance, since most of the eliminated fixed points are spurious states.

The calculation uses techniques developed to count the number of metastable states in a spin glass at finite temperature (Bray and Moore, 1980) and to study the distribution of spurious states in the state space of a neural network of binary neurons (Gardner, 1986). More recently, these techniques have been applied to a variety of neural network models (Treves and Amit, 1988; Fukai, 1990; Fukai and Shiino, 1990; Kepler, 1991). To conform with these references, we will adopt a slightly different notation in this section, making the transformations $F(x) \rightarrow F(\beta x)$ for the transfer function and $T_{ij} \rightarrow \alpha^{-1/2} T_{ij}$ for the Hebb-rule interconnection matrix of Eq. (23).

We have seen already that the dynamical behavior of an analog network can depend on the updating rule, and we might therefore expect that the number of fixed points will also. However, continuous-time and discrete-time networks with the same interconnection matrices have the same fixed points. This can be seen by setting $dx_i(t)/dt = 0$ in Eq. (1) and $x_i(t + 1) = x_i(t)$ in Eq. (2) to find the fixed-point condition for both updating rules:

$$H_i \equiv F_i^{-1}(x_i) - \beta \sum_j T_{ij} x_j - \beta I_i = 0 \quad (35)$$

Since we are interested only in the number of these fixed points, differences in the dynamics away from fixed points resulting from different updating rules are irrelevant.

The calculation, which involves using the replica method to compute the determinant in Eq. (36), is presented in detail in Waugh et al. (1991). The number N_{fp} of fixed points for a single realization of the interconnection matrix is found by integrating a product of delta functions of the

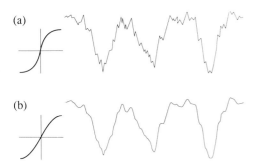

Fig. 10.6. Schematic Liapunov landscape for (a) high-gain and (b) low-gain neuron transfer functions, showing how the Liapunov function (ordinate) varies as a function of an abstract state space coordinate (abscissa). Decreasing neuron gain smoothes the Liapunov landscape, greatly reducing the number of spurious states. (From Waugh et al., 1990, with permission.).

quantities H_i over state space:

$$N_{fp} = \int \prod_i [d\rho(x_i)] \prod_i [\delta(H_i)] |\det \mathbf{A}| \quad (36)$$

The measure $d\rho(x_i)$ equals 1 over the range of the neuron transfer function and equals 0 otherwise. The quantity $|\det \mathbf{A}|$ is the Jacobian normalizing the delta functions. The elements of $\mathbf{A}$ are given by

$$A_{ij} = \frac{\partial H_i}{\partial x_j} = a(x_i)\delta_{ij} - \beta T_{ij} \quad (37)$$

where δ_{ij} equals 1 for $i = j$ and equals 0 otherwise and where

$$a(x_i) \equiv \frac{d}{dx}[F^{-1}(x)]_{x=x_i} \quad (38)$$

For the transfer function $F(x) = \tanh(x)$, Eq. (38) gives $a(x) = (1 - x^2)^{-1}$.

In the limit of large N, the average number of fixed points varies exponentially with N, with an exponent determined by saddle point methods as an extremum over four order parameters q, λ, Δ, and B:

$$\langle N_{fp} \rangle = \max_{q,\lambda,\Delta} \min_{B} \exp[N\tilde{g}(\alpha,\beta, q, \lambda, \Delta, B) \quad (39)$$

The function $\tilde{g}$ depends on the storage capacity α, the neuron gain β, and the four order parameters:

$$\tilde{g} = -\frac{\sqrt{\alpha}}{\beta}(B + \Delta) + \frac{\alpha}{2}\ln\left[\frac{\alpha C}{(B/\beta - \sqrt{\alpha})^2}\right]$$

$$+ \ln\left(\frac{I}{\beta\sqrt{2\pi q}}\right) \quad (40)$$

where

$$C = \frac{1}{\alpha}[(\Delta/\beta + \sqrt{\alpha})^2 - 2\lambda q] \quad (41)$$

$$I = \int d\tilde{\rho}(x)[a(x) + B]$$

$$\times \exp\left[-\frac{(F^{-1}(x) - \Delta x)^2}{2\beta^2 q} + \lambda x^2\right] \quad (42)$$

and the measure $d\tilde{\rho}(x)$ equals 1 on that part of the range of F where $(a(x) + B) > 0$ and equals 0 otherwise.

Extrema of Eq. (39) with respect to q, λ, B, and Δ are found by setting partial derivatives of $\tilde{g}$ equal to zero, which gives a set of four saddle-

point equations:

$$q = C\langle\langle x^2 \rangle\rangle \quad (43)$$

$$\lambda = -\frac{C}{2q}\left(1 - \frac{1}{\beta^2 q}\langle\langle(F^{-1}(x) - \Delta x)^2\rangle\rangle\right) \quad (44)$$

$$C = 1 + \frac{C}{\beta q\sqrt{\alpha}}\langle\langle xF^{-1}(x)\rangle\rangle \quad (45)$$

$$B = \left(\frac{\beta B}{\sqrt{\alpha}} - \beta^2\right)\langle\langle(a(x) + B)^{-1}\rangle\rangle \quad (46)$$

In Eqs. (43) to (46), the double angled brackets $\langle\langle\cdots\rangle\rangle$ indicate a weighted average with the weight function $W(x)$ given by the integrand of I:

$$\langle\langle\cdots\rangle\rangle \rightarrow \frac{\int d\tilde{\rho}(x)\, W(x)(\cdots)}{\int d\tilde{\rho}(x)\, W(x)} \quad (47)$$

$$W(x) = [a(x) + B]\exp\left[-\frac{(F^{-1}(x) - \Delta x)^2}{2\beta^2 q} + \lambda x^2\right] \quad (48)$$

For given values of α and β, self-consistent solutions for q, λ, Δ, and B are found by solving Eqs. (43) to (46) numerically. The solutions are then inserted into Eqs. (40) to (42) to yield a value for the quantity of interest, the scaling exponent $g(\alpha, \beta)$. Values of $g(\alpha, \beta)$ are plotted in Fig. 10.7 for the transfer function $F(x) = \tanh(x)$. The result shows that for any value of the storage fraction α, the function $g(\alpha, \beta)$ decreases as the neuron gain is lowered. Since the number of fixed points depends exponentially on the product $Ng(\alpha, \beta)$, even small changes in $g(\alpha, \beta)$

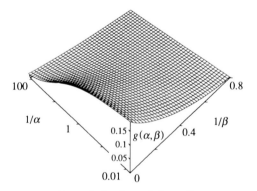

Fig. 10.7. Theoretical values of the scaling exponent $g(\alpha, \beta)$ as a function of inverse storage ratio $1/\alpha$ and inverse neuron gain $1/\beta$, for the neuron transfer function $F(x) = \tanh(\beta x)$. The expected number of fixed points in the network is $e^{Ng(\alpha,\beta)}$. (Adapted from Waugh et al., 1990, with permission.)

dramatically affect the number of fixed points, especially for large N.

As an example, consider the effect of lowering the neuron gain from $\beta = 100$ to $\beta = 10$ in an analog associative memory network with storage fraction $\alpha = 0.1$. The computed values $g(0.1, 100) = 0.059$ and $g(0.1, 10) = 0.040$ imply that the average number of fixed points will be reduced by approximately 97% for $N = 200$ and by eight orders of magnitude for $N = 1000$. As the fixed points eliminated are all spurious states, these results imply a dramatic increase in recall performance.

The results depicted in Fig. 10.7 have been tested numerically for the transfer function $F(x) = \tanh(x)$ by counting the stable fixed points in small computer-generated analog neural networks (Waugh et al., 1991). The number of fixed points in each network was counted by choosing random initial conditions $x_i(0) = \pm 1$ and iterating the discrete-time network of Eq. (2) until a fixed point was reached. The iteration was carried out sequentially rather than in parallel in order to eliminate oscillatory states. (Sequentially updated networks with symmetric interconnections and no self-connections also obey the fixed-point condition (35) but converge to a fixed point for all values of gain.) Weighted least-squares fits to the data were made to find numerical results for the scaling exponent g. As seen in Fig. 10.8, the resulting values show good agreement with the theoretical curves for $\alpha = 10$, 1, and 0.1, especially for large α and β.

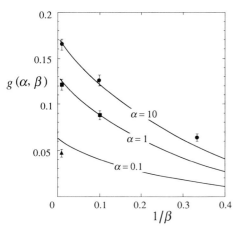

Fig. 10.8. Comparison of theoretical and numerical results for the scaling exponent $g(\alpha, \beta)$. Numerical results for $\alpha = 10$ (circles), $\alpha = 1$ (squares), and $\alpha = 0.1$ (triangles) agree well with theoretical results (curves) for these values of α as calculated from Eqs. (40 to 42). (From Waugh et al., 1991, with permission.)

Reducing neuron gain is not the only way to eliminate spurious fixed-point attractors. Adding asymmetry to the interconnection matrix has been shown to reduce the number of fixed points and the total number of attractors in zero-temperature spin glasses with parallel updating (Gutfreund et al., 1988), and similar effects have been discussed for neural networks (Parisi, 1986; Crisanti and Sompolinski, 1987). For symmetric interconnections, a time-dependent self-connection has been used that sends parallel-update associative memories to an oscillatory state, rather than to a spurious fixed point, whever the network has sufficiently low overlap with a stored pattern (Fontanari and Köberle, 1988b). Unlike these methods, deterministic annealing eliminates spurious states without introducing other non-fixed-point attractors.

10.6. MULTIPLE TIME-STEP UPDATING

In this section, we consider another method for improving the recall capability of discrete-time analog associative memories. The idea is to modify the updating rule of Eq. (2) so that the input of each neuron is averaged over several time steps (Marcus and Westervelt, 1990). The payoff for using multiple time-step updating is that the criterion for guaranteeing convergence to a fixed point allows a greater range of neuron gain in direct proportion to the number of time steps used in the averaging. At the same time, other properties such as the storage capacity are unchanged.

With multiple-time-step updating, the updating rule becomes

$$x_i(t + 1) = F_i\left[\sum_j T_{ij}z_j(t) + I_i\right] \quad i = 1, \ldots, N$$
(49)

where $z_j(t)$ is the output of neuron j averaged over M previous time steps:

$$z_j(t) = \frac{1}{M}\sum_{\tau=0}^{M-1} x_j(t - \tau)$$

$$j = 1, \ldots, N \quad M \in \{1, 2, 3, \ldots\} \quad (50)$$

As before, the updating of the neuron states $x_i(t)$ as well as the quantities $z_i(t)$ is done in parallel and is fully deterministic. The interconnection matrix T_{ij} is real-valued and symmetric, and we again require that the neuron transfer functions F_i be monotonically increasing and that they rise slower than linearly when their arguments are large in magnitude. We will refer to this system as a multistep neural network. Note that, when $M = 1$, the multistep network reduces to the discrete-time network of Eq. (2).

Multistep networks have only fixed-point attractors when the neuron gains all satisfy the criterion

$$\frac{1}{\beta_i} > -\frac{\lambda_{\min}}{M} \qquad (51)$$

where, as before, $\lambda_{\min}$ is the minimum eigenvalue of the interconnection matrix. This result is obtained by showing that the function

$$L_m(t) = -\frac{1}{2}\sum_{i,j} T_{ij}z_i(t)z_j(t)$$

$$+ \sum_i \frac{1}{M} \sum_{\tau=0}^{M-1} [G_i(x_i(t-\tau)) - I_i x_i(t-\tau)] \qquad (52)$$

is a Liapunov function for the M-step multistep network when (51) is satisfied (Marcus and Westervelt, 1990). The proof closely follows that appearing in Section 10.3 for the case $M = 1$. Comparing (51) to the result (17) for $M = 1$, we see that, for a given interconnection matrix, multistep updating increases the range of neuron gains over the network and is guaranteed to converge to a fixed point. It has been proven (Herz, 1991) that, when (51) is violated, the only other attractors besides fixed points are limit cycles with period $M + 1$ and its integer divisors.

For analog dissociative memories that store fixed-point patterns, (51) implies that the recall region for multistep updating is larger than for standard $M = 1$ updating. The effect of multistep updating on the size of the recall region for Hebb-rule interconnections is shown in Fig. 10.9.

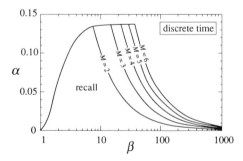

Fig. 10.9. The recall region for the discrete-time Hebb-rule associative memory for various values of M in the multistep updating rule of Eqs. (49) and (50). Increasing the number of time steps over which the neuron input is averaged extends the recall region to higher gain.

More complicated multistep networks can store limit cycles as well as fixed points (Herz, 1991). In these networks, the interconnection matrix T_{ij} and external inputs I_i depend on the time delay τ. The recall ability of associative memories that store sequences with a Hebb-like rule has been investigated in the high-gain limit and found to be comparable to that for fixed point storage (Herz et al., 1991). The recall of pattern sequences has also been addressed by a number of other authors (Kleinfeld, 1986; Sompolinsky and Kanter, 1986; Dehaene et al., 1987; Guyon et al., 1988; Mori et al., 1989).

10.7. SUMMARY

We have seen in this chapter now networks of analog neurons can operate reliably as associative memories, with high operating speed due to parallel updating and improved recall capability due to the reduction of spurious states. Perhaps just as importantly, we have seen how to analyze the behavior of these networks by combining techniques of nonlinear dynamics and the statistical mechanics of disordered systems, an approach that is still very much in its infancy.

The topics we have addressed only begin to probe the capabilities of analog neural networks. Other topics of recent research include stochastic networks of threshold-linear neurons (Amit and Treves, 1989; Treves and Amit, 1989; Treves, 1990), networks with diluted, asymmetric interconnections (Mertens, 1991), and networks in which analog values are coded in the frequency rather than the magnitude of neuron outputs (Murray, 1991). A number of outstanding problems, such as using other pattern storage rules or other, nonsigmoidal, transfer functions in analog networks can be approached using the methods discussed in this chapter. Other outstanding problems will require new techniques. Among these, one issue with great importance for implementations is how to reduce the excessive wiring requirements of pattern storage rules with all-to-all interconnections. When interconnections have geometrical structure, new dynamical phenomena can occur, such as the formation of domains of neuron activity the sizes of which are determined by neuron gain (Noest, 1989, 1990; Coolen, 1990; Marcus et al., 1991). These phenomena may be useful in building associative memories that detect patterns or textures in visual scenes (Noest, 1989).

REFERENCES

Alspector, J., Gupta, B., and Allen, R. B. (1989). "Performance of a Stochastic Learning Microchip," in *Proc. Conf. Neural Information Processing Systems*, Denver, CO, 1987, D. S. Touretsky, ed. San Mateo, Morgan Kaufmann, pp. 748–760.

Alspector, J., Gannett, J. W., Haber, S., Parker, M. B., and Chu, R. (1991). "A VLSI-efficient Technique for Generating Multiple Uncorrelated Noise Sources and its Application to Stochastic Neural Networks," *IEEE Trans. Circuits and Systems*, **38**, 109–123.

Amit, D. J. (1989). *Modeling Brain Function: The World of Attractor Neural Networks*. Cambridge, Cambridge University Press.

Amit, D. J. and Treves, A. (1989). "Associative Memory Neural Network with Low Temporal Spiking Rates" *Proc. Nat. Acad. Sci. U.S.A.*, **86**, 7871–7875.

Amit, D. J., Gutfreund, H., and Sompolinsky, H. (1985a). "Spin-glass Models of Neural Networks," *Phys. Rev. A*, **32**, 1007–1018.

Amit, D. J., Gutfreund, H., and Sompolinsky, H. (1985b). "Storing Infinite Numbers of Patterns in a Spin-Glass Model of Neural Networks," *Phys. Rev. Lett.*, **55**, 1530–1533.

Amit, D. J., Gutfreund, H., and Sompolinsky, H. (1987). "Statistical Mechanics of Neural Networks Near Saturation," *Ann. Phys. N.Y.*, **173**, 30–67.

Andreou, A. G., Boahen, K. A., Pouliquen, P. O., Pavasovic, A., Jenkins, R. E., and Strohbehn, K. (1991). "Current-mode Subthreshold MOS Circuits for Analog VLSI Neural Systems," *IEEE Trans. Neural Networks*, **2**, 205–213.

Binder, K. and Young, A. P. (1986). "Spin Glasses: Experimental Facts, Theoretical Concepts, and Open Questions," *Rev. Mod. Phys.*, **58**, 801–976.

Blake, A. and Zisserman, A. (1987). *Visual Reconstruction*. MIT Press, Cambridge, MA.

Bray, A. J. and Moore, M. A. (1980). "Metastable States in Spin Glasses," *J. Phys. C*, **13**, L469–L476.

Buhmann, J. (1989). "Oscillations and Low Firing Rates in Associative Memory Neural Networks," *Phys. Rev. A*, **40**, 4145–4148.

Caianiello, E. R. (1961). "Outline of a Theory of Thought and Thinking Machines," *J. Theor. Biol.*, **1**, 204–235.

Cohen, M. A. and Grossberg, S. (1983). "Absolute Stability of Global Pattern Formation and Parallel Memory Storage by Competitive Neural Networks," *IEEE Trans.*, **SMC-13**, 815–826.

Coolen, A. C. C. (1990). "Ising-spin Neural Networks with Spatial Structure," in *Statistical Mechanics of Neural Networks: Proc. XIth Sitges Conference*, Sitges, Barcelona, Spain, June 1990, L. Garrido, ed. Springer-Verlag, New York, pp. 381–396.

Cragg, B. G. and Temperley, H. N. V. (1954). "The Organization of Neurones: a Cooperative Analogy," *Electroenceph. Clin. Neuro.*, **6**, 85–92.

Crisanti, A. and Sompolinsky, H. (1987). "Dynamics of Spin Systems with Randomly Asymmetric Bonds: Langevin Dynamics and a Spherical Model," *Phys. Rev. A*, **36**, 4922.

Dehaene, S., Changeux, J. P., and Nadal, J. P. (1987).

"Neural Networks That Learn Temporal Sequences by Selection," *Proc. Nat. Acad. Sci. U.S.A.*, **84**, 2727–2731.

Diederich, S. and Opper, M. (1987). "Learning of Correlated Patterns in Spin-glass Networks by Local Learning Rules," *Phys. Rev. Lett.*, **58**, 949–952.

Domany, E., van Hemmen, J. L., and Schulten, K. (Eds.) (1991). *Models of Neural Networks*. Springer-Verlag, New York.

Durbin, R. and Willshaw, D. (1987). "An Analogue Approach to the Travelling Salesman Problem Using an Elastic Net Method," *Nature*, **326**, 689–691.

Fischer, K. H. and Hertz, J. A. (1991). *Spin Glasses*. Cambridge University Press, Cambridge.

Fisher, W. A., Fujimoto, R. J., and Smithson, R. C. (1991). "A Programmable Neural Network Processor," *IEEE Trans. Neural Networks*, **2**, 222–228.

Fogelman-Soulie, F., Mejia, C., Goles, E., and Martinez, S. (1989). "Energy Functions in Neural Networks with Continuous Local Functions," *Complex Systems*, **3**, 269–293.

Fontanari, J. F. and Köberle, R. (1987). "Information Storage and Retrieval in Synchronous Neural Networks," *Phys. Rev. A*, **36**, 2475–2477.

Fontanari, J. F. and Köberle, R. (1988a). "Information Processing in Synchronous Neural Networks," *J. Phys. (France)*, **49**, 13–23.

Fontanari, J. F. and Köberle, R. (1988b). "Neural Networks with Transparent Memory," *J. Phys. A*, **21**, L259–L262.

Fukai, T. (1990). "Metastable States of Neural Networks Incorporating the Dale Hypothesis," *J. Phys. A*, **23**, 249–258.

Fukai, T. and Shiino, M. (1990). "Large Suppression of Spurious States in Neural Networks of Nonlinear Analog Neurons," *Phys. Rev. A*, **42**, 7459–7466.

Gardner, E. J. (1986). "Structure of Metastable States in the Hopfield Model," *J. Phys. A*, **19**, L1047–L1052.

Gardner, E. J. (1988). "The Space of Interactions in Neural Network Models," *J. Phys. A*, **21**, 257–270.

Garrido, L. (Ed.) (1990). *Statistical Mechanics of Neural Networks: Proc. XIth Sitges Conference*, Sitges, Barcelona, Spain, June 1990. Springer-Verlag, New York.

Geman, S. (1980). "A Limit Theorem for the Norm of Random Matrices," *Ann. Prob.*, **8**, 252–261.

Golden, R. M. (1986). "The 'Brain-State-in-a-Box' Neural Model is a Gradient Descent Algorithm," *J. Math. Psychol.*, **30**, 73–80.

Goles, E. and Vichniac, G. Y. (1986). "Lyapunov Functions for Parallel Neural Networks," in *Neural Networks for Computing*, J. S. Denker, ed., AIP Conf. Proc. 151. American Institute of Physics, New York, pp. 165–181.

Goles-Chacc, E., Fogelman-Soulie, F., and Pellegrin, D. (1985). "Decreasing Energy Functions as a Tool for Studying Threshold Networks," *Disc. Appl. Math.*, **12**, 261–277.

Golomb, D., Rubin, N., and Sompolinsky, H. (1990). "Willshaw Model: Associative Memory with Sparse Coding and Low Firing Rates," *Phys. Rev. A*, **41**, 1843–1854.

Graf, H. P., Jackel, L. D., Howard, R. E., Straughn, B.,

Denker, J. S., Hubbard, W., Tennant, D. M., and Schwartz, D. (1986). "VLSI Implementation of a Neural Network Memory with Several Hundreds of Neurons," in *Neural Networks for Computing*, J. S. Denker, ed., AIP Conf. Proc. 151, American Institute of Physics, New York, pp. 182–187.

Grossberg, S. (1967). "Nonlinear Difference-Differential Equations in Prediction and Learning Theory," *Proc. Nat. Acad. Sci. U.S.A.*, **58**, 1329–1334.

Grossberg, S. (1968). "Some Nonlinear Networks Capable of Learning a Spatial Pattern of Arbitrary Complexity," *Proc. Nat. Acad. Sci. U.S.A.*, **59**, 368.

Gutfreund, H., Reger, J. D., and Young, A. P. (1988). "The Nature of Attractors in an Asymmetric Spin Glass with Deterministic Dynamics," *J. Phys. A*, **21**, 2775–2797.

Guyon, I., Personnaz, L., Nadal, J. P., and Dreyfus, G. (1988). "Storage and Retrieval of Complex Sequences in Neural Networks," *Phys. Rev. A*, **38**, 6365–6372.

Hebb, D. O. (1949). *The Organization of Behavior*. Wiley, New York.

Hertz, J. A., Krogh, A. S., and Palmer, R. G. (1991). *Introduction to the Theory of Neural Computation*. Addison-Wesley, Reading, MA.

Herz, A. V. M. (1991). "Global Analysis of Parallel Analog Networks with Retarded Feedback," *Phys. Rev. A*, **44**, 1415–1418.

Herz, A. V. M., Li, Z., and van Hemmen, J. L. (1991). "Statistical Mechanics of Temporal Association in Neural Networks with Transmission Delays," *Phys. Rev. Lett.*, **66**, 1370–1373.

Holler, M., Tam, S., Castro, H., and Benson, R. (1989). "An Electrically Trainable Artificial Neural Network (ETANN) with 10240 'Floating Gate' Synapses," in *Proc. Int. Joint Conference Neural Networks*, Washington DC, pp. 191–196.

Hopfield, J. J. (1982). "Neural Networks and Physical Systems with Emergent Collective Computational Abilities," *Proc. Nat. Acad. Sci. U.S.A.*, **79**, 2554–2558.

Hopfield, J. J. (1984). "Neurons with Graded Response Have Collective Computational Properties Like Those of Two-State Neurons," *Proc. Nat. Acad. Sci. U.S.A.*, **81**, 3088–3092.

Hopfield, J. J. and Tank, D. W. (1985). "'Neural' Computation of Decisions in Optimization Problems," *Biological Cybernetics*, **52**, 141–152.

Hopfield, J. J. and Tank, D. W. (1986). "Computing with Neural Circuits: A Model," *Science*, **233**, 625.

Kanter, I. and Sompolinsky, H. (1987). "Associative Recall of Memory Without Errors," *Phys. Rev. A*, **35**, 380–392.

Kepler, T. B. (1991). "Domains of Attraction and the Density of Static Metastable States in Single-pattern Iterated Neural Networks," *J. Phys. A*, **24**, 1089–1092.

Kirkpatrick, S., Gelatt, C. D. Jr., and Vecchi, M. P. (1983). "Optimization by Simulated Annealing," *Science*, **220**, 671–680.

Kleinfeld, D. (1986). "Sequential State Generation by Model Neural Networks," *Proc. Nat. Acad. Sci. U.S.A.*, **83**, 9469–9473.

Koch, C., Marroquin, J., and Yuille, A. (1986). "Analog 'Neuronal' Networks in Early Vision," *Proc. Nat. Acad. Sci. U.S.A.*, **83**, 4263–4267.

Kohring, G. A. (1990). "Performance Enhancement of Willshaw Type Networks Through the Use of Limit Cycles," *J. Phys. (France)*, **51**, 2387–2393.

Köhler, H., Diederich, S., Kinzel, W., and Opper, M. (1990). "Learning Algorithm for a Neural Network with Binary Synapses," *Z. Phys. B—Condensed Matter*, **78**, 333–342.

Kohonen, T. (1974). "An Adaptive Associative Memory Principle" *IEEE Trans. Computers*, **C-23**, 444–445.

Kühn, R. (1990). "Statistical Mechanics for Networks of Analog Neurons," in *Statistical Mechanics of Neural Networks: Proc. XIth Sitges Conference*, Sitges, Barcelona, Spain, June 1990, L. Garrido, ed. Springer-Verlag, New York, pp. 19–32.

Kühn, R., Bös, S., and van Hemmen, J. L. (1991). "Statistical Mechanics for Networks of Graded-response Neurons," *Phys. Rev. A*, **43**, 2084–2087.

Le Cun, Y., Kanter, I., and Solla, S. A. (1991). "Eigenvalues of Covariance Matrices: Application to Neural-network Learning," *Phys. Rev. Lett.*, **66**, 2396–2399.

Little, W. A. (1974). "The Existence of Persistent States in the Brain," *Math. Biosci.*, **19**, 101–120.

Maher, M. A. C., Deweerth, S. P., Mahowald, M. A., and Mead, C. A. (1989). "Implementing Neural Architectures Using Analog VLSI Circuits," *IEEE Trans. Circuits and Systems*, **36**, 643–652.

Marcus, C. M. and Westervelt, R. M. (189a). "Stability of Analog Neural Networks with Delay," *Phys. Rev. A*, **39**, 347–359.

Marcus, C. M. and Westervelt, R. M. (1989b). "Dynamics of Iterated-map Neural Networks," *Phys. Rev. A*, **40**, 501–504.

Marcus, C. M. and Westervelt, R. M. (1990). "Stability and Convergence of Analog Neural Networks with Multiple-Time-Step Parallel Dynamics," *Phys. Rev. A*, **42**, 2410–2417.

Marcus, C. M., Waugh, F. R., and Westervelt, R. M. (1990). "Associative Memory in an Analog Iterated-map Neural Network," *Phys. Rev. A*, **41**, 3355–3364.

Marcus, C. M., Waugh, F. R., and Westervelt, R. M. (1991). "Connection Topology and Dynamics in Lateral-inhibition Networks," in *Advances in Neural Information Processing Systems*, Vol. 3. Morgan Kaufman, San Mateo, CA, pp. 98–104.

McCulloch, W. and Pitts, W. (1943). "A Logical Calculus of Ideas Immanent in Nervous Activity," *Bull. Math. Biophys.* **5**, 115–133. (Reprinted in *Brain Theory, Reprint Volume, Advanced Series in Neuroscience*, Vol. 1, G. L. Shaw and G. Palm, eds. World Scientific, Singapore, 1988.)

Mead, C. A. (1989). *Analog VLSI and Neural Systems*. Addison-Wesley, Reading, MA.

Mertens, S. (1991). "An Extremely Diluted Asymmetric Network with Graded Response Neurons," *J. Phys. A*, **24**, 337–351.

Mezard, M., Parisi, G., and Virasoro, M. A. (1987). *Spin Glass Theory and Beyond*. World Scientific, Singapore.

Mori, Y., Davis, P., and Shigetoshi, N. (1989). "Pattern Retrieval in an Asymmetric Neural Network with Embedded Limit Cycles," *J. Phys. A*, **22**, L525–L532.

Murray, A. F. (1991). *"Silicon Implementations of Neural Networks," IEE Proc. F*, **138**, 3–13.

Noest, A. J. (1989). "Domains in Neural Networks with Restricted-range Interactions," *Phys. Rev. Lett.*, **63**, 1739–1742.

Noest, A. J. (1990). "Semi-local Signal Processing in the Visual System" in *Statistical Mechanics of Neural Networks: Proc. XIth Sitges Conference*, Sitges, Barcelona, Spain, June 1990, L. Garrido, ed. Springer-Verlag, New York, pp. 303–316.

Parisi, G. (1986). "Asymmetric Neural Networks and the Process of Learning," *J. Phys. A*, **19**, L675–L680.

Personnaz, L., Guyon, I., and Dreyfus, G. (1985). "Information Storage and Retrieval in Spin-glass-like Neural Networks," *J. Phys. Lett. (Paris)*, **46**, L359–L365.

Personnaz, L., Guyon, I., and Dreyfus, G. (1986a). "Collective Computational Properties of Neural Networks: New Learning Mechanisms," *Phys. Rev. A*, **34**, 4217–4228.

Personnaz, L., Guyon, I., Dreyfus, G., and Toulouse, G. (1986b). "A Biologically Constrained Learning Mechanism in Networks of Formal Neurons," *J. Statist. Phys.*, **43**, 411–422.

Sherrington, D. and Kirkpatrick, S. (1975). "Solvable Model of a Spin Glass," *Phys. Rev. Lett.*, **35**, 1792–1796.

Shiino, M. and Fukai, T. (1990). "Replica-symmetric Theory of the Nonlinear Analogue Neural Networks," *J. Phys. A*, **23**, L1009–L1017.

Sompolinsky, H. and Kanter, I. (1986). "Temporal Association in Asymmetric Neural Networks," *Phys. Rev. Lett.*, **57**, 2861–2864.

Treves, A. (1990). "Threshold-linear Formal Neurons in Auto-associative Nets," *J. Phys. A*, **23**, 2631–2650.

Treves, A. and Amit, D. J. (1988). "Metastable States in Asymmetrically Diluted Hopfield Networks," *J. Phys. A*, **21**, 3155–3169.

Treves, A. and Amit, D. J. (1989). "Low Firing Rates: an Effective Hamiltonian for Excitatory Neurons," *J. Phys. A*, **22**, 2205–2226.

Waugh, F. R., Marcus, C. M., and Westervelt, R. M. (1990). "Fixed-point Attractors in Analog Neural Computation," *Phys. Rev. Lett.*, **64**, 1986–1989.

Waugh, F. R., Marcus, C. M., and Westervelt, R. M. (1991). "Reducing Neuron Gain to Eliminate Fixed-point Attractors in an Analog Associative Memory" *Phys. Rev. A*, **43**, 3131–3142.

11

Dynamics and Stability Analysis of the Brain-State-in-a-Box (BSB) Neural Models

STEFEN HUI, WALTER E. LILLO, and STANISLAW H. ŻAK

11.1. INTRODUCTION

A possible function of the Brain-State-in-a-Box (BSB) model proposed by Anderson et al. (1977) is to recognize a pattern given a noisy version of the pattern (see Anderson's Chapter 4 of this book for further discussion of the BSB models). For this reason it is often referred to as an associative memory. The stable equilibrium points should represent the complete stored patterns and the noisy patterns are represented by points in the basins of attraction of the stable equilibrium points. The dynamics of the BSB model is characterized by the difference equation

$$\mathbf{x}(k + 1) = g(\mathbf{x}(k) + \alpha \mathbf{W}\mathbf{x}(k))$$

and the initial condition

$$\mathbf{x}(0) = \mathbf{x}_0$$

where $\mathbf{x}(k) \in \mathbb{R}^n$ is the state of the system at time k, α is the step size and $\mathbf{W} \in \mathbb{R}^{n \times n}$ is a symmetric weight matrix, determined during the training process, and the function $g: \mathbb{R}^n \to \mathbb{R}^n$ is a vector-valued function whose ith component, $g_i(\cdot)$, is defined as follows:

$$g_i(y) = \begin{cases} 1 & \text{if } y \geq 1 \\ y & \text{if } -1 < y < 1 \\ -1 & \text{if } y \leq -1 \end{cases}$$

The BSB model gets its name from the fact that the state of the system is constrained to be in the hypercube $H_n = [-1, 1]^n$. Normally only the stable equilibrium points should be located at the vertices of the hypercube, although it is not necessary that every vertex be a stable equilibrium point.

The BSB and related neural models have been studied extensively by a number of researchers. In particular Golden (1986) showed that the BSB model can be viewed as a gradient descent algorithm. Later Greenberg (1988) studied the BSB model where the weight matrix is diagonally

dominant but is no longer required to be symmetric. More recently, Hui and Żak (1991, 1992), and Golden (1992) proposed a generalized BSB model. The dynamics of this generalized BSB model is described by

$$\mathbf{x}(k + 1) = g(\mathbf{x}(k) + \alpha(\mathbf{W}\mathbf{x}(k) + \mathbf{b})) \quad \mathbf{x}(0) = \mathbf{x}_0$$

Li et al. (1989) examined a class of linear dynamical systems operating on a closed hypercube, which they referred to as a "linear system in a saturated mode" (LSSM). The LSSM can be viewed as a continuous version of the BSB model proposed by Hui and Żak (1991, 1992) and Golden (1992). Under various assumptions on the weight matrix $\mathbf{W}$ and the vector $\mathbf{b}$, Li et al. (1989) were able to locate all the equilibrium points of their network and determine their stability properties.

In this chapter we examine the dynamics of the generalized BSB model. The original BSB model can just be viewed as a special case of the generalized BSB model where $\mathbf{W}$ is symmetric and $\mathbf{b} = \mathbf{0}$. In our examination of the generalized BSB model we will make various assumptions concerning the weight matrix $\mathbf{W}$ and the vector $\mathbf{b}$, and determine the effect of these assumptions on the dynamics of the model. First we consider the case where $\mathbf{W}$ is symmetric. Then we relax the symmetry conditions but place the restrictions on $\mathbf{W}$ proposed by Hui and Żak (1991, 1992).

This chapter is divided into five sections. Section 11.2 contains background results. Section 11.3 discusses the stability of the generalized BSB model under the assumption that the weight matrix $\mathbf{W}$ is symmetric. Section 11.4 examines the case where the weight matrix $\mathbf{W}$ is diagonally dominant. Section 11.5 is a chapter summary.

11.2. BACKGROUND RESULTS

In this section we present some basic definitions and preliminary results which are used in the remaining sections. For convenience we introduce the following notation. Let

$$L(\mathbf{x}) = (\mathbf{I}_n + \alpha\mathbf{W})\mathbf{x} + \alpha\mathbf{b}$$

where $\mathbf{I}_n$ is the $n \times n$ identity matrix, and

$$T(\mathbf{x}) = g(L(\mathbf{x}))$$

Throughout this chapter we are concerned with the stability or instability of equilibrium points of the generalized BSB model. For this reason we present a brief review of stability analysis of dynamical systems modeled by the equation

$$\mathbf{x}(k + 1) = f(\mathbf{x}(k)) \qquad \mathbf{x}(0) = \mathbf{x}_0$$

where $\mathbf{x} = \mathbf{x}(k) \in \mathbb{R}^n$ and f is a continuous vector-valued function having components

$$f_i(x_1, x_2, \ldots, x_n) \qquad i = 1, \ldots, n$$

For more information on the stability of dynamic systems modeled by difference equations, the reader is referred to Hahn (1958) or LaSalle (1986). Note that in the case of the generalized BSB model $f(\mathbf{x}) = T(\mathbf{x})$.

Definition 2.1 (Equilibrium Point). A point $\mathbf{x}$ is an equilibrium point of $\mathbf{x}(k + 1) = f(\mathbf{x}(k))$ if $f(\mathbf{x}) = \mathbf{x}$.

Definition 2.2 (Stability in the Sense of Liapunov). An isolated equilibrium point $\mathbf{x}^*$ is said to be stable in the sense of Liapunov if for any scalar $\varepsilon > 0$ there is a $\delta > 0$ such that

$$\|\mathbf{x}(0) - \mathbf{x}^*\| < \delta \text{ implies } \|\mathbf{x}(k) - \mathbf{x}^*\| < \varepsilon \quad \forall k \geq 0$$

where $\|\mathbf{x}\|$ is the Euclidean norm of the vector $\mathbf{x}$, and the symbol $\forall$ means "for all."

Definition 2.3 (Asymptotic Stability). An isolated equilibrium point $\mathbf{x}^*$ is said to be asymptotically stable if in addition to being stable in the sense of Liapunov there exists $\delta > 0$ so that $\mathbf{x}(k) \to \mathbf{x}^*$ as $k \to \infty$ if $\|\mathbf{x}(0) - \mathbf{x}^*\| < \delta$.

Definition 2.4 (Positive Definite Function). Let $\Omega \subseteq \mathbb{R}^n$ be an open neighborhood of the origin of $\mathbb{R}^n$. A continuous function $V: \mathbb{R}^n \to \mathbb{R}$ is said to be positive definite over the set $\Omega \subseteq \mathbb{R}^n$ if $V(\mathbf{x}) > 0 \ \forall \mathbf{x} \in \Omega, \mathbf{x} \neq \mathbf{0}$, and $V(\mathbf{0}) = 0$.

In the following theorems we assume, without loss of generality, that the equilibrium point of interest has been translated to the origin of $\mathbb{R}^n$.

THEOREM 2.1 (Liapunov Stability Theorem). *An isolated equilibrium point $\mathbf{x}^* = \mathbf{0}$ is stable in the sense of Liapunov if for some neighborhood Ω about $\mathbf{x}^*$, there exists a positive definite function $V(\mathbf{x})$ so that $V(f(\mathbf{x})) \leq V(\mathbf{x}) \ \forall \mathbf{x} \in \Omega$.*

THEOREM 2.2 (Asymptotic Stability Theorem). *An isolated equilibrium point $\mathbf{x}^* = 0$ is asymptotically stable if for some neighborhood Ω about $\mathbf{x}^*$, there exists a positive definite function $V(\mathbf{x})$ so that $V(f(\mathbf{x})) < V(\mathbf{x}) \ \forall \mathbf{x} \in \Omega, \mathbf{x} \neq \mathbf{0}$.*

In analyzing the stability of equilibrium points it is useful in some cases to be able to determine if a point is a local or global minimizer of a function. For this reason we have included the following definitions and theorems. For more information on the subject of convex functions the reader is referred to Luenberger (1984).

Definition 2.5. A subset D of $\mathbb{R}^n$ is said to be convex if for any two points $\mathbf{x}, \mathbf{y} \in D$ the line segment connecting $\mathbf{x}$ and $\mathbf{y}$ is contained in D. This condition can be expressed as

$$\{\mathbf{z} \in \mathbb{R}^n | \mathbf{z} = \lambda\mathbf{x} + (1 - \lambda)\mathbf{y}, \mathbf{x}, \mathbf{y} \in D, 0 < \lambda < 1\} \subset D$$

Definition 2.6. A function $f: D \to \mathbb{R}$ defined over the convex set $D \subseteq \mathbb{R}^n$ is said to be convex if for all $\mathbf{x}, \mathbf{y} \in D$ and $0 \leq \lambda \leq 1$,

$$f(\lambda\mathbf{x} + (1 - \lambda)\mathbf{y}) \leq \lambda f(\mathbf{x}) + (1 - \lambda)f(\mathbf{y})$$

A function is said to be strictly convex if, for all $\mathbf{x}, \mathbf{y} \in D, \mathbf{x} \neq \mathbf{y}$, and $0 < \lambda < 1$,

$$f(\lambda\mathbf{x} + (1 - \lambda)\mathbf{y}) < \lambda f(\mathbf{x}) + (1 - \lambda)f(\mathbf{y})$$

A function f is concave (strictly concave) if $-f$ is convex (strictly convex).

If we make some assumptions about the smoothness of a function then we have the following theorems.

THEOREM 2.3. *If f is a differentiable function and D is a convex set then f is a convex function if and only if $f(\mathbf{x}) \geq f(\mathbf{y}) + \nabla f(\mathbf{y})^T(\mathbf{x} - \mathbf{y}) \ \forall \mathbf{x}, \mathbf{y} \in D$. It is strictly convex if and only if $f(\mathbf{x}) > f(\mathbf{y}) + \nabla f(\mathbf{y})^T(\mathbf{x} - \mathbf{y}) \ \forall \mathbf{x}, \mathbf{y} \in D, \mathbf{x} \neq \mathbf{y}$.*

THEOREM 2.4. *If f is a twice continuously differentiable function and D is a convex set containing an interior point, then f is convex over D if the Hessian matrix of f is positive semifinite throughout D. The function f is strictly convex if its Hessian matrix is positive definite.*

If we assume the domain D of a function f is a convex set, then an important tool in developing a useful characterization of a minimizer is the concept of feasible directions. For a feasible point $\mathbf{x}$ a direction $\mathbf{d}$ is said to be feasible if there exists a scalar $\eta^* > 0$ such that for $0 \leq \eta \leq \eta^*$, the vector $\mathbf{x} + \eta\mathbf{d} \in D$.

THEOREM 2.5. *Let f be continuously differentiable on the convex set D. If there is a point $\mathbf{x}^* \in D$ such that $\forall \mathbf{y} \in D$, $\nabla f(\mathbf{x}^*)^T(\mathbf{y} - \mathbf{x}^*) \geq 0$, then $\mathbf{x}^*$ is a global minimizer of f over D.*

THEOREM 2.6. *For a convex function defined over a convex set D, any local minimizer is a global minimizer.*

It follows from the above theorem that if a function is strictly convex over a closed convex set then there is exactly one minimizer and that is a global minimizer.

THEOREM 2.7. *Let f be a concave function defined over a closed, bounded, convex set D. If f has a minimum over D, then it is achieved at an extreme point of D.*

It can be shown that if f is strictly concave then a minimum cannot occur at any point which is not an extreme point.

Definition 2.7 (Row Diagonal Dominance). A matrix $\mathbf{W} \in \mathbb{R}^{n \times n}$ is said to be row diagonal dominant if

$$w_{ii} \geq \sum_{\substack{j=1 \\ j \neq i}}^{n} |w_{ij}| \qquad i = 1, \ldots, n$$

It is said to be strongly row diagonal dominant if

$$w_{ii} > \sum_{\substack{j=1 \\ j \neq i}}^{n} |w_{ij}| \qquad i = 1, \ldots, n$$

THEOREM 2.8 (Lévy-Desplanques Theorem). *A complex $n \times n$ matrix $\mathbf{W}$ with the property*

$$|w_{ii}| > \sum_{\substack{j=1 \\ j \neq i}}^{n} |w_{ij}| \qquad i = 1, \ldots, n$$

is nonsingular.

For a proof of the above theorem see Marcus and Minc (1964).

11.3. ANALYSIS OF THE GENERALIZED BSB MODEL WITH A SYMMETRIC WEIGHT MATRIX

In this section we examine the dynamics of the generalized BSB model under the assumption that the weight matrix $\mathbf{W}$ is symmetric. As a consequence of this assumption, the generalized BSB model can be viewed as a gradient descent algorithm, with a step size α, when operating in the open hypercube H_n^o (Golden, 1992). In the interior of the hypercube the dynamics can be described by the difference equation:

$$\mathbf{x}(k + 1) = L(\mathbf{x}(k)) = \mathbf{x}(k) - \alpha\nabla E(\mathbf{x}(k))$$

where $E(\mathbf{x})$ is the energy function given by

$$E(\mathbf{x}) = -\tfrac{1}{2}\mathbf{x}^T\mathbf{W}\mathbf{x} - \mathbf{x}^T\mathbf{b}$$

We now show, in a manner similar to Golden (1992), that the energy function $E(\mathbf{x})$ evaluated on the trajectories of the generalized BSB model is monotonically nonincreasing. In order to prove this statement we need the following Definition and Lemma. Let, as in Golden (1992),

$$\gamma(i, k) = \begin{cases} 0 & \text{if } (\nabla E(\mathbf{x}(k)))_i = 0 \\ \dfrac{(T(\mathbf{x}(k)))_i - x_i(k)}{-\alpha(\nabla E(\mathbf{x}(k)))_i} & \text{otherwise} \end{cases}$$

Note that

$$(T(\mathbf{x}(k)))_i = x_i(k) - \alpha\gamma(i, k)(\nabla E(\mathbf{x}(k)))_i$$

LEMMA 3.1. *For the generalized BSB model with a symmetric weight matrix $\mathbf{W}$ the variable $\gamma(i, k)$ satisfies*

$$0 \leq \gamma(i, k) \leq 1$$

PROOF. This clearly holds when $(\nabla E(\mathbf{x}(k)))_i = 0$. We then have two cases.

Case 1: $(T(\mathbf{x}(k)))_i = (L(\mathbf{x}(k)))_i$. Then

$$\gamma(i, k) = \frac{(T(\mathbf{x}(k)))_i - x_i(k)}{-\alpha(\nabla E(\mathbf{x}(k)))_i}$$

$$= \frac{x_i(k) - \alpha(\nabla E(\mathbf{x}(k)))_i - x_i(k)}{-\alpha(\nabla E(\mathbf{x}(k)))_i} = 1$$

Case 2: $(T(\mathbf{x}(k)))_i \neq (L(\mathbf{x}(k)))_i$. Then without loss of generality consider the case where $(T(\mathbf{x}(k)))_i = 1$. This implies that $(T(\mathbf{x}(k)))_i < (L(\mathbf{x}(k)))_i$ and thus

$$1 < x_i(k) - \alpha(\nabla E(\mathbf{x}(k)))_i$$

Since $|x_i(k)| \leq 1$, we have $-\alpha(\nabla E(\mathbf{x}(k)))_i > 0$. Hence

$$\gamma(i, k) = \frac{(T(\mathbf{x}(k)))_i - x_i(k)}{-\alpha(\nabla E(\mathbf{x}(k)))_i}$$

$$< \frac{x_i(k) - \alpha(\nabla E(\mathbf{x}(k)))_i - x_i(k)}{-\alpha(\nabla E(\mathbf{x}(k)))_i} = 1$$

In addition, since $(T(\mathbf{x}(k)))_i - x_i(k) \geq 0$, and $-\alpha(\nabla E(\mathbf{x}(k)))_i > 0$, we have $\gamma(i, k) \geq 0$.

The case where $(T(\mathbf{x}(k)))_i = -1$ is similar. $\square$

We now prove Golden's (1992) BSB Energy Minimization Theorem.

THEOREM 3.1 (Golden, 1992). *Suppose* $\mathbf{W} = \mathbf{W}^T$ *is positive definite or* $\alpha < (2/|\lambda_{\min}(\mathbf{W})|)$, *where* $\lambda_{\min}(\mathbf{W})$ *is the smallest eigenvalue of the symmetric matrix* $\mathbf{W}$. *Then the energy function E evaluated on the trajectories of the generalized BSB model is monotonically decreasing unless it encounters an equilibrium point.*

PROOF

Case 1: $\mathbf{x}(k) = \mathbf{x}(k + 1)$. In this case $\mathbf{x}(k)$ is an equilibrium point and we have

$$E(\mathbf{x}(k + 1)) = E(\mathbf{x}(k))$$

Case 2: $\mathbf{x}(k) \neq \mathbf{x}(k + 1)$. We are going to show that

$$E(\mathbf{x}(k)) - E(\mathbf{x}(k + 1)) > 0$$

For convenience we introduce the vector $\mathbf{d}$ whose ith element, d_i, is defined by

$$d_i = x_i(k + 1) - x_i(k)$$

Then from the definition of $\gamma(i, k)$ we have

$$d_i = -\alpha\gamma(i, k)(\nabla E(\mathbf{x}(k)))_i$$

Utilizing the Taylor series expansion, the definition of $\mathbf{d}$, and the fact that

$$\lambda_{\min}(\mathbf{W})\|\mathbf{d}\|^2 \leq \mathbf{d}^T\mathbf{W}\mathbf{d}$$

for $\mathbf{W} = \mathbf{W}^T$, we obtain

$E(\mathbf{x}(k)) - E(\mathbf{x}(k + 1))$

$= \frac{1}{2}\mathbf{d}^T\mathbf{W}\mathbf{d} - \nabla E(\mathbf{x}(k))^T\mathbf{d}$

$\geq \frac{1}{2}\lambda_{\min}(\mathbf{W})\mathbf{d}^T\mathbf{d} - \nabla E(\mathbf{x}(k))^T\mathbf{d}$

$= \frac{1}{2}\lambda_{\min}(\mathbf{W})\|\mathbf{d}\|^2 + \alpha \sum_{i=1}^{n} \gamma(i, k)((\nabla E(\mathbf{x}(k)))_i)^2$

Note that if $\mathbf{W}$ is positive definite then $\lambda_{\min}(\mathbf{W}) > 0$ and thus $E(\mathbf{x}(k)) - E(\mathbf{x}(k + 1)) > 0$. Otherwise we can use the definition of $\mathbf{d}$ to obtain

$E(\mathbf{x}(k)) - E(\mathbf{x}(k + 1))$

$= \alpha \sum_{i=1}^{n} \gamma(i, k)\left(-|\lambda_{\min}(\mathbf{W})|\frac{\alpha}{2}\gamma(i, k) + 1\right)$

$\times ((\nabla E(\mathbf{x}(k)))_i)^2$

Since $0 \leq \gamma(i, k) \leq 1$, we have

$$\alpha < \frac{2}{|\lambda_{\min}(\mathbf{W})|} \leq \frac{2}{\gamma(i, k)|\lambda_{\min}(\mathbf{W})|}$$

Thus if

$$\alpha < \frac{2}{|\lambda_{\min}(\mathbf{W})|}$$

we have $E(\mathbf{x}(k) - E(\mathbf{x}(k + 1)) > 0$. $\square$

We assume the hypothesis of Theorem 3.1 to be satisfied for the rest of this section.

THEOREM 3.2. *An isolated equilibrium point is asymptotically stable if and only if it is a strict local minimizer of the energy function E.*

PROOF. Sufficiency: Let $\mathbf{x}^*$ be a strict minimizer of the function E. For simplicity we assume that a coordinate transformation has been applied so that the point $\mathbf{x}$ is now at the origin of $\mathbb{R}^n$. Consider the Liapunov function candidate $V(\mathbf{x}) = E(\mathbf{x}) - E(\mathbf{x}^*)$. Note that V is positive definite in a neighborhood of $\mathbf{x}^*$. By Theorem 3.1, the trajectory of the generalized BSB model is such that the function $E(\mathbf{x}(k))$ is monotonically decreasing for $\mathbf{x} \neq \mathbf{x}^*$. It is clear that $V(T(\mathbf{x})) - V(\mathbf{x}) < 0$ for $\mathbf{x} \neq \mathbf{x}^*$. Thus $\mathbf{x}^*$ is an asymptotically stable equilibrium point by Theorem 2.2.

Necessity: Let $\mathbf{x}^*$ be an asymptotically stable equilibrium point of the generalized BSB model. We show, by contraposition, that $\mathbf{x}^*$ is a strict local minimizer of E. If $\mathbf{x}^* \in H_n$ is not a strict local minimizer of the energy function over the hypercube H_n then in any neighborhood of the point $\mathbf{x}^*$ inside the hypercube, there are points where the energy function is of equal or lower value. By Theorem 3.1, the energy function E evaluated on any trajectory of the model was monotonically decreasing if $\mathbf{x}(k + 1) \neq \mathbf{x}(k)$. Thus, if the initial state of the model corresponded to one of these points, then the trajectory would never converge to $\mathbf{x}^*$. This would contradict the fact that $\mathbf{x}^*$ is an asymptotically stable equilibrium point. $\square$

COROLLARY 3.1. *An equilibrium point* $\mathbf{x}^*$ *is stable in the sense of Liapunov if and only if it is a local minimizer of the function E.*

Li et al. (1989) gave conditions under which all equilibrium points of the LSSM could be found. In a manner similar to that of Li et al. (1989), we present a more general method for locating all equilibrium points of the generalized BSB model assuming only that the matrix $\mathbf{W}$ is symmetric. We then introduce more restrictive conditions, similar to those of Li et al. (1989), under which the stability or instability of points can be determined. In their analysis, Li et al. (1989) divided the hypercube H_n into 3^n regions. The regions correspond to the permutations of the three possible conditions listed below for the components of the state variable $\mathbf{x}$. They are:

1. $x_i = 1$, positive border
2. $-1 < x_i < 1$, interior
3. $x_i = -1$, negative border

Having defined these regions we now attempt to answer the following questions:

(a) Are there any equilibrium points in a given region?
(b) Is an equilibrium point in a region stable?

The first region we consider is the open hypercube $H_n^o = (-1, 1)^n$. If a point $\mathbf{y} \in H_n^o$ is an equilibrium point in this region, then

$$T(\mathbf{y}) = L(\mathbf{y}) = \mathbf{y}$$

Thus

$$L(\mathbf{y}) = (\mathbf{I} + \alpha\mathbf{W})\mathbf{y} + \alpha\mathbf{b} = \mathbf{y}$$

or equivalently,

$$\mathbf{W}\mathbf{y} + \mathbf{b} = \mathbf{0}$$

If $\mathbf{W}$ is full rank this condition is equivalent to

$$\mathbf{y} = -\mathbf{W}^{-1}\mathbf{b}$$

and $\mathbf{y}$ is the only equilibrium point in H_n^o if $\mathbf{y} \in H_n^o$. If $\mathbf{W}$ is not of full rank then $\mathbf{b}$ must still lie in the column space of $\mathbf{W}$ for a solution to $\mathbf{W}\mathbf{y} = -\mathbf{b}$ to exist. In this case there is an infinite number of equilibrium points in H_n^o. These points may be characterized by the set

$$X_e = \{\mathbf{z} \mid \mathbf{z} \in H_n^o, \mathbf{z} \in \mathbf{y} + N(\mathbf{W})\}$$

where $N(\mathbf{W})$ denotes the null space of $\mathbf{W}$. In a similar manner it can be shown that there is no equilibrium point in H_n^o if:

1. $\mathbf{W}$ is of full rank and $-\mathbf{W}^{-1}\mathbf{b} \notin H_n^o$.
2. $\mathbf{W}$ is not full rank and $\mathbf{b} \notin R(\mathbf{W})$, where $R(\mathbf{W})$ is the range of $\mathbf{W}$.
3. $\mathbf{W}$ is not full rank and $\mathbf{b} \in R(\mathbf{W})$, however the set X_e is the empty set, i.e., $(\mathbf{y} + N(\mathbf{W})) \cap H_n^o = \phi$.

Having examined the case where all the components of the state vector are in the interior of the hypercube, we now consider the regions where one or more components of the state vector are at the border of the hypercube H_n. We refer to this region as an edge.

Definition 3.2. We say that S is an edge of H_n

if there is a nonempty subset J of the set $\{1, 2, \ldots, n\}$ such that

$$S = \{[s_1, \ldots, s_n]^T \mid -1 \le s_j \le 1 \text{ for } j \notin J$$

$$\text{and } s_j = v_j \text{ for } j \in J, \text{ where } v_j \text{ is fixed at } \pm 1\}$$

We define the dimension of S as the number of nonfixed components, that is

$$\dim(S) = n - |J|$$

The interior of S, or the open edge S^o, is defined by

$$S^o = \{[s_1, s_2, \ldots, s_n]^T \in S \mid -1 < s_j < 1 \text{ for } j \notin J\}$$

We now illustrate the notion of an edge on the following two examples.

Example 3.1
Let $n = 3$, $J = \{2, 3\}$, and $x_2 = v_2 = 1$, $x_3 = v_3 = -1$. Then

$$S = \{\mathbf{x} \in H_3 \mid -1 \le x_1 \le 1, x_2 = 1, x_3 = -1\}$$

Note that since $|J| = 2$ the set S is a one-dimensional hypercube.

Example 3.2
Let $n = 3$, $J = \{2\}$, and $x_2 = v_2 = -1$. Then

$$S = \{\mathbf{x} \in H_3 \mid -1 \le x_1 \le 1, x_2 = -1,$$

$$-1 \le x_3 \le 1\}$$

In this example $|J| = 1$, and hence S is a two-dimensional hypercube (a face of H_3).

We now consider the $3^n - 2^n - 1$ cases where one or more components of the state vector is on a border of H_n, but not all of them. We do this by considering an arbitrary edge S, where $0 < |J| < n$. In order for an equilibrium point to exist on an open edge S^o there must be a point $\mathbf{x} \in S^o$ such that

$$(T(\mathbf{x}))_i = x_i \qquad i = 1, \ldots, n$$

We can break the above condition down into two cases corresponding to the components on the border and those in the interior. For the border case we have

$$(L(\mathbf{x}))_i x_i \ge 1 \qquad \forall i \in J$$

For the interior case we have

$$(L(\mathbf{x}))_i = x_i \qquad \forall i \notin J$$

where $|x_i| < 1 \; \forall i \notin J$, since S^o is an open edge. This is equivalent to the condition

$$\sum_{j=1}^{n} w_{ij}x_j + b_i = 0 \qquad \forall i \notin J$$

or

$$\sum_{j \notin J} w_{ij} x_j + \sum_{j \in J} w_{ij} v_j + b_i = 0 \quad \forall i \notin J$$

We define $\tilde{\mathbf{W}} \in \mathbb{R}^{(n-|J|) \times (n-|J|)}$ to be the matrix $\mathbf{W}$ with rows and columns which correspond to indices which are in the set J deleted. In addition, we define the vector $\tilde{\mathbf{b}} \in \mathbb{R}^{(n-|J|)}$ to have components

$$\tilde{b}_i = \sum_{j \in J} w_{ij} v_j + b_i \quad \forall i \notin J$$

Utilizing the above definitions we observe that in order for an equilibrium point to exist on an edge S° there must exist a point $\mathbf{x} \in \mathbb{R}^n$ whose border components satisfy the condition

$$(L(\mathbf{x}))_i v_i \geq 1 \quad \forall i \in J$$

and the interior components must satisfy the condition

$$\tilde{\mathbf{W}} \tilde{\mathbf{x}} = -\tilde{\mathbf{b}}$$

where $\tilde{\mathbf{x}} \in \mathbb{R}^{(n-|J|)}$ is a vector whose components correspond to the interior components of $\mathbf{x}$. For the interior of a given edge the following procedure can be used to determine the equilibrium points:

Step 1. In a manner similar to the case where we were analyzing the interior of the hypercube, it should be determined if there are any solutions to $\tilde{\mathbf{W}} \tilde{\mathbf{x}} = -\tilde{\mathbf{b}}$ which are in the open hypercube $H^\circ_{(n-|J|)}$. If there is no suitable point found, then there is no equilibrium point in the interior of that edge, and the remaining step is unnecessary.

Step 2. Check if any of the points on the open edge S°, whose nonborder components correspond to the appropriate components of the $\tilde{\mathbf{x}}$, satisfy the condition $(L(\mathbf{x}))_i v_i \geq 1$ $\forall i \in J$.

Any point satisfying conditions given in steps (1) and (2) is an equilibrium point. For the 2^n cases where all components of the state vector are on a border (i.e. $\mathbf{x}$ is a vertex of the hypercube H_n), the condition for the existence of an equilibrium point becomes

$$(L(\mathbf{x}))_i v_i \geq 1 \quad i = 1, \ldots, n$$

Having given conditions whereby the position of equilibrium points can be determined for a generalized BSB model with a symmetric weight matrix $\mathbf{W}$, we now present conditions whereby the stability or instability of an equilibrium point can be determined.

THEOREM 3.3. *If the matrix* $\mathbf{W} = \mathbf{W}^T$ *is positive definite, then any stable equilibrium point of the*

model must be located at a vertex of the hypercube H_n.

PROOF. Since the matrix $\mathbf{W}$ is positive definite, the energy function is strictly concave by Theorem 2.4. From Theorem 2.7 we know that a strictly concave function defined over a convex set cannot have a local minimizer at a point which is not an extreme point of the set. From Corollary 3.1 it follows that a stable equilibrium point must correspond to a local minimizer of the energy function E. Thus if an equilibrium point is stable it must be an extreme point of the hypercube H_n. The extreme points of H_n are precisely the vertices. $\qquad \square$

In order to prove our next theorem we need the following lemma.

LEMMA 3.2. *If* $\mathbf{W} = \mathbf{W}^T$ *is negative definite, then any equilibrium point* $\mathbf{x}^*$, *is a strict local minimizer of the function* E.

PROOF. Since $\mathbf{W}$ is negative definite, the function E is convex by Theorem 2.4. Let $\mathbf{x}^* \in H_n$ be an equilibrium point. We treat the cases of $\mathbf{x}^* \in H^\circ_n$ and $\mathbf{x}^* \notin H^\circ_n$ separately.

Case 1: $\mathbf{x}^* \in H^\circ_n$. Since $\mathbf{x}^*$ is an interior point, it follows that $\nabla E(\mathbf{x}^*) = 0$ (Luenberger, 1984, p. 169). With this in mind we consider the Taylor series expansion of the function E about the point $\mathbf{x}^*$ in an arbitrary direction $\mathbf{d}$. We have

$$E(\mathbf{x}^* + \eta \mathbf{d}) - E(\mathbf{x}^*) = -\frac{\eta^2}{2} \mathbf{d}^T \mathbf{W} \mathbf{d} > 0$$

Thus $\mathbf{x}^*$ is a strict local minimizer of the energy function E.

Case 2: $\mathbf{x}^* \notin H^\circ_n$. Since $\mathbf{x}^* \notin H^\circ_n$, $\mathbf{x}^*$ must lie on some edge S of the hypercube H_n. We define the set G to be the set of indices corresponding to the components of $\mathbf{x}^*$ whose values are ± 1. We also assume that $\tilde{\mathbf{W}}, \tilde{\mathbf{b}}$, and $\tilde{\mathbf{x}}$ are as defined earlier. Thus, since $\mathbf{x}^*$ is an equilibrium point, it follows that

$$(L(\mathbf{x}^*))_i x_i^* = (x_i^* - \alpha (\nabla E(\mathbf{x}^*))_i) x_i^* \geq 1 \quad \forall i \in G$$

This implies that

$$-(\nabla E(\mathbf{x}^*))_i x_i^* \geq 0 \quad \forall i \in G$$

If we let $\mathbf{d}$ be an arbitrary feasible direction, it follows that for any $i \in G$, either $d_i = 0$ or $\mathrm{sgn}(d_i) = -x_i^*$. Thus

$$(\nabla E(\mathbf{x}^*))_i d_i \geq 0 \quad \forall i \in G$$

Now consider the components of $\mathbf{x}^*$ which are not ± 1. Since $\mathbf{x}^*$ is an equilibrium point, it is clear that

$$(\nabla E(\mathbf{x}^*))_i = 0 \qquad \forall i \notin G$$

Taking the second-order Taylor series expansion of E about $\mathbf{x}^*$ yields

$$E(\mathbf{x}^* + \eta\mathbf{d}) - E(\mathbf{x}^*) = \eta\nabla E(\mathbf{x}^*)^\mathrm{T}\mathbf{d} - \frac{\eta^2}{2}\mathbf{d}^\mathrm{T}\mathbf{W}\mathbf{d} > 0$$

since $\mathbf{W} = \mathbf{W}^\mathrm{T} < 0$. Thus $\mathbf{x}^*$ is a strict local minimizer of the energy function E, and hence by Theorem 3.2, $\mathbf{x}^*$ is an asymptotically stable equilibrium point.   □

THEOREM 3.4. *If $\mathbf{W}$ is negative definite, then there is a unique asymptotically stable equilibrium point of the generalized BSB model.*

PROOF. From Lemma 3.2, it follows that any equilibrium point $\mathbf{x}^*$ is a strict local minimizer of E. Furthermore, since $\mathbf{W}$ is negative definite the energy function E is a strictly convex function. Thus by Theorem 2.6 any local minimizer is a strict global minimizer of the function E. Obviously there cannot be more than one strict global minimizer. Therefore $\mathbf{x}^*$ must be the only equilibrium point of the model and thus by Theorem 3.2, $\mathbf{x}^*$ is an asymptotically stable equilibrium point.   □

COROLLARY 3.2. *Let $\mathbf{x}^*$ be an equilibrium point on an edge S where $\tilde{\mathbf{W}}$ is negative definite. Then $\mathbf{x}^*$ is the only equilibrium point on the edge S.*

PROOF. We consider the energy function E over the set S. Since we are only considering the edge S of the hypercube H_n, we can treat the components of the state vector $\mathbf{x}$ whose components are in the set J as fixed, i.e., $x_i = \pm 1$ $\forall i \in J$. Thus the energy function on the edge S can be expressed as follows:

$$E(\tilde{\mathbf{x}}) = -\tfrac{1}{2}\tilde{\mathbf{x}}^\mathrm{T}\tilde{\mathbf{W}}\tilde{\mathbf{x}} - \tilde{\mathbf{x}}^\mathrm{T}\tilde{\mathbf{b}} - C$$

where

$$C = \frac{1}{2}\sum_{i \in J}\sum_{j \in J} w_{ij}v_i v_j + \sum_{i \in J} b_i v_i$$

Then, since the Hessian matrix of $E(\tilde{\mathbf{x}})$, is positive definite we can apply Theorem 3.4 and deduce that $\mathbf{x}^*$ is the only equilibrium point of the $(n - |J|)$-dimensional generalized BSB model. This in turn implies that $\mathbf{x}^*$ is the only equilibrium point of the original model on the edge S, since if a point on the edge S of the original model is an equilibrium point then it is an equilibrium point of the lower-dimensional model.   □

In our next theorem we concern ourselves with equilibrium points on an edge S, but not at a vertex.

THEOREM 3.5. *Let $\mathbf{x}^*$ be an equilibrium point in the interior of an edge S where $\tilde{\mathbf{W}}$ is negative definite. If*

$$(L(\mathbf{x}^*))_i x_i^* > 1 \qquad \forall i \in J$$

then $\mathbf{x}^*$ is a strict local minimizer of the energy function E, and an asymptotically stable equilibrium point.

PROOF. Since $L(\mathbf{x})$ is continuous, there exists a neighborhood $N(\mathbf{x}^*)$ about $\mathbf{x}^*$ such that for any point $\mathbf{x} \in N(\mathbf{x}^*)$,

$$(L(\mathbf{x}))_i x_i > 1 \qquad \forall i \in J$$

Thus for any point $\mathbf{x}(k) \in N(\mathbf{x}^*)$, $\mathbf{x}(k + 1) = T(\mathbf{x}(k)) \in S$. Therefore there is no equilibrium point in the region defined by

$$N(\mathbf{x}^*) \cap (H_n \backslash S)$$

In addition, since $\tilde{\mathbf{W}}$ is negative definite, it follows from Corollary 3.2 that there is no other equilibrium point on the edge S. Hence there is no equilibrium point other than $\mathbf{x}^*$ in the neighborhood $N(\mathbf{x}^*)$. Thus we have shown that $\mathbf{x}^*$ is an isolated equilibrium point. We now show that $\mathbf{x}^*$ is a strict local minimizer of E. Consider the Taylor series expansion of E about the point $\mathbf{x}^*$ in an arbitrary feasible direction $\mathbf{d}$. We have

$$E(\mathbf{x}^* + \eta\mathbf{d}) - E(\mathbf{x}^*) = \eta\nabla E(\mathbf{x}^*)^\mathrm{T}\mathbf{d} - \frac{\eta^2}{2}\mathbf{d}^\mathrm{T}\mathbf{W}\mathbf{d}$$

In order for a direction $\mathbf{d}$ to be feasible it is necessary that

$$d_i = 0 \quad \text{or} \quad \mathrm{sgn}(d_i) = -x_i^* \qquad \forall i \in J$$

This combined with the fact that

$$(L(\mathbf{x}^*))_i x_i^* = (x_i^* - \alpha(\nabla E(\mathbf{x}^*))_i)x_i^* > 1 \qquad \forall i \in J$$

implies that if $d_i \neq 0$ and $i \in J$ then

$$(\nabla E(\mathbf{x}^*))_i d_i > 0$$

If we consider the components of $\mathbf{x}^*$ which are not ± 1, then since $\mathbf{x}^*$ is an equilibrium point, it is clear that

$$(\nabla E(\mathbf{x}^*))_i = 0 \qquad \forall i \notin J$$

Therefore, unless $d_i = 0 \ \forall i \in J$,

$$\nabla E(\mathbf{x}^*)^\mathrm{T}\mathbf{d} > 0$$

Thus for a sufficiently small η it follows that

$$E(\mathbf{x}^* + \eta\mathbf{d}) - E(\mathbf{x}^*) = \eta\nabla E(\mathbf{x}^*)^\mathrm{T}\mathbf{d} + o(\eta) > 0$$

where $o(\eta)/\eta \to 0$ as $\eta \to 0$. If $d_i = 0 \; \forall i \in J$, then

$$\nabla E(\mathbf{x}^*)^T \mathbf{d} = 0$$

and the second order Taylor series expansion of E about $\mathbf{x}^*$ yields

$$E(\mathbf{x}^* + \eta \mathbf{d}) - E(\mathbf{x}^*) = -\frac{\eta^2}{2} \mathbf{d}^T \mathbf{W} \mathbf{d} > 0$$

Recall that $d_i = 0 \; \forall i \in J$. Thus if we define $\tilde{\mathbf{d}}$ to be the vector with the components whose indices are in the set J deleted, then the above expression can be rewritten as

$$E(\mathbf{x}^* + \eta \mathbf{d}) - E(\mathbf{x}^*) = -\frac{\eta^2}{2} \tilde{\mathbf{d}}^T \tilde{\mathbf{W}} \tilde{\mathbf{d}} > 0$$

Thus $\mathbf{x}^*$ is a strict local minimizer of the energy function E, and an isolated equilibrium point. Therefore by Theorem 3.2, $\mathbf{x}^*$ is an asymptotically stable equilibrium point of the generalized BSB model. □

COROLLARY 3.3. *Let* $\mathbf{x}^*$ *be a vertex. If*

$$(L(\mathbf{x}^*))_i x_i^* > 1 \qquad i = 1, \ldots, n$$

then $\mathbf{x}^*$ *is an asymptotically stable equilibrium point.*

PROOF. Since $L(\mathbf{x})$ is continuous there exists a neighborhood $N(\mathbf{x}^*)$ of $\mathbf{x}^*$ such that $\forall \mathbf{x} \in N(\mathbf{x}^*)$

$$(L(\mathbf{x}))_i x_i > 1 \qquad i = 1, \ldots, n$$

Thus $\forall \mathbf{x}(k) \in N(\mathbf{x}^*)$ we have

$$\mathbf{x}(k + 1) = T(\mathbf{x}(k)) = \mathbf{x}^*$$

which implies that $\mathbf{x}^*$ is an asymptotically stable equilibrium point. □

We now illustrate the above results by the following numerical example.

Example 3.3
For $n = 2$ let

$$\mathbf{W} = \begin{bmatrix} 1.2 & -0.4 \\ -0.4 & 1.8 \end{bmatrix} \quad \mathbf{b} = \begin{bmatrix} -0.9 \\ 0 \end{bmatrix} \quad \alpha = 0.3$$

Note that $\mathbf{W} = \mathbf{W}^T > 0$. In this example we have three asymptotically stable equilibrium points corresponding to the three vertices of H_2. They are $\mathbf{e}^{(2)} = [-1, 1]^T$, $\mathbf{e}^{(3)} = [-1, -1]^T$, and $\mathbf{e}^{(4)} = [1, -1]^T$. This follows from Corollary 3.3. The vertex $\mathbf{e}^{(1)} = [1, 1]^T$ is not even an equilibrium point since for $i = 1$ we have

$$(L(\mathbf{e}^{(1)}))_i v_1 = 0.97 < 1$$

In Fig. 11.1 we depict the basins of attraction for

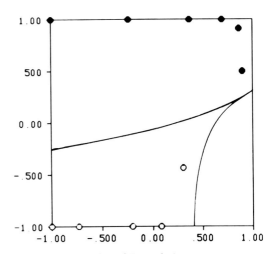

Fig. 11.1. Illustration of Example 1.

the three asymptotically stable equilibrium points. The unstable equilibrium points are located at $\mathbf{x}^{(1)} = [0.81, 0.18]$, $\mathbf{x}^{(2)} = [0.42, -1]^T$, and $\mathbf{x}^{(3)} = [-1, -0.22]^T$ (see Theorem 3.3). We have also shown two trajectories in Fig. 11.1. The trajectory marked with filled circles starts at $\mathbf{x}(0) = [0.9, 0.5]^T$. The trajectory denoted by empty circles starts at $\mathbf{x}(0) = [0.4, -0.25]^T$.

11.4. ANALYSIS OF THE GENERALIZED BSB MODEL WITH A ROW DIAGONAL DOMINANT WEIGHT MATRIX

In this section we consider the case where the weight matrix $\mathbf{W}$ satisfies the property

$$w_{ii} \geq \sum_{\substack{j=1 \\ j \neq i}}^{n} |w_{ij}| + |b_i| \qquad i = 1, \ldots, n$$

Note that the weight matrix $\mathbf{W}$ is no longer required to be symmetric, therefore the results of the last section do not necessarily apply here. The analysis of the generalized BSB model under these conditions is similar to that of Hui and Żak (1991, 1992).

LEMMA 4.1. *If a matrix* $\mathbf{W} \in \mathbb{R}^{n \times n}$ *and a vector* $\mathbf{b} \in \mathbb{R}^n$ *satisfy*

$$w_{ii} > \sum_{\substack{j=1 \\ j \neq i}}^{n} |w_{ij}| + |b_i| \qquad i = 1, \ldots, n$$

then $\mathbf{W}^{-1}\mathbf{b} \in H_n^o$.

PROOF. By Theorem 2.8 the matrix $\mathbf{W}$ is invertible. Let $\mathbf{a} = \mathbf{W}^{-1}\mathbf{b}$. If $\mathbf{b} = \mathbf{0}$, then $\mathbf{a} = \mathbf{0} \in H_n^o$.

Suppose $\mathbf{b} \neq \mathbf{0}$. Let a_i be the component such that

$$|a_i| = \max_{1 \le j \le n} |a_j|$$

Then $|a_i| > 0$ since $\mathbf{W}$ is nonsingular. From the assumption, we have

$$w_{ii} - \sum_{\substack{j=1 \\ j \neq i}}^{n} |w_{ij}| > |b_i|$$

Hence

$$|b_i| = \left| \sum_{j=1}^{n} w_{ij} a_j \right| = \left| w_{ii} a_i + \sum_{\substack{j=1 \\ j \neq i}}^{n} w_{ij} a_j \right|$$

$$= |a_i| \left| w_{ii} + \sum_{\substack{j=1 \\ j \neq i}}^{n} w_{ij} \frac{a_j}{a_i} \right|$$

$$\ge |a_i| \left| w_{ii} - \sum_{\substack{j=1 \\ j \neq i}}^{n} |w_{ij}| \right| > |a_i| |b_i|$$

We conclude that $b_i \neq 0$ and $|a_i| < 1$. Consequently $|a_i| < 1 \; \forall j = 1, \ldots, n$. $\qquad \square$

Using Lemma 4.1 and the arguments of the previous section, one can show that the point $-\mathbf{W}^{-1}\mathbf{b}$ is the only equilibrium point in H_n^o. Hui and Żak (1991, 1992) showed that this point is unstable, as follows from the following theorem.

THEOREM 4.1. *If* $\|(\mathbf{I}_n + \alpha\mathbf{W})\mathbf{x}\| > \|\mathbf{x}\|$ *for* $\mathbf{x} \neq \mathbf{0}$, *then for any point* $\mathbf{y} \in H_n^o$, $\mathbf{y} \neq -\mathbf{W}^{-1}\mathbf{b}$, *either* $T(\mathbf{y})$ *lies on the boundary of* H_n *or* $\|T(\mathbf{y}) + \mathbf{W}^{-1}\mathbf{b}\| > \|\mathbf{y} + \mathbf{W}^{-1}\mathbf{b}\|$.

PROOF. Let $\mathbf{y} \in H_n^o$, $\mathbf{y} \neq -\mathbf{W}^{-1}\mathbf{b}$, and $\mathbf{\Theta} = \mathbf{y} + \mathbf{W}^{-1}\mathbf{b}$. Suppose that $T(\mathbf{y}) \in H_n^o$. Then

$$T(\mathbf{y}) = L(\mathbf{y}) = (\mathbf{I}_n + \alpha\mathbf{W})\mathbf{y} + \alpha\mathbf{b}$$

$$= (\mathbf{I}_n + \alpha\mathbf{W})(\mathbf{\Theta} - \mathbf{W}^{-1}\mathbf{b}) + \alpha\mathbf{b}$$

$$= (\mathbf{I}_n + \alpha\mathbf{W})\mathbf{\Theta} - \mathbf{W}^{-1}\mathbf{b}$$

Therefore, since by assumption $\|(\mathbf{I}_n + \alpha\mathbf{W})\mathbf{x}\| > \|\mathbf{x}\| \; \forall \mathbf{x} \neq \mathbf{0}$, we have

$$\|T(\mathbf{y}) + \mathbf{W}^{-1}\mathbf{b}\| = \|(\mathbf{I}_n + \alpha\mathbf{W})\mathbf{\Theta}\| > \|\mathbf{\Theta}\|$$

$$= \|\mathbf{y} + \mathbf{W}^{-1}\mathbf{b}\| \qquad \square$$

The above theorem states that for this class of weight matrices the distance between the state and the equilibrium point $-\mathbf{W}^{-1}\mathbf{b}$ increases with time until the boundary of H_n is reached. It should be noted that the hypothesis of Theorem 4.1 holds true for any nonsingular weight matrix $\mathbf{W}$ such that $\mathbf{W} = \mathbf{W}^T > 0$ or $(\mathbf{W}^T + \mathbf{W})$ is positive

definite. To show this, consider

$$\|(\mathbf{I}_n + \alpha\mathbf{W})\mathbf{x}\|^2$$

$$= \mathbf{x}^T(\mathbf{I}_n + \alpha\mathbf{W}^T)(\mathbf{I}_n + \alpha\mathbf{W})\mathbf{x}$$

$$= \mathbf{x}^T(\mathbf{I}_n + \alpha(\mathbf{W}^T + \mathbf{W}) + \alpha^2\mathbf{W}^T\mathbf{W})\mathbf{x}$$

Since from the assumption $\mathbf{W}^T + \mathbf{W} > 0$ and $\mathbf{W}^T\mathbf{W} > 0$, thus we have $\lambda_{\min}(\mathbf{W}^T + \mathbf{W}) > 0$ and $\lambda_{\min}(\mathbf{W}^T\mathbf{W}) > 0$. Hence

$$\|(\mathbf{I}_n + \alpha\mathbf{W})\mathbf{x}\|^2 \ge (1 + \alpha\lambda_{\min}(\mathbf{W}^T + \mathbf{W})$$

$$+ \alpha^2\lambda_{\min}(\mathbf{W}^T\mathbf{W}))\|\mathbf{x}\|^2$$

$$> \|\mathbf{x}\|^2$$

Therefore, if $\mathbf{W} = \mathbf{W}^T > 0$ or $\mathbf{W} + \mathbf{W}^T > 0$, then $\|(\mathbf{I}_n + \alpha\mathbf{W})\mathbf{x}\| > \|\mathbf{x}\|$.

THEOREM 4.2. *If* $w_{ii} \ge \sum_{j=1, j \neq i}^{n} |w_{ij}| + |b_i|$ *for* $i = 1, \ldots, n$, *then all edges are invariant, i.e., once a component of the vector* $\mathbf{x}(k)$ *reaches a value of* ± 1 *it will remain at that value.*

PROOF. Let the ith coordinate of $\mathbf{x}(k)$ be 1. Then the ith coordinate of $(\mathbf{I}_n + \alpha\mathbf{W})\mathbf{x}(k) + \alpha\mathbf{b}$ is

$$(1 + \alpha w_{ii}) + \alpha \sum_{\substack{j=1 \\ j \neq i}}^{n} w_{ij}x_j(k) + \alpha b_i$$

$$\ge 1 + \alpha \left(w_{ii} - \sum_{\substack{j=1 \\ j \neq i}}^{n} |w_{ij}| - |b_i| \right) \ge 1$$

The last inequality follows from the assumption. Thus $x_i(k + 1) = (T(\mathbf{x}(k)))_i = 1$. Similarly we have the result for $x_i(k) = -1$. $\qquad \square$

As a consequence of the above theorem we observe that all the extreme points of the hypercube are equilibrium points. We now show that this is only the case when the weight matrix $\mathbf{W}$ satisfies the conditions specified in this section.

THEOREM 4.3. *If all vertices of the hypercube are equilibrium points, then the weight matrix of the generalized BSB model must satisfy the condition*

$$w_{ii} \ge \sum_{\substack{j=1 \\ j \neq i}}^{n} |w_{ij}| + |b_i| \qquad i = 1, \ldots, n$$

PROOF. By contraposition. Assume that every vertex of H_n is an equilibrium point and there exists an i such that

$$w_{ii} < \sum_{\substack{j=1 \\ j \neq i}}^{n} |w_{ij}| + |b_i|$$

Consider the vertex given by $\mathbf{e} = [e_1, \ldots, e_n]^{\mathrm{T}}$, where

$$e_j = \begin{cases} -\operatorname{sgn}(b_i) & \text{if } j = i \text{ or } w_{ij} = 0 \\ \operatorname{sgn}(w_{ij}) \operatorname{sgn}(b_i) & \text{if } j \neq i \text{ and } w_{ij} \neq 0 \end{cases}$$

where $\operatorname{sgn}(w_{ij}) = w_{ij}/|w_{ij}| = |w_{ij}|/w_{ij}$. Without loss of generality we will assume that $\operatorname{sgn}(b_i) = -1$, Then

$$(L(\mathbf{e}))_i = ((\mathbf{I}_n + \alpha\mathbf{W})\mathbf{e})_i + \alpha b_i$$

$$= 1 + \alpha w_{ii} e_i + \alpha \sum_{\substack{j=1 \\ j \neq i}}^{n} w_{ij} e_j + \alpha b_i$$

$$= 1 + \alpha w_{ii} - \alpha \sum_{\substack{j=1 \\ j \neq i}}^{n} |w_{ij}| - \alpha |b_i|$$

$$= 1 + \alpha\left(w_{ii} - \sum_{\substack{j=1 \\ j \neq i}}^{n} |w_{ij}| - |b_i| \right)$$

$$< 1$$

Thus $\mathbf{e}$ is a vertex that is not an equilibrium point. □

Our next theorem is an extension of Greenberg's (1988) result.

THEOREM 4.4. *If* $w_{ii} > \sum_{j=1, j \neq i}^{n} |w_{ij}| + |b_i|$, $i = 1, \ldots, n$, *then the vertices of the hypercube* H_n *are asymptotically stable equilibrium points.*

PROOF. Let $\mathbf{e}$ be an arbitrary vertex of H_n. Then for an arbitrary index i,

$$(L(\mathbf{e}))_i e_i = 1 + \alpha e_i((\mathbf{W}\mathbf{e})_i + b_i)$$

$$= 1 + \alpha\left(w_{ii} + \sum_{\substack{j=1 \\ j \neq i}}^{n} w_{ij} e_j e_i + b_i e_i \right)$$

$$\geq 1 + \alpha\left(w_{ii} - \sum_{\substack{j=1 \\ j \neq i}}^{n} |w_{ij}| - |b_i| \right)$$

$$> 1$$

Since $(L(\mathbf{x}))_i \mathbf{x}_i$ is a continuous function there exists a neighborhood $N(\mathbf{e})$ about $\mathbf{e}$ such that

$$(L(\mathbf{y}))_i y_i > 1 \qquad \forall i = 1, \ldots, n, \text{ and } \mathbf{y} \in N(\mathbf{e})$$

Therefore for any $\mathbf{y} \in N(\mathbf{e})$ we have

$$T(\mathbf{y}) = \mathbf{e} \qquad \square$$

We now illustrate the above results by the following two numerical examples.

Example 4.1

Let $n = 2$ and

$$\mathbf{W} = \begin{bmatrix} 1.2 & -0.4 \\ -0.4 & 1.8 \end{bmatrix} \qquad \mathbf{b} = \begin{bmatrix} -0.5 \\ -0.9 \end{bmatrix} \qquad \alpha = 0.3$$

Note that the weight matrix $\mathbf{W}$ in this example is the same as in Example 3.3. However, the vector $\mathbf{b}$ has been modified to meet the conditions of Theorem 4.4. As a result of this modification we now have four asymptotically stable equilibrium points located at the vertices of H_2. In Fig. 11.2 we depicted the basins of attraction of the asymptotically stable equilibrium points, and three sample trajectories. The three trajectories shown in Fig. 11.2 start from $[0.7, 0.7]^{\mathrm{T}}$, $[0, 0.3]^{\mathrm{T}}$, and $[0.7, 0.5]^{\mathrm{T}}$.

Example 4.2

In this example we have a nonsymmetric weight matrix. Specifically

$$\mathbf{W} = \begin{bmatrix} 2 & -0.8 \\ -0.5 & 1 \end{bmatrix} \qquad \mathbf{b} = \begin{bmatrix} 0.4 \\ -0.2 \end{bmatrix} \qquad \alpha = 0.3$$

Note that since $\mathbf{W}$ satisfies the condition of Theorem 4.4, we have four asymptotically stable equilibrium points located at the vertices of H_2. In Fig. 11.3 the basin of attraction of the asymptotically stable equilibrium points are shown. In addition, we have shown two sample trajectories. The trajectory marked by empty circles starts at $[-0.25, 0.1]^{\mathrm{T}}$. The trajectory marked by filled circles starts at $[0, 0]^{\mathrm{T}}$.

Hui and Żak (1991, 1992) gave a bound on the number of steps required to hit a boundary of H_n for a given initial condition $\mathbf{x}(0)$.

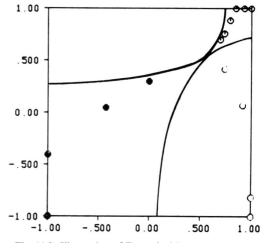

Fig. 11.2. Illustration of Example 4.1.

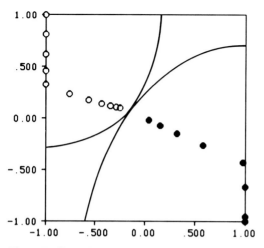

Fig. 11.3. Illustration of Example 4.2.

THEOREM 4.5. *Given a generalized BSB model whose weight matrix* $\mathbf{W}$ *satisfies*

$$w_{ii} > \sum_{\substack{j=1 \\ j \neq i}}^{n} |w_{ij}| + |b_i| \qquad i = 1, \ldots, n,$$

and

$$\mathbf{W} + \mathbf{W}^T > 0$$

Let $\mathbf{x}(0) \in H_n^o$, $\mathbf{x}(0) \neq -\mathbf{W}^{-1}\mathbf{b}$, *and* $\xi = \mathbf{x}(0) + \mathbf{W}^{-1}\mathbf{b}$. *Then* $\mathbf{x}(k)$ *lies on the boundary of the hypercube* H_h *for*

$$k \geq \frac{2 \log\left(\dfrac{2\sqrt{n}}{\|\xi\|}\right)}{\log(1 + \alpha\lambda_{\min}(\mathbf{W} + \mathbf{W}^T) + \alpha^2\lambda_{\min}(\mathbf{W}^T\mathbf{W}))}$$

where $\lambda_{\min}(\mathbf{P})$ *denotes the smallest eigenvalue of the symmetric matrix* $\mathbf{P}$.

PROOF. Since $\mathbf{W} + \mathbf{W}^T$ and $\mathbf{W}^T\mathbf{W}$ are positive definite matrices, thus $\lambda_{\min}(\mathbf{W} + \mathbf{W}^T) > 0$ and $\lambda_{\min}(\mathbf{W}^T\mathbf{W}) > 0$.
If $\mathbf{x}(i) \in H_n^o$, $i = 0, \ldots, k$, then we can express $\mathbf{x}(k)$ as

$$\mathbf{x}(k) = (\mathbf{I}_n + \alpha\mathbf{W})^k \xi - \mathbf{W}^{-1}\mathbf{b}$$

Thus

$$\mathbf{x}(k) + \mathbf{W}^{-1}\mathbf{b} = (\mathbf{I}_n + \alpha\mathbf{W})^k \xi$$

and

$$\|\mathbf{x}(k) + \mathbf{W}^{-1}\mathbf{b}\|^2 = \|(\mathbf{I}_n + \alpha\mathbf{W})^k \xi\|^2$$

Define

$$\xi_k = (\mathbf{I}_n + \alpha\mathbf{W})^k \xi$$

Then for $k \geq 1$,

$$\xi_k = (\mathbf{I}_n + \alpha\mathbf{W})\xi_{k-1}$$

We have

$$\|\xi_k\|^2 = \|(\mathbf{I}_n + \alpha\mathbf{W})\xi_{k-1}\|^2$$
$$= \xi_{k-1}^T(\mathbf{I}_n + \alpha(\mathbf{W}^T + \mathbf{W}) + \alpha^2\mathbf{W}^T\mathbf{W})\xi_{k-1}$$
$$\geq \|\xi_{k-1}\|^2(1 + \alpha\lambda_{\min}(\mathbf{W}^T + \mathbf{W})$$
$$+ \alpha\lambda_{\min}(\mathbf{W}^T\mathbf{W}))$$

Iteratively applying this relation yields

$$\|\mathbf{x}(k) + \mathbf{W}^{-1}\mathbf{b}\|^2 = \|\xi_k\|^2 = \|(\mathbf{I}_n + \alpha\mathbf{W})^k\xi\|^2$$
$$\geq \|\xi\|^2(1 + \alpha\lambda_{\min}(\mathbf{W}^T + \mathbf{W}) + \alpha^2\lambda_{\min}(\mathbf{W}^T\mathbf{W}))^k$$

Thus

$$\|\mathbf{x}(k) + W^{-1}\mathbf{b}\|$$
$$\geq \|\xi\|(1 + \alpha\lambda_{\min}(\mathbf{W}^T + \mathbf{W}) + \alpha^2\lambda_{\min}(\mathbf{W}^T\mathbf{W}))^{k/2}$$

Since $-\mathbf{W}^{-1}\mathbf{b}$, $\mathbf{x}(k) \in H_n^o$, it follows that

$$\|\mathbf{x}(k) + \mathbf{W}^{-1}\mathbf{b}\| < 2\sqrt{n}$$

where $2\sqrt{n}$ is the length of the diagonal of the hypercube H_n. Thus a necessary condition for the trajectory to be inside the hypercube is

$$2\sqrt{n} > \|\xi\|(1 + \alpha\lambda_{\min}(\mathbf{W}^T + \mathbf{W}) + \alpha^2\lambda_{\min}(\mathbf{W}^T\mathbf{W}))^{k/2}$$

Solving for k, we can rewrite this condition as follows

$$k < \frac{2 \log\left(\dfrac{2\sqrt{n}}{\|\xi\|}\right)}{\log(1 + \alpha\lambda_{\min}(\mathbf{W} + \mathbf{W}^T) + \alpha^2\lambda_{\min}(\mathbf{W}^T\mathbf{W}))}$$

It follows therefore that the trajectory has hit, and by Theorem 4.1 will stay on the boundary of the hypercube H_n^o for all k such that

$$k \geq \frac{2 \log\left(\dfrac{2\sqrt{n}}{\|\xi\|}\right)}{\log(1 + \alpha\lambda_{\min}(\mathbf{W} + \mathbf{W}^T) + \alpha^2\lambda_{\min}(\mathbf{W}^T\mathbf{W}))}$$

$\square$

Having shown that each edge of the hypercube is invariant and having given a bound on the number of iterations necessary for a trajectory to reach an edge, we now turn our attention to studying the dynamics of the model when the trajectory is constrained to an edge S of the hypercube H_n. Since the edges of the hypercube are invariant we only need to consider the components x_j of the state vector with $|x_j| < 1$. Recall that the dynamics of the model for any component of the state variable $\mathbf{x}$ is given by

$$x_i(k + 1) = (T(\mathbf{x}(k)))_i$$

$$= g_i(x_i(k) + \alpha(\mathbf{Wx}(k))_i + \alpha b_i)$$

$$= g_i\left(x_i(k) + \alpha\left(\sum_{j=1}^{n} w_{ij}x_j(k) + b_i\right)\right)$$

$$= g_i\left(x_i(k) + \alpha\left(\sum_{j \in J} w_{ij}x_j(k)\right.\right.$$

$$\left.\left. + \sum_{j \notin J} w_{ij}x_j(k) + b_i\right)\right)$$

$$= g_i\left(x_i(k) + \alpha\left(\sum_{j \in J} w_{ij}v_j\right.\right.$$

$$\left.\left. + \sum_{j \notin J} w_{ij}x_j(k) + b_i\right)\right)$$

$$= g_i\left(x_i(k) + \alpha\left(\tilde{b}_i + \sum_{j \notin J} w_{ij}x_j(k)\right)\right)$$

Recalling the definitions for $\tilde{\mathbf{W}}$, $\tilde{\mathbf{x}}$, and $\tilde{\mathbf{b}}$, we note that when constrained to an edge, the dynamic behavior of the model can be described by the equation

$$\tilde{\mathbf{x}}(k + 1) = g((\mathbf{I}_{n-|J|} + \alpha\tilde{\mathbf{W}})\tilde{\mathbf{x}}(k) + \alpha\tilde{\mathbf{b}})$$

Observe that the form of the above equation is simply an $(n - |J|)$-dimensional generalized BSB. In addition, the lower-dimensional model inherits the diagonal dominance properties of the higher-dimensional model as is shown below. Indeed

$$\tilde{w}_{ii} - \sum_{\substack{j=1 \\ j \neq i}}^{n} |\tilde{w}_{ij}| - |\tilde{b}_i| = w_{ii} - \sum_{\substack{j \notin J \\ j \neq i}} |w_{ij}|$$

$$- \left|b_i + \sum_{j \in J} w_{ij}v_j\right|$$

$$\geq w_{ii} - \sum_{\substack{j \notin J \\ j \neq i}} |w_{ij}|$$

$$- \sum_{j \in J} |w_{ij}| - |b_i|$$

$$= w_{ii} - \sum_{\substack{j=1 \\ j \neq i}}^{n} |w_{ij}| - |b_i|$$

$$> 0$$

Since the lower-dimensional model has the same form as the higher-dimensional model, we can iteratively apply the previous results. This leads us to the following result from Hui and Żak (1991, 1992).

THEOREM 4.6. *For a generalized BSB model*

with *the property*

$$w_{ii} > \sum_{\substack{j=1 \\ j \neq i}}^{n} |w_{ij}| + |b_i| \qquad i = 1, \ldots, n$$

there are 3^n equilibrium points and only the 2^n extreme points are stable equilibrium points.

PROOF. Previously we noted that the extreme points of the hypercube are asymptotically stable equilibrium points. We also noted that the equilibrium point on the interior of the hypercube is unstable. Since for the interior of an edge, the trajectory is determined by the equation

$$\tilde{\mathbf{x}}(k + 1) = g((\mathbf{I}_{n-|J|} + \alpha\tilde{\mathbf{W}})\tilde{\mathbf{x}}(k) + \alpha\tilde{\mathbf{b}})$$

we can show that there exists one equilibrium point on the interior of that edge, and its nonborder components are given by $-\tilde{\mathbf{W}}^{-1}\tilde{\mathbf{b}}$. In addition it is clear that this point will be unstable. Thus the number of equilibrium points is

$$1 + \sum_{j=1}^{n-1} \binom{n}{j} 2^j + 2^n = 3^n \qquad \square$$

11.5. SUMMARY

In this chapter we have examined the stability of the generalized BSB neural model. We studied the generalized BSB model first under the assumption that the weight matrix $\mathbf{W}$ was symmetric. Under this assumption it was noted that the model was a gradient algorithm. It was then shown that if the step size was sufficiently small then the model trajectory would act in such a way that the energy function would decrease until the model reached an equilibrium point. We then characterized the set of equilibrium points in the system and gave conditions to determine if they are stable or asymptotically stable. Next we examined the case where the weight matrix $\mathbf{W}$ was strongly diagonally dominant, and gave conditions which were necessary and sufficient for the model to have stable equilibrium points at every vertex of the hypercube. It was then shown that each edge was invariant under the conditions, and a bound for the time needed for a trajectory to reach an edge was given. Finally for this case it was shown that there were 3^n equilibrium points and only 2^n of them were stable.

REFERENCES

Anderson, J. A., Silverstein, J. W., Ritz, S. A., and Jones, R. S. (1977). "Distinctive Features, Categorical

Perception, and Probability Learning: Some Applications of a Neural Model," *Psychol. Rev.*, **84**, 413–451.

Golden, R. M. (1986). "The Brain-State-in-a-Box Neural Model is a Gradient Descent Algorithm," *J. Math. Psychol.*, **30**, 73–80.

Golden, R. M. (1992). "Stability and Optimization Analysis of the Generalized Brain-State-in-a-Box Neural Network Model," *J. Math. Psychol.* to appear.

Greenberg, H. J. (1988). "Equilibria of the Brain-State-in-a-Box Neural Model," *Neural Networks*, **1**, 323–324.

Hahn, W. (1958). "Über die Anwendung der Methode von Ljapunov auf Differenzengleichungen" *Mathematische Annalen*, **136**, 430–441.

Hui, S. and Żak, S. H. (1991). "On the Brain-State-in-a-Box (BSB) Neural Models," *Proc. Int. AMSE Conference on Neural Networks*, San Diego, May 29–31, 1991, Vol. 1, pp. 35–46.

Hui, S. and Żak, S. H. (1992). "Dynamical Analysis of the Brain-State-in-a-Box (BSB) Neural Models," *IEEE Trans. Neural Networks*, **3**, 86–94.

LaSalle, J. P. (1986). *The Stability and Control of Discrete Processes*. Springer-Verlag, New York.

Li, J.-H., Michel, A. N., and Porod, W. (1989). "Analysis and Synthesis of a Class of Neural Networks: Linear Systems Operating on a Closed Hypercube," *IEEE Trans. Circuits and Systems*, **36**, 1405–1422.

Luenberger, D. G. (1984). *Linear and Nonlinear Programming*, 2nd ed. Addison-Wesley, Reading, MA.

Marcus, M. and Minc, H. (1964). *A Survey of Matrix Theory and Matrix Inequalities*. Allyn and Bacon, Boston, MA.

12

Feature and Memory-Selective Error Correction in Neural Associative Memory[1]

GIRISH PANCHA AND SANTOSH S. VENKATESH

12.1. INTRODUCTION

In this chapter we describe models of auto-associative memory based upon densely inter-connected recurrent networks of McCulloch–Pitts neurons (McCulloch and Pitts, 1943). The model neurons in these networks are linear threshold elements that compute the sign of a linear form of their inputs. In a recurrent network, a collection of these neurons communicate with each other through linear synaptic weights and each neuron changes state on the basis of the net synaptic potential from all the neurons in the network. The instantaneous state of the neural network is described by the collective states of the individual neurons, and the choice of synaptic weights and the neuron updating rule determine the nature of flow in the state space of the network. As in any dynamical system, the fixed points of the network play a critical role in determining its computational properties. In particular, such networks can be used for encoding a set of prescribed items identified with states $\mathbf{u}$ as fixed points of the network, i.e., states $\mathbf{u}$ which are fixed under the dynamics of the network. In addition, some form of error correction is desired, so that states $\mathbf{u} + \delta\mathbf{u}$ in the vicinity of $\mathbf{u}$ are mapped into $\mathbf{u}$. When the item is to be retrieved from the network, the search can then be done on either the whole item or on part of it. Such searches by data association are especially useful in pattern recognition applications such as speech and vision. *We will describe methods of specifying in a probabilistic sense the conditions under which such error correction occurs.*

We now consider a fully interconnected network of n McCulloch–Pitts neurons. Each neuron is capable of assuming two values: 1 (firing) and -1 (not firing). The instantaneous binary output of each neuron is fed back as an input to the network. At epoch t, if $u_1[t]$, $u_2[t], \ldots, u_n[t]$ are the outputs of each of the n neurons in the network, then at epoch $t + 1$, the ith neuron updates itself according to the following threshold rule:

$$u_i[t + 1] = f\left(\sum_{j=1}^{n} w_{ij} u_j[t] - w_{i0} \right)$$

where

$$f(x) = \begin{cases} +1 & \text{if } x \geq 0 \\ -1 & \text{if } x < 0 \end{cases} \quad (1)$$

Based on the preceding firing rule, each neuron is characterized by a set of n real synaptic weights, and a real threshold value, and the network as a whole is characterized by a matrix of n^2 weights w_{ij} and n thresholds w_{i0}. Without loss of generality, we can confine our analysis to zero thresholds as thresholds are easily subsumed by the simple expedient of adding a constant input of -1 to each neuron.

The instantaneous state of the network is an n-tuple $\mathbf{u} = (u_1, \ldots, u_n) \in \mathbb{B}^n$, where $\mathbb{B} = \{-1, 1\}$, and u_i, the ith component of $\mathbf{u}$ is the output value of the ith neuron. Our goal is to encode items (states in $\mathbb{B}^n$) that are to be stored as fixed points of the recurrent network by an appropriate choice of weight matrix. We label these prescribed items $\mathbf{u}^{(1)}, \ldots, \mathbf{u}^{(m)}$, and hereafter refer to them as memories to distinguish them from other states of the network.

12.1.1. Liapunov Functions and Error Correction

The network can operate under different modes of updating. If all neurons are simultaneously updated, the mode of operation is said to be synchronous. If at most one neuron is updated at each epoch, the network is said to operate in

[1] This work was supported in part by the Air Force Office of Scientific Research under grant AFOSR 89-0523.

an asynchronous mode. It has been shown that both modes of operation lead to very similar associative behavior in neural networks. Note that in this synchronous mode case, we have

$$\mathbf{u}[t + 1] = F(\mathbf{W}\mathbf{u}[t])$$

where $F: \mathbb{R}^n \to \mathbb{B}^n$ is an n-ary pointwise threshold operator whose ith component $f_i(x)$ is as in (1).

Given the arbitrarily prescribed set of m memories, we are interested in specifying patterns of interconnectivity. The nature of flow in state space of the network is completely determined once the matrix of neural interconnection weights is computed. In order for our network to act as an associative memory, we require that the memories be stable. As described earlier, a memory $\mathbf{u}^{(\alpha)}$ is stable if all subsequent mappings return $\mathbf{u}^{(\alpha)}$, i.e., $\mathbf{u}^{(\alpha)}[t + 1] = F(\mathbf{W}\mathbf{u}^{(\alpha)}[t])$ for all t. Furthermore, we required that the memories exercise a region of influence around themselves, i.e., states close to or similar to memories should map to the corresponding memories in the network, and thereby exhibit error-correcting properties. The Euclidean distance between a state $\mathbf{u}$ and a memory $\mathbf{u}^{(\alpha)}$ is given by

$$d_E(\mathbf{u}, \mathbf{u}^{(\alpha)}) = \left[\sum_{i=1}^{n} (u_i^{(\alpha)} - u_i)^2 \right]^{1/2}$$

In $\mathbb{B}^n$, $(u_i^{(\alpha)} - u_i) = \pm 2$ if the components are mismatched, and 0 otherwise. Therefore $d_E = 2\sqrt{d_H}$ where the Hamming distance $d_H(\mathbf{u}, \mathbf{u}^{(\alpha)})$ is defined to be the number of mismatched components between the two states. We shall use the average Hamming distance from a memory over which error corrections are exhibited as a natural measure of attraction, and call it the *attraction radius*, ρ, of the memory.

From the theory of dynamical systems, we can expect attraction behavior in neural networks if we can find functions on the systems that are bounded and monotone nonincreasing along trajectories in the state space. Such functions are called Liapunov functions. If such a function exists, the stable points of the system reside at minima of the function. If the memories are programmed to be at these minima, we can achieve the desired attraction behavior around the memories.

Two such Liapunov functions are the Hamiltonian Energy ($E(\mathbf{u})$) function and the Manhattan Norm ($G(\mathbf{u})$) function, where

$$E(\mathbf{u}) = -\frac{1}{2} \sum_{i=1}^{n} \sum_{j=1}^{n} w_{ij} u_i u_j$$

and

$$G(\mathbf{u}) = -\sum_{j=1}^{n} \left| \sum_{j=1}^{n} w_{ij} u_j \right|$$

These functions act as Liapunov functions for certain classes of weight matrices under particular modes of operation (cf. Hopfield, 1982; Goles and Vichniac, 1986; Venkatesh and Psaltis, 1989).[1]

PROPOSITION 1. $E(\mathbf{u})$ *is nonincreasing in asynchronous mode if* $\mathbf{W}$ *is symmetric and has nonnegative diagonal elements;* $E(\mathbf{u})$ *is noncreasing in any mode if* $\mathbf{W}$ *is symmetric and nonnegative definite,*

PROPOSITION 2. $G(\mathbf{u})$ *is noncreasing in synchronous mode if* $\mathbf{W}$ *is symmetric.*

While the existence of Liapunov functions indicates attraction behavior, the lack of one does not necessarily indicate that the network will not function as desired. In the following sections, *we shall discuss a family of near-optimal algorithms to compute the weight matrix* $\mathbf{W}$. All but one of these algorithms result in a symmetric weight matrix, and all of them exhibit both desired properties of stability and attraction. In the following sections, we will establish the relationship between the various algorithms, and evaluate their performance.

12.1.2. Capacity and Complexity

Two measures characteristic of any algorithm are the *algorithmic capacity* and *algorithmic complexity*. We will look at these measures for each of the algorithms in the succeeding sections.

Capacity is the maximal number of memories that can be stored with high probability. It is useful to define capacity as a rate of growth rather than an exact number. Specifically, a sequence of numbers $\{C(n), n \geq 1\}$ is a sequence of capacities if and only if for every $\lambda \in (0, 1)$, as $n \to \infty$ the probability that each of the memories is stable approaches 1 whenever $m \leq (1 - \lambda)C(n)$, and approaches 0 whenever $m \geq (1 + \lambda)C(n)$. We note that a consequence of this is that a sequence of capacities, if it exists, is not unique, but rather determines an equivalence of sequences $\mathscr{C}$ where, $C(n)$ and $C'(n)$ are sequences in $\mathscr{C}$ iff $C(n) \sim C'(n)$ as $n \to \infty$.[2]

---

[1] Proofs of propositions in the main text of the chapter are deferred to Appendix A.

[2] On asymptotic notation. If $\{x(n)\}$ and $\{y(n)\}$ are any two sequences, we denote: $n_n = O(y_n)$ if there exists a constant K such that $|x(n)| \leq K|y(n)|$ for every n; $x(n) = o(y(n))$ if $|x(n)|/|y(n)| \to 0$ as $n \to \infty$; $x(n) \sim y(n)$ if $x(n)/y(n) \to 1$ as $n \to \infty$; and $x(n) = \omega(y(n))$ if $|x(n)|/|y(n)| \to \infty$ as $n \to \infty$.

We define complexity to be the number of elementary operations required to compute the matrix of weights. For the purposes of this discussion, the elementary operations are multiplication and addition of two real values. We are therefore interested in determining the complexity of an algorithm given m memories in n-space. In some cases, an algorithm may have a recursive definition which may result in a reduced algorithmic complexity from a practical standpoint when memories are added to the network one at a time.

12.2. OUTER PRODUCT ALGORITHM

Let $\mathbf{u}^{(1)}, \ldots, \mathbf{u}^{(m)}$ be a selection of m memories. The outer product algorithm prescribes that the weight matrix $\mathbf{W}^{\mathrm{op}}$ be chosen in the following manner (see Chapter 1 for additional details):

$$\mathbf{W}^{\mathrm{op}} = \mathbf{U}\mathbf{U}^{\mathrm{T}} \qquad (2)$$

where $\mathbf{U} = [\mathbf{u}^{(1)}, \mathbf{u}^{(2)} \cdots \mathbf{u}^{(m)}]$.

This scheme uses the sum of outer products of the memory vectors as correlation between the memories to form a weight matrix $\mathbf{W}^{\mathrm{op}}$ so that an input vector close to a memory vector will pick out that memory vector. The functioning of the algorithm as an efficient associative memory has been well documented (cf. Hopfield, 1982), and theoretical results on the capacity have been derived (McEliece et al., 1987; Komlós and Paturi, 1988).[3] We will now review some results for the scheme.

12.2.1. Error Correction

From the definition it follows that $\mathbf{W}^{\mathrm{op}}$ is symmetric, nonnegative definite. Therefore, the algorithmic flow in state space is towards the minimization of bounded functionals (the Manhattan Norm G in synchronous mode and the Energy E in any mode).[4] The trajectories therefore will tend to terminate in stable states that are local minima of the functionals. If these stable states correspond to the stored memories, the outer product algorithm satisfies the requirements of a physical associative memory. To examine its efficacy, however, we need to estimate its storage capacity, and the algorithmic complexity of computing the weights.

[3] See also Chapter 9.

[4] In some variations a zero diagonal is enforced for the matrix $\mathbf{W}^{\mathrm{op}}$ in (2). The matrix is then symmetric, with nonnegative diagonal elements so that the energy is non-increasing in asynchronous operation.

We first consider the effect of $\mathbf{W}^{\mathrm{op}}$ on a memory $\mathbf{u}^{(\alpha)}$. We have

$$
\begin{aligned}
[\mathbf{W}^{\mathrm{op}}\mathbf{u}^{(\alpha)}]_i &= \sum_{j=1}^{n} w_{ij}^{\mathrm{op}} u_j^{(\alpha)} \\
&= \sum_{j \neq i} \sum_{\beta=1}^{m} u_i^{(\beta)} u_j^{(\beta)} u_j^{(\alpha)} \\
&= (n-1) u_i^{(\alpha)} + \sum_{j \neq i} \sum_{\beta \neq \alpha} u_i^{(\beta)} u_j^{(\beta)} u_j^{(\alpha)} \\
&= (n-1) u_i^{(\alpha)} + \delta u_i^{(\alpha)} \qquad (3)
\end{aligned}
$$

We see that, in effect, there is a "signal" term and a "noise" term. Assuming that the memories are chosen randomly from a sequence of symmetric Bernoulli trials, the noise term $\delta u_i^{(\alpha)}$ has a mean of 0 and a standard deviation of $\sqrt{[(n-1)(m-1)]}$. The mean of the absolute value of the signal term $(n-1) u_i^{(\alpha)}$ is $(n-1)$. Thus, if $m = o(n)$, the signal term dominates the error term, and we can write

$$\mathbf{W}^{\mathrm{op}}\mathbf{U} = (n-1)\mathbf{U} + \delta\mathbf{U} \approx (n-1)\mathbf{U}$$

It will follow that $F(\mathbf{W}^{\mathrm{op}}\mathbf{U}) = \mathbf{U}$ with high probability if $m = o(n)$.

12.2.2. Capacity and Complexity

The memories $\mathbf{u}^{(\alpha)}$, $\alpha = 1, \ldots, m$ can be identified as *pseudo-eigenvectors* of the linear operator $\mathbf{W}^{\mathrm{op}}$ with *pseudo-eigenvalues* $n - 1$. When randomly chosen, they are stable in a probabilistic sense only if the mean to standard deviation given by $\sqrt{[(n-1)/(m-1)]}$ is large. More precisely, the following assertion holds (cf. McEliece et al., 1987; Komlós and Paturi, 1988). Chapter 9 contains more details.

PROPOSITION 3. *The (stable state) capacity of the outer product algorithm is $n/(4 \log n)$.*

We sketch one side of the proof in Appendix A to illustrate some of the ideas involved. The reader is also invited to delve into Chapter 1 for similar derivations.

Somewhat more can be shown than asserted above. In fact, if $\rho \in [0, 1/2)$ is any fixed quantity, and random probes are generated at a distance ρn from each of the memories, then all the errors in all the probes are corrected in one synchronous step with high probability if the number of memories m increases no more rapidly with n than $(1 - 2\rho)^2 n/(4 \log n)$. In particular, this result implies that within capacity each of the memories has (asymptotically) an identically sized ball of attraction of radius ρn. From a physical viewpoint this implies that all the memories are treated equivalently by the outer product algorithm as are all the features (memory components).

If we label the number of elementary operations required to compute $\mathbf{W}^{op}$ for m memories as N^{op}, then by counting the number of operations needed for matrix multiplication, by considering that the weight matrix $\mathbf{W}^{op}$ is symmetrical, and by noting that the diagonal elements are trivially specified, we find that $N^{op} = (mn^2 - mn)/2$. In addition, by inspection of (3), we notice that the outer product matrix can be recursively computed via

$$\mathbf{W}^{op}[\alpha] = \mathbf{W}^{op}[\alpha - 1] + \mathbf{u}^{(\alpha)}(\mathbf{u}^{(\alpha)})^T \qquad \alpha \geq 1$$

where $\mathbf{W}^{op}[\alpha]$ denotes the outer product weight matrix generated by the first α memories, and $\mathbf{W}^{op}[0] = \mathbf{0}$. This means that the incremental complexity $N^{op}[\alpha] = (n^2 - n)$, and the cost of computing the weight matrix incrementally is twice the cost of computing it from scratch for any given set of m memories.

12.3. MEMORY-SELECTIVE ALGORITHMS

In this section, we will discuss schemes to generate the weight matrix to yield a larger capacity than the outer product scheme. In addition, *these schemes will enable us to selectively increase attraction radii around specified memories.* (Compare with the outer product algorithm, where uniform attraction balls around the memories obtain.) The constructions are an extension of the outer product scheme to make the memories true eigenvectors of the linear operator $\mathbf{W}$, and then specification of the eigenvalues of the memories.

Construction 1. Define the interconnection matrix $\mathbf{W}^s$ as follows:

$$\mathbf{W}^s = \mathbf{U}\Lambda(\mathbf{U}^T\mathbf{U})^{-1}\mathbf{U}^T \qquad (4)$$

where $\Lambda = \mathbf{dg}[\lambda^{(1)}, \ldots, \lambda^{(m)}]$ is the $m \times m$ diagonal matrix of positive eigenvalues $\lambda^{(1)}, \ldots, \lambda^{(m)} > 0$, and $\mathbf{U} = [\mathbf{u}^{(1)} \mathbf{u}^{(2)} \cdots \mathbf{u}^{(m)}]$ is the $n \times m$ matrix of memory column vectors.

Construction 2. Given $\mathbf{u}^{(1)}, \ldots, \mathbf{u}^{(m)}$, choose any $(n - m)$ vectors $\mathbf{u}^{(m+1)}, \ldots, \mathbf{u}^{(n)} \in \mathbb{B}^n$ such that $\mathbf{u}^{(1)}, \ldots, \mathbf{u}^{(n)}$ are linearly independent. Define the interconnection matrix $\mathbf{W}^s$ as follows:

$$\mathbf{W}^s = \mathbf{U}_a\Lambda_a\mathbf{U}_a^{-1}$$

where the augmented matrices $\mathbf{U}_a$ and Λ_a are

defined as

$$\Lambda_a = \mathbf{dg}[\lambda^{(1)}, \ldots, \lambda^{(m)}, 0, \ldots, 0]$$
$$\mathbf{U}_a = [\mathbf{u}^{(1)} \cdots \mathbf{u}^{(n)}]$$

We note that

$$\mathbf{W}^s\mathbf{U} = \mathbf{U}\Lambda \qquad \mathbf{W}^s\mathbf{U}_a = \mathbf{U}_a\Lambda_a \qquad (5)$$

so that $\mathbf{u}^{(1)}, \ldots, \mathbf{u}^{(m)}$ are eigenvectors of $\mathbf{W}^s$ and Λ is the spectrum of $\mathbf{W}^s$ (Personnaz et al., 1985; Venkatesh and Psaltis, 1989). Therefore, we are guaranteed to have stable memories as long as $\mathbf{W}^s$ is well defined.

Hybrids of the two methods described above can also be used where the matrix $\mathbf{U}$ is partially augmented and then the pseudo-inverse is computed.

12.3.1. Error Correction

For the case of an m-fold degenerate spectrum, $\lambda^{(1)}, \ldots, \lambda^{(m)} = \lambda > 0$, we see that the matrix $\mathbf{W}^s$ is symmetric with nonnegative eigenvalues, i.e., it is nonnegative definite. Therefore there exist Liapunov functions in this case. In fact, consider the energy function $E(\mathbf{u}) = -\frac{1}{2}\langle \mathbf{u}, \mathbf{W}\mathbf{u} \rangle$.

For each memory, $\mathbf{u}^{(\alpha)}$, the energy is hence given by

$$E(\mathbf{u}^{(\alpha)}) = -\frac{1}{2}\langle \mathbf{u}^{(\alpha)}, \mathbf{W}\mathbf{u}^{(\alpha)} \rangle = -\frac{\lambda n}{2}$$

Let $\mathbf{u} \in \mathbb{B}^n$ be arbitrary. We can write $\mathbf{u}$ in the form

$$\mathbf{u} = \sum_{\alpha=1}^{m} c^{(\alpha)}\mathbf{u}^{(\alpha)} + \mathbf{u}_\perp = \mathbf{u}_\parallel + \mathbf{u}_\perp$$

where $\mathbf{u}_\parallel$ is the projection of $\mathbf{u}$ into the linear span of the m memories, $\mathbf{u}^{(1)}, \mathbf{u}^{(2)}, \ldots, \mathbf{u}^{(m)}$, and $\mathbf{u}_\perp$ is a vector in the orthogonal subspace. Then we have $\langle \mathbf{u}_\parallel, \mathbf{u}_\perp \rangle = 0$, and by the Pythagorean theorem,

$$\|\mathbf{u}\|^2 = \|\mathbf{u}_\parallel\|^2 + \|\mathbf{u}_\perp\|^2 = \left\| \sum_{\alpha=1}^{m} c^{(\alpha)}\mathbf{u}^{(\alpha)} \right\|^2 + \|\mathbf{u}_\perp\|^2$$

The energy is then given by

$$E(\mathbf{u}) = -\frac{1}{2}\langle \mathbf{u}, \mathbf{W}\mathbf{u} \rangle$$
$$= -\frac{1}{2}\left\langle \sum_{\alpha=1}^{m} c^{(\alpha)}\mathbf{u}^{(\alpha)} + \mathbf{u}_\perp, \sum_{\alpha=1}^{m} \lambda c^{(\alpha)}\mathbf{u}^{(\alpha)} \right\rangle$$
$$= -\frac{\lambda}{2}\left\| \sum_{\alpha=1}^{m} c^{(\alpha)}\mathbf{u}^{(\alpha)} \right\|^2$$
$$\geq -\frac{\lambda}{2}\|\mathbf{u}\|^2 = -\frac{\lambda n}{2}$$

It follows that the stored memories form *global* energy minima.

For the general spectral matrix in (4), exact Liapunov functions are hard to come by. The signal-to-noise ratio, however, serves as a good ad hoc measure of attraction capability. Consider synchronous operations with $\mathbf{W}^s$ on a state vector $\mathbf{u} = \mathbf{u}^{(\alpha)} + \delta\mathbf{u} \in \mathbb{B}^n$. We have

$$\mathbf{W}^s\mathbf{u} = \mathbf{W}^s(\mathbf{u}^{(\alpha)} + \delta\mathbf{u}) = \mathbf{W}^s\mathbf{u}^{(\alpha)} + \mathbf{W}^s\,\delta\mathbf{u}$$

Once again, there exists a "signal" term, $\mathbf{W}^s\mathbf{u}^{(\alpha)}$, and a "noise" term, $\mathbf{W}^s\,\delta\mathbf{u}$. We anticipate that the greater the signal-to-noise ratio, the greater the attraction around $\mathbf{u}^{(\alpha)}$. Let the Hamming distance between $\mathbf{u}$ and $\mathbf{u}^{(\alpha)}$, $d_H(\mathbf{u}, \mathbf{u}^{(\alpha)})$, equal d, i.e., $\|\delta\mathbf{u}\| = 2\sqrt{d}$. The (strong) norm of the matrix $\mathbf{W}^s$ is defined as

$$\|\mathbf{W}^s\| = \sup_{\mathbf{x}} \frac{\|\mathbf{W}^s\mathbf{x}\|}{\|\mathbf{x}\|} \qquad \|\mathbf{x}\| \neq 0$$

It follows (cf. Strang, 1980) that $\|\mathbf{W}^s\| = \sqrt{k}$, where k is the largest eigenvalue of the matrix $(\mathbf{W}^s)^T\mathbf{W}^s$. For the case of the degenerate spectrum,

$$\lambda^{(1)}, \ldots, \lambda^{(m)} = \lambda > 0,$$

$\mathbf{W}^s$ is symmetric, and $(\mathbf{W}^s)^T\mathbf{W}^s = (\mathbf{W}^s)^2$. Therefore the maximum eigenvalue of $(\mathbf{W}^s)^T\mathbf{W}^s = k = \lambda^2$, and the signal-to-noise ratio (SNR) is given by

$$\text{SNR} = \frac{\|\mathbf{W}^s\mathbf{u}^{(\alpha)}\|}{\|\mathbf{W}^s\,\delta\mathbf{u}\|} \geq \frac{\lambda^{(\alpha)}\sqrt{n}}{(\sqrt{k}(2\sqrt{d})} = \frac{1}{2}\sqrt{\frac{n}{d}}$$

Thus, we would expect the attraction sphere around $\mathbf{u}^{(1)}, \ldots, \mathbf{u}^{(m)}$ to increase as n increases for the m-fold degenerate spectral scheme. For the general nondegenerate case, we expect that by varying the size of $\lambda^{(\alpha)}$, the SNR, and hence the attraction capability, will be proportionately increased or decreased for the αth memory $\mathbf{u}^{(\alpha)}$ (Fig. 12.1).

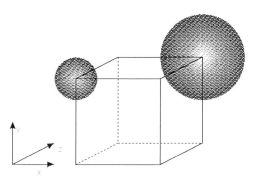

Fig. 12.1. Schematic representation of the directional attraction space around two memories with different eigenvalues with spectral algorithms.

12.3.2. Capacity and Complexity

To determine the capacity of these schemes, we use the following proposition (Komlós, 1967).

PROPOSITION 4. *Let m increase with n such that $m \leq n$. Then the probability that a randomly chosen set of n-tuples $\mathbf{u}^{(1)}, \ldots, \mathbf{u}^{(m)} \in \mathbb{B}^n$ is linearly independent approaches 1 as $n \to \infty$.*

It follows that the probability that $\mathbf{W}^s$ is well-defined approaches 1 as $n \to \infty$. Since a linear transformation has at most n eigenvalues, the static capacity of the spectral scheme is n. (Dynamic capacities pose greater problems in evaluation here because of the added dependency structure. For some theoretical results in this regard, see Dembo (1989); for numerical simulations see Venkatesh and Psaltis (1989) and Chapter 1 of this volume.)

Let N^s denote the number of elementary operations required to compute the weight matrix $\mathbf{W}^s$ directly from the m memories to be stored. Then using the fact that $(\mathbf{U}^T\mathbf{U})^{-1}$ is symmetric, we can use the Cholesky decomposition to compute its inverse. This along with the rest of the matrix multiplications yields $N^s = mn^2 + m^2n + m^3/2 + O(n^2)$.

When the eigenvalues $\lambda^{(\alpha)}$ are m-fold degenerate, Greville's algorithm can be used to recursively compute the pseudo-inverses, which in turn results in a recursive construction for the weight matrices $\mathbf{W}^s[\alpha]$. Here $\mathbf{W}^s[\alpha]$ denotes the spectral weight matrix corresponding to the first α memories. In fact, let $\lambda^{(\alpha)} = \lambda > 0, \alpha \geq 1$. For each $\alpha \geq 1$ let $\mathbf{e}^{(\alpha)}$ be the n-vector defined by

$$\mathbf{e}^{(\alpha)} = (\lambda\mathbf{I} - \mathbf{W}^s[\alpha - 1])\mathbf{u}^{(\alpha)}$$

where we define $\mathbf{W}^s[0] = \mathbf{0}$. Then it is easy to verify by induction that

$$\mathbf{W}^s[\alpha] = \mathbf{W}^s[\alpha - 1] + \frac{\mathbf{e}^{(\alpha)}(\mathbf{e}^{(\alpha)})^T}{(\mathbf{u}^{(\alpha)})^T\mathbf{e}^{(\alpha)}} \qquad \alpha \geq 1 \quad (6)$$

Now let $N^s[\alpha]$ denote the number of elementary operations needed to compute the update of the weight matrix according to the recursion (6). Again counting the number of multiplications (the number of additions is of the same order), we get the cost estimate $N^s[\alpha] = 2n^2 + 2n, \alpha \geq 1$. Note that for all choices of $m < n$, we have $mN^s[\alpha] \gtrsim 2N^s$, so that, especially for large n, the recursive construction of $\mathbf{W}^s$ through the updates (6) is computationally about twice as expensive as the direct estimation of $\mathbf{W}^s$. Note that the cost is only about four times more than that of the simple outer product algorithm.

12.4. FEATURE-SELECTIVE ALGORITHMS

12.4.1. Dual Spectral Algorithm

The following scheme, formally related to the outer product and spectral algorithms, was introduced by Maruani et al. (1987). Let $\mathbf{U} = [\mathbf{u}^{(1)}\,\mathbf{u}^{(2)}\cdots\mathbf{u}^{(m)}]$ be the matrix of memories as before. Let $\mathbf{x}^{(\beta)}$, $\beta = 1, \ldots, n - m$, be a set of linearly independent vectors in $\mathbb{R}^n$ which are individually orthogonal to each of the memories, i.e., $\mathbf{X}^{\mathrm{T}}\mathbf{U} = \mathbf{0}$, where we define the $n \times (n - m)$ matrix $\mathbf{X} = [\mathbf{x}^{(1)}\,\mathbf{x}^{(2)}\cdots\mathbf{x}^{(n-m)}]$. Define a weight matrix $\mathbf{W}$ with weights w_{ij} given by

$$w_{ij} = \begin{cases} -\sum_{\beta=1}^{n-m} x_{i\beta} x_{j\beta} & \text{if } i \neq j \\ 0 & \text{if } i = j \end{cases}$$

where $x_{k\beta}$ is the kth component of $\mathbf{x}^{(\beta)}$. If we define $\hat{\mu}_i = \sum_{\beta=1}^{n-m} x_{i\beta}^2$, $i = 1, \ldots, n$, we see that

$$\mathbf{W} = \hat{\mathbf{M}} - \mathbf{X}\mathbf{X}^{\mathrm{T}} \qquad (7)$$

where $\hat{\mathbf{M}} = \mathbf{dg}[\hat{\mu}_1, \ldots, \hat{\mu}_n]$. Thus,

$$\mathbf{W}\mathbf{U} = \hat{\mathbf{M}}\mathbf{U} - \mathbf{X}\mathbf{X}^{\mathrm{T}}\mathbf{U}$$
$$= \hat{\mathbf{M}}\mathbf{U} \qquad (8)$$

12.4.2. Error Correction

Comparing (5) and (8) we see that the spectral and dual spectral algorithms exhibit an interesting duality.

In the spectral scheme, the eigenvectors of $\mathbf{W}^s$ are the memories, so that the column space of $\mathbf{W}^s$ is given by the span of the memories. Therefore, if the memories are far enough from each other and the initial state vector $\mathbf{u}$ is close enough to a memory, $\mathbf{W}^s$ combined with the thresholding operation *projects* $\mathbf{u}$ onto the memory.

In the dual spectral scheme, since the parameters $\hat{\mu}_i$ are positive for each choice of i, it follows that

$$f(\mathbf{W}\mathbf{u}^\alpha)_i = f(\hat{\mu}_i u_i^\alpha) = u_i^\alpha$$
$$\text{for each } i = 1, \ldots, n, \ \alpha = 1, \ldots, m$$

So the memories $\mathbf{u}^{(1)}, \ldots, \mathbf{u}^{(m)}$ are fixed points in the scheme as well. $\mathbf{W}$ as defined in (7) is a zero-diagonal symmetric matrix. Thus, we know that there exist Liapunov functions in both modes of operation and that the network will exhibit some form of attraction behavior. The weight matrix $\mathbf{W}^d$ is obtained by taking the correlation of vectors that are orthogonal to the memories and then setting the diagonal elements to be 0. In creating the zero diagonal, we essentially add

perturbations to the left nullspace of $\mathbf{U}$ in the directions of the memories. The *strength* of the perturbations along any component, i, is proportional to $\hat{\mu}_i$. Thus, each of the $\hat{\mu}_i$ corresponds to a directional distortion, and we expect the SNR of the dual spectral scheme to vary from direction to direction proportionately with the value of $\hat{\mu}_i$. We therefore expect that the larger the $\hat{\mu}_i$, the more the information that is lost if the ith bit is flipped and, hence, the smaller the attraction in the ith direction.

As an illustration, let us consider the case where $n = 3$, and $\mu_y \gg \mu_x, \mu_z$. Each memory $\mathbf{u}$ would be preferentially attracted in the x and z direction, indicated schematically by an attraction spheroid in Fig. 12.2; i.e., a vector with a different y component is less likely to map back to $\mathbf{u}$ but vectors with different x and/or z components will probably be within the attraction region of $\mathbf{u}$. In other words,

$$P\left[\begin{pmatrix} u_x \\ -u_y \\ u_z \end{pmatrix} \to \begin{pmatrix} u_x \\ u_y \\ u_z \end{pmatrix}\right] < P\left[\begin{pmatrix} u_x \\ u_y \\ -u_z \end{pmatrix} \to \begin{pmatrix} u_x \\ u_y \\ u_z \end{pmatrix}\right],$$

$$P\left[\begin{pmatrix} -u_x \\ u_y \\ u_z \end{pmatrix} \to \begin{pmatrix} u_x \\ u_y \\ u_z \end{pmatrix}\right]$$

12.4.3. Constructing Feature-Selective Weights

In the previous section, the orthogonal basis, $\mathbf{X}$, was chosen arbitrarily and therefore resulted in some lack of control in specifying attraction capability. As we argued above, the $\hat{\mu}_i$'s essentially control directional attraction and we have no means of specifying these under the above approach. We now suggest a few *schemes for constructing the weight matrices that specify the μ-values and thereby achieve direction-specific*

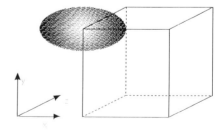

Fig. 12.2. Schematic representation of the directional attraction space in the dual spectral scheme with $\mu_y \gg \mu_x, \mu_z$.

attraction. Specifically, for a *prescribed set* $\mu_1, \ldots, \mu_n > 0$ of directional attraction strengths, and $\mathbf{M} = \mathbf{dg}[\mu_1, \ldots, \mu_n]$, we require a weight matrix $\mathbf{W}^d$ such that

$$\mathbf{W}^d \mathbf{U} = \mathbf{MU} \qquad (9)$$

We define $\mathbf{W}^d$ such that

$$w_{ij}^d = \begin{cases} -\sum_{\beta=1}^{n-m}(x_{i\beta}b_\beta)(x_{j\beta}b_\beta) & \text{if } i \neq j \\ 0 & \text{if } i = j \end{cases} \qquad (10)$$

where $x_{i\beta}$ is the ith component of the basis vector $\mathbf{x}^{(\beta)}$ as defined earlier, and b_β is the βth component of a vector which we will specify shortly. Thus, given $\mu_1, \ldots, \mu_n$ we need to find a vector $\mathbf{b}$ such that with $\mathbf{Y} = \mathbf{Xb}$

$$\mathbf{W}^d = \mathbf{M} - \mathbf{YY}^\mathsf{T} \qquad (11)$$

(Note that the columns of $\mathbf{Y}$, in general, are not orthogonal.)

Assuming that $\mathbf{W}^d$ has the form given in (10), let us now consider the effect of $\mathbf{W}^d$ on the ith element of a memory $\mathbf{u}^{(\alpha)}$:

$$[\mathbf{W}^d \mathbf{u}^{(\alpha)}]_i = \sum_{j=1}^{n} w_{ij}^d u_j^{(\alpha)}$$

$$= -\sum_{j \neq i} \sum_{\beta=1}^{n-m} b_\beta^2 x_{i\beta} x_{j\beta} u_j^{(\alpha)}$$

$$= -\sum_{j=1}^{n} \sum_{\beta=1}^{n-m} b_\beta^2 x_{i\beta} x_{j\beta} u_j^{(\alpha)} + \sum_{\beta=1}^{n-m} b_\beta^2 x_{i\beta}^2 u_i^{(\alpha)}$$

$$= \sum_{\beta=1}^{n-m} b_\beta^2 x_{i\beta}^2 u_i^{(\alpha)}$$

We require from (9) that $[\mathbf{W}^d \mathbf{u}^{(\alpha)}]_i = \mu_i u_i^{(\alpha)}$ where $\mu_i > 0$. By inspection, we obtain the relationship

$$\mu_i = \sum_{\beta=1}^{n-m} x_{i\beta}^2 b_\beta^2$$

Define $a_{i\beta} = x_{i\beta}^2$, and $c_\beta = b_\beta^2$. Then we require $\mathbf{Ac} = \mathbf{M}_\mu$, where $\mathbf{A}$ is a known $n \times (n-m)$ matrix with nonnegative elements $a_{i\beta} = x_{i\beta}^2$, $\mathbf{c}$ is an unknown $(n-m)$-dimensional vector with $c_\beta = b_\beta^2$ constrained to be nonnegative, and $\mathbf{M}_\mu$ is a specified n-dimensional vector with positive components $\mu_1, \ldots, \mu_n$.

We notice that this is an overspecified system of n equations with $(n-m)$ unknowns, where both $\mathbf{c}$ and $\mathbf{M}_\mu$ are constrained to have nonnegative elements. Linear programming techniques can be used to solve this system of equations. We can choose the μ-values in a variety of ways. A few representative methods are:

1. Specify $\mu_1, \ldots, \mu_k$, $k \leq n-m$, let $\mu_k + 1, \ldots, \mu_n - m < \varepsilon$, and minimize ε.

2. Specify $\mu_1, \ldots, \mu_n$, and
   (a) Minimize the mean-square error given by

$$\|\mathbf{Ac} - \mathbf{M}_\mu\|^2$$

$$= \sum_{i=1}^{n} (a_{i,1}c_1 + \cdots + a_{i,n-m}c_{n-m} - \mu_i)^2$$

subject to the constraints $\mathbf{M}_\mu > 0$, $\mathbf{c} > 0$.
   (b) Minimize the largest absolute error, c_0, given by

$$\max(|\varepsilon_1|, \ldots, |\varepsilon_n|)$$

where ε_i, the error in μ_i, is

$$\mu_i - (a_{i,1}c_1 + \cdots + a_{i,n-m}c_{n-m})$$

$$i = 1, \ldots, n$$

For simplicity, we consider algorithms employing the first linear programming approach outlined above. We have modified the initial basis for the nullspace of $\mathbf{U}$ using the results of the simplex method such that

$$\mathbf{W}^d = \mathbf{M} - \mathbf{YY}^\mathsf{T}$$

where $\mathbf{M} = \mathbf{dg}[\mu_1, \mu_2, \ldots, \mu_n]$ with $\mu_1, \ldots, \mu_k > 0$ specified by us,

$$0 < \mu_{k+1}, \ldots, \mu_n \leq \varepsilon < \min(\mu_1, \ldots, \mu_k)$$

and $\mathbf{Y} = \mathbf{Xb}$ is a set of basis vectors for the left nullspace of $\mathbf{U}$. Since μ_i, $i = 1, \ldots, n$, are positive, we see that all the memories are strictly stable in the dual spectral scheme as long as the memories, $\mathbf{u}^{(1)}, \ldots, \mathbf{u}^{(m)}$, are linearly independent, and we are able to find the vector $\mathbf{c}$ in the system (12) through linear programming. In Appendix B we outline the linear programming approach in some detail.

12.4.4. Capacity and Complexity

By Komlós's result (Proposition 4), we are guaranteed that almost all choices of n memories or fewer are linearly independent, so that for almost all choices of $n-1$ memories there is an orthogonal subspace of dimension 1, while almost all choices of n memories span the space $\mathbb{R}^n$ and therefore the orthogonal subspace is of dimension 0. The storage capacity of the dual spectral scheme of (10) is directly $n-1$, since $n-1$ is the number of memories for which we can still specify a left nullspace $\mathbf{X}$.

To find an n-dimensional vector under constraints, the simplex method iterates from one feasible solution to another until it finds an optimal feasible solution. In the worst case, we need to test each vertex of the feasible region, which is an n-sided polyhedron, leading to $2^n - 1$ iterations. However, such cases are rare and require careful specification of the constraints designed solely to approach the worst-case behavior. In practice, it has been widely reported (Chvátal, 1983; Murty, 1983) that the number of iterations is almost always between 1 and 3 times the number of constraints. Thus, for the case of specifying k values of $\mathbf{M}_\mu$, we would expect at the most $3n$ iterations. For the (revised) simplex method, a good estimate of the average cost of each iteration in our scheme is $52n - 10m - 10k + 10$, while for the standard simplex method, a good estimate is $2n^2 - mn - kn + n)/4$ (cf. Chvátal, 1983, p. 113). Thus, we estimate that the total cost of specifying k values of $\mathbf{M}_\mu$ is $O(n^2)$ (using the revised simplex method). The cost of finding a basis for the nullspace of $\mathbf{U}$ (through Gram–Schmidt orthogonalization) includes finding $(\mathbf{U}^T\mathbf{U})^{-1}$ and two matrix multiplications and is given by $mn^2 + (m^2n)/2 - m^3/2 + O(n^2)$. Finally, the cost of finding $\mathbf{W}^d$ from $\mathbf{c}$ and $\mathbf{X}$ is $n^3 - n^2m + O(n^2)$. So, we can say that on the average

$$N^d = n^3 + \tfrac{1}{2}m^2n - mn^2 - \frac{m^3}{2} + O(n^2)$$

where N^d is the number of elementary operations needed to compute $\mathbf{W}^d$.

12.5. COMPOSITE ALGORITHMS

In Section 12.3 we saw ways of increasing the radii of attraction-spheres around memories. In Section 12.4.3 we saw ways of specifying increased attraction in certain directions around each of the memories. A natural extension of these schemes is to create a composite scheme (Venkatesh et al., 1990) with weight matrix $\mathbf{W}^c$ given by

$$\mathbf{W}^c = \mathbf{W}^s + \mathbf{W}^d$$

Since $\mathbf{W}^c$ is a linear combination of $\mathbf{W}^s$ and $\mathbf{W}^d$, we would expect memories to be stable in the composite scheme for reasons described in the previous sections. The idea of the composite scheme is to specify both memory-specific attraction by specifying λ for each memory, and direction-specific attraction μ for the individual directions (Fig. 12.3).

Here, the spectrum of $\mathbf{W}^s$ is no longer degenerate, and $\mathbf{W}^c$, consequently, is no longer symmetric. As the composite algorithm combines the memory-specific spectral algorithm, and the direction-specific dual spectral algorithm, it works effectively in shaping the attraction regions as desired. It should be noted that the relative values of the $\lambda^{(1)}, \ldots, \lambda^{(m)}$, compared to the $\mu_1, \ldots, \mu_n$, need to be considered in order not to lose the effects of one of the two parts of the composite scheme.

Note that the capacity of the composite scheme is $n - 1$. The algorithm complexity of the composite scheme is the sum of the complexities of the spectral and dual spectral schemes, except that we need not find $(\mathbf{U}^T\mathbf{U})^{-1}$ twice. Therefore the complexity, N^c, is given by $3n^3 + O(n^2)$ for $m \lesssim n$.

12.6. SIMULATIONS

There are a number of open questions involved with the schemes outlined above. In this section, we discuss trends observed in computer simulations that validate some of our conjectures. All memories were chosen randomly using a binomial pseudo-random generator. Test input vectors at specified Hamming distances from the memories were generated by reversing the signs of randomly chosen components for the outer product and spectral schemes. In the case of the dual spectral schemes, test input vectors were generated by reversing the signs of randomly chosen components with the specified or unspecified μ values.

Analytical bounds are difficult to arrive at for the attraction radius as a function of the number of memories and the dimension of the state space. Figure 12.4 plots the attraction radius for the

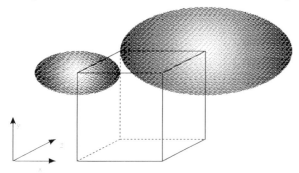

Fig. 12.3. Schematic representation of the joint memory-specific and direction-specific attraction space for two memories in the composite scheme.

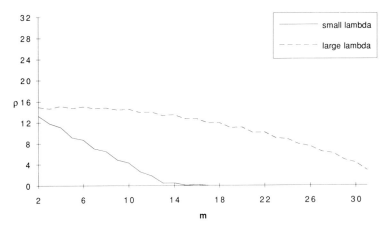

Fig. 12.4. Attraction radii in the spectral scheme.

spectral scheme. The attraction radius was estimated by averaging the maximum Hamming distance of error correction around stable memories over several independent runs. The memories were divided into two equal-sized groups, one group with eigenvalue λ(large) $= 3n$, and the other group with eigenvalue λ(small) $= n$. The respective attraction radii of the λ(large) memories and the λ(small) memories are plotted against the number of memories m.

In the dual spectral schemes, there is the question of the number of directions, k, that can be specified given a set of m memories and n neurons, arising from the nature of the construction of the $\mathbf{W}^d$ matrix. It is obvious from previous discussions about the dimensions of $\mathbf{A}$ and $\mathbf{c}$ that we can surely specify no more than

$n - m$ directions. However, there is a possibility (albeit small) that there exist no feasible solutions for pathological cases where $k < n - m$. Another quantity we are interested in is the size of ε, the largest of the unspecified μ's, compared to the size of the specified μ's, since we have conjectured that this will affect directional attraction.

While there exists little theory for the simplex method that will enable us to gauge these parameters, simulations show that ε is typically small (Fig. 12.5) compared to μ_i for the specified directions ($< 0.5 \mu_i$), and k is typically of the order of $n/4$ in the ranges simulated. We conjecture that this behavior continues to hold for large n. Figures 12.6a and 12.6b plot directional attraction for the dual spectral scheme. Attraction data for a given direction were generated by investigating

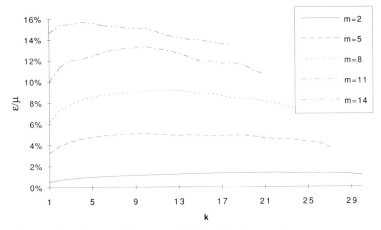

Fig. 12.5. Variation of ε, the largest of the unspecified directional parameters, $\mu_{k+1}, \ldots, \mu_n$, as a ratio of the specified directional parameters with k, the number of directions specified.

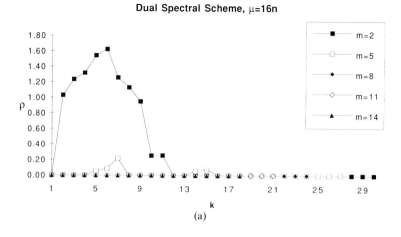

(a)

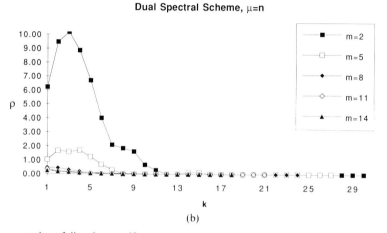

(b)

Fig. 12.6. Demonstration of direction-specific attraction in the dual spectral scheme. (a) Large μ and (b) small μ.

probe vectors at various Hamming distances from a memory with the component of the probe in the direction being investigated being chosen to be opposite in sign to the corresponding component of the memory. Flipping a bit in an important (large μ) direction almost always reduced the attraction to the memory compared to an unimportant (small μ) direction.

Figures 12.7a and 12.7b plot the results for the composite scheme. The memories were divided into two groups, one group corresponding to a "large" eigenvalue, $\lambda_{lg} = 3$, and the other group corresponding to a "small" eigenvalue, $\lambda_{sm} = 1$. Directional attraction parameters for k directions were set equal to $\mu_{lg} = 16$. Attraction radii were determined for the large μ (specified) and small μ (unspecified) directions as a function of k, for both the large-eigenvalue memories and the small-eigenvalue memories. As can be seen, there

is degradation of the behavior when compared to the memory-specific and feature-specific schemes. However, the attraction behavior is in keeping with our expectations, and we conjecture that the behavior of these networks improves as n increases.

12.7. SUMMARY

The recurrent neural network paradigm for associative memory is attractive in its simplicity and computational tractability. The classical outer product algorithm, for instance, has very low implementation complexity and yet exhibits near-linear memory storage capacities with correction of a linear number of random errors uniformly in balls around the memories. In this chapter we have shown how it is possible to

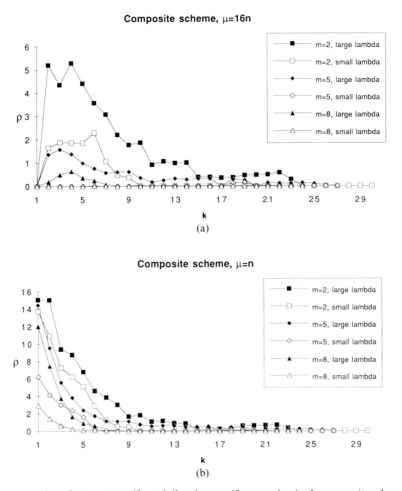

Fig. 12.7. Demonstration of memory-specific and direction-specific attraction in the composite scheme. (a) Large μ and (b) small μ.

exercise a macroscopic degree of nonuniform error correction around the stored memories. In particular, both memory- and feature-selective error correction is feasible using a composite algorithm which exploits the spectral characteristics of the interconnectivity weight matrix. These spectral-based approaches are near-optimal in character and, in particular, are characterized by low implementation complexities and linear storage capacities. Extensions of these approaches are possible to higher-order neural networks (where the model neurons compute the sign of polynomial forms of their inputs) with concomitant increases in the storage capacities of the networks (cf. Venkatesh and Baldi, 1991a, b).

APPENDIX A: Propositions

PROPOSITION 1. $E(\mathbf{u})$ *is nonincreasing in asynchronous mode if* $\mathbf{W}$ *is symmetric and has nonnegative diagonal elements;* $E(\mathbf{u})$ *is nonincreasing in any mode if* $\mathbf{W}$ *is symmetric and nonnegative definite.*

PROOF. (a) Consider either a synchronous or an asynchronous mode of operation. For any state $\mathbf{u}$, the algorithm results in a flow in state space defined by $\mathbf{u} \mapsto \mathbf{u} + \delta\mathbf{u}$. We note that $\delta\mathbf{u}$ is an n-tuple with each component taking on one of the values 0, -2, or 2. The change in E is given by

$$\delta E = E(\mathbf{u} + \delta\mathbf{u}) - E(\mathbf{u})$$

$$= -\tfrac{1}{2}[\langle \delta\mathbf{u}, \mathbf{W}\mathbf{u} \rangle + \langle \mathbf{u}, \mathbf{W}\,\delta\mathbf{u} \rangle + \langle \delta\mathbf{u}, \mathbf{W}\,\delta\mathbf{u} \rangle]$$

Since $\mathbf{W}$ is symmetric, $\langle \delta\mathbf{u}, \mathbf{W}\mathbf{u}\rangle = \langle \mathbf{u}, \mathbf{W}\,\delta\mathbf{u}\rangle$. Hence, we have

$$\delta E = -\langle \delta\mathbf{u}, \mathbf{W}\mathbf{u}\rangle - \tfrac{1}{2}\langle \delta\mathbf{u}, \mathbf{W}\,\delta\mathbf{u}\rangle$$

We note that the nature of the algorithm is such that the sign of each component of $\delta\mathbf{u}$ is the same as that of the corresponding component of $\mathbf{W}\mathbf{u}$. Thus, the inner product $\langle \delta\mathbf{u}, \mathbf{W}\mathbf{u}\rangle \geq 0$ for every state vector $\mathbf{u}\in \mathbb{B}^n$. Furthermore, if $\mathbf{W}$ is nonnegative definite, the quadratic form $\langle \delta\mathbf{u}, \mathbf{W}\,\delta\mathbf{u}\rangle \geq 0$. Thus $\delta E \leq 0$ and E is a monotone nonincreasing function.

(b) In the synchronous mode, assume that the kth neuron updates itself at epoch t. We therefore have

$$u_i[t+1] - u_i[t] = \begin{cases} 0 & \text{if } i \neq k \\ -2u_k[t] & \text{if } i = k \end{cases}$$

The change in energy δE is given by

$$\delta E = -\delta u_k \left(\sum_{j=1}^{n} w_{kj}u_j\right) - \tfrac{1}{2}(\delta u_k)^2 w_{kk}$$

Once again, the algorithm ensures that δu_k is of the same sign as $\sum_{j=1}^{n} w_{kj}u_j$, and since w_{kk} is nonnegative, we see that $\delta E \leq 0$, and E is monotone nonincreasing. $\square$

PROPOSITION 2. $G(\mathbf{u})$ *is nonincreasing in synchronous mode if* $\mathbf{W}$ *is symmetric.*

PROOF. In the synchronous mode of operation the change in the Manhattan Norm is given by

$$\delta G = G(\mathbf{u}[t+1]) - G(\mathbf{u}[t])$$

$$= -\frac{1}{2}\sum_{i=1}^{n}\sum_{j=1}^{n} u_i[t+1]w_{ij}u_j[t]$$

$$+ \frac{1}{2}\sum_{p=1}^{n}\sum_{q=1}^{n} u_p[t]w_{pq}u_q[t-1]$$

As $\mathbf{W}$ is symmetric, we have $w_{pq} = w_{qp}$ and

$$\sum_{p=1}^{n} u_p[t]w_{qp} = \sum_{j=1}^{n} w_{ij}u_j[t].$$

So

$$\delta G = -\frac{1}{2}\sum_{i=1}^{n}(u_i[t+1] - u_i[t-1])\sum_{j=1}^{n} w_{ij}u_j[t]$$

Let I be the set of indices for which $u_i[t+1] = -u_i[t-1]$. (Note that I can be empty.) Then

$$\delta G = -\sum_{i\in I} u_i[t+1]\sum_{j=1}^{n} w_{ij}u_j[t]$$

Once again we note that the nature of the algorithm guarantees us that the two sums in the above equation are of the same sign. Thus,

$$\delta G = -\sum_{i\in I}\left|\sum_{j=1}^{n} w_{ij}u_j[t]\right| \leq 0$$

and G is a monotone nonincreasing function along trajectories in state space. $\square$

PROPOSITION 3. *The (stable state) of the outer product algorithm is* $n/4 \log n$.

PROOF. We will prove only that $n/4 \log n$ is a lower bound for the stable state capacity of the outer product algorithm. Consider the random sums

$$X_n^{(i,\alpha)} = u_i^{(\alpha)}\sum_{j=1}^{n} w_{ij}^{\text{op}}u_j^{(\alpha)}$$

$$= n + m - 1 + \sum_{j\neq i}\sum_{\beta\neq\alpha} u_i^{(\alpha)}u_i^{(\beta)}u_j^{(\alpha)}u_j^{(\beta)}$$

$$1 \leq i \leq n \qquad 1 \leq \alpha \leq m$$

Fix i and α, and write simply X_n instead of $X_n^{(i,\alpha)}$. We can now write

$$X_n = n + m - 1 + \sum_{\beta\neq\alpha}\sum_{j\neq i} Z_j^{(\beta)}$$

where, for fixed i and α, we define the random variables $Z_j^{(\beta)} = u_i^{(\alpha)}u_i^{(\beta)}u_j^{(\alpha)}u_j^{(\beta)}$. Note that the random variables $Z_j^{(\beta)}$, $j\neq i$, $\beta\neq\alpha$ are i.i.d., symmetric, ± 1 random variables.[5] Now recall that a simple application of Chebyshev's inequality yields that for any random variable Y and any $t \geq 0$,

$$P\{Y \leq -t\} \leq \inf_{r\geq 0} e^{-rt}E(e^{-rY})$$

Applying this result here we obtain

$$P\{X_n \leq 0\} = P\left\{\sum_{\beta\neq\alpha}\sum_{j\neq i} Z_j^{(\beta)} \leq -n - m + 1\right\}$$

$$\leq \inf_{r\geq 0} e^{-r(n+m-1)}E\left\{e^{-r\sum_{\beta\neq\alpha}\sum_{j\neq i} Z_j^{(\beta)}}\right\}$$

$$= \inf_{r\geq 0} e^{-r(n+m-1)}E\left\{\prod_{\beta\neq\alpha}\prod_{j\neq i} e^{-rZ_j^{(\beta)}}\right\}$$

The terms in the product, $e^{-rZ_j^{(\beta)}}$, $\beta\neq\alpha$, $j\neq i$ are independent random variables as the random variables $Z_j^{(\beta)}$ are independent. The expectation of the product of random variables above can, hence, be replaced by the product of expectations. Accordingly, denoting by Z a random variable which takes on values -1 and 1 only, each with probability $1/2$, we have

$$P\{X_n \leq 0\} \leq \inf_{r\geq 0} e^{-r(n+m-1)}[E(e^{-rZ})]^{(m-1)(n-1)}$$

$$= \inf_{r\geq 0} e^{-r(n+m-1)}(\cosh r)^{(m-1)(n-1)}$$

Now, for every $r\in\mathbb{R}$ we have $\cosh r \leq e^{r^2/2}$. Hence

$$P\{X_n \leq 0\} \leq \inf_{r\geq 0}\exp\left(\frac{r^2(m-1)(n-1)}{2} - r(n+m-1)\right)$$

$$= \exp\left(-\frac{(n+m-1)^2}{2(n-1)(m-1)}\right)$$

We hence obtain that the probability that any given

---

[5] The critical fact here is that each random variable $Z_j^{(\beta)}$ has a distinct multiplicative term $u_j^{(\beta)}$ which occurs solely in the expression for $Z_j^{(\beta)}$.

component of a memory is not stable is bounded by

$$P\{X_n^{(1,\alpha)} \leq 0\} \leq e^{-n/2m}$$

for large enough n provided m grows so that $m = o(n)$ and $m = \omega(\sqrt{n})$.

Denote the event

$$\mathscr{E}_n = \left\{ \bigcup_{i=1}^{n} \bigcup_{\alpha=1}^{m} \{X_i^{(i,\alpha)}\} \leq 0 \right\}$$

that one or more of the nm memory components is not stable. A simple application of Boole's inequality now yields

$$P\{\mathscr{E}_n\} \leq \sum_{i=1}^{n} \sum_{\alpha=1}^{m} P\{X_n^{(i,\alpha)} \leq 0\} \leq nm \exp\left\{-\left(\frac{n}{2m}\right)\right\}$$

For a choice of

$$m = \frac{n}{4 \log n} \left[1 + \frac{\log \log n + \log 4\varepsilon}{2 \log n} - o\left(\frac{\log \log n}{(\log n)^2}\right)\right]$$

we hence obtain $P\{\mathscr{E}_n\} \leq \varepsilon$ as $n \to \infty$. As the probability that each of the memories is stable is exactly $1 - P\{\mathscr{E}_n\}$, this establishes that the capacity is at least $n/4 \log n$. $\square$

PROPOSITION 4. *Let m increase with n such that $m \leq n$. Then the probability that a randomly chosen set of n-tuples $\mathbf{u}^{(1)}, \ldots, \mathbf{u}^{(m)} \in \mathbb{B}^n$ is linearly independent approaches 1 as $n \to \infty$.*

This result follows directly from a result of Komlós (1967) asserting that the probability that a random $n \times n \pm 1$ matrix is nonsingular approaches 1 as $n \to \infty$.

APPENDIX B: Linear Programming

1. Specify $\mu_1, \ldots, \mu_k$, $k \leq n - m$, let

$$\mu_k + 1, \ldots, \mu_n - m < \varepsilon$$

and minimize ε. The canonical form of the linear programming problem that the simplex method solves is: Minimize the *goal function* $\mathbf{c}^T \mathbf{y}$ subject to the constraints

$$\mathbf{A}\mathbf{y} = \mathbf{b}$$

where the vector $\mathbf{y}$ is unknown and $\mathbf{y} > \mathbf{0}$.

In this case, we specify k positive values of $\mathbf{M}_\mu$ and minimize the maximum of the $(n - k)$ unspecified values of $\mathbf{M}_\mu$ subject to the constraints $\mu_{k+1}, \ldots, \mu_n > 0$, and $c_1, \ldots, c_{n-m} > 0$. In other words, we have the following equations:

$$a_{1,1}c_1 + \cdots + a_{1,n-m}c_{n-m} = \mu_1$$
$$\vdots$$
$$a_{k,1}c_1 + \cdots + a_{k,n-m}c_{n-m} = \mu_k$$
$$a_{k+1,1}c_1 + \cdots + a_{k+1,n-m}c_{n-m} \leq \varepsilon$$
$$\vdots$$
$$a_{n,1}c_1 + \cdots + a_{n,n-m}c_{n-m} \leq \varepsilon$$

where $c_i \geq 0$, $\varepsilon > 0$, and we want to find $\mathbf{c}$ which minimizes ε.

To convert the $n - k$ inequalities to equalities, we subtract ε from both sides of the equation and add *slack variables* $z_1, \ldots, z_{n-k}$ to give us the following $n - k$ equations:

$$a_{k+1,1}c_1 + \cdots + a_{k+1,n-m}c_{n-m} - \varepsilon + z_1 = 0$$
$$\vdots$$
$$a_{n,1}c_1 + \cdots + a_{n,n-m}c_{n-m} - \varepsilon + z_{n-k} = 0$$

in addition to the first k equations. Now we have n equations with $2n - m - k$ unknown quantities $(c_1, \ldots, c_{n-m}, z_1, \ldots, z_{n-k})$. Let us label ε as c_0. By inspection, we see that the goal function to be minimized is c_0, subject to the constraints $\mathbf{A}'\mathbf{c}' = \mathbf{M}'_\mu$ where $\mathbf{c}'$ is a $(2n - m - k + 1)$-dimensional vector, $\mathbf{M}'_\mu$ is an n-dimensional vector, and $\mathbf{A}'$ is an

$$n \times (2n - m - k + 1)$$

matrix; i.e., we require to solve

$$
\begin{pmatrix}
0 & & 0 & \\
\vdots & & & \ddots \\
0 & & & 0 \\
-1 & \mathbf{A} & 1 & \\
\vdots & & & \ddots \\
-1 & & 1 &
\end{pmatrix}
\begin{pmatrix}
c_0 \\
c_1 \\
\vdots \\
c_{n-m} \\
z_1 \\
\vdots \\
z_{n-k}
\end{pmatrix}
=
\begin{pmatrix}
\mu_1 \\
\vdots \\
\mu_k \\
0 \\
\vdots \\
0
\end{pmatrix}
$$

(12)

and $c_i, z_i \geq 0$. This is in the canonical form for the simplex method.

2. Specify $\mu_1, \ldots, \mu_n$, and (a) minimize the mean-square error given by

$$\|\mathbf{A}\mathbf{c} - \mathbf{M}_\mu\|^2 = \sum_{i=1}^{n} (a_{i,1}c_1 + \cdots + a_{i,n-m}c_{n-m} - \mu_i)^2$$

subject to the constraints $\mathbf{M}_\mu > 0$, $\mathbf{c} > 0$.

This is a quadratic programming problem. However, this problem can be reformulated as a simplex method problem and can be solved using a variation of the traditional simplex method called Wolfe's method (Wolfe, 1959).

(b) Minimize the largest absolute error, c_0, given by

$$\max(|\varepsilon_1|, \ldots, |\varepsilon_n|)$$

where ε_i, the error in μ_i, is

$$\mu_i - (a_{i,1}c_1 + \cdots + a_{i,n-m}c_{n-m}) \qquad i = 1, \ldots, n$$

Our problem now is to minimize c_0 subject to $c_0, c_1, \ldots, c_{n-m} \geq 0$. To solve this problem,[6] we note that we have n pairs of inequality constraints of the

[6] This is known as Chebyshev's approximation (Franklin, 1980, p. 8).

of the form

$$-c_0 + a_{i,1}c_1 + \cdots + a_{i,n-m}c_{n-m} \le \mu_i$$

$$-c_0 - a_{i,1}c_1 - \cdots - a_{i,n-m}c_{n-m} \le -\mu_i$$

The addition of slack variables puts the problem in canonical form.

REFERENCES

Chvátal, V. (1983). *Linear Programming*. W. H. Freeman, New York.

Dembo, A. (1989). "On the Capacity of Associative Memories with Linear Threshold Functions," *IEEE Trans. Information Theory*, **35**, 709–720.

Franklin, J. (1980). *Methods of Mathematical Economics*. Springer-Verlag, New York.

Goles, E. and Vichniac, G. Y. (1986). "Lyapunov Functions for Parallel Neural Networks," in *Neural Networks for Computing*, J. Denker, ed. AIP, New York, No. 151, pp. 165–181.

Hopfield, J. J. (1982). "Neural Networks and Physical Systems with Emergent Collective Computational Abilities," *Proc. Nat. Acad. Sci. U.S.A.*, **79**, 2554–2558.

Komlós, J. (1967). "On the Determinant of (0, 1) Matrices," *Studia Scientarum Mathematicarum Hungarica*, **2**, 7–21.

Komlós, J. and Paturi, R. (1988). "Convergence Results in an Associative Memory Model," *Neural Networks*, **1**, 239–250.

Maurani, A. D., Chevalier, R. C., and Sirat, G. (1987). "Information Retrieval in Neural Networks. I. Eigenproblems in Neural Networks," *Rev. Phys. Appl.*, **22**, 1321–1325.

McCulloch, W. W. and Pitts, W. (1943). "A Logical Calculus of the Ideas Immanent in Neural Activity," *Bull. Math. Biophys.*, **5**, 115–133.

McEliece, R. J., Posner, E. C., Rodemich, E. R., and Venkatesh, S. S. (1987). "The Capacity of the Hopfield Associative Memory," *IEEE Trans. Information Theory*, **33**, 461–482.

Murty, K. G. (1983). *Linear Programming*. Wiley, New York.

Personnaz, L., Guyon, I., and Dreyfus, G. (1985). "Information Storage and Retrieval in Spin-glass-like Networks," *J. Physique Lett.*, **46**, L359–365.

Strang, G. (1980). *Linear Algebra and Its Applications*. Academic Press, New York.

Venkatesh, S. S. and Baldi, P. (1991a). "Programmed Interactions in Higher-order Neural Networks: Maximal Capacity," *J. Complexity*, **7**, 316–337.

Venkatesh, S. S. and Baldi, P. (1991b). "Programmed Interactions in Higher-order Neural Networks: The Outer Product Algorithm," *J. Complexity*, **7**, 443–479.

Venkatesh, S. S. and Psaltis, D. (1989). "Linear and Logarithmic Capacities in Associative Neural Networks," *IEEE Trans. Information Theory*, **35**, 558–568.

Venkatesh, S. S., Pancha, G., Psaltis, D., and Sirat, G. (1990). "Shaping Attraction Basins in Neural Networks," *Neural Networks*, **3**, 613–623.

Wolfe, P. (1959). "The Simplex Method for Quadratic Programming," *Econometrica*, **27**, 282–298.

13

Analysis of Dynamics and Capacity of Associative Memory Using a Nonmonotonic Neuron Model

SHUJI YOSHIZAWA, MASAHIKO MORITA, and SHUN-ICHI AMARI

13.1. INTRODUCTION

In the long history of studies on neural network models of associative memory from the early 1970s (see for example, Nakano, 1972; Kohonen, 1972; Anderson, 1972; Amari, 1972), researchers' attention has mainly been directed toward the improvement of memory storage, for example, the generalized-inverse memory matrix (Kohonen and Ruohonen, 1973; Amari, 1977), the optimal capacity using a general memory matrix (Gardner, 1988), the sparsely encoded associative memory (Willshaw et al., 1969; Palm, 1980; Amari, 1989; Meunier et al., 1991), introduction of sparsity in connections (Yanai et al., 1991), and introduction of excitatory and inhibitory neurons (Shinomoto, 1987). (Also refer to Chapters 1 and 5–7 in this book.)

Recently, using computer simulation, Morita et al. (1990b) have shown that the performance of the Hopfield-type associative memory is improved remarkably by replacing the usual sigmoid neuron with a nonmonotonic neuron. An increase in memory capacity and the disappearance of spurious memories are the main improvements of this modification. These results are reviewed in Section 13.2.

In this chapter we adopt a piecewise linear model of the nonmonotonic neuron and investigate properties of the recall process such as the location and the stability of equilibrium states, and derive two kinds of theoretical estimates for the memory capacity (Yoshizawa et al., 1992).

13.2. DYNAMICS OF AUTOCORRELATION ASSOCIATIVE MEMORY AND ITS IMPROVEMENT

In this section we give a short summary of the autocorrelation associative memory of the Hopfield type and describe Morita's improvement of its recall process.

We consider a neural network of n neurons. The memorized patterns for the network are n-dimensional random vectors whose elements take the values ± 1 randomly with equal probability and are denoted by $\mathbf{s}^{(\mu)} = (s_1^{(\mu)} \, s_2^{(\mu)} \cdots s_n^{(\mu)})^{\mathrm{T}}$ ($\mu = 1, 2, \ldots, m$), where T denotes the transpose operator. By using these patterns, a memory matrix is constructed as follows,

$$\mathbf{W} = \frac{1}{n}\mathbf{S}\mathbf{S}^{\mathrm{T}} - a\mathbf{E}_n \qquad (1)$$

where

$$\mathbf{S} = [\mathbf{s}^{(1)} \, \mathbf{s}^{(2)} \cdots \mathbf{s}^{(m)}] \qquad (2)$$

$\mathbf{E}_n$ is the n-dimensional unit matrix, and a is the memory ratio defined by

$$a = \frac{m}{n} \qquad (3)$$

Since $\mathbf{s}^{(\mu)}$ is a random vector and $a < 1$, we can assume that the covariance matrix $\mathbf{S}^{\mathrm{T}}\mathbf{S}$ is nonsingular. Without loss of generality we define $\mathbf{s}^{(1)}$ as

$$\mathbf{s}^{(1)} = (1, \ldots, 1)^{\mathrm{T}} \qquad (4)$$

In the original Hopfield associative memory, the discrete-time recall process

$$\mathbf{z}(t + 1) = f(\mathbf{W}\mathbf{z}(t)) \qquad (5)$$

is used (Hopfield, 1982), where

$$\mathbf{z}(t) = (z_1(t), z_2(t), \ldots, z_n(t))^{\mathrm{T}}$$

represents the state of the network at time t and f is the output function of the neuron that operates on each element of its argument vector. On the other hand, we adopt the continuous-time recall dynamics as follows:

$$\frac{d\mathbf{u}}{dt} = -\mathbf{u} + \mathbf{W}f(\mathbf{u}) \qquad (6)$$

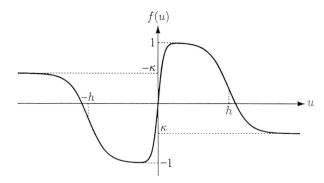

Fig. 13.1. Nonmonotonic output function.

where $\mathbf{u} = (u_1, u_2, \ldots, u_n)^T$ is an n-dimensional vector which represents the internal potential of neurons.

Conventional associative memories use a sigmoid output function; however, the Morita model uses a nonmonotonic function defined by

$$f(u) = \frac{1 - \exp(-cu)}{1 + \exp(-cu)} \cdot \frac{1 + \kappa \exp c'(|u| - h)}{1 + \exp c'(|u| - h)} \quad (7)$$

which is depicted in Fig. 13.1 (Morita et al., 1990a, b) for parameter values $c = 50$, $c' = 15$, $h = 1$, and $\kappa = -0.5$.

The recall process of an associative memory of the Hopfield type is as follows. For a given input pattern $\mathbf{u}_0$ we evaluate Eq. (6) with initial condition $\mathbf{u}_0$, and if the solution converges to an equilibrium state $\mathbf{u}_\infty$ the recalled pattern is determined by the signum of $\mathbf{u}_\infty$: $\mathrm{sgn}(\mathbf{u}_\infty)$. If the solution does not converge, we cannot determine a recalled pattern and we deem the recall failed.

Comparisons of the recall properties of associative memories with the conventional and the nonmonotonic neurons are shown in Figs. 13.2 and 13.3. Figure 13.2 shows time evolutions of the recall process with a memory ratio of 0.32, where abscissa and ordinate are the time and the direction cosine between the signum of $\mathbf{u}(t)$ and $\mathbf{s}^{(1)}$:

$$\frac{1}{n} \sum_{i=1}^{n} \{\mathrm{sgn}(u_i(t)) s_i^{(1)}\} = \frac{1}{n} \sum_{i=1}^{n} \mathrm{sgn}(u_i(t))$$

Figure 13.2a shows that the direction cosines for the conventional neuron converge to values far less than 1 even if their initial values are set close to 1. This means that the memory pattern $\mathbf{s}^{(1)}$ is unstable for this memory ratio. For a nonmonotonic neuron (Fig. 13.2b), however, it can be seen that there is a basin of attraction around $\mathbf{s}^{(1)}$; for this special example, the critical direction cosine is about 0.44. Figure 13.3 is a summary of the relation between the memory ratio and the critical direction cosine, from which

it can be seen that the capacity of the non-monotonic model (curve B) is more than $0.5n$ and that of the sigmoid model (curve A) is $0.15n$ at most. For the benefit of comparison, a result for the sigmoid neuron with the generalized-inverse matrix (refer to Chapter 1 for details on the generalized inverse-recorded associative memory) is also shown by curve C.

Lastly, we remark that the time evolution of $\mathbf{u}$ in Fig. 13.2b does not converge to an equilibrium state when memorized pattern $\mathbf{s}^{(1)}$ cannot be recalled. This suggests that spurious memories are rare for the nonmonotonic-neuron model.

13.3. PIECEWISE LINEAR MODEL AND EQUILIBRIUM STATE ANALYSIS

To study theoretically the above-mentioned properties of the nonmonotonic model we replace the nonmonotonic function shown in Fig. 13.1 by the piecewise linear function shown in Fig. 13.4. By doing so we can obtain the maximum value of the memory ratio for which Eq. (6) has a stable equilibrium in each quadrant of the memorized pattern. We call this maximum memory ratio the (absolute) capacity of the associative memory.

The recall dynamics for the associative memory using the piecewise linear neuron is

$$\left. \begin{array}{l} \dfrac{d\mathbf{u}}{dt} = -\mathbf{u} + \mathbf{W}\mathbf{x}(\mathbf{u}) \\[2mm] \mathbf{x}(\mathbf{u}) = \mathrm{sgn}(\mathbf{u}) - k\mathbf{u} \end{array} \right\} \quad (8)$$

We assume that

$$k = \frac{1}{a} \quad (9)$$

Then we have the following theorem.

THEOREM 1. *Let $\mathbf{u}$ be an equilibrium state of Eq. (8) in the quadrant of the memory pattern*

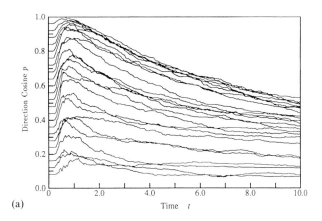

(a)

Fig. 13.2. Time sequences of recall process ($n = 1000$, $a = 0.32$). The abscissa and the ordinate are the time and the direction cosine between $\mathrm{sgn}(\mathbf{u}(t))$ and $\mathbf{s}^{(1)}$. Accordingly, the larger the direction cosine is, the closer to $\mathbf{s}^{(1)}$ the recalled pattern is. The direction cosines for the conventional neuron (a) converge to values far less than 1 even if their initial values are set close to 1. This fact means that the memory pattern $\mathbf{s}^{(1)}$ is unstable. For a nonmonotonic neuron (b), there is a basin of attraction around $\mathbf{s}^{(1)}$; for this special example, the critical direction cosine is about 0.44.

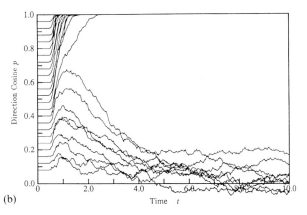

(b)

$\mathbf{s}^{(1)}$ and $\mathbf{x} = -k\mathbf{u} + \mathrm{sgn}(\mathbf{u})$. *Then $\mathbf{x}$ is characterized by the following three conditions,*

$$(a) \quad \sum_{i=1}^{n} x_i s_i^{(\mu)} = 0 \quad (\mu = 2, 3, \ldots, m)$$

$$(b) \quad \sum_{i=1}^{n} x_i s_i^{(1)} = \sum_{i=1}^{n} x_i = m \qquad (10)$$

$$(c) \quad x_i < 1 \quad (i = 1, 2, \ldots, n)$$

PROOF. The equilibrium state of Eq. (8) is determined by

$$-\mathbf{u} + \mathbf{W}\mathbf{x}(\mathbf{u}) = 0 \qquad (11)$$

In the quadrant of $\mathbf{s}^{(1)}$ it holds that $\mathrm{sgn}(\mathbf{u}) = \mathbf{s}^{(1)}$. By substituting this relation, the second equation of Eq. (8) and Eq. (1) into Eq. (11), we obtain

$$\frac{1}{k}\mathbf{s}^{(1)} - \left(\frac{1}{k} - a\right)\mathbf{x} = \frac{1}{n}\mathbf{S}\mathbf{S}^{\mathrm{T}}\mathbf{x}$$

By Eq. (9) the above equation is reduced to

$$\frac{1}{k}\mathbf{s}^{(1)} = \frac{1}{n}\mathbf{S}\mathbf{S}^{\mathrm{T}}\mathbf{x} \qquad (12)$$

This is rewritten as

$$\mathbf{S}\left(\frac{1}{k} - \frac{1}{n}\sum_{i=1}^{n} x_i s_i^{(1)} \quad -\frac{1}{n}\sum_{i=1}^{n} x_i s_i^{(2)} \quad \cdots \right.$$

$$\left. -\frac{1}{n}\sum_{i=1}^{n} x_i s_i^{(m)}\right)^{\mathrm{T}} = 0$$

Taking into account that the covariance matrix $\mathbf{S}^{\mathrm{T}}\mathbf{S}$ is nonsingular, we have (a) and (b) of Eq. (10). The condition (c) is equivalent to condition that $\mathbf{x}$ is in the quadrant of $\mathbf{s}^{(1)}$. □

The following geometrical interpretation of the problem described by Eq. (10) is useful for further analysis. Vectors $\mathbf{x}$ which satisfy condition (a) are contained in the ortho-complement of the subspace spanned by $\mathbf{s}^{(\mu)}$ ($\mu = 2, 3, \ldots, m$). Call this ortho-complement subspace Q. On the other hand, the vectors that satisfy conditions (b) and (c) form an n-polygon which is the intersection of the $(n-1)$-dimensional hyperplane defined by condition (b) and the semi-infinite region determined by $x_i \leq 1$ ($i = 1, 2, \ldots, n$). This region is the hatched area labeled B in Fig. 13.5. Let R be

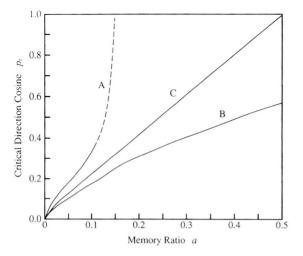

Fig. 13.3. The relation between the memory ratio and the critical direction cosine. It can be seen that the nonmonotonic model (curve B) is superior to the sigmoid model (curve A). For comparison, a result for the sigmoid neuron with the generalized-inverse matrix is also shown by curve C.

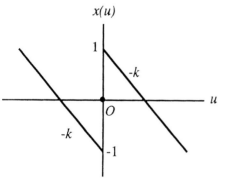

Fig. 13.4. Piecewise linear approximation to a non-monotonic output function.

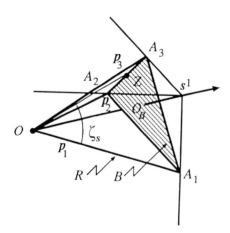

Fig. 13.5. Geometrical representation of an equilibrium state. The hatched area illustrates the region of $\mathbf{x}$ which is determined by $\sum_{i=1}^{n} x_i s_i^1 = m$, and $x_i < 1$.

an n-simplex whose vertices are those of B and the origin. Then the existence of an $\mathbf{x}$ that satisfies Eq. (10) is equivalent to the existence of an intersection of R and Q.

Since the $\mathbf{s}^{(\mu)}$ ($\mu = 1, 2, \ldots, m$) are assumed to be random vectors, the simplex R takes a random direction with respect to the subspace Q and we can determine the probability of existence of an $\mathbf{x}$ that satisfies Eq. (10). The capacity of the associative memory is given by the minimum value of a for which the probability of existence of such an $\mathbf{x}$ becomes 0 as the number n of neurons increases to ∞.

If it is possible to approximate the simplex R by a cone whose vertex is at the origin, then we can calculate the above probability by Theorem 2. Our calculation depends on the assumption that the $\mathbf{s}^{(\mu)}$ are distributed in every direction uniformly, though in the original formulation there are only a finite number of possible directions.

THEOREM 2. *Let us fix an* $(n - m + 1)$-*dimensional subspace* Q *in* n-*dimensional Euclidean space. If an* n-*dimensional cone* R *with its vertex at the origin and with vertex angle* $2\bar{\theta}$ *is selected in a random direction, then the probability* P *of nonintersection of* Q *and* R *(other than the trivial one at the origin) is given by*

$$P = \frac{\int_0^\Theta \sin^{n-m} \theta \cos^{m-2} \theta \, d\theta}{\int_0^{\pi/2} \sin^{n-m} \theta \cos^{m-2} \theta \, d\theta} \tag{13}$$

where

$$\Theta = \frac{\pi}{2} - \bar{\theta} \tag{14}$$

PROOF. See Appendix A.

13.4. ESTIMATES OF THE MEMORY CAPACITY

In this section, we show that when n is large it is possible to approximate the simplex R by a cone as far as the existence of the intersection of R and the subspace Q is concerned. Also, two kinds of estimates for the vertex angle of the cone are given and the corresponding estimates for the memory capacity are calculated.

13.4.1. Estimate of the Vertex Angle by the Angle Between an Edge and its Facing Plane

We show that when n increases, the angle A_iOZ (denoted ζ_s) approaches a constant value, where point A_i is any vertex of the n-polygon B, point Z is any point on the facing plane of A_i and point O is the origin (see Fig. 13.5).

The coordinates of these points can be written as follows.

$$
\left.
\begin{aligned}
A_1: \quad & \mathbf{p}_1 = (m - n + 1, 1, 1, \ldots, 1)^T \\
A_2: \quad & \mathbf{p}_2 = (1, m - n + 1, 1, \ldots, 1)^T \\
& \cdots \\
A_n: \quad & \mathbf{p}_n = (1, \ldots, 1, 1, m - n + 1)^T
\end{aligned}
\right\} \quad (15)
$$

$$
\left.
\begin{aligned}
& \mathbf{Z} = (z_1, z_2, \ldots, z_n)^T \\
& z_1 = 1 \\
& z_i = \sum_{j=2}^{n} \lambda_j p_{ij} \quad (i = 2, \ldots, n) \\
& \lambda_j \geq 0 \quad \sum_{j=2}^{n} \lambda_j = 1
\end{aligned}
\right\} \quad (16)
$$

where p_{ij} is the ith element of vector $\mathbf{p}_j$.

With these notations we have the following theorem.

THEOREM 3

$$
\lim_{n \to \infty} \zeta_s = \frac{\pi}{2} \quad (17)
$$

PROOF. In the notations of (15) and (16) the angle ζ_s satisfies the following relation,

$$
\cos \zeta_s = \frac{1}{\sqrt{\left(1 + \sum_{i=2}^{n} z_i^2\right)}} \frac{(m - n + 1)z_1 + \sum_{j=2}^{n} z_i}{\sqrt{[(m - n + 1)^2 + n - 1]}}
$$

$$(18)$$

Note that z_i satisfies

$$
\sum_{i=2}^{n} z_i = \sum_{j=2}^{n} \lambda_j \sum_{i=2}^{n} p_{ij} = m - 1 \quad (19)
$$

and

$$
\sum_{i=2}^{n} z_i^2 \geq \frac{\left(\sum_{i=2}^{n} z_i\right)^2}{n - 1} = \frac{(m - 1)^2}{n - 1} \quad (20)
$$

Then from relations (18), (19), and (20) we have

$$
|\cos \zeta_s| \leq \frac{1}{\sqrt{\left(1 + \frac{(m - 1)^2}{n - 1}\right)}}
$$

$$
\times \frac{m - 1}{\sqrt{[(m - n + 1)^2 + n - 1]}}
$$

If we put $m = an$ then

$$
|\cos \zeta_s| \simeq \left| \frac{2a - 1}{(1 - a)a\sqrt{n}} \right| \to 0 \text{ as } n \to \infty
$$

Hence, we get $\zeta_s \to \pi/2$. $\qquad \square$

On the other hand, the coordinate of a point on the plane $A_1A_2 \cdots A_{n-1}$ which contains A_1 can be written

$$
\mathbf{m}_\lambda = \sum_{j=1}^{n-1} \lambda_j \mathbf{p}_j \quad \lambda_j \geq 0 \quad \sum_{j=1}^{n-1} \lambda_j = 1
$$

Therefore, we have

$$
\cos \zeta \simeq \frac{\lambda_1}{\sqrt{(\sum \lambda_j^2)}}
$$

and ζ takes any value between 0 and $\pi/2$.

From the above theorem we see that the angle of simplex R at the origin O is $\pi/2$ for any direction. So we can approximate simplex R by a cone with vertex angle ζ_s and adopt this value as $2\bar{\theta}$ in Eq. (14).

The following facts are helpful to understand the shape of R. For the circumscribed cone of R, it holds that

$$
\cos \frac{\zeta}{2} = \frac{a}{(1 - a)\sqrt{n}} \left(1 + O\left(\frac{1}{n}\right)\right)
$$

and so $\zeta \to \pi$ as $n \to \infty$. On the other hand, for the inscribed cone it holds that

$$
\cos \frac{\zeta}{2} = 1 + O\left(\frac{1}{\sqrt{n}}\right)
$$

and $\zeta \to 0$ as $n \to \infty$. Hence both cones do not provide us with any information about the limiting value of a.

13.4.2. Estimate of the Vertex Angle by a Cone that has the same Solid Angle as Simplex R

In the previous section we observed that when n is large the inscribed cone shrinks into a line and the circumscribed cone flattens into a plane but the angle between an edge and its facing plane is $\pi/2$. Thus edges of the simplex thrust out sharply and so it will be worth adopting a more averaged criterion than the angle between an edge and its facing plane. The most reasonable quantity from this point of view will be the vertex angle of a cone with a solid angle equal to that of the simplex.

Using Amari's method we can calculate solid angle spanned by the simplex R (Amari, 1990).

THEOREM 4. *Let v_0 be the value of v at which the function*

$$\Psi(v, \sigma) = \frac{v^2}{2\sigma^2} - \log\{1 - \Phi(v)\} \quad (21)$$

takes its minimum, where $\Phi(v)$ is the error integral defined by

$$\Phi(v) = \int_{-\infty}^{v} \frac{1}{\sqrt{2\pi}} \exp\left\{-\frac{u^2}{2}\right\} du$$

Then, when the number n of neurons is sufficiently large, the solid angle K_n spanned by the simplex R at the origin is given by

$$K_n = C(v_0, \sigma)S_n \exp\left\{-n\left[\frac{v_0^2}{2\sigma^2} - \log\{1 - \Phi(v_0)\}\right]\right\} \quad (22)$$

where

$$\sigma = \frac{\sqrt{(1 - 2a)}}{a} \qquad 0 \le a \le \tfrac{1}{2} \quad (23)$$

S_n is the surface of the n-dimensional unit sphere and $C(v, \sigma)$ is defined by

$$C(v, \sigma) = \frac{\sigma}{\sqrt{[\sigma^2 + (\sigma^2 + 1)v^2]}}$$

PROOF. Let us use notation (15) again and define a region R^∞ which is the positive convex hull of $\mathbf{p}_i$ $(i = 1, \ldots, n)$ defined by,

$$R^\infty = \left\{\mathbf{x} \,|\, \mathbf{x} = \mu \sum_{i=1}^{n} \lambda_i \mathbf{p}_i, \sum_{i=1}^{n} \lambda_i = 1, \lambda_i \ge 0, \mu \ge 0\right\}$$

Moreover, let us introduce an n-dimensional random vector $\mathbf{u} = (u_1, \ldots, u_n)^\mathrm{T}$ whose components u_i are mutually independent and obey a normal distribution $N(0, 1)$. We will denote P_{R^∞}

the probability that the random vector $\mathbf{u}$ is contained in the region R^∞.

The solid angle K_n spanned by the simplex R is equal to the probability P_{R^∞} multiplied by the surface area of the n-dimensional unit sphere.

Vectors $\mathbf{e}_i$ which are normalization of $\mathbf{p}_i$ in (15) have the following direction cosines

$$\cos(\mathbf{e}_i, \mathbf{e}_j) = \frac{2m - n}{(m - n + 1)^2 + n - 1}$$

$$= \frac{1}{n} \frac{2a - 1}{(1 - a)^2}\left(1 + o\left(\frac{1}{n}\right)\right) \quad (24)$$

Hence, we can orthogonalize them by introducing the following transformation

$$\tilde{\mathbf{e}}_i = \mathbf{e}_i - \frac{c}{n} \sum_{i=1}^{n} \mathbf{e}_i \quad (25)$$

$$c = \begin{cases} \dfrac{2a - 1}{a} & (<0) \ (0 < a < \tfrac{1}{2}) \\[2mm] \dfrac{1}{a} & (>0) \ (a > \tfrac{1}{2}) \end{cases} \quad (26)$$

so that

$$\tilde{\mathbf{e}}_i \cdot \tilde{\mathbf{e}}_j = \delta_{ij}$$

Under the transformation (25) the random variables u_i are transformed to

$$\tilde{u}_i = u_i - \frac{c}{n} \sum_{i=1}^{n} u_i \quad (27)$$

These new random variables $\tilde{u}_i$ are not mutually independent and have correlations

$$E(\tilde{u}_i, \tilde{u}_j) = \frac{(c - 2)c}{n}$$

$$= \frac{1 - 2a}{na^2} \quad (28)$$

Thus $\tilde{u}_i$ can be written as

$$\tilde{u}_i = u_i + \frac{1}{\sqrt{n}} v \quad (29)$$

where v is a normal random variable with mean 0 and variance $\sigma = \sqrt{(1 - 2a)}/a$ (where the case $a > 1/2$ is not treated here) and so

$$v \in N(0, \sigma^2)$$

$$\sigma^2 = \frac{1 - 2a}{a^2}$$

After the transformation (25) the probability that $\mathbf{u} = (u_1 \cdots u_n)^\mathrm{T}$ is contained in R^∞ becomes the probability that $\tilde{\mathbf{u}} = (\tilde{u}_1, \ldots, \tilde{u}_n)^\mathrm{T}$ is contained in the positive convex hull of $\tilde{e}_i$ $(i = 1, \ldots, n)$, namely $\tilde{u}_i > 0$ for all i. Therefore we have

$$\frac{K_n}{S_n} = P_{R^\infty} = \Pr(\tilde{u}_i > 0; \forall i)$$

$$= E_v\left(\prod_{i=1}^n \left(1 - \Phi\left(\frac{v}{\sqrt{n}} \right) \right) \right)$$

$$= \frac{\sqrt{n}}{\sqrt{2\pi}\sigma} \int_{-\infty}^\infty$$

$$\times \exp\left\{ -n\left[\frac{v^2}{2\sigma^2} - \log\{1 - \Phi(v)\} \right] \right\} dv \quad (30)$$

By the application of the saddle point approximation method to the last integral of Eq. (30) we have

$$\frac{K_n}{S_n} = \frac{\sqrt{n}}{\sqrt{2\pi}\,\sigma} \exp\left\{ -n\left[\frac{v_0^2}{2\sigma^2} - \log\{1 - \Phi(v_0)\} \right] \right\}$$

$$\times \sqrt{\left(\frac{2\pi}{n\Psi''(v_0, \sigma)} \right)} \quad (31)$$

where $''$ denotes the second derivative. Thus we get Eq. (22). $\qquad\square$

The following corollary determines the vertex angle of the cone which has the same solid angle as the simplex R.

COROLLARY 5. If $2\theta_c$ is the vertex angle of a cone whose solid angle is equal to that given in Theorem 4, then

$$\sin \theta_c = \exp\left\{ -\left[\frac{v_0^2}{2\sigma^2} - \log\{1 - \Phi(v_0)\} \right] \right\} \quad (32)$$

where v_0 is the same as in Theorem 4.

PROOF. See Appendix B.

In order to see the explicit relation between the memory ratio a and the vertex angle θ_c we have to solve Eq. (32) numerically. The following two values are easily obtained,

$$a = 0 \rightarrow \sigma = \infty \rightarrow \theta_c = \frac{\pi}{2}$$

$$a = \tfrac{1}{2} \rightarrow \sigma = 0 \rightarrow \theta_c = \frac{\pi}{6}$$

13.4.3. Estimates of Memory Capacity

Now we calculate the capacity of the associative memory based on the above obtained vertex

angles. Let us remark that the integrand of Eq. (13),

$$\sin^{n-m} \theta \cos^{m-2} \theta$$

is positive in $0 \le \theta \le \pi/2$ and takes its maximum value at

$$\tilde{\theta}(a) = \arctan\sqrt{\left(\frac{1-a}{a} \right)} \quad (33)$$

Also the denominator of Eq. (13) can be split into two terms as

$$\int_0^\Theta \sin^{n-m} \theta \cos^{m-2} \theta \, d\theta$$

$$+ \int_\Theta^{\pi/2} \sin^{n-m} \theta \cos^{m-2} \theta \, d\theta$$

Thus using the saddle point approximation we can rewrite Eq. (13) as follows,

$$P = \begin{cases} 0 & \Theta(a) < \tilde{\theta} \\ 1 & \Theta(a) > \tilde{\theta} \end{cases} \quad (34)$$

In the case of the vertex angle of the n-simplex, Theorem 3 gives the following Θ:

$$\Theta = \frac{\pi}{2} - \bar{\theta} = \frac{\pi}{2} - \frac{\pi}{4} = \frac{\pi}{4}$$

Hence we have

$$\tilde{\theta}(a) > \Theta \quad \text{for } 0 \le a < \tfrac{1}{2}$$

and memory capacity is estimated as 0.5.

On the other hand, in the case of the solid angle of the n-simplex, we have to solve Eq. (32) numerically to get a function $\Theta = \Theta(a)$ which is compared with $\tilde{\theta}(a)$ as in Fig. 13.6.

These results are summarized in the following corollary.

COROLLARY 6. In the case of $k = a^{-1}$, an upper bound on memory capacity is given by $0.5n$ and the average value is estimated at $0.398n$.

In numerical calculations of Eq. (8) for $n = 300$, we could get a correct recall result for $a = 0.5$ by selecting proper initial values. But for $n = 1000$ the upper limit of a for correct recall was between 0.4 and 0.41.

13.5. STABILITY OF EQUILIBRIUM SOLUTIONS

In our piecewise linear model, the stability of equilibrium solutions is easily determined if their locations are known. Namely, the variational equation at equilibrium solution $\mathbf{x} = \mathbf{x}_0$ is

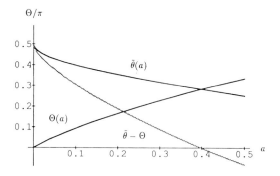

Fig. 13.6. Capacity estimated by a cone with the same solid angle. The value of a at which $\bar{\theta} - \Theta$ is equal to 0 gives the memory capacity.

given by

$$\frac{d\mathbf{z}}{dt} = -\frac{k_0 - k}{k_0}\mathbf{z} - \frac{k}{n}\mathbf{S}\mathbf{S}^T\mathbf{z} \qquad (35)$$

where $\mathbf{z} = \mathbf{x} - \mathbf{x}_0$ and $k_0 = 1/a$. Introduce the following coordinate transformation,

$$\mathbf{T} = [\mathbf{S}\mathbf{U}]^{-1} \qquad \xi = \mathbf{T}\mathbf{x} \qquad (36)$$

where $\mathbf{S}$ is the memory matrix in Eq. (2) and $\mathbf{U}$ is a matrix which is composed of the basis vectors $\mathbf{u}^v$ ($v = 1, \ldots, n - m$) of the ortho-complement of the subspace spanned by $\mathbf{s}^\mu$ ($\mu = 1, \ldots, m$). Then Eq. (35) reduces to

$$\frac{d\xi}{dt} = -\frac{k_0 - k}{k_0}\xi - \frac{k}{n}\begin{bmatrix} \mathbf{S}^T\mathbf{S} & \mathbf{0} \\ \mathbf{0} & \mathbf{0} \end{bmatrix}\xi \qquad (37)$$

Taking into consideration that $\mathbf{S}^T\mathbf{S}$ is positive definite, we have the following results.

1. $\mathbf{x}_0$ is stable when $k < k_0$.
2. $\mathbf{x}_0$ is neutrally stable when $k = k_0$.
3. $\mathbf{x}_0$ is unstable when $k > k_0$. (This implies that directions of nonmemorized vectors are unstable.)

13.6. SUMMARY

In this chapter we investigate the recall process and capacity of an autocorrelation associative memory whose neuron element has nonmonotonic output function. We derive theoretical results on capacity which show good agreement with numerical experiments. The absolute capacity of this memory is equal to $0.4n$ which contrasts favorably with the theoretical absolute capacity of $n/(2 \log n)$ (Weisbuch, 1985; McEliece et al., 1987), and the relative capacities of $0.15n$ (obtained by computer simulations; Hopfield, 1982), 0.14n (obtained by the replica method; Amit et al., 1985a, b), and $0.16n$ (obtained by the statistical neurodynamical method; Amari and Maginu, 1988) of Hopfield-type memories employing monotonic neurons.

The following two facts are worth mentioning here: (1) A similar capacity can be proved for the case of $k \neq 1/a$ if k is close enough to $1/a$. (2) The probability of existence of spurious memories decreases as k approaches $1/a$.

Our consideration in this chapter was restricted to the existence of equilibrium solutions and local stability. It will be interesting to investigate the size of the basins of attraction, and the behavior with clustered memory patterns for such memories.

REFERENCES

Amari, S. (1972). "Learning Patterns and Pattern Sequences by Self-Organizing Nets of Threshold Elements," *IEEE Trans. Comput.*, **C-21**(11), 1197–1206.

Amari, S. (1977). "Neural Theory of Association and Concept-Formation," *Biological Cybernetics*, **26**, 175–185.

Amari, S. (1989). "Characteristics of Sparsely Encoded Associative Memory," *Neural Networks*, **2**, 451–457.

Amari, S. (1990). "Mathematical Foundations of Neurocomputing," *Proc. IEEE*, **78**(9), 1443–1463.

Amari, S. and Maginu, K. (1988). "Statistical Neurodynamics of Associative Memory," *Neural Networks*, **1**, 63–73.

Amit, D. J., Gutfreund, H., and Sompolinsky, H. (1985a). "Spin-Glass Models of Neural Networks," *Phys. Rev. A*, **32**, 1007–1018.

Amit, D. J., Gutfreund, H., and Sompolinsky, H. (1985b). "Storing Infinite Numbers of Patterns in a Spin-Glass Model of Neural Networks," *Phys. Rev. Lett.*, **55**(14), 1530–1533.

Anderson, J. A. (1972). "A Simple Neural Network Generating an Interactive Memory," *Math. Biosc.*, **14**, 197–220.

Gardner, E. (1988). "The Space of Interaction in Neural Network Models," *J. Phys. A: Math. Gen.*, **21**, 257–270.

Hopfield, J. J. (1982). "Neural Network and Physical Systems with Emergent Collective Computational Abilities," *Proc. Nat. Acad. Sci. U.S.A.*, **79**, 2554–2558.

Kohonen, T. (1972). "Correlation Matrix Memories," *IEEE Trans. Computers*, **C-21**(4), 353–359.

Kohonen, T. and Ruohonen, M. (1973). "Representation of Associated Data by Matrix Operators," *IEEE Trans. Computers*, **C-22**(7), 701–702.

McEliece, R. J., Posner, E. C., Rodemich, E. R., and Venkatesh, S. S. (1987). "The Capacity of the Hopfield Associative Memory," *IEEE Trans. Information Theory*, **IT-33**(4), 461–482.

Meunier, C., Yanai, H., and Amari, S. (1991). "Sparsely Coded Associative Memories: Capacity and Dynamical Properties," *Network*, **2**(4), 469–487.

Morita, M., Yoshizawa, S., and Nakano, K. (1990a). "Analysis and Improvement of the Dynamics of Autocorrelation Associative Memory," *Trans. Institute Electronics, Information and Communication Engrs.*, **J73-D-III**(2), 232–242.

Morita, M., Yoshizawa, S., and Nakano, K. (1990b). "Memory of Correlated Patterns by Associative Neural Networks with Improved Dynamics," in *Proc. INNC'90*, Paris, Vol. 2, pp. 868–871.

Nakano, K. (1972). "Associatron—A Model of Associative Memory," *IEEE Trans. Systems, Man, and Cybernetics*, **SMC-2**, 381–388.

Palm, G. (1980). "On Associative Memory," *Biological Cybernetics*, **36**, 19–31.

Shinomoto, S. (1987). "A Cognitive and Associative Memory," *Biological Cybernetics*, **57**, 197–206.

Weisbuch, G. (1985). "Scaling Laws for the Attractors of Hopfield Networks," *J. Phys. Lett.*, **46**(14).

Willshaw, D. J., Buneman, O. P., and Longuet-Higgins, H. C. (1969). "Non-Holographic Associative Memory," *Nature*, **222**, 960–962.

Yanai, H., Sawada, Y., and Yoshizawa, S. (1991). "Dynamics of an Auto-Associative Neural Network Model with Arbitrary Connectivity and Noise in the Threshold," *Network*, **2**(3), 295–314.

Yoshizawa, S., Morita, M., and Amari, S. (1992). *Capacity of Autocorrelation-Type Associative Memory Using a Non-Monotonic Neuron Model.* Technical Report METR92-01, Dept. Mathematical Engineering and Information Physics, University of Tokyo.

APPENDIX A: Proof of Theorem 2

First, two notations are introduced.

$S_n(r)$ The surface area of an n-dimensional sphere of radius r ($\sum_{i=1}^n x_i^2 \leq r^2$).

$S_{n,m}^\theta$ The solid angle covered by the center of a cone C whose vertex is at the origin and whose vertex angle is $2\bar{\theta}$, when the cone C moves in such a way that it has a nontrivial intersection with an $(n-m+1)$-dimensional subspace Q.

LEMMA A1. *The surface area of an n-dimensional sphere with radius r can be expressed as*

$$S_n(r) = \int \frac{r}{\sqrt{(r^2 - |\mathbf{x}|^2)}} dx_2 \cdots dx_n \qquad (A.1)$$

where

$$|\mathbf{x}|^2 = \sum_{i=2}^n |x_i|^2$$

PROOF. It is well known that $S_n(r)$ is expressed as

$$S_n(r) = r^{n-1} \int_0^\pi \cdots \int_0^\pi \int_0^{2\pi} \sin^{n-2}\theta_1 \sin^{n-3}\theta_2$$
$$\cdots \sin\theta_{n-2}\, d\theta_1\, d\theta_2 \cdots d\theta_{n-1}$$

By changing from polar coordinates into orthogonal coordinates,

$$x_1 = r\cos\theta_1$$
$$x_2 = r\sin\theta_1\cos\theta_2$$
$$\cdots$$
$$x_{n-1} = r\sin\theta_1 \cdots \sin\theta_{n-2}\cos\theta_{n-1}$$
$$x_n = r\sin\theta_1 \cdots \sin\theta_{n-2}\sin\theta_{n-1}$$

the above integral can be rewritten as

$$S_n(r) = \int \frac{1}{\cos\theta_1} dx_2 \cdots dx_n$$

since the Jacobian for the coordinate change is

$$J = (r^{n-1}\sin^{n-2}\theta_1 \cdots \sin\theta_{n-2}\cos\theta_1)^{-1}$$

From this we have Eq. (A.1). ☐

Now let us denote the $(m-1)$-dimensional subspace which is spanned by $\mathbf{s}^\mu$ ($\mu = 2,\ldots,m$) by $\mathbf{y} = (x_1,\ldots,x_{m-1})^{\mathsf{T}}$ and its ortho-complement Q which is an $(n-m+1)$-dimensional subspace, by

$$\mathbf{x} = (x_m, x_{m+1}, \ldots, x_n)^{\mathsf{T}}$$

Then noting that the solid angle $S_{n,m}^{\bar{\theta}}$ is the surface area of that part of the unit sphere whose distance from Q is less than $\sin\bar{\theta}$, we can write

$$S_{n,m}^{\bar{\theta}} = \int_{|\mathbf{x}|=\cos\bar{\theta}}^1 dx \int_{(1-|\mathbf{x}|^2)\geq|\mathbf{y}|^2} \frac{dy}{\sqrt{[(1-|\mathbf{x}|^2)-|\mathbf{y}|^2]}}$$

where $|\mathbf{x}|^2 = \sum_{i=m}^n x_i^2$, $|\mathbf{y}|^2 = \sum_{i=2}^{m-1} x_i^2$, and $dx = dx_m \cdots dx_n$, $dy = dx_2 \cdots dx_{m-1}$. By Using Eq. (A.1) we can rewrite the above as

$$S_{n,m}^{\bar{\theta}} = \int \frac{dx}{\sqrt{(1-|\mathbf{x}|^2)}} \int \frac{\sqrt{(1-|\mathbf{x}|^2)}\, dy}{\sqrt{[(1-|\mathbf{x}|^2)-|\mathbf{y}|^2]}}$$

$$= \int \frac{dx}{\sqrt{(1-|\mathbf{x}|^2)}} S_{m-1}(\sqrt{(1-|\mathbf{x}|^2)})$$

$$= \int (1-|\mathbf{x}|^2)^{(m-3)/2}\, dx\, S_{m-1}(1) \qquad (A.2)$$

By changing $\mathbf{x}$ into polar coordinates $(\rho, \varphi_1, \ldots, \varphi_{n-m})$ we can rewrite the last integral of the above as

$$\int_{\rho=\cos\theta}^1 (1-\rho^2)^{(m-3)/2}\rho^{n-m}\, d\rho \int_0^\pi$$
$$\cdots \int_0^\pi \int_0^{2\pi} \sin^{n-m-1}\varphi_1 \cdots \sin\varphi_{n-m}\, d\varphi$$

Transformation $\rho = \cos \xi$ in this integral yields

$$\int_0^{\bar{\theta}} \sin^{m-2} \xi \cos^{n-m} \xi \, d\xi \, S_{n-m+1}(1) \qquad (A.3)$$

Then from Eqs. (A.2), (A.3), and (14), Theorem 2 follows. ☐

APPENDIX B: Proof of Corollary 5

The solid angle of a cone with vertex angle equal to $2\theta_c$ satisfies the following equation:

$$\int_0^{\theta_c} \int_0^{\pi} \cdots \int_0^{\pi} \int_0^{2\pi} \sin^{n-2} \theta_1 \cdots \sin \theta_{n-1} \, d\theta_1 \cdots d\theta_{n-1} \qquad (B.1)$$

$$= S_{n-1} \int_0^{\theta_c} \sin^{n-2} \theta_1 \, d\theta_1 \qquad (B.2)$$

Therefore, the vertex angle $2\theta_c$ of a cone whose solid angle is equal to that given in Theorem 4 is obtained from the relation

$$\exp\left\{ -n\left[\frac{v_0^2}{2\sigma^2} - \log\{1 - \Phi(v_0)\} \right] \right\}$$

$$= \frac{S_{n-1}}{S_n} \frac{1}{C(v_0, \sigma)} \int_0^{\theta_c} \sin^{n-2} \theta_1 \, d\theta_1 \qquad (B.3)$$

where v_0 and $C(v, \sigma)$ are defined in Theorem 4.

Now let us note that we can construct a sequence of C^{∞} functions $s_k(\theta)$ such that

$$s_k(\theta) = \begin{cases} \sin \theta & \left| \theta - \dfrac{\theta_c}{2} \right| \leq \dfrac{\theta_c}{2} - \dfrac{1}{k} \\[2mm] 0 & \left| \theta - \dfrac{\theta_c}{2} \right| \geq \dfrac{\theta_c}{2} \end{cases} \qquad (B.4)$$

$$0 < s_k(\theta) < \sin \theta \qquad 0 < \theta < \frac{1}{k}$$

or

$$\theta_c - \frac{1}{k} < \theta < \theta_c$$

and with $s_k''(0)$ of polynomial order k, and

$$s_k(\theta) \to s(\theta) = \begin{cases} \sin \theta & 0 \leq \theta \leq \theta_c \\ 0 & \theta_c < \theta \leq \pi/2 \end{cases} \qquad (B.5)$$

as $k \to \infty$. For example we can use

$$s_k(\theta) = \frac{\rho(\theta_c \theta - \theta^2) \sin \theta}{\rho(\theta_c \theta - \theta^2) + \rho(\theta^2 + \theta_c \theta + (\theta_c/k) - (1/k)^2)} \qquad (B.6)$$

where ρ is defined by

$$\rho(t) = \begin{cases} \exp(-1/t) & t > 0 \\ 0 & t \leq 0 \end{cases}$$

Then, the integral of the right-hand side of Eq. (B.3) is approximated by

$$\int_0^{\pi/2} s_k^{n-2}(\theta) \, d\theta = \int_0^{\pi/2} \exp\{(n-2) \log s_k(\theta)\} \, d\theta \qquad (B.7)$$

Since $s_k(\theta)$ has its maximum at a θ (which is denoted θ_0) in the interval $(\theta_c - 1/k, \theta_c)$, we can apply the saddle point approximation method and the value of the integral (B.7) is given by

$$s_k^{n-2}(\theta_0) \sqrt{\left(\frac{2\pi}{-(n-2) s_k''(\theta_0)} \right)} \qquad (B.8)$$

Lastly, if we note that the maximum point θ_0 approaches θ_c and hence $s_k(\theta_0)$ approaches $\sin \theta_c$, and that the factor S_{n-1}/S_n satisfies

$$\log \frac{S_{n-1}}{S_n} = o(n)$$

we have Eq. (32) by taking log of Eq. (B.3).

14

Fault-Tolerance of Optical and Electronic Hebbian-type Associative Memories

PAU-CHOO CHUNG and THOMAS F. KRILE

14.1. INTRODUCTION

Hebbian-type associative memories (HAMs) have been applied to various problems and appear to have very attractive performance characteristics. The future generations of HAMs will employ hundreds of neurons with high interconnectivity to achieve useful results. One of the formidable problems with such large networks is to determine how the HAMs are affected by interconnection matrix faults. Neural networks are known to have the capability of fault-tolerance because of their distributed information storage. However, to what degree can a network tolerate interconnection faults? How will various faults cause different performance degradations? These are some major concerns to be determined prior to designing large physical implementations. Once the reliability properties of HAMs are determined, the effects of network design trade-offs can be studied before proceeding to implementation.

A similar issue has been discussed by Amit (1989) and Sompolinsky (1986). Using the spin glass concept, they have examined noisy HAMs with diluted interconnections. The dilution model used by Amit and Sompolinsky was obtained by deleting (or reducing) some of the interconnection values symmetrically, and then adding these values back to the remaining weights in an average sense. They also discussed the input noise effects on network performance degradation. The noise model used is normally distributed with a zero mean value. In contrast, the case considered here is one where some of the interconnections are damaged while the others remain unchanged. The damage may be caused by incorrect control processes, wafer contamination or finite component lifetimes (in the case of VLSI implementations), or faulty pixels or pixel addressing structures of optical spatial light modulator masks (in the case of optical implementations). Symmetry in the distribution of faulty weights

is not restricted here. The failed interconnections can be considered to behave like either open-circuited or short-circuited connections and they can be either randomly distributed in the interconnection matrix or clustered within a small area.

This chapter is organized as follows. First, VLSI and optical implementations are introduced and the distributions of failed interconnections are discussed. After that, the open-circuit and the short-circuit effects of failed interconnections are analyzed. Then the discussion is extended to the case where network attraction radius is considered. Theoretical equations for estimating the probability of direct one-step convergence, P_{dc}, for networks with both types of failed interconnections are established as a function of the ratio of failed interconnections to the total number of connections. The effect of clustered failed interconnections is also discussed and compared to that of uniformly distributed failed interconnections. Finally, fault-tolerance capabilities of quadratic HAMs are introduced and compared with those of linear HAMs.

14.2. OPTICAL AND VLSI IMPLEMENTATIONS

Instead of performing a program of instructions sequentially, neural networks perform a computation by simultaneously using massively parallel nets composed of many simple computational elements. The real promise for applications of neural networks lies in specialized hardware. Today optoelectronics and VLSI techniques are two major technologies proposed for implementing neural networks. In optical implementations, a hologram or a spatial light modulator (SLM), with optical or electronic addressing of cells, is used for implementing the interconnection matrix. The input vectors can be represented by a one-dimensional array of light

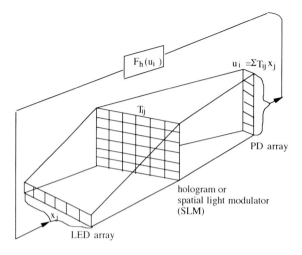

Fig. 14.1. A common optical implementation of associative memory.

sources (e.g., LED array). An array of photodiodes (PD) can be used to detect the output vector. One example showing an optical implementation of an associative memory is depicted in Fig. 14.1. This is a recursive structure where the output is thresholded by $F_h(\cdot)$ and fed back to the input. As we can see, the interconnection matrix $\mathbf{T}$ is the major part of this implementation. Defects in the interconnection matrix may result from damage to the mask or the failure of address control circuits so that the hologram (or SLM) cannot be properly programmed. These defects could be randomly distributed or clustered within specific areas, as shown in Fig. 14.2, depending on the effects causing the defects and the environment the circuit has experienced.

In VLSI implementations, networks can be either analog or digital. Digital implementations are known to have higher noise-immunity and provide higher-resolution weight representations; but more control logic and multiplier-adder circuits are required, resulting in a large circuit even for a modest number of neurons. Using analog computation, we can achieve a multiplication with a single resistor, and the summing of currents is accomplished "for free" on the input wire of the amplifier. However, a large silicon

area is required in order to achieve high resolution in interconnection strengths. Moreover, designing large integrated analog circuits is a difficult task and there is a tendency to avoid analog computation and do everything digitally. Even so, the high robustness (or fault tolerance) and relatively low precision needed in neural networks make analog designs popular.

The interconnection matrix is implemented, basically, by variable resistors. Junction field-effect transistors (JFETs) and metal–oxide–semiconductor field-effect transistors (MOSFETs) are the simplest active devices for simulating the variable resistors. Today, the CMOS switch based on a pair of complementary MOSFETs is considered as the most practical design, and is currently used (Foo et al., 1990; Masaki et al., 1990). No matter what technique has been used for implementation, the interconnections tend to be laid out in a regular pattern on a wafer along with the addressing and amplifier circuitry as shown in Fig. 14.3. Defects in interconnections can come from wafer contamination, incorrect process control, and finite component lifetime. They can be categorized as point defects or area defects according to their size. Point defects may come from small-particle contamination or finite

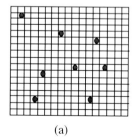

(a)

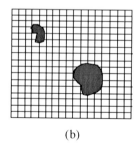

(b)

Fig. 14.2. Defects in optically-implemented interconnection masks: (a) randomly distributed; (b) clustered.

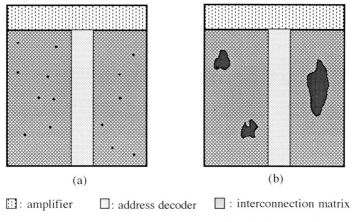

(a)   (b)

⊞ : amplifier   ☐ : address decoder   ▨ : interconnection matrix

Fig. 14.3. Interconnection defects in VLSI design: (a) randomly distributed; (b) clustered.

component lifetime. Clustered (or area) defects can result from the contamination of groups of particles. The clustering of particles is believed to be caused by aggregates of particles that have collected in the manufacturing machinery. Typical randomly distributed (point) and clustered (area) defects are shown in Fig. 14.3. Defects in the interconnections could result in open-circuit or short-circuit situations.

14.3. OPEN-CIRCUIT EFFECTS

The autoassociative memory model is constructed by interconnecting a large number of simple processing units, as shown in Fig. 14.1. For a network consisting of N processing units, or neurons, each neuron i, $1 \leq i \leq N$, receives input from neuron j, $1 \leq j \leq N$, through a connection T_{ij}. Let $\mathbf{x}^k = [x_1^k, x_2^k, \ldots, x_N^k]$, $1 \leq k \leq M$, be M vectors stored in the network with each $x_i^k = -1$ or $+1$. Then T_{ij} is obtained as

$$T_{ij} = \sum_{k=1}^{M} x_i^k x_j^k \qquad (1)$$

As a probe vector $\mathbf{x}^f(t) = [x_1^f(t), x_2^f(t), \ldots, x_N^f(t)]$, similar to some stored vector $\mathbf{x}^q$, is applied at time t, the network evolves according to

$$x_i(t + 1) = F_h\left(\sum_{j=1}^{N} T_{ij} x_j^f(t) \right) \qquad (2)$$

where $x_i(t + 1)$ is the value of neuron i after iteration t and $F_h(\ )$ is the hard-limiting function defined as $F_h(z) = 1$ if $z \geq 0$ and -1 if $z < 0$.

Thus the components of the current vector are updated to produce a new vector at each iteration. Here the reliability of a HAM within

a one-step update is considered, called direct (one-step) convergence. Connections to neuron i are $T_{i1}, T_{i2}, \ldots, T_{iN}$. Further, assume that among these interconnections $T_{im1}, T_{im2}, \ldots, T_{imn}$ are failed and the fraction of failed connections is $p = n/N$. We define the set of failed interconnections to neuron i as $A = \{m1, m2, \ldots, mn\}$. If we assume that the failed interconnections result in a disconnected (open-circuit) state, then Eq. (2) can be rewritten as

$$\begin{aligned}
x_i(t + 1) &= F_h\left(\sum_{\substack{j=1 \\ j \notin A}}^{N} \sum_{k=1}^{M} x_i^k x_j^k x_j^f(t) \right) \\
&= F_h\left(x_i^q \sum_{\substack{j=1 \\ j \notin A}}^{N} x_j^q x_j^f(t) \right. \\
&\quad \left. + \sum_{\substack{j=1 \\ j \notin A}}^{N} \sum_{\substack{k=1, \\ k \neq q}}^{M} x_i^k x_j^k x_j^f(t) \right) \qquad (3)
\end{aligned}$$

If $i \notin A$, i.e., the self-connection of neuron i is a good interconnection at iteration t, the second term of Eq. (3) can further be separated into the cases $j = i$ and $j \neq i$. Hence, when $i \notin A$, Eq. (3) can be rewritten as

$$\begin{aligned}
x_i(t + 1) &= F_h\left(x_i^q \sum_{\substack{j=1 \\ j \notin A}}^{N} x_j^q x_j^f(t) + (M - 1) x_i^f(t) \right. \\
&\quad \left. + \sum_{\substack{j=1 \\ j \notin A \cup \{i\}}}^{N} \sum_{\substack{k=1, \\ k \neq q}}^{M} x_i^k x_j^k x_j^f(t) \right) \qquad (4)
\end{aligned}$$

When $i \notin A$, the first term of Eq. (3) is the signal term and the second term of Eq. (3) is the noise term. When $i \in A$, the summation of the first and second terms of Eq. (4) is considered as the signal value and the third term of Eq. (4) is the noise term. Assume that there is no error bit in the probe vector. When only the first iteration is considered, $x_i^f(t)$ represents the ith bit in the applied probe vector. In this case, the signal, S, is calculated as $x_i^q(N - pN + M - 1)$ when $i \notin A$ and $x_i^q(N - pN)$ when $i \in A$. Assume that the stored patterns are sampled from a distribution such that each element x_i^k has $P(x_i^k = +1) = 0.5$ and $P(x_i^k = -1) = 0.5$. Then, each term of $x_i^k x_j^k x_j^f(t)$ in the noise terms is an identical independently distributed (i.i.d.) random variable with mean 0 and variance 1. Therefore, as NM gets large, using the central limit theorem, the noise term can be approximated as having a Gaussian distribution with mean 0 and variance σ^2, where σ^2 is obtained as $(N - pN - 1)(M - 1)$ if $i \notin A$ and $(N - pN)(M - 1)$ if $i \in A$. A parameter, defined as the average ratio of signal to the standard deviation of noise, similar to the parameter defined in previous work for characterizing a HAM (Wang et al., 1990), is obtained as

$$C = \frac{p(N - pN)}{\sqrt{[(N - pN)(M - 1)]}}$$
$$+ \frac{(1 - p)(N - pN + M - 1)}{\sqrt{[(N - pN - 1)(M - 1)]}} \quad (5)$$

This parameter uniquely defines the probability of a bit remaining in an incorrect state no matter what its previous state is. The larger C is, the higher the probability of a neuron remaining in a correct state. How the C parameter is used in deriving the probability of a bit remaining in an incorrect state is described in detail in Wang et al. (1990). The same procedure can be applied to our modified C parameter to give the following results.

Given a fixed C, the probability of a bit remaining in an incorrect state is $\phi(C)$, where

$\phi(C)$ is the standard error function, i.e.,

$$\phi(x) = \frac{1}{\sqrt{2\pi}} \int_x^\infty e^{-t^2/2} \, dt \quad (6)$$

Since the activity of each neuron is considered to be independent of any other neurons (synchronous update), the probability that every neuron in the network is correct after the first iteration is

$$P_{dc} = (1 - \phi(C))^N \quad (7)$$

Theoretical and simulation results for P_{dc} versus p are shown in Fig. 14.4, where (N, M) is chosen to be (50, 6). Simulation results in this paper are obtained from the ensemble average of 5000 randomly sampled vectors. Each element, x_i^k, in the stored vectors is sampled from a distribution with $P(x_i^k = +1) = 0.5$ and $P(x_i^k = -1) = 0.5$. The number of correct recalls from these 5000 stored vectors in one state updating is then used to compute the P_{dc}. Note that both synchronous and asynchronous neuron updates are closely modeled by Eq. (7).

Network fault-tolerance capabilities for variously sized networks are plotted in Fig. 14.5. As can be seen, as the networks grow larger the effect of the same percentage of failed interconnections becomes more severe. Moreover, as the fraction of failed interconnections increases, the larger networks are affected more than the smaller networks. This is an important result in looking at the scaling characteristics of neural networks, especially in physical implementations where real networks are built with thousands of neurons.

The capacity of Hebbian-type networks has been extensively studied (McEliece et al., 1987; Kuh and Dickinson, 1989; Venkatesh and Psaltis, 1989; Dembo, 1989; Abu-Mostafa and Jacques, 1985; Newman, 1988). An early study was made by Abu-Mostafa and Jacques (1985). More recently, McEliece et al. (1987), gave a very

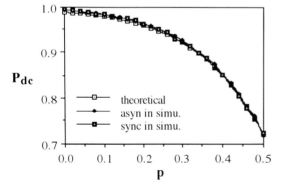

Fig. 14.4. Network reliabilities obtained from theoretical and simulation results.

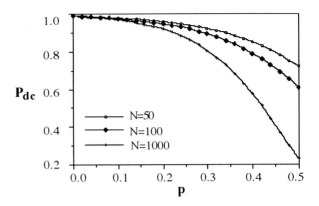

Fig. 14.5. Comparison of network reliability with various N.

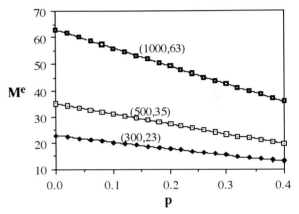

Fig. 14.6. Theoretical decrease in network capacity (M^e) of networks with various (N, M^c) and $P_{dc} = 0.989$.

thorough description. Their results predict a storage capacity of $N/(4 \log N)$ for a network with N neurons, where log is the natural logarithm. However, an increase in the number of failed interconnections decreases the probability of correct recall. Hence, the number of stored vectors must be decreased in order to maintain network performance above a specific limit, say $P_{dc} > 0.98$, even when the fraction of failed interconnections increases up to a certain value. The new number of stored vectors M^e is related to the original network capacity M^c. Assume that a network has the capacity for storing M^c vectors and has a P_{dc} equal to a required value when no interconnection faults exist. Since the P_{dc} must be maintained above the required value even when a fraction p of interconnections are disconnected, the number of vectors that can now be stored in the network, M^e, is related to the original memory capacity, M^c, as

$$M^e = \frac{(N + M^c)^2}{2M^c(1 - p)} - N - \sqrt{\left[\frac{(N + M^c)^2}{2M^c(1 - p)}\right]} \times \left(\frac{(N + M^c)^2}{2M^c(1 - p)} - 2N\right)\right] \quad (8)$$

This equation can be derived by setting the ratios of signal to standard deviation of noise of these two networks equal (Chung, 1991). Equation (8) indicates that, in order to keep the same P_{dc}, the number of vectors stored in the network is decreased from M^c to M^e, when the fraction of failed interconnections increases from 0 to p. Given that the required P_{dc} is kept above 0.989, Fig. 14.6 shows how M^e decreases as p increases for networks with various (N, M^c).

14.4. SHORT-CIRCUIT EFFECTS

Network reliability for HAMs with short-circuit-type failed interconnections is analyzed in this section. By short-circuit-type failed interconnections we mean here that a large signal magnitude is added into the network through this interconnection even when a small signal is applied to the input lead, similar to a short circuit in an electronic circuit. For an optical implementation, it means that a reflective or transmissive weight has a value of unity. In the analysis followed in this section, a value GS_{ij},

$G > 0$, $S_{ij} = \text{sign}(T_{ij})$, is assumed for the short-circuit-type failed interconnections, where $\text{sign}(\ )$ is defined as $\text{sign}(x) = 1$ if $x > 0$, $\text{sign}(x) = 0$ if $x = 0$, and $\text{sign}(x) = -1$ if $x < 0$. In this case, the state of neuron i iterates as

$$x_i(t + 1) = \left(\sum_{\substack{j=1 \\ j \notin A}}^{N} \sum_{k=1}^{M} x_i^k x_j^k x_j^f(t) + \sum_{j \in A} GS_{ij}x_j^f(t) \right) \tag{9}$$

This equation is the same as Eq. (3) except that an extra term $\sum GS_{ij}x_j^f(t)$ is added. Again, analysis can be separated into two cases where $i \in A$ and $i \notin A$. In the case $i \in A$, $\sum GS_{ij}x_j^f(t)$ is further expanded to

$$\sum_{j \in A} GS_{ij}x_j^f(t) = Gx_i^f(t) + G \sum_{j \in A\setminus\{i\}} S_{ij}x_j^f(t) \tag{10}$$

where "\" denotes set subtraction. $A\setminus\{i\}$ represents a set which consists of elements of A from which element i has been removed. S_{ij} is related to the summation of the products of x_i^k and x_j^k for $1 \leq k \leq M$, which includes the pattern to be retrieved, x_i^q, as

$$S_{ij} = \text{sign}(x_i^1 x_j^1 + x_i^2 x_j^2 + \cdots + x_i^q x_j^q$$
$$+ \cdots + x_i^M x_j^M) \tag{11}$$

Define $[x]$ as the smallest integer that is greater than or equal to x and $C_x^M = M!/x!(M - x)!)$. After some manipulation, when $x^f(t) = x^q$, it can be proved that $S_{ij}x_j^f(t)$ is a random variable with mean $\mu_s x_i^q$ and variance $1 - (\mu_s)^2$ (Chung, 1991), where μ_s is defined as

$$\mu_s = \left(\frac{1}{2} \right)^{M-1} C_{[(M-1)/2]}^{M-1} \tag{12}$$

The $S_{ij}x_j^f(t)$ are identical independently distributed (i.i.d.) random variables. As the network size gets large, by the central limit theorem, $\sum S_{ij}x_j^f(t)$ can be approximated as a normally distributed random variable. The mean and the variance of $\sum S_{ij}x_j^f(t)$ are $(pN - 1)\mu_s x_i^q$ and $(pN - 1)(1 - \mu_s^2)$, respectively, when $i \in A$. Hence $G \sum S_{ij}x_j^f(t)$, with summation index $j \in A\setminus\{i\}$, is approximately normal with mean $G(pN - 1)\mu_s x_i^q$ and variance $G^2(pN - 1)(1 - \mu_s^2)$. Since $x_i^k x_j^k x_j^f(t)$ for $j \notin A$ and $S_{ij}x_j^f(t)$ for $j \in A\setminus\{i\}$ are independent and the summation of two independent normal random variables is still a normal random variable with its mean being the summation of the two means and variance being the summation of the two variances (Papoulis, 1985; Hogg and Craig, 1978), the distribution of the noise in the short-circuit failed interconnections is also normal with mean $G(pN - 1)\mu_s x_i^q$ and variance, σ^2,

obtained as

$$\sigma^2 = (N - pN)(M - 1) + G^2(pN - 1)(1 - \mu_s^2) \tag{13}$$

The mean value of the noise term adds a signal component to a network. Hence, the overall signal value of a network is obtained as

$$x_i^q(N - pN + G + G\mu_s(pN - 1))$$

when $i \in A$. Similarly, when $i \notin A$, the signal and the variance of noise are obtained as

$$x_i^q(N - pN + M - 1 + G\mu_s pN)$$

and

$$(N - pN - 1)(M - 1) + G^2 pN(1 - \mu_s^2)$$

respectively. The probabilities of $i \in A$ and $i \notin A$ are p and $1 - p$, respectively. Hence, the average ratio of signal to the standard deviation of noise for networks with short-circuit-type failed interconnections, C_s, is obtained as

$$C_s = \frac{p(N - pN + G + G\mu_s(pN - 1))}{\sqrt{[(N - pN)(M - 1) + G^2(pN - 1)(1 - \mu_s^2)]}}$$
$$+ \frac{(1 - p)(N - pN + M - 1 + G\mu_s pN)}{\sqrt{[(N - pN - 1)(M - 1) + G^2 pN(1 - \mu_s^2)]}} \tag{14}$$

and the probability of direct convergence in this case is obtained as $(1 - \phi(C_s))^N$.

From Eq. (12), as M increases, μ_s decreases and σ^2 increases. The increase of noise variance, especially for a physical system when $N \to \infty$ and $M \to \infty$, decreases network performance drastically as G increases. Hence, short-circuit-type failed interconnections affect a network much more than the open-circuit type, especially when both M and G are large. This phenomenon is supported by the simulation results as shown in Figs. 14.7 and 14.8, where G is assumed to be the highest interconnection value, M. In 14.7, it is shown that for a network with $(N, M) = (100, 9)$, network P_{dc} decreases from 0.993 when $p = 0$ to 0.79 when $p = 0.4$ for short-circuit-type failed interconnections. But for a network with $(N, M) = (500, 35)$, network P_{dc} decreases to 0.2 when p increases only up to 0.08, as shown in Fig. 14.8. The results also show that a network with open-circuit failed interconnections is more robust than a network with short-circuit failed interconnections.

In order to demonstrate network characteristics when the fraction of failed interconnections increases up to the extreme limit 1.0, network

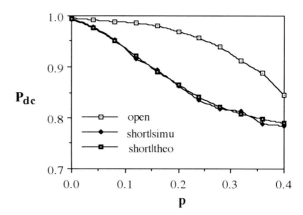

Fig. 14.7. Network reliability with open- and short-circuit failed synapses, where $(N, M) = (100, 9)$.

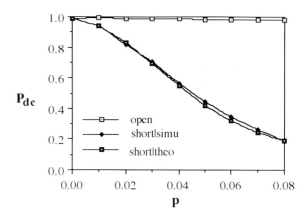

Fig. 14.8. Network reliability with open- and short-circuit failed synapses, where $(N, M) = (500, 35)$.

reliabilities with various values of G are shown in Fig. 14.9. The curves in Fig. 14.9 indicate that, as the ratio of failed interconnections increases continuously, P_{dc} for a short-circuit network can first decrease and then increase. On the other hand, when $G = 0$, i.e., with open-circuit failed interconnections, as p increases up to around 0.36 network performance starts to decrease at a fast rate. The curves for $G = 0$ and $G = 35$ intersect at p equal to about 0.56. From Fig. 14.9, we can

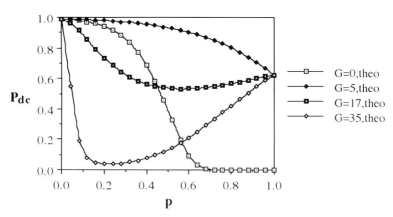

Fig. 14.9. Theoretical network reliabilities with various G, when $(N, M) = (500, 35)$.

also see that all the curves, except $G = 0$, coincide at the same point where $P_{dc} = 0.624$, when $p = 1.0$. This result can be obtained intuitively. When $G \neq 0$, as $p = 1.0$, all the interconnections are replaced with their signums multiplied by one scalar G. But in network retrieval each neuron decides its state according to the sign (either positive or negative) of the summation of its weighted inputs. Hence, as long as G is not equal to 0, no matter what value it is, the network performs exactly the same way. This phenomenon is supported in Amit (1989) and Sompolinsky (1986), where Amit and Sompolinsky indicated that even when all the interconnections in a linear associative memory are clipped to their signums, $+1$ or -1, the network still retains its storage capacity to a certain extent.

It is also found that some relatively small values of G affect the network much less than the very large values of G, e.g., $G = 35$, or the very small values of G, e.g., $G = 0$ or 1. From the theoretical equations, through trial and error, it is possible to see that as $G = 5$ the (500, 35) network has the greatest robustness. From intuition, the optimal G, which gives the least change in network performance, is obtained by $G_{opt} = E\{|T_{ijk}|\}$ where $E\{\cdot\}$ is the expectation operator. The actual value obtained for this G_{opt} is 4.75, which is quite close to the best value, 5. This value can be used inside a fault-detection and replacement mechanism to improve network performance when interconnection faults occur. Actually, from both theoretical and simulation results, as long as G is around G_{opt}, networks have almost the same P_{dc} curves with various values of G, i.e., performance is not very sensitive to G.

14.5. SYNCHRONOUS NETWORK THEORY

Radius of attraction is an important parameter in associative memories. It indicates how many input error bits a network can tolerate and still give us an acceptably high probability of correct recall. Synchronous and asynchronous networks have different network performance, particularly in the probability of direct one-step convergence, when attraction radius is considered. Associative memories are known to have the capability of error correction, especially when a network does not exceed its capacity. Hence, the number of incorrect input bits in an asynchronous network is reduced when state updating proceeds from one neuron to another and the probability of a neuron changing to a correct state is increased. This causes the asynchronous network to have a

performance superior to the synchronous network.

For a synchronous network, the state updating of one particular neuron is not affected by the updating of other neurons. During each iteration cycle, all neurons have the same input patterns. When the probability of direct one-step convergence is considered, the input pattern is the probe vector applied to the network input leads. Assume that the probe vector, $\mathbf{x}^f(t)$, is different from the pattern to be retrieved, $\mathbf{x}^q$, by b bits. When the updating neuron, say neuron i, has a disconnected self-connection weight, i.e., $i \in A$, referring to Eq. (3), the signal term can be obtained as

$$S = x_i^q(1 - p)(N - 2b) \qquad (15)$$

However, when $i \notin A$, Eq. (4) should be used in obtaining the signal value. In this case, the signal for correct bits, i.e., $x_i^f(t) = x_i^q$, and incorrect bits, i.e., $x_i^f(t) \neq x_i^q$, are different. By summing the first and second terms of Eq. (4), the signal for a correct bit is obtained as

$$S = x_i^q((1 - p)(N - 2b) + M - 1) \qquad (16)$$

On the other hand, if neuron i is an incorrect bit, i.e., $x_i^f(t) \neq x_i^q$, the signal is obtained as

$$S = x_i^q((1 - p)(N - 2b) - (M - 1)) \qquad (17)$$

Inserting the ratios of signal to the standard deviation of noise for correct and incorrect bits into Eq. (7), the probability of correct recall is calculated as

$$P_{dc} = (1 - \phi(C_1))^{N-b}(1 - \phi(C_2))^b \qquad (18)$$

where C_1 and C_2 are the average ratios of signal to the standard deviation of noise for correct and incorrect bits, respectively. The probability of $i \in A$ is p and that of $i \notin A$ is $1 - p$. Hence C_1 and C_2 are obtained as

$$C_1 = \frac{p(1 - p)(N - 2b)}{\sqrt{[(N - pN)(M - 1)]}}$$
$$+ \frac{(1 - p)((1 - p)(N - 2b) + (M - 1))}{\sqrt{[(N - pN - 1)(M - 1)]}} \qquad (19)$$

and

$$C_2 = \frac{p(1 - p)(N - 2b)}{\sqrt{[(N - pN)(M - 1)]}}$$
$$+ \frac{(1 - p)((1 - p)(N - 2b) - (M - 1))}{\sqrt{[(N - pN - 1)(M - 1)]}} \qquad (20)$$

Given the required probability of correct recall, an increase in the fraction of failed interconnections decreases network attraction radius b.

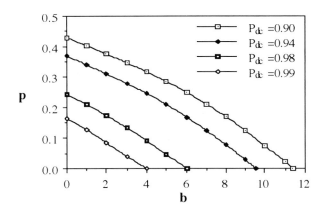

Fig. 14.10. Effect of fraction of failed interconnections on attraction radius for given P_{dc}.

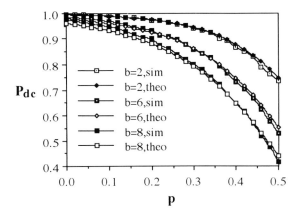

Fig. 14.11. Network P_{dc} for various b in synchronous operation.

Figure 14.10 shows this effect, when $(N, M) = (100, 8)$, given that the probability of correct recall is at least 0.9. A similar diagram where curves of P_{dc} versus p (fraction of failed interconnections) are plotted for various b is shown in Fig. 14.11. Obviously, an increase in failed interconnections decreases the network attraction radius and thus decreases the generalization capability of the network.

14.6. ASYNCHRONOUS NETWORK THEORY

In asynchronous operation the updating sequence and the positions of the input error bits greatly affect network performance because of the capability of error correction in neural networks. Networks analyzed in this section are assumed to have a P_{dc} approaching 1. Given that the network updating sequence proceeds from neuron $1 \rightarrow 2 \rightarrow 3 \rightarrow \cdots \rightarrow N$ and the b error bits are located in the first b neurons, after state updating of the first neuron, it is almost certain that the state of neuron 1 has been corrected. Hence, in

updating the state for neuron 2, there are only $b - 1$ incorrect input bits. During state updating for neurons $b + 1$ to N, there is a high probability that there are no input error bits remaining. This capability of error correction results in asynchronous operation giving superior performance when the network attraction radius is considered.

Assume that an asynchronous network updates its neuron states going from neuron $1 \rightarrow 2 \rightarrow 3 \rightarrow \cdots \rightarrow N$. Further assume that the b error bits in the input probe vector $\mathbf{x}^f(t)$, $t = 0$, are randomly distributed across the N neurons. There are C_b^N combinations in which these b error bits are distributed, with each combination having probability $1/C_b^N$. As N increases, the number of combinations is extremely large. It is impossible to consider all of the combinations individually. However, note that, for these C_b^N combinations, there is a total of NC_b^N bits. Among these NC_b^N bits, there are

$$N_b = C_{b-1}^{N-1} + 2C_{b-1}^{N-2} + 3C_{b-1}^{N-3}$$
$$+ \cdots + (N + 1 - b)C_{b-1}^{b-1} \qquad (21)$$

situations where neurons have n input error bits

before their state update. This equation can be obtained by looking into all of the possible combinations. If the first input is an incorrect bit, neuron 1 has b input error bits when computing its state update. After neuron 1 has been updated, it is almost certain that this incorrect bit has been corrected. The number of combinations where the first input will be an incorrect bit is C_{b-1}^{N-1}. For those combinations where the first input is correct and the second input is incorrect, both neuron 1 and neuron 2 have b input error bits in computing their state update. There are a total of C_{b-1}^{N-2} combinations when neuron 1 is correct and neuron 2 is incorrect, resulting in $2C_{b-1}^{N-2}$ neurons having b input error bits during their state update. Following this argument, Eq. (21) is obtained.

Now we will look into those neurons having $b-1$ input error bits in their state update. First, assume that the network is within its storage capacity, so that a neuron, no matter whether it has a correct or an incorrect initial state, will turn to a correct state after update. Based on this assumption, a neuron has $b-1$ input error bits during its state update if and only if there is exactly one error bit located in the neurons preceding it. In other words, within those neurons which have been updated before the neuron of interest, there is exactly one neuron which has an incorrect initial state. This result can be clarified by looking at Fig. 14.12, where the network size $N = 5$ and the number of error bits $b = 3$. Each row in Fig. 14.12 represents one input combination. The X's show where the three input error bits, and consequently the three erroneous initial states, are located. The blocks marked with an encircled 2 represent those neurons having $b-1 = 2$ input error bits during their state update. As we can see from the figure, for each block with a 2 marked inside, there is exactly one X in front of it. The total number of neurons having $b-1$ input error bits can be computed by taking the error bits starting from the first bit plus the error bits starting from the second bit, up to the error bits starting from the last bth bit, as follows.

1. The error bits start from the first bit. For example, suppose neuron 1 has an incorrect initial state.

When neuron 1 is assigned as one of the neurons having an incorrect initial state, the remaining $b-1$ incorrect bits can be distributed within the rest of the $N-1$ neurons. Therefore, the number of neurons having $b-1$ input error bits, in this case, can be obtained from Eq. (21) with N replaced by $N-1$ and b replaced by

comb. \ neu.	1	2	3	4	5
1	③ X	② X	① X	⓪	⓪
2	③ X	② X	①	① X	⓪
3	③ X	② X	①	①	① X
4	③ X	②	② X	① X	⓪
5	③ X	②	② X	①	① X
6	③ X	②	②	② X	① X
7	③	③ X	② X	① X	⓪
8	③	③ X	② X	①	① X
9	③	③ X	②	② X	① X
10	③	③	③ X	② X	① X

Fig. 14.12. Error bit mapping in asynchronous operation. X represents the error bit location, the number inside the circle represents the number of input error bits in the state update.

$b-1$ as

$$C_{b-2}^{N-2} + 2C_{b-2}^{N-3} + 3C_{b-2}^{N-4} + \cdots + (N+1-b)C_{b-2}^{b-2}$$

2. The error bits start from the second bit. For example, suppose neuron 1 has a correct initial state and neuron 2 has an incorrect initial state.

Since neurons 1 and 2 are already assigned as having a correct and incorrect initial state, respectively, there are only $N-2$ neurons to which the remaining $b-1$ errors can be distributed. Hence, within this case, the number of neurons having $b-1$ incorrect input bits is obtained from Eq. (21) with N replaced by $N-2$ and b replaced by $b-1$ as

$$C_{b-2}^{N-3} + 2C_{b-2}^{N-4} + 3C_{b-2}^{N-5} + \cdots + (N-b)C_{b-2}^{b-2}$$

The same argument can be followed up to when neurons $1, 2, \ldots, N+1-b$ have correct initial states and neurons $N-b$ to N have incorrect initial states. In this last case, the number of neurons having $b-1$ input error bits is

$$C_{b-2}^{b-2} = 1$$

Adding all of these cases together, the total number of neurons having $b-1$ input error bits in their state update is obtained as

$$N_{b-1} = \{C_{b-2}^{N-2} + 2C_{b-2}^{N-3} + 3C_{b-2}^{N-4} + \cdots$$

$$+ (N + 1 - b)C_{b-2}^{b-2}\}$$

$$+ \{C_{b-2}^{N-3} + 2C_{b-2}^{N-4} + 3C_{b-2}^{N-5} + \cdots$$

$$+ (N - b)C_{b-2}^{b-2}\} + \cdots + \{C_{b-2}^{b-2}\} \quad (22)$$

Using the fact that

$$C_x^N = \sum_{i=1}^{N-x+1} C_{x-1}^{N-i} \quad (23)$$

it is clear that $N_b = N_{b-1}$. The same procedure can be used to prove that $N_b = N_{b-1} = \cdots = N_1$ where N_b is equal to Eq. (21). It can also be shown that the number of neuron occurrences having no input error bits in their state update is equal to

$$N_0 = C_{b-1}^{N-2} + 2C_{b-1}^{N-3} + \cdots + (N - b)C_{b-1}^{b-1} \quad (24)$$

Obviously, N_0 is smaller than

$$N_b = N_{b-1} = \cdots = N_1$$

From this, given a random combination of input bit errors, the average number of neuron occurrences having $b, b-1, \ldots, 0$ input error bits in their state update can be obtained as

$$n_b = n_{b-1} = \cdots = n_1 = \frac{N_b}{NC_b^N} N = \frac{N_b}{C_b^N} \quad (25)$$

and

$$n_0 = \frac{N_0}{C_b^N} \quad (26)$$

One example showing how to calculate N_b, $N_{b-1}, \ldots, N_1$ and N_0 is as follows. Given that the network size $N = 5$ and the number of error bits, b, is equal to 3, it is easy to find that $C_3^5 = 10$. So there are 10 combinations in which the 3 error bits can be distributed within these 5 input positions. There are 50 bits total in these 10 input combinations. Applying Eqs. (21) and (24), N_3, N_2, N_1, and N_0 are calculated to be

$$N_3 = C_2^4 + 2C_2^3 + 3C_2^2 = 15$$

$$N_2 = 15 \qquad N_1 = 15$$

and

$$N_0 = C_2^3 + 2C_2^2 = 5$$

These numbers can be verified by looking at Fig. 14.12, where there are 15 circles with numbers, 3, 2, and 1, respectively, and 5 circles with number 0.

Before using Eqs. (18) to (20) for calculating network convergence probability, we need to predict whether the state of the neuron itself is correct or not. From Fig. 14.12, it can be seen that within each $N_b, N_{b-1}, \ldots, N_1$ there are C_b^N neurons having incorrect inputs (and initial states). Thus, we have $C_b^N: N_b = B: n_b$, where B represents the number of neurons within each n_b, $n_{b-1}, \ldots, n_1$ which have incorrect initial states. Hence, it is reasonable to expect that $B = n_b C_b^N / N_b = 1$. This result coincides with the intuition that if there are N_x neurons having x error bits in their state update, these N_x neurons must be consecutive and the last of them must be the only neuron which has an incorrect initial state. Now the same procedure used for analyzing the synchronous operation can be used here. The probability of convergence is then

$$P_{dc} = \left\{ \prod_{i=1}^{b} (1 - \phi(C_i^c)) \right\}^{n_b - 1}$$

$$\times \left\{ \prod_{i=1}^{b} (1 - \phi(C_i^e)) \right\} \{1 - \phi(C_0^c)\}^{n_0} \quad (27)$$

where

$$C_i^e = \frac{p(1-p)(N-2i)}{\sqrt{[(N-pN)(M-1)]}}$$

$$+ \frac{(1-p)((1-p)(N-2i) - (M-1))}{\sqrt{[(N-pN-1)(M-1)]}} \quad (28)$$

$$C_i^c = \frac{p(1-p)(N-2i)}{\sqrt{[(N-pN)(M-1)]}}$$

$$+ \frac{(1-p)((1-p)(N-2i) + (M-1))}{\sqrt{[(N-pN-1)(M-1)]}} \quad (29)$$

and C_0^c represents C_i^c with i equal to zero.

The performance of asynchronous operation compared with synchronous operation, when $(N, M) = (100, 8)$ and $b = 6$, is shown in Fig. 14.13. Theoretical estimates using Eqs. (18) to (20) and (27) to (29) are also plotted in this figure.

14.7. CLUSTERED FAILED INTERCONNECTIONS

In some cases, when the interconnection damage results from a large particle or some aggregate of contaminants, interconnection faults can be clustered within a small area, rather than evenly distributed over the network. In this section, the failed interconnections are assumed to be open circuits which are located within a small square block. The reason for choosing a square area is

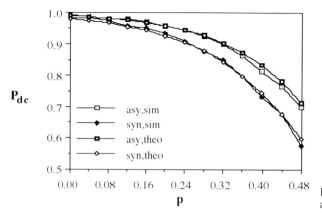

Fig. 14.13. Network direct convergence for asynchronous and synchronous operations.

that analysis is easier. Also, we assume that the interconnections are orderly in space such that T_{ij} is next to $T_{i+1,j}$ and T_{ij+1}. Given p, the ratio of failed interconnections, there are a total of pN^2 failed interconnections in a network. Assume that all of these failed interconnections are clustered within a square block, with the cluster center randomly located over the interconnection matrix. Each side of the square block is $N(p)^{1/2}$. This means that among these N neurons, $N(p)^{1/2}$ neurons have failed interconnections in their inputs, with each neuron having $N(p)^{1/2}$ failed interconnections. All the failed interconnections in the input leads of a specific neuron occur in consecutive synapses. Therefore, the equation for estimating network performance is modified to

$$P_{dc} = (1 - \phi(C_1))^{N\sqrt{p}}(1 - \phi(C_2))^{N - N\sqrt{p}} \quad (30)$$

where C_1 and C_2 are the ratios of signal to standard deviation of noise from Eq. (5) with the ratios of failed interconnections being $(p)^{1/2}$ and 0, respectively. Network reliability comparisons with square clustered failed interconnections and with uniformly distributed failed interconnections

are shown in Fig. 14.14 for the case of open-circuit type failures and $(N, M) = (50, 6)$. We note that clustered failures are more damaging than uniform failures.

From the analyses in the previous sections, equations for estimating performance for various geometric structures of failed interconnections can also be derived, as long as the number of failed interconnections in each neuron is known. For example, given an arbitrary geometric structure, if x_{di} represents the number of open-circuit failed interconnections in the input of neuron i, the probability of network convergence is obtained as

$$P_{dc} = \prod_{i=1}^{N} (1 - \phi(C|p = x_{di}/N)) \quad (31)$$

where $\phi(C|p = x_{di}/N)$ represents the standard error function with parameter C obtained by using $p = x_{di}/N$. The analysis described in this section can also be extended to obtain an analytical expression for P_{dc} for networks with clustered short-circuit failed interconnections.

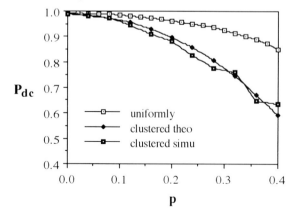

Fig. 14.14. Network reliability with uniform and clustered failed interconnections.

14.8. QUADRATIC HEBBIAN-TYPE ASSOCIATIVE MEMORIES (QHAMs)

Higher-order associative memories were proposed mainly for enhancing storage capacity. The increase in memory capacity is achieved primarily through the high-order correlation between neurons, which leads to an increase in network nonredundant parameters. The correlation between neurons in a quadratic associative memory is extended from two-neuron interactions into three-neuron interactions, resulting in an increase in the number of interconnections from $O(N^2)$ to $O(N^3)$, where N is the number of neurons. This section briefly discusses the fault-tolerance capabilites of QHAMs. As will be shown shortly, not only the memory capacity but also the tolerance to interconnection faults is increased dramatically in a quadratic memory. However, its fault-tolerance capability in terms of noise input patterns is very low, i.e., generalization capability is reduced.

Because of the correlation among different neurons, a dependent component within the noise term needs to be carefully separated out in analyzing a QHAM (Chung, 1991). Following that, the C parameter for an open-circuited QHAM can be obtained as

$$C = \frac{(N-1)(N-2)(1-p)}{\sqrt{[(M-1)(N-1)(N-2)(2-3p+p^2)]}}$$

(32)

For the same M and N, this number is much higher than that obtained in a HAM in Eq. (5). In other words, given the same N, the M value in a quadratic HAM can be raised to a very high value and the network still has the same value of C. Hence, the capacity of a quadratic network is increased. For instance, given the same probability of correct recall, the number of vectors stored in a HAM with $N = 50$ is 6, but for a QHAM with $N = 46$, the M can be as high as 69. Comparing Eqs. (5) and (32), we see that the fraction of failed interconnections, p, decreases the C parameter of a quadratic HAM much more slowly than for the linear case. Hence, reliability characteristics of a quadratic network are superior to those of a linear network. When network attraction radius is considered, the C parameters given a fraction p of interconnection faults are computed as

$$C_1 = \frac{((N-2b)(N-2b-1)-2(N-b-1))(1-p)}{\sqrt{[(M-1)(N-1)(N-2)(2-3p+p^2)]}}$$

(33)

and

$$C_2 = \frac{((N-2b)(N-2b+1)-2(b-1))(1-p)}{\sqrt{[(M-1)(N-1)(N-2)(2-3p+p^2)]}}$$

(34)

for the correct and incorrect bits, respectively, during synchronous operation. Examining these two equations, we can see that an increase in the number of input error bits significantly decreases network performance. For instance, from our theoretical calculations, a network with $N = 42$ and $b = 0$ can store 69 vectors and still have a P_{dc} equal to 0.99. If 3 error bits are allowed in the probe vectors, the network can only store 38 vectors in order to keep a P_{dc} as high as 0.99. If the number of input error bits is increased to 6, only 17 vectors are allowed for the same P_{dc}. The dramatic drop in network capacity indicates that a slight increase in the number of error bits decreases the network P_{dc} considerably. From this, it is also predictable that the ratio of failed interconnections will not affect network attraction radius much when p is small.

14.9. SUMMARY

The interconnection faults in a resistive weight matrix could result in open-circuit or short-circuit situations. Equivalent fault models can also be applied to optical neural network implementations. With the same fraction of interconnection faults, the short-circuit failed interconnections decrease network performance more than the open-circuit type do, especially when short circuit gain G is large. This result also implies that an implementation technique that would cause an open-circuit situation in failed interconnections is better than a technique that would cause a short-circuit situation. Certain values of G are found to have only a mild impact on network performance degradation, a fact which can be used for improving network reliability. When network attraction radius is considered, an asynchronous network presents a higher performance than a synchronous network. Clustered faults result in a greater rate of performance degradation than uniformly distributed faults. Finally, quadratic HAMs provide higher fault-tolerance capability than ordinary (or linear type) HAMs.

Based on "Characteristics of Hebbian-Type Associative Memories" by Pau-Choo Chung and Thomas F. Krile, which appeared in *IEEE Transactions on Neural Networks*, 3(6), 969–980, November 1992. Copyright © 1992 IEEE.

REFERENCES

Abu-Mostafa, Y. S. and Jacques, J. M. (1985). "Information Capacity of the Hopfield Model," *IEEE Trans. Information Theory*, **IT-31**, 461–464.

Amit, D. J. (1989). *Modeling Brain Function: The World of Attractor Neural Networks*, Cambridge University Press, Cambridge.

Chung, P. C. (1991). "Reliability Characteristics of Neural Networks Having Faulty Interconnections," Ph.D. dissertation, Dept. E.E., Texas Tech University, August.

Dembo, A. (1989). "On the Capacity of Associative Memories with Linear Threshold Functions," *IEEE Trans. Information Theory*, **35**, 709–719.

Foo, S. Y., Anderson, L. R., and Takefuji, Y. (1990). "Analog Components for the VLSI Implementation of Neural Networks," *Circuits & Devices*, **6**, 18–26.

Hogg, R. V. and Craig, A. T. (1978). *Introduction to Mathematical Statistics*, 4th edn. Macmillan, New York.

Kuh, A. and Dickinson, B. W. (1989). "Information Capacity of Associative Memories," *IEEE Trans. Information Theory*, **35**, 59–68.

Masaki, A., Hirai, M., and Yamada, M. (1990). "Neural Networks in CMOS: a Case Study," *Circuits & Devices*, **6**, 12–17.

McEliece, R. J., Posner, E. C., Rodemich, E. R., and Venkatesh, S. S. (1987). "The Capacity of the Hopfield Associative Memory," *IEEE Trans. Information Theory*, **IT-33**, 461–482.

Newman, C. M. (1988). "Memory Capacity in Neural Network Models: Rigorous Lower Bounds," *Neural Networks*, **1**, 223–238.

Papoulis, A. (1985). *Probability, Random Variables, and Stochastic Processes*, 3rd ed. McGraw-Hill, New York.

Sompolinsky, H. (1986). "The Theory of Neural Networks: The Hebb Rule and Beyond," in *Heidelberg Colloquium on Glassy Dynamics*, J. L. Van Hermmen and I. Morgenstern, eds. Springer-Verlag, New York.

Venkatesh, S. S. and Psaltis, D. (1989). "Linear and Logarithmic Capacities in Associative Neural Networks," *IEEE Trans. Information Theory*, **IT-35**, 558–568.

Wang, J. H., Krile, T. F., and Walkup, J. F. (1990). "Determination of Hopfield Associative Memory Characteristics Using a Single Parameter," *Neural Networks*, **3**, 319–331.

PART IV

IMPLEMENTATION

15

Analog Implementation of an Associative Memory: Learning Algorithm and VLSI Constraints

MICHEL VERLEYSEN, JEAN-DIDIER LEGAT, and PAUL JESPERS

15.1. INTRODUCTION

Artificial neural networks can be classified by their architectures; the main classes of models are feedback networks (as the Hopfield net), layered networks (as the multilayer perceptron), and auto-organized networks (as the Kohonen map). All these networks realize associative memory processing; the Hopfield model corresponds to an autoassociative memory, perceptrons correspond to heteroassociative memories, and Kohonen maps to classifiers, although this distinction is quite artificial. Most current models of artificial neural networks are based on one of these three architectures; many differences in the structure, in the learning rule, or in the way the data are represented can, however, exist in other models. Some combine these concepts, for example, by adding feedback to multilayer perceptrons. The diversity of all models makes it difficult to build a complete list with their full characteristics.

What is important here is to point out the common features in all these models. The first characteristic relies on the common operations involved in nets. In the Hopfield model, a matrix product is performed between the input vector (or the output of the previous computation step) and a matrix of weights. In layered nets of perceptrons, a similar matrix–vector product is performed in each layer; they differ from the Hopfield model by the absence of feedback. In Kohonen maps, the first layer is devoted to the measure of distances between the input vector and the stored ones, which is also a matrix–vector product.

The second important common aspect between these models involves the precautions that must be taken regarding the precision and the dynamics of the values involved in the operations (synaptic weights, activation levels, ...), and

the dimensions and cascadability of chips. Except for some particular networks, interesting properties of neural nets can only be found when their size (number of neurons and synapses) is large. Real parallel computing can then be performed, and the speed of the computations can become significant in comparison to classical serial computing methods. The advantages of analog VLSI, in particular situations, will be discussed in a later section, but it will be seen that the problems mentioned above, especially the cascadability, are not so simple to handle with analog design; again, the solutions proposed for one model can easily be transposed to others, and the topological layout of the net is of less importance.

This chapter emphasizes the Hopfield model. It has been chosen for several reasons. First, historically, it was one of the most used networks for associative memories; applications were lacking for self-organizing maps, and multiple-layer nets of perceptrons suffered from the complexity of their learning algorithms. Second, the simplicity of the Hopfield model makes it easier to study compared to more complex networks. Finally, as mentioned above, VLSI constraints for the implementation of neural networks are similar in the different models; it is thus natural to experiment with a simple model, keeping in mind that the results can easily be transposed to more complex ones.

15.2. HOPFIELD NETWORK

The Hopfield model is a simple recurrent network, as shown in Fig. 15.1. Refer to Chapter 1 for operation and details of this architecture.

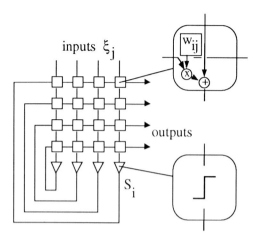

inputs ξ_j

w_{ij}

outputs

S_i

Fig. 15.1. Hopfield network.

15.3. LEARNING RULES FOR HOPFIELD NETWORKS

15.3.1. Hebb's Rule

The Hebb rule (Hebb, 1949) for storing p bipolar patterns is given by

$$
\left.
\begin{array}{l}
w_{ij} = \dfrac{1}{N} \displaystyle\sum_{k=1}^{p} \xi_i^k \xi_j^k \quad \text{if } i \neq j \\[2ex]
w_{ii} = 0
\end{array}
\right\} \tag{1}
$$

where w_{ij} is the weight of the synapse connecting neurons i and j, and ξ_i^k is bit i of pattern k to be memorized.

For random stored patterns, it is generally agreed that a capacity limit of $0.15N$ vectors must not be exceeded, to keep performances acceptable. This corresponds to the theoretical limits found by McEliece (1987) and Abu-Mostafa (1985). To integrate this learning rule in VLSI circuits, where the precision and dynamics of the coefficients are limited, the clipping of weights must be examined. Equation (1) shows that each pattern to be stored contributes to each weight a value $+1$ or -1. If p patterns are stored, it is thus obvious that $2p + 1$ different values are allowed for the weights. The number of bits required for storing the weights is then given by

$$
B = \log_2(2p + 1) \tag{2}
$$

The smallest integer value greater than B is the number of digital memory points necessary for storing the weight.

Clipping the Hebb rule to a one-bit accuracy is possible by applying a sign function to the weight obtained by (1). The capacity, simply

measured as the number of random patterns that it is possible to store without error, is then reduced to about $0.11N$ vectors (compared to $0.15N$ for the unclipped Hebb rule). Morgenstern (1987) proposed a variant of this method, the zero model, which consists of setting the weight to zero if its absolute value is lower than some threshold Z, and applying the sign function for the other cases. The capacity can then be raised to a mean value of $0.12N$.

15.3.2. Projection Rule

The projection rule (generalized-inverse rule) is covered and discussed in Chapter 1 (see also Personnaz, 1985). It is given by

$$
W = \Psi \cdot \Psi^+ \tag{3}
$$

where

$$
\Psi^+ = (\Psi^T \cdot \Psi)^{-1} \cdot \Psi^T
$$

and

$$
\Psi = [\xi^1, \xi^2, \ldots, \xi^p] \tag{4}
$$

Clipping the weights obtained with Eq. (3) leads to disastrous results. The reason is the matrix inversion involved in the rule; pseudo-inverse is indeed a generalization of matrix inversion, which involves many divisions in its computation. Depending on the vectors, the results can theoretically have infinite dynamics, and all the information contained in this dynamics is completely lost with any type of truncation. Clipping the projection rule will thus not be considered here.

15.3.3. Optimization Rule

The projection rule guarantees that patterns ξ^k are fixed points as long as these patterns are linearly independent; but the method does not speak about the degree of stability of such patterns. Basins of attraction can be represented as in Fig. 15.2 for all patterns memorized with the projection rule; circles around each pattern define the minimum basins guaranteed by the rule. It must be mentioned that Fig. 15.2 is only a schematic 2-D representation of the attractiveness of vectors; the reality would require an N-dimensional representation with vectors situated on hypercube vertices, and Hamming distances instead of 2-D Euclidian ones.

Good behavior of the algorithm is shown for patterns situated on the left part of Fig. 15.2; however, when the size of the basins of attraction guaranteed by the rule is small in comparison with the distance between patterns, the limits of an attraction domain can be close to a pattern;

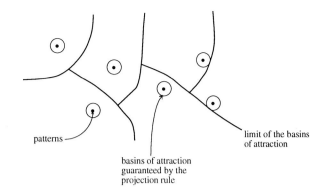

Fig. 15.2. Basins of attraction with the projection rule.

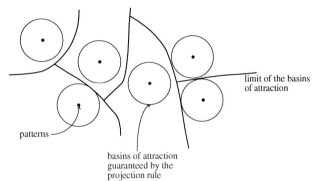

Fig. 15.3. Basins of attraction with the optimization rule.

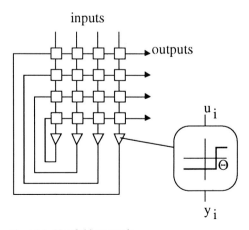

Fig. 15.4. Hopfield network.

this is nonoptimal, as shown in the right part of Fig. 15.2. We propose here an original algorithm (Verleysen et al., 1989) which maximizes the size of the basins of attraction for a set of given patterns (Fig. 15.3). This learning rule will be termed the "optimization rule."

If the Hopfield network of Fig. 15.4 is examined, it can be seen that a measure of the stability of a pattern is the difference between the activation of a neuron and its threshold Θ. The activation u_i is defined by

$$u_i = \sum_{j=1}^{N} w_{ij} y_j \qquad (5)$$

A stability factor S is defined here for a particular bit i in a particular pattern ξ^k as

$$S_i^k = |u_i - \Theta| \qquad (6)$$

Without loss of generality, Θ will be set to 0 in the following, since a threshold can be formally replaced by a supplementary fixed input.

The proposed optimization rule relies on the following intuitive view. A change in one bit of the vector **y** (for example -1 instead of $+1$) can induce a change of a maximum of 2 (or -2) in the activation value u_i (assuming that the weights w_{ij} are in $[-1, +1]$). If the stability factor S is constant for all bits of all stored patterns, and is set to 1 as in the projection rule, it means that one wrong bit in the pattern **y** can lead to a wrong calculation of any other bit in the pattern during the next iteration of the network. A remedy to this problem would be to increase the stability factor S, again for all bits i and patterns k. If S is set to C, a maximum number of $C/2$ erroneous bits will be corrected at the next iteration, since the signs of all y_i are not affected by these wrong bits. Formally, for a stable state ξ^k,

$$|u_i| = \left| \sum_{j=1}^{N} w_{ij}\xi_j^k \right| \geq C \qquad (7)$$

If $\mathbf{y}$ differs from ξ^k by D bits,

$$|u_i| = \left| \sum_{j=1}^{N} w_{ij}y_j \right| \geq C - 2D \qquad (8)$$

Thus if $D < C/2$, the sign of u_i in (8) will be identical to the sign of u_i in (5), and this confirms that $\mathbf{y}$ will evolve to ξ^k at the next iteration. S is thus a good measure for the attractiveness of the patterns in one cycle; it is assumed here that a good attractiveness factor S measured in one-cycle convergence will also lead to good attractiveness in several cycles.

The idea of the optimization rule is thus to maximize the value of S. Of course, S cannot be increased indefinitely; there exists a maximum value of S for which the system can be solved, with w_{ij} in $[-1, +1]$. The problem is the dependence of this maximum on the value of patterns ξ^k. Intuitively, it is obvious that a weak correlation between the patterns ξ^k will permit a higher value for S than if the patterns are strongly correlated; this is, however, extremely difficult to compute analytically. It is thus proposed to rewrite the system as an optimization problem. The objective is now to maximize S,

$$S_i^k = \left| \sum_{j=1}^{N} w_{ij}\xi_j^k \right| \qquad (9)$$

Forcing the same S for all bits i and all patterns k is, however, too restrictive. In order to have only one variable to maximize, the minimum of all S_i^k will be considered. The problem is thus

$$\text{Maximize } S \quad \text{where } S = \min_{i,k}(S_i^k) \qquad (10)$$

and

$$S_i^k = \left| \sum_{j=1}^{N} w_{ij}\xi_j^k \right| = \sum_{j=1}^{N} w_{ij}\xi_j^k\xi_i^k \qquad (11)$$

It can be seen in Fig. 15.4 and in equations (5) to (11) that the maximizations are independent of i. In order to simplify the algorithm that will solve the problem, different S_i will be considered for the different bits. The complete problem can then be defined by:

$$\text{Maximize } S_i \quad \text{where } S_i = \min_k(S_i^k) \qquad (12)$$

with

$$S_i^k = \sum_{j=1}^{N} w_{ij}\xi_j^k\xi_i^k - E_i^k \qquad (13)$$

under the constraints

$$E_i^k \geq 0 \qquad (14)$$

$$-1 \leq w_{ij} \leq 1 \qquad (15)$$

the constraints being valid for all patterns k and all bits j. The coefficients E_i^k add free degrees to the system, so that the S_i^k can be different. The problem (12)–(15) must be repeated N times for the N bits i, and can easily be solved by a standard simplex procedure (Lasdon, 1970).

This rule can be used in problems where on-line computation of weights is not necessary. Indeed the complexity of the computations, and the restriction that learning is nonadaptive (all weights must be recomputed each time information is added into the network), make the rule best suited to systems where the learning can be computed by an external computer before using the associative memory in its recall phase.

15.3.4. Clipping the Optimization Rule

The problem of clipping is quite different with this learning rule. Equation (15) limits the dynamics of the coefficients to the range $[-1, +1]$; it could be limited to any other symmetric range without changing any result, because of the above-mentioned scaling property. Only the precision must then be known to evaluate the number of necessary bits in each memory point.

The algorithm applied to one column of the synaptic matrix can be seen as an optimization problem with N variables (the N synaptic weights) and p constraints (the p equations 13) to maximize simultaneously); $N - p$ variables will then directly take the values -1 or 1, while the remaining p may take any value in $[-1, +1]$. In most problems, however, N is much greater than p; furthermore, if this were not the case, it can be seen that the performance of the memory would be too poor for efficient applications. The optimization method thus guarantees that most of the weights can be coded into one bit.

Equation (9) shows the quantity to maximize for each of the patterns. It can intuitively be seen that the only effect of any further restriction on the variables (the synaptic weights) is to decrease S in (9) and (10). Since S is the quantity to be maximized, these restrictions have the same effect as a nonoptimal convergence of the optimization process; the solution will not be the optimum, but will be near it. Clipping the weights to the values 1 or -1, or eventually 0, is such a restriction. It is, however, difficult to appreciate the difference between S in the optimal solution and S in the solution obtained with these restrictions, since it depends on the absolute value of S, which itself depends on the patterns to be memorized. However, since the number of weights where this restriction is applied is less than p, with a small ratio p/N, the quantity S will not decrease drastically.

Our simulations show that there is only a very small decrease in memory performance if the weights are truncated to $\{-1, 0, +1\}$ or $\{-1, +1\}$ or $\{-1, +1\}$; the reader is referred to Chapter 1 for additional discussion of the effects of weight accuracy on memory retrieval.

This concept becomes important when the associative memory is realized with analog or digital VLSI circuits. The number of bits will indeed determine the complexity of each synapse, i.e., their size on the chip. If only one or two bits are needed, the number of synapses that can be implemented on a single chip can be increased, enhancing the performance of the memory for a fixed area. Circuits using low synaptic weight accuracy will be developed in the rest of this chapter.

15.4. VLSI ASSOCIATIVE MEMORIES

Modern computers are based on up-to-date VLSI techniques, and their performance and reliability depend on the quality of the technological process, the quality of the design, and the use of appropriate architectures. Neural networks can be viewed as a way to realize defined functions and objectives by using adaptive nonconventional methods that differ from conventional processing by their lack of need for deterministic algorithms. It is, however, obvious that neural networks cannot replace von Neumann's architecture. Present artificial neural nets are most effective in specialized processing, in the same way that mathematical coprocessors increase the efficiency of CPUs in mathematical problems. All functions that are, or will be, realized with neural networks can also be realized with any kind of personal computer or minicomputer by programming it with sequential languages. This is usually called simulation, although the term is not appropriate: what is the limit between simulation and realization when speaking about neural networks? Neurons and synapses are indeed biological cells, and their electronic or optical realization is already a "simulation" of their biological behavior; an implementation of the algorithm on a PC is a "simulation" of the behavior of the electronic circuit, and the first

study of the algorithm is a "simulation" of its behavior when it is used for real applications.

VLSI is thus only one way among many others of realizing artificial neural networks, but it is a way adapted to the requirements of the problems intended to be solved with these architectures. Neural networks are structures where many simple processing elements are connected together, and where the computational power comes from the collective behavior of the network. VLSI is therefore an excellent candidate to their realization, since it is easy to implement the simple processing elements involved, and since the recent advances in VLSI technology allow the integration of large numbers of transistors and devices, and thus of the large number of neuronal cells, needed for a collective behavior.

Other implementations could also be considered, such as optical ones (see the chapters 18 and 19 in this book). Neural networks are certainly not the panacea, and obviously cannot be used for the resolution of all problems in engineering fields. In their present form, neural networks are only interesting when connected to general-purpose computing machines; and the mixing of electronics and optical processing is still an emerging technology that will not be fully used for decades! The only valid method of research for short-term use of neural networks thus seems to be VLSI implementation, although other solutions are not excluded for the future.

15.4.1. Perspectives and Objectives

An ideal associative memory would be a device that can be used in any association problem, of any size, and that finds the correct solution after a negligible delay. Such an ideal artificial neural network of course does not exist, and any VLSI realization will always be a compromise between size, performance, and speed of computation. These three concepts can be detailed in terms of VLSI cells.

15.4.2. Size

The collective behavior of neural networks necessitates large numbers of neurons and synapses. Hopfield networks, for example, require N neurons and N^2 synapses to handle patterns of N bits of information. This number is even larger in layered nets of perceptrons. Size is thus one of the most important features in VLSI implementations of neural networks, and will determine the complexity allowed for each cell, since the total number of transistors allowed on a chip cannot be increased indefinitely. This number depends on the technology used, and on the percentages of defects per unit area. The number of cells that can be implemented on a single chip will thus be directly determined by the number of transistors in each cell, and the goal here is to simplify these cells as much as possible.

15.4.3. Performance

The performance of a defined associative memory structure is an important aspect to consider. This is normally affected by the precision of the weights, the activation values (signals fed into activation nodes), and the state values (output of activation nodes). For example, it is quite different working with one- or two-bit weights in Hopfield networks than with 16-bit weights. Since precision is area-consuming, the goal here will be to consider the constraint of weight precision in the VLSI network. In later sections of this chapter, reduced precision architectures as well as a compromise between network precision and speed will be considered.

15.4.4. Computation Speed

Speed is of course one of the goals of VLSI implementations of associative memories. If no speed requirements exist in a given neural network application, it is not worth realizing it in electronics since conventional computers can be utilized effectively. However, in many applications, neural networks are used as signal-processing devices where the time constants depend on physical constraints, such as in speech or image processing. Speed will thus also be an important aspect of the realization, which goes in the same way as the simplicity of the elements.

15.4.5. Analog and Digital Implementations

The advantages and drawbacks of analog and digital VLSI implementations of neural networks have already been the theme of many papers and articles (Verleysen and Jespers, 1991a). However, it seems important to clarify some ideas before going further in the description of analog implementations.

Digital cells are powerful for the implementation of complex functions used in some algorithms; the precision of the computations can be extended depending on the requirements of the algorithm by increasing the number of bits in each cell of the processors. However, this greatly

increases the size of the cells, and it rapidly becomes impossible to realize the massive interconnections required by a given algorithm.

The connections between the neurons occupy most of the silicon area on a chip; these synapses contain the memory points, and the elementary operations such as product and/or sums. The neurons, however, contain only nonlinear functions (in the case of Hopfield networks or multilayer perceptrons); even if the neurons are larger than the synapses, due to the complexity of their operations, the number of synapses is generally much larger than the number of neurons, and most of the area will be occupied by synapses.

It is not unusual to handle networks with many thousands of synapses; the complexity of digital cells makes it hard to implement all of them on a single chip. Multiplexing is then the solution: the same cells are sequentially used for several operations. This of course reduces the processing speed in the same way that the silicon area is reduced.

If speed is not to be compromised, analog cells can be used, since the area occupied can be much less than the area occupied by digital cells for similar realized operations. Drawbacks of the use of analog cells are the restricted precision of the computations that can be handled, and cascadability problems. The dynamical range of analog values is indeed limited due to mismatch between components on the same chip. This normally restricts bit accuracy to 5 or 6 bits, unless elaborate matching techniques are employed.

15.5. CURRENT-MODE APPROACH

There are generally three ways to represent data in VLSI circuits. First, the information can be represented by a voltage at a node on a line; its transmission is then made in a natural way along the line. Here, the resistance of the wire can affect the voltages between two distant points. Currents may also be employed to represent data; in this case, the information is represented as a current flow in a circuit. Leakage currents of junctions can affect the value of a current, but the transmission of the information between two points is generally realized better than with voltages. The third way is to represent the data in a temporal way, with frequencies or widths of pulses; although this obviously offers some advantages for the regeneration of corrupted information, the circuitry needed to handle the pulses generally requires more area on silicon, and will therefore not be considered in this chapter.

Most analog VLSI neural networks are based on the following principle: a product is realized at a synapse, between its input and a stored weight. The output value of the synapse is converted into current with a current source, and all resulting currents of synapses connected to a single neuron are summed (for free!) on a current line.

The neuron connected to the end of the current line will then apply a nonlinear operation to the input current. The information at the input of the neuron is called the "activation," while the one at the output is called the "state value," or "state" of the neuron. For the sake of simplicity, the relation between activation and state values is assumed to be a step function with zero threshold. A nonzero value for the threshold can be modeled by adding one synapse connected to the neuron, with a fixed input equal to 1, and with a connection weight equal to the negative of the threshold. The step function can be replaced by a sigmoid by varying the gain of the amplifier in the neuron.

The goal is to realize large sizes of synaptic arrays; it is thus necessary to design the network in such a way that large numbers of synapses can be connected together with sufficient precision that the neuron can make accurate decisions. If the current flowing from each synapse to the current line is binary (one current unit or no current), the neuron must be able to count the number of current units flowing to the current line. If the precision on the current units is infinite, adequate design of the neuron will avoid any decision problem in the neuron. However, a finite precision on the current units limits the number of synapses that can be connected together without inducing any wrong decision in the neuron.

The property of fault-tolerance of neural networks can compensate for some imperfections. However, it is usually agreed that the fault-tolerance relies only on random errors that could occur in the synapses, for example their being stuck to 1 or 0 at their output (see Chapter 14 of this volume by Chung and Krile for a detailed analysis of the effects of hardware faults on memory performance). Systematic errors in the synapses or in the neurons are more difficult to handle, since they may change the behavior of the circuit, and its convergence process. Wrong decisions taken by the neuron on the number of active synapses that are connected are of the second class of errors, the systematic ones. The values of the synaptic currents indeed depend on

physical parameters like the oxide thickness of the transistor gate, but the errors remain constant once the circuit is realized; it is then not correct to speak about fault-tolerance for this kind of error and the precision of the currents at the output of a synapse will be of importance.

15.6. TWO-LINE SYSTEM

In the beginning of this chapter, it was shown that learning rules for Hopfield networks can be adapted to the constraints imposed by VLSI, keeping only one- or two-bits precision for weights. The realization of the synapses is then greatly facilitated: products are digital ones and can be realized with logic gates, and memorization of weights in static registers is not too area-expensive since only one or two memory points are needed.

Figure 15.5 shows the first current-mode architecture proposed by Graf and De Vegvar (1987a, b). The synaptic weight is contained in two memory points, and its possible values are +1, 0 and −1. The input value of a synapse is either 1 or −1, and is materialized by a positive voltage on either "in" or "inb."

The excitatory state of the synapse will inject current into the current line through transistors T1 and T2, while the inhibitory state will sink current from the line through transistors T3 and T4. The content of memory points M1 and M2 will be respectively 0 and 0 for a positive connection, 1 and 1 for a negative connection, and 1 and 0 for no connection. The state with M1 = 0 and M2 = 1 is not allowed.

In Graf's circuit, the excitatory and inhibitory

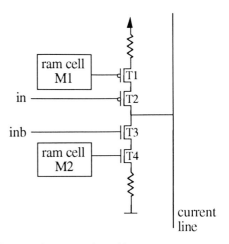

Fig. 15.5. Current-mode architecture.

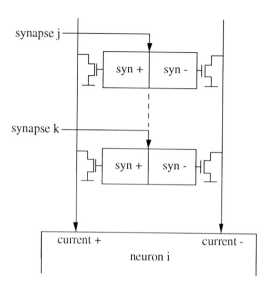

Fig. 15.6. Two-lines architecture.

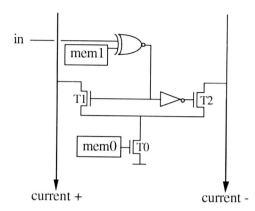

Fig. 15.7. Synapse.

15.6.1. Synapse

Each synapse in Fig. 15.6 is a programmable current source controlled by a differential pair (see Fig. 15.7). Three connection values are allowed in each synapse. If mem0 = 1, current is sunk to one of the two lines; mem1 and the input determines to which of them the current is sunk. If mem0 = 0, no connection exists between neurons i and j, and no current flows to either the excitatory or the inhibitory line. This synapse is designed in accordance with the optimization learning rule described previously, where only 2-bit precision is necessary in the synaptic weights; the synapse input is here a logic (1-bit) value.

15.6.2. Neuron

Depending on the state of the XOR function (see Fig. 15.7), the current may be sunk either from the line current + or from the line current −. The comparison between the two total currents must be achieved in the neuron; this is done by means of the current mirror shown in Fig. 15.8. The currents on the two lines are converted into voltages across transistors T3 and T4; these

currents flow through N-type and P-type current sources, respectively. The mismatch between the two types of transistors makes it impossible to obtain exactly the same currents for the two states of the synapses; as a consequence, systematic errors occur in the neuron decision function, and the convergence process of the network can be altered. Mismatch between N-type and P-type current sources come from the mobility differences between the two types of transistors. The size of the sources can of course be adjusted to compensate for these differences, but the exact ratio between the mobilities depends on the technological process, and cannot be known exactly before chip fabrication. Size compensation is thus not sufficient to ensure correct behavior of the circuit.

This mismatch problem can be avoided by using a two-line system with only one type of current source (Verleysen et al., 1989). The architecture is shown in Fig. 15.6. Its purpose is to use only one type of current source, in order to avoid the mismatching effects between N- and P-type transistors. However, even with one type of source, the currents will be slightly different one from another; this will be discussed later. As described in Fig. 15.6, all excitatory currents are summed on one current line, and all inhibitory currents are summed on another one. The role of the neuron will be to compare the two total currents, and to apply the nonlinear function to the difference.

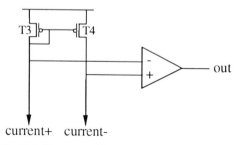

Fig. 15.8. Neuron architecture.

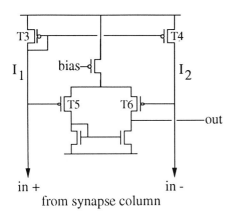

from synapse column

Fig. 15.9. Neuron amplifier.

voltages themselves are compared in the operational amplifier shown in the figure.

To obtain digital values at the output of a neuron, its gain must be very large; the opamp will thus be a simple amplifier with no feedback, as shown in Fig. 15.9.

15.7. PRECISION IN NEURON ELEMENTS

The design of the neuron clearly determines the sum-of-products precision achieved. Since each synaptic current is relatively small, the current mirror, and the other components in the neuron, have to be designed accurately to avoid computation errors. Several mismatch errors can be quantified in the neuron of Fig. 15.9 (Verleysen et al., 1989). First, threshold voltages of the two transistors in the mirror (T3 and T4) can differ (Verleysen et al., 1991b). For first order, the effect of this difference is directly related to the transconductance of a transistor in saturated mode:

$$\Delta I_{tm} \approx \sqrt{\left(2\mu C_{ox}\frac{W}{L}I_1\right)}\Delta V_{tm} \qquad (16)$$

where $\Delta I_t{}^m$ is the current error, μ is the mobility of carriers in the transistors, C_{ox} is their oxide capacitance, W/L is their size, and ΔV_{tm} is the difference of threshold voltages between T3 and T4. This leads to

$$\Delta I_{tm} \approx \frac{2}{V_{gs} - V_t}\Delta V_{tm} I_1 \qquad (17)$$

where V_{gs} is the gate-to-source voltage of transistor T3, and V_t its threshold voltage.

A second effect is the variation of β factors between the two transistors. The resulting current difference is expressed by

$$\Delta I_{\beta} = \frac{\Delta\beta}{\beta}I_1 \qquad (18)$$

A third matching error in the mirror is due to the difference of the Early voltages between the two transistors. Taking the Early effect into consideration, and neglecting the Early effect of the N-type current sources, the expression for the current in the transistors is

$$I = \mu C_{ox}\frac{W}{L}\frac{(V_{gs} - V_t)^2}{2}\left(1 - \frac{V_{ds} - V_{dsat}}{V_{EA}}\right) \qquad (19)$$

where V_{ds} is the drain-to-source voltage of the transistor, and V_{dsat} is its pinch-off voltage. If we define

$$\lambda_p = \frac{1}{V_{EAp}} \qquad (20)$$

then the current in the two transistors becomes

$$I_{ds1} = \mu C_{ox}\frac{W}{L}\frac{(V_{gs1} - V_t)^2}{2}[1 - \lambda_{p1}(V_{ds1} - V_{dsat})] \qquad (21)$$

$$I_{ds2} = \mu C_{ox}\frac{W}{L}\frac{V_{gs1} - V_t)^2}{2}[1 - \lambda_{p2}(V_{ds2} - V_{dsat})] \qquad (22)$$

Equation (22) leads to

$$V_{ds2} = \frac{1}{\lambda_{p2}}\left(1 + \frac{2I_{ds2}}{\mu C_{ox}\frac{W}{L}(V_{gs1} - V_t)^2}\right) \qquad (23)$$

As proposed by Hoekstra (1990), the derivative of V_{gs1} can be computed by differentiating (21) implicitly with respect to λ_{p1}. This leads to

$$0 = \frac{\partial V_{gs1}}{\partial\lambda_{p1}}[2 - \lambda_{p1}(V_{gs1} - V_t)] - (V_{gs1} - V_t)V_{gs1} \qquad (24)$$

$$\frac{\partial V_{gs1}}{\partial\lambda_{p1}} = \frac{(V_{gs1} - V_t)V_{gs1}}{2 - \lambda_{p1}(V_{gs1} - V_t)} \qquad (25)$$

Differentiating V_{ds2} in (23) with respect to λ_{p2} results in

$$\frac{\partial V_{ds2}}{\partial\lambda_{p2}} = \frac{V_{ds2}}{\lambda_{p2}} \qquad (26)$$

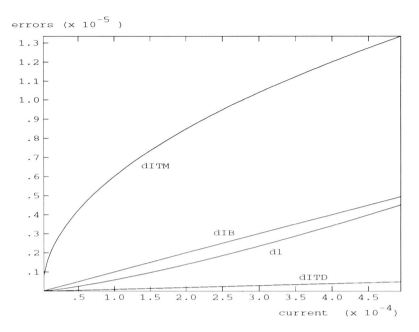

Fig. 15.10. Errors in a current mirror.

Substituting these values in (23) for a fixed $\Delta\lambda_p$ equal for the two transistors but with opposite signs (worst case) leads to a resulting difference between V_{ds2} and V_{ds1}:

$$V_{ds2} - V_{gs1} = \left(\frac{V_{ds2}}{\lambda_{p2}} - \frac{(V_{gs1} - V_t)V_{gs1}}{2 - \lambda_{p1}(V_{gs1} - V_t)} \right) \Delta\lambda_p \tag{27}$$

which reported in terms of current difference between I_1 and I_2 gives approximately

$$\Delta I_\lambda = \left[\frac{V_{ds2}}{\lambda_{p2}} - \frac{(V_{gs1} - V_t)V_{gs1}}{2 - \lambda_{p1}(V_{gs1} - V_t)} \right] \Delta\lambda_p \frac{I_1}{V_{EA}} \tag{28}$$

A fourth and last error to consider is the mismatch between the two transistors at the input stage of the differential amplifier which measures the difference between the gate-to-source voltages of the two transistors in the mirror (i.e., the threshold voltage difference between T5 and T6 in Fig. 15.9). This mismatch, reported in terms of current difference, is given by

$$\Delta I_{td} = \Delta V_{td} \frac{I_1}{V_{EA}} \tag{29}$$

In order to compare these four errors, realistic values are chosen: I from 0 to 500 μA; $\mu_p C_{ox} = 1.5 \times 10^{-5}$ A/V^2 (standard CMOS process); and W/L must be chosen to cope with the maximum current (500 μA). If $(V_{gs} - V_t)$ can be up to

1.5 V, then

$$\frac{W}{L} = \frac{I_{max}}{\mu_p C_{ox}(V_{gs} - V_t)^2/2} \approx 30 \tag{30}$$

with $V_t = 0.8$ V, $V_{EA} = 20$ V, $\Delta V_{tm} = 10$ mV, $\Delta\beta/\beta = 0.01$, $\Delta V_{td} = 10$ mV, and $\Delta\lambda_p = 0.01$ V^{-1}. These values can be reached with careful design of the mirrors and comparators. The errors are illustrated in Fig. 15.10.

The error generated by the threshold voltage difference in the current mirror is obviously the most significant one, especially for small currents; this results from the fact that the error is proportional to the square root of the size of the transistors in the mirror. Even for small currents, the error is thus significant because the two transistors must be large enough to drive the maximum current (here 500 μA). A solution to this problem would be to connect several mirrors in parallel, each of them being active only when necessary to drive the total current.

15.8. SUMMARY

This chapter presents a general method for implementing neural networks with VLSI circuits. It shows how basic learning algorithms for associative memory can be adapted to the requirements of VLSI circuits. The basic principles of analog VLSI architectures are presented, together with most precision limitations encountered with such technology. It is also shown how to overcome such limitations by appropriate design of the neuron.

The principles explained in this chapter are illustrated for the Hopfield memory network. They are, however, more general, and can be used in most neural network implementations where the basic operation is the sum-of-product (i.e., the product of a matrix and a vector), and where binary weights may be used.

This chapter also presents a learning rule for the Hopfield memory adapted to such circuits; the rule shows good performance even when the weights' dynamic range is restricted to one or two bits.

REFERENCES

Abu-Mostafa Y. S. and St. Jacques, J.-M. (1985). "Information Capacity of the Hopfield Model," *IEEE Trans. Information Theory*, **31**(4), 461–464.

Graf, H. P. and De Vegvar, P. (1987a). "A CMOS Associative Memory Chip Based on Neural Networks, in *Proceedings of the IEEE International Solid-State Circuits Conference*, New York.

Graf, H. P. and De Vegvar, P. (1987b). "A CMOS Implementation of a Neural Network Model," in Losleben, P., ed. *Proceedings of the 1987 Stanford Conference on Advanced Research in VLSI*. MIT Press, Cambridge, MA.

Hebb, D. O. (1949). *The Organization of Behaviour*. Wiley, New York.

Hertz, J., Krogh, A., and Palmer, R. G. (1991). *Introduction to the Theory of Neural Computation*. Addison-Wesley, Redwood City, CA.

Hopfield, J. J. (1982). "Neural Networks and Physical Systems with Emergent Collective Computational Abilities," *Proc. Nat. Acad. Sci., USA*, **79**, 2554–2558.

Lasbon, L. (1970). *Optimization Theory for Large Systems*. Collier-Macmillan, London.

McEliece, R. L., Posner, E. C., Rodemich, E. R., and Venkatesh, S. S. (1987). "The Capacity of the Hopfield Associative Memory," *IEEE Trans. Information Theory*, **33**(4), 461–482.

Morgenstern, I. (1987). "Neural Networks and Chip Design," *Condensed Matter*, **67**, 265–270.

Hoekstra, A. (1990). "An Associative Memory Based on Neural Networks in CMOS VLSI Circuits," Master Thesis, Faculty of Electrical Engineering, University of Twente.

Personnaz, L., Guyon, I., and Dreyfus, G. (1985). "Information Storage and Retrieval in Spin-glass-like Neural Networks," *J. de Physique Lett.*, **46**, 359–365.

Verleysen, M. and Jespers, P. (1991a). "VLSI Chips for Neural Networks," in *Progress in Neural Networks*, Vol. 1, O. M. Omidvar, ed. Ablex, Norwood, NJ.

Verleysen, M. and Jespers, P. (1991b). "Precision of Computations in Analog Neural Networks," in *VLSI Design of Neural Networks*, Ramacher, U. and Rückert, U., eds. Kluwer Academic Publishers, Boston.

Verleysen, M., Sirletti, B., Vandemeulebroecke, A., and Jespers, P. (1989). "Neural Networks for High-storage Content-addressable Memory: VLSI Circuit and Learning Algorithm," *IEEE J. Solid-State Circuits*, **24**(3), 562–569.

16

Recurrent Correlation Associative Memories and their VLSI Implementation

TZI-DAR CHIUEH and RODNEY M. GOODMAN

16.1. INTRODUCTION

Unlike traditional computer memory (e.g. ROM, RAM), associative memories have error-correcting capability in that they can generate accurate responses even with partially incorrect input patterns. Associative recall operation is found in many human information-processing tasks, such as knowing names from faces, recalling tunes and lyrics, recollecting telephone numbers, etc. Because of their potential usage in many information processing systems, associative memories have attracted much attention. Since the seminal work of Hopfield (1982, 1984), there has been much interest in building associative memories using neural-network approaches. The storage capacity of the Hopfield memory has been found, both empirically (Hopfield, 1982) and theoretically (McEliece et al., 1987), to scale less than linearly with the number of components in memory patterns. Other researchers have proposed new architectures that utilize nonlinear circuits and correlations between memory patterns and the input pattern (Psaltis and Park, 1986; Sayeh and Han 1987; Dembo and Zeitouri, 1988a, b; see also Chapter 6 by Dembo in this book). Previously, we also proposed a new associative-memory model that adopts the exponentiation function (Chiueh and Goodman, 1988). These models can all be implemented by a two-layer recurrent network: The first layer computes the correlations of the current-state pattern and all the memory patterns, followed by some nonlinear weighting function; the second layer calculates a weighted sum of all memory patterns and thresholds that sum to produce the next-state pattern. Since these recurrent neural-network associative memories are based on the correlation measure, we call them recurrent correlation associative memories (RCAM).

The organization of the rest of this chapter is as follows. In Section 16.2 a model for the RCAM is presented. Also, some known associative memories are shown to be instances of the RCAM model. Section 16.3 deals with the convergence property of the RCAM. By defining a Liapunov ("energy") function and demonstrating that it never increases, the RCAM is shown to be asymptotically stable in both synchronous and asynchronous update modes if its weighting functions are continuous and monotone non-decreasing. Section 16.4 concentrates on a particular model called the exponential correlation associative memory (ECAM). The relationship between the storage capacity and the attraction radius of the ECAM is investigated. If all state patterns inside a hypersphere of some attraction radius centered at a memory pattern are to be attracted, in one iteration, to that memory pattern with very high probability, then as N (the number of components in memory patterns) approaches infinity, the storage capacity (the maximum number of memory patterns) is proportional to c^N. The constant c is greater than 1 and it decreases as the attraction radius increases. More importantly, we find that under certain conditions the asymptotic storage capacity of the ECAM meets the ultimate upper bound for the capacity of associative memories derived by sphere-packing arguments in Chou (1988). We also find that the asymptotic storage capacity of the ECAM is proportional to the dynamic range of its exponentiation circuits if that dynamic range is limited. In Section 16.5, we present the results of some simulation experiments of the ECAM that confirm the theoretical findings about the asymptotic storage capacity of the ECAM, even though N is not excessively large. VLSI implementation of the ECAM and the associated circuits are described in Section 16.6. Test results and an application of the ECAM chip to a binary-image vector-quantization problem are discussed in Section 16.7. Section 16.8 is a summary of the chapter.

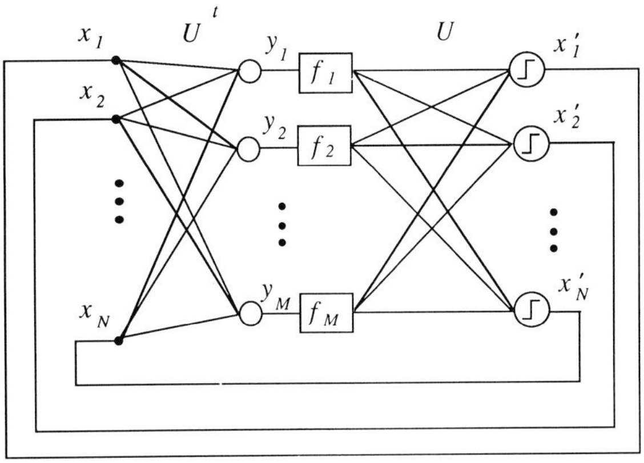

Fig. 16.1. Architecture of the recurrent correlation associative memories. Matrix U is an $M \times N$ matrix made up of M N-bit bipolar ($+1$ or -1) memory patterns, $\mathbf{u}^{(k)}$, $k = 1, 2, \ldots, M$. $\mathbf{x}$ and $\mathbf{x}'$ are the current-state and the next-state patterns, respectively. The f_k are the weighting functions.

16.2. A MODEL FOR RECURRENT CORRELATION ASSOCIATIVE MEMORIES

Let $\mathbf{x}$ and $\mathbf{w}$ be two N-bit bipolar patterns whose components are either $+1$ or -1, and denote the correlation between $\mathbf{x}$ and $\mathbf{w}$ by

$$\langle \mathbf{x}, \mathbf{w} \rangle \equiv \sum_{j=1}^{N} x_j w_j \tag{1}$$

Also suppose that $\mathbf{u}^{(1)}, \mathbf{u}^{(2)}, \ldots, \mathbf{u}^{(M)}$ are N-bit bipolar ($+1$ or -1) memory patterns to be stored in this associative memory. The RCAM computes its N-bit bipolar next state pattern $\mathbf{x}'$ according to the motion equation

$$\mathbf{x}' = \text{sgn}\left\{ \sum_{k=1}^{M} f_k(\langle \mathbf{u}^{(k)}, \mathbf{x} \rangle) \cdot \mathbf{u}^{(k)} \right\} \tag{2}$$

where $\mathbf{x}$ is the current-state pattern, f_k are called weighting functions, and sgn is the vector operation that produces signs of each component. The RCAM is in principle an evolutonary system, moving from one state pattern to another until it reaches a fixed point, which is the response to the initial input pattern. As in the Hopfield model, if the system starts with a noise-corrupted pattern, the RCAM will very likely return that memory pattern except that with the additional nonlinearity the RCAM can either tolerate more severe noise or store more memory patterns than the Hopfield memory does. Figure 16.1 illustrates the architecture of the recurrent correlation associative memories. Matrix U is an $M \times N$ matrix made up of the M memory patterns $\mathbf{u}^{(1)}, \mathbf{u}^{(2)}, \ldots, \mathbf{u}^{(M)}$. The signal y_k is equal to the correlation of the input pattern, $\mathbf{x}$, and $\mathbf{u}^{(k)}$. The second interconnection matrix then computes weighted sums of components of all memory patterns and then thresholds these sums individually to generate the next state pattern, $\mathbf{x}'$. Let us now describe how some known neural-network associative memories can be expressed as instances of the RCAM.

Correlation-Matrix Associative Memory. This model is essentially equivalent to the Hopfield memory except that the diagonal of the connection weight matrix is not zeroed (Kohonen, 1972; Anderson, 1972). The connection weight matrix is given by

$$T_{ij} = \sum_{k=1}^{M} u_i^{(k)} \cdot u_j^{(k)}$$

It is easily shown that the correlation-matrix associative memory is an instance of the RCAM with

$$f_k(t) \equiv t$$

High-Order Correlation Associative Memory. In this type of associative memories (Psaltis and Park, 1986) the evolution equation is Eq. (2) with

$$f_k(t) \equiv (N + t)^q$$

where q is an integer great than 1. The storage capacity of the high-order correlation associative memory is asymptotically proportional to N^q.

Potential-Function Correlation Associative Memory. Dembo and Zeitouni (1988a, b) and Sayeh and Han (1987) independently introduced similar models that utilize a potential-type function. Originally, they are continuous-time systems with real-valued patterns. Nonetheless, it is straightforward to express these models in discrete-time formulation with bipolar patterns. The evolution equation then takes the form of

Eq. (2) with

$$f_k(t) = (N - t)^{-L}$$

where $L \geq 1$. The storage capacity of this model grows exponentially with the number of components in memory patterns (Dembo and Zeitouni, 1988b; see also Chapter 6).

Exponential Correlation Associative Memory (ECAM). We have introduced the exponential correlation associative memory (ECAM), which is an instance of the RCAM with all f_k equal to an exponentiation function (Chiueh and Goodman, 1988, 199); i.e.,

$$f_k(t) \equiv a^t$$

where $a > 1$. The storage capacity of the ECAM will be explored in Section 16.4.

Other Recurrent Correlation Associative Memories. In principle, as soon as one specifies a set of weighting functions $f_k(\cdot)$, one builds an RCAM. However, the important thing is to find weighting functions that are easy to implement, and that produce an RCAM that is asymptotically stable and has a large storage capacity. In the next section, we will present a condition on the weighting functions $f_k(\cdot)$ that is sufficient for the asymptotic stability of that RCAM.

16.3. THE CONVERGENCE PROPERTY OF THE RCAM

Since the RCAM is based on a recurrent network structure, understanding its asymptotic behavior is important. Hopfield (1982) proved that his model is asymptotically stable when running in the asynchronous update mode (when only one neuron in the output layer updates itself at a time). At first, he introduced a Liapunov ("energy") function of the system, and went on to demonstrate that the Liapunov function either decreases or stays the same after each iteration. Moreover, he showed that the energy function has a lower bound and that the system cannot stay at the same energy level forever. These facts imply that the Hopfield memory will eventually reach a stable state with minimum "energy" level. However, if the Hopfield memory is running in the synchronous update mode (all neurons in the output layer update themselves at the same time), it may not converge to a fixed point, but may instead become oscillatory between two states (Bruck and Goodman, 1987). In this section, we prove that the first four RCAMs in the previous section are all asymptotically stable in both synchronous and asynchronous update modes.

To begin with, let us introduce a lemma.

LEMMA 1. *Let $f(t)$ be continuous and monotone nondecreasing over $[-N, N]$; then the RCAM with the following evolution equation,*

$$\mathbf{x}' = \text{sgn}\left\{ \sum_{k=1}^{M} f(\langle \mathbf{u}^{(k)}, \mathbf{x} \rangle) \cdot \mathbf{u}^{(k)} \right\}$$

is asymptotically stable in both synchronous and asynchronous (sequential) update modes.

PROOF. See Chiueh and Goodman (1991).

THEOREM 2. *The first four RCAMs in the previous section are all asymptotically stable in both synchronous and asynchronous (sequential) update modes.*

PROOF. First of all, all four weighting functions in the previous section are continuous. Furthermore, for any $t_1 > t_2$, one has

$$t_1 \geq t_2$$
$$(N + t_1)^q \geq (N + t_2)^q$$
$$(N - t_1)^{-L} \geq (N - t_2)^{-L}$$
$$a^{t_1} \geq a^{t_2}$$

where $q > 1$, $L \geq 1$, and $a > 1$. Consequently, Lemma 1 can be applied and the theorem proved.
□

The significance of Lemma 1 is that it ensures that one can employ any continuous, monotone nondecreasing weighting function, and the resulting RCAM will be asymptotically stable in both synchronous and asynchronous update modes. This proves to be very helpful when it comes to hardware implementation of the RCAM. Since almost all physical devices exhibit a certain amount of fluctuation from their ideal characteristics, any real implementation of an RCAM is bound to encounter deviation in the nonlinearity. With Lemma 1 one can rest assured that an RCAM is asymptotically stable as long as the real response of the nonlinear devices stays continuous and monotone nondecreasing, although in this case the performance in storage capacity and error-correction ability may suffer.

16.4. THE CAPACITY AND THE ATTRACTION RADIUS OF THE ECAM

The exponential correlation associative memory seems to be the RCAM that is most amenable

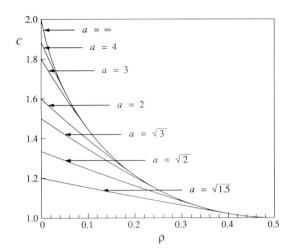

Fig. 16.2. Relationship of the base constant c and the two parameters, a and ρ. The ECAM has an exponential storage capacity that is proportional to c^N.

to VLSI implementation; therefore, this section is devoted to the exploration of the storage capacity and the attraction radius of the ECAM. Our definition of the storage capacity is somewhat similar to that of McEliece et al.'s (1987). Suppose we choose $M = M(N)$ N-bit memory patterns at random, program an ECAM with those M memory patterns and initialize that ECAM with an input pattern r ($r \equiv \rho N$, and $0 \le \rho < \frac{1}{2}$) bits away from the nearest memory pattern. We then ask, as $N \to \infty$, what is the greatest rate of growth of $M(N)$ so that after one iteration the bit-error probability (the probability that a bit in the next-state is different from the corresponding bit in the nearest memory pattern) is less than $(4T)^{-1/2}e^{-T}$, where T is a fixed and large number? By adjusting T, one can make a trade-off between the bit-error probability and the storage capacity of an ECAM.

To begin with, assume that all M N-bit memory patterns $\mathbf{u}^{(k)}$, $k = 1, 2, \ldots, M$ are randomly chosen; in other words, each bit in any of the M memory patterns is the outcome of a Bernoulli trial ($+1$ or -1). Let us now present the theorem about the storage capacity of the ECAM.

THEOREM 3. *Suppose an ECAM is loaded with*

$$M(N) = \begin{cases} [a^4/(4T)]2^{N(1-\mathscr{H}(\rho'))} + 1 \\ \qquad \text{if } \rho' \ge (1 + a^2)^{-1} \\ [a^4/(4T)]2^N[(1 + a^{-2})a^{2\rho'}]^{-N} + 1 \\ \qquad \text{if } \rho' < (1 + a^2)^{-1} \quad (3) \end{cases}$$

N-bit memory patterns, where $\rho' = \rho + (1/N)$, $0 \le \rho < \frac{1}{2}$, and $\mathscr{H}(\rho')$ is the binary information entropy of ρ'. If the current-state pattern $\mathbf{x}$ is ρN

bits away from the nearest memory pattern, then as $N \to \infty$, the bit-error probability (P_e) is less than $(4\pi T)^{-1/2}e^{-T}$.

The proof of Theorem 3 can be found in Chiueh and Goodman (1991) and is omitted here. From Theorem 3 we conclude that the ECAM has a storage capacity that scales *exponentially* with N—the number of bits in the memory patterns. In other words, the ECAM can store c^N memory patterns—all N-bits wide—and will be capable of some error correction. The base constant c is a relatively complex function of two parameters, a and ρ. Refer to Fig. 16.2 to see how c decreases with smaller a and larger ρ. Also note that c is never less than 1. More importantly, in the case when $\rho' \ge (1 + a^2)^{-1}$, one has, as $N \to \infty$,

$$[\log_2 M(N)]/N = 1 - \mathscr{H}(\rho')$$
$$+ [4\log_2(a) - \log_2(4T)]/N$$
$$\simeq 1 - \mathscr{H}(\rho')$$
$$\simeq 1 - \mathscr{H}(\rho)$$

Hence, when $\rho' \ge (1 + a^2)^{-1}$ the asymptotic storage capacity of the ECAM meets the ultimate upper bound for the capacity of associative memories (Chou, 1988), the $a = \infty$ curve in Fig. 16.2.

This exponential capacity is very attractive; however, the dynamic range required of the exponentiation circuit also grows exponentially with N. In any real implementation, this requirement is very difficult to meet, if not impossible. In a typical CMOS VLSI process, a transistor operating in the subthreshold region working as an exponentiation circuit has a dynamic range of approximately 10^5 to 10^7 (Glasser and Dopperpuhl, 1985). Thus, we need to study how the storage capacity of the ECAM changes if the dynamic range of its exponentiation circuits is limited.

Suppose the dynamic range (D) of the exponentiation circuits is fixed and

$$D \equiv a^N$$

Then as N increases, a will decrease and M will no longer scale exponentially with N. We now concentrate on the case when N approaches infinity. Since N is very large and D is fixed, a will be near 1. Let

$$a \equiv 1 + \mu$$

where μ is a small positive number; then

$$\log D = N \log a = N \log(1 + \mu) \simeq N\mu$$

As a approaches 1, ρ' will be less than $(1 + a^2)^{-1}$ in practically all cases (remember that $\rho < 1/2$ and $\rho' = \rho + 1/N$); therefore, only the second formula in Eq. (3) needs to be considered. It follows that with fixed D and as N approaches infinity,

$$\begin{aligned} M(N) &= [a^2/(4T)]2^N[(1 + a^{-2})a^{2\rho'}]^{-N} \\ &\simeq [a^4/(4T)]2^N[(2 - 2\mu)(1 + 2\rho'\mu)]^{-N} \\ &\simeq [a^4/(4T)](1 + [N\mu(1 - 2\rho')]/N)^N \\ &\simeq [a^4/(4T)]\,e^{N\mu(1 - 2\rho)} \\ &\simeq [a^4/(4T)]D^{1 - 2\rho} \end{aligned} \qquad (4)$$

From the above equation, one sees that the asymptotic storage capacity of the ECAM is proportional to the dynamic range (D) when the required attraction radius (ρ) is zero. However,

as the attraction radius is increased, the storage capacity decreases. These findings are not at all discouraging, since they say that the ECAM can be only as good as one of its components—the exponentiation circuit.

16.5. SIMULATION RESULTS

A few simulations have been conducted in order to confirm the theoretical results about the storage capacity of the ECAM. We set $a = 2$ and randomly choose 10 sets of M N-bit memory patterns. For each set of M memory patterns, program an ECAM with these M patterns. For each ECAM, 100 initial-state patterns are generated by randomly picking a memory pattern and flipping d bits. They are then fed to the ECAM and the ECAM is allowed to run until it reaches a fixed point. The resulting fixed point is then compared with the original memory pattern, and the run is called a *success* if they match, a *failure* otherwise. The number of successes out of 1000 runs is then collected. If this number is greater than 998, we say that loaded with M memory patterns, the ECAM can tolerate d errors. The largest d for a fixed M is called the *attraction radius* (r).

In Figure 16.3, the normalized attraction radius $(\rho = r/N)$ is plotted against the number of memory patterns (M) for various N. Note that if a horizontal line is drawn across the plot such that it intersects the four curves in the figure, the four intersection points are approximately equidistantly apart. Since the four curves correspond

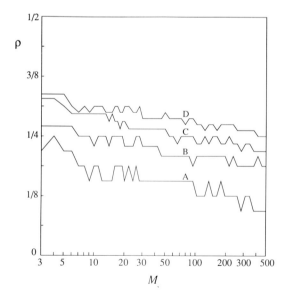

Fig. 16.3. Normalized attraction radius $(\rho = r/N)$ vs. number of stored memory patterns (M) in the ECAM, Curve A, $N = 32$; curve B, $N = 48$; curve C, $N = 64$; curve D, $N = 80$.

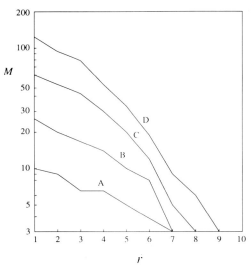

Fig. 16.4. Number of stored memory patterns (M) vs. attraction radius (r) of the ECAM with $N = 32$ and fixed dynamic range, D. Curve A, $D = 2^4$; curve B, $D = 2^5$; curve C, $D = 2^6$; curve D, $D = 2^7$.

to the cases where N increases linearly, this observation implies that for a fixed ρ the storage capacity of the ECAM scales exponentially with N, confirming Theorem 3.

Next we fix N at 32 and vary the dynamic range of the exponentiation circuits. Figure 16.4 illustrates how the relationship between the attraction radius (r) and the number of loaded memory patterns (M) changes for different dynamic ranges when $N = 32$. As one can easily see, the curves intersect the vertical axis ($r = 1$) at four points, each of which is approximately twice as large as the point before. Since the dynamic ranges of these four curves double successively, the storage capacity of the ECAM is thus proportional to the dynamic range of the exponentiation circuits for fixed N. Furthermore, if one draws a vertical line at larger r, it again intersects the four curves at points equidistantly apart, though with a smaller interval than in the previous case. Another simulation with $N = 64$ also gives similar results. Therefore, we conclude that the previous result about the storage capacity of the ECAM with fixed-dynamic-range exponentiation circuits (i.e., Eq. (4)) is valid.

16.6. VLSI IMPLEMENTATION OF THE ECAM

In the previous sections we have introduced a model for the recurrent correlation associative memories. We addressed, in particular, the case when the weighting functions are exponential, namely, the ECAM. The evolution equation of the ECAM is given by

$$\mathbf{x}' = \mathrm{sgn}\left\{\sum_{k=1}^{M} a^{\langle \mathbf{u}^{(k)}, \mathbf{x}\rangle}\mathbf{u}^{(k)}\right\} \qquad (5)$$

where a is a constant greater than 1.

The ECAM chip we designed is *programmable*; that is, one can change the set of memory patterns stored in an ECAM chip at will. To perform an associative recall, one first loads a set of memory patterns onto the chip. The chip is then switched to the associative-recall mode, and an input pattern is presented to the ECAM chip. The ECAM chip then computes the next-state pattern according to Eq. (5) and presents it at the output port of the chip. No clock signal is necessary since, after the internal circuits have settled, the components of the next-state pattern appear in parallel at the output port. Feedback is easily incorporated by connecting the output port to the input port, in which case the chip will cycle until a fixed point is reached.

From the evolution equation of ECAM, one notices that there are essentially three operations that need to be carried out:

$\langle \mathbf{u}^{(k)}, \mathbf{x}\rangle$, correlation computation

$$\sum_{k=1}^{M} a^{\langle \mathbf{u}^{(k)}, \mathbf{x}\rangle}\mathbf{u}^{(k)}, \text{exponentiation, multiplication,}$$

and summation

$\mathrm{sgn}(\cdot)$, thresholding.

Now let us describe each circuit, present its design, and finally integrate all these circuits to get the complete design of the ECAM chip.

16.6.1. Correlation Computation Circuit

In Fig. 16.5, we illustrate a voltage-divider type circuit consisting of NMOS transistors working as controlled resistors (linear resistors or open circuits). This circuit computes the correlation between the input pattern $\mathbf{x}$ and a memory pattern $\mathbf{u}^{(k)}$. If the ith components of these two patterns are the same, the corresponding XOR gate outputs a "0" and there is a connection from the node $V_{ux}^{(k)}$ to V_{BB}; otherwise, there is a connection from $V_{ux}^{(k)}$ to ground. Hence the output voltage will be proportional to the number of positions at which $\mathbf{x}$ and $\mathbf{u}^{(k)}$ match. The maximum output voltage is controlled by an externally supplied bias voltage V_{BB}. Normally, V_{BB} is set to a voltage lower than the threshold voltage of NMOS transistors (V_{RH}) for a reason to be explained later.

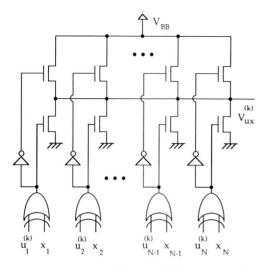

Fig. 16.5. Diagram of the correlation computation circuit.

The conductance of an NMOS transistor in the ON mode is not fixed, but rather depends on its gate-to-source voltage and its drain-to-source voltage. Thus, some nonlinearity is bound to occur in the correlation computation circuit. We have simulated a correlation computation circuit with $N = 64$ using SPICE. In Fig. 16.6, we illustrate the SPICE output voltage $V_{ux}^{(k)}$ and the ideal linear response, i.e., the case when ON transistors are replaced by linear resistors and OFF transistors by open circuit. As shown in

Fig. 16.6, there is only slight deviation from the ideal response throughout the whole operating range—from 0 V to V_{BB}. One therefore concludes that the proposed correlation computation circuit is good enough for the purpose of the ECAM chip.

16.6.2. Exponentiation, Multiplication, and Summation Circuit

Figure 16.7 depicts a circuit that computes the exponentiation of $V_{ux}^{(k)}$, the product of $u_i^{(k)}$ and the exponential, and the sum of all M products.

The exponentiation function is implemented by an NMOS transistor whose gate voltage is $V_{ux}^{(k)}$. Since V_{BB}, the maximum value that $V_{ux}^{(k)}$ can assume, is set to be lower than the threshold voltage (V_{TH}), the NMOS transistor is guaranteed to be in the subthreshold region, where its drain current depends exponentially on its gate-to-source voltage (Mead, 1989). For the time being, ignore the two transistors controlled by $u_i^{(k)}$ or the complement of $u_i^{(k)}$; then the current flowing through the exponentiation transistor associated with $V_{ux}^{(k)}$ scales exponentially with $V_{ux}^{(k)}$. Thus the exponentiation function is properly computed.

Since the multiplier $u_i^{(k)}$ assumes either $+1$ or -1, the multiplication can be done by forming two branches, each made up of a pass transistor in series with an exponentiation transistor whose gate voltage is $V_{ux}^{(k)}$. One of the two pass transistors is controlled by $u_i^{(k)}$, and the other by the complement of $u_i^{(k)}$. Consequently, when

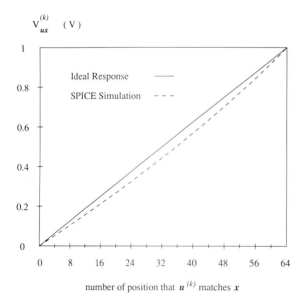

Fig. 16.6. Comparison of the SPICE-simulation output voltage of a correlation computation circuit with $N = 64$ and the ideal linear response.

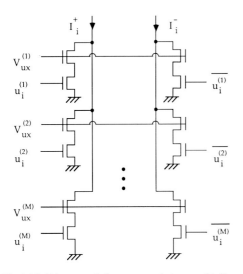

Fig. 16.7. Diagram of the exponentiation, multiplication, and summation circuit.

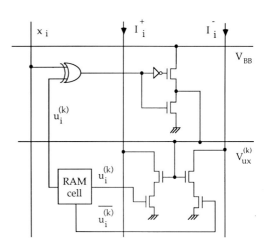

Fig. 16.9. Circuit diagram of the basic ECAM cell.

depicts the top half of a simple differential amplifier, which can be integrated with the circuit in Fig. 16.7 to decide x'_i.

16.6.4. ECAM Cell

For easy VLSI implementation, we have designed a basic ECAM cell that realizes all the afore-mentioned computation. The idea is to draw the correlation computation circuit, the exponentiation, multiplication, and summation circuit, and then extract a basic repeating block. This block, together with a static RAM cell, makes up the basic ECAM cell as illustrated in Fig. 16.9. The major part of an exponential correlation associative memory that holds M N-bit memory patterns can be obtained by replicating the basic ECAM cell M times in the vertical direction and N times in the horizontal direction.

16.7. TESTING RESULTS OF THE ECAM CHIP

The complete ECAM chip, made up of 32×24 ECAM cells, holds 32 memory patterns each 24 bits wide. We have tested the ECAM chip using a specialized VME-bus host computer. The testing procedure is to first generate 32 memory patterns randomly and program the ECAM chip with these 32 patterns. We then pick a memory pattern, flip a specified number of bits randomly, and feed the resulting pattern to the ECAM as an input pattern ($\mathbf{x}$). The output pattern ($\mathbf{x}'$) can then be fed back to the input side of the ECAM chip. This iteration continues until the pattern on the input bus is the same as that on the

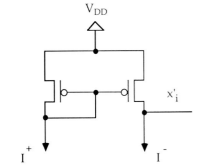

Fig. 16.8. Diagram of the threshold circuit.

$u_i^{(k)} = 1$, the positive branch will carry a current that scales exponentially with the correlation of input $\mathbf{x}$ and the kth memory pattern $\mathbf{u}^{(k)}$, while the negative branch is essentially an open circuit, and vice versa.

Summation of the M terms in the evolution equation is done by current summing. The final results are two currents, I_i^+ and I_i^-, to be compared by the threshold circuit in order to determine the sign of the ith bit of the next state pattern, x'_i.

16.6.3. Threshold Circuit

The function of the threshold circuit is to decide whether or not I_i^+ is greater than I_i^-. Thus, any differential amplifier is sufficient. Figure 16.8

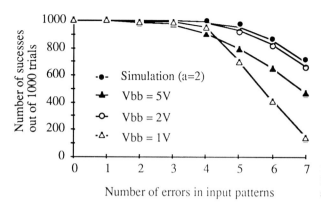

Fig. 16.10. Error-correcting ability of the ECAM chip with different V_{BB} and an ECAM simulation with $a = 2$.

output bus, at which time the ECAM chip is said to have reached a stable state. We select 10 sets of 32 memory patterns and for each set we run the ECAM chip on 100 trial input patterns with a fixed number of errors. Altogether, the test consists of 1000 trials.

16.7.1. Results and Discussions

In Fig. 16.10 we illustrate the test results. The number of successes is plotted against the number of errors in input patterns for the following four cases: (1) The ECAM chip with $V_{BB} = 5$ V; (2) $V_{BB} = 2$ V; (3) $V_{BB} = 1$ V; and (4) a simulated ECAM with a equals 2. It is apparent from Fig. 16.10 that as the number of errors increases, the number of successes decreases, which is expected. Also, one notices that the simulated ECAM is by far the best one, which is again not unforeseen because the ECAM chip is, after all, only an approximation of the ECAM model and thus will definitely do worse.

What is unexpected is that the best performance occurs when $V_{BB} = 2$ V rather than when $V_{BB} = 1$ V (V_{TH} in this CMOS process). This phenomenon arises due to two contradicting effects brought about by increasing V_{BB}. On the one hand, increasing V_{BB} increases the dynamic range of the exponentiation transistors in the ECAM chip. Suppose that the correlations of two memory patterns $\mathbf{u}^{(l)}$ and $\mathbf{u}^{(k)}$ with the input pattern $\mathbf{x}$ are t_l and t_k, respectively, where $t_l > t_k$; then

$$V_{ux}^{(l)} = \frac{(t_l + N) V_{BB}}{2N} \qquad V_{ux}^{(k)} = \frac{(t_k + N) V_{BB}}{2N}$$

Therefore, as V_{BB} increases, so does the difference between $V_{ux}^{(l)}$ and $V_{ux}^{(k)}$, and $\mathbf{u}^{(l)}$ becomes more dominant than $\mathbf{u}^{(k)}$ in the weighted sum of the evolution equation. Hence, as V_{BB} increases, the

erorr-correcting ability of the ECAM chip should improve. On the other hand, as V_{BB} increases beyond the threshold voltage, the exponentiation transistors leave the subthreshold region and may enter saturation, where the drain current is approximately proportional to the *square* of the gate-to-source voltage. Since a second-order correlation associative memory in general possesses a smaller storage capacity than an ECAM, one expects that, with a fixed number of loaded memory patterns, the ECAM should have a better error-correcting power than the second-order correlation associative memory does. To sum up, two opposing effects are occurring as V_{BB} is raised; one tends to enhance the performance of the ECAM chip, while the other tends to degrade it. A compromise between these two effects is reached, and the best performance is achieved when $V_{BB} = 2$ V.

For the case when $V_{BB} = 2$ V, the drain current versus gate-to-source voltage characteristic of the exponentiation transistors is actually a hybrid of a square function and an exponentiation function: At the bottom it is of an exponential form, and it gradually flattens out to a square function once the gate-to-source voltage becomes larger than the threshold voltage. Therefore, the ECAM chip with $V_{BB} = 2$ V is a mixture of the second-order correlation associative memory and the ECAM. According to the convergence theorem for recurrent correlation associative memories and the fact that the weighting function in the ECAM chip with $V_{BB} = 2$ V is still monotonically nondecreasing, the ECAM chip is still asymptotically stable when $V_{BB} = 2$ V.

16.7.2. A Vector-Quantization Example

We have tested the speed of the ECAM chip using binary-image vector quantization as an example

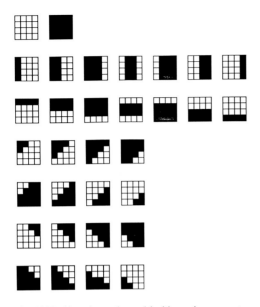

Fig. 16.11. 32 codewords used in binary image vector quantization.

problem. Vector quantization is a means of data compression (source coding) on information to be transmitted or stored, e.g, speech waveforms, images, etc. (Gray, 1984). In principle, given a set of codewords and an input, a vector quantizer should find the nearest codeword to the input. Then only the index of the nearest codeword is transmitted or stored instead of the information itself. Usually, the number of possible codewords is much smaller than the number of possible information patterns, hence reducing the required transmission/storage bandwidth.

Each pixel in the test images is either black or white. At first, input images are partitioned into 4×4 blocks, and each block is vector-quantized by the ECAM chip. A set of 32 codewords is chosen, and they correspond to all-white, all-black, horizontal-edge, vertical-edge, and diagonal-edge blocks (see Fig. 16.11). The choice of these codewords is totally heuristic and is not optimal in any way since the objective of this experiment is to apply the ECAM chip to solve some real problem and measure the speed of the chip. The ECAM chip is programmed with these codewords, and 4×4 blocks from a binary image are fed to the ECAM chip one at a time. The nearest codeword to each input block appears at the output of the ECAM chip. The indices of those codewords can then be transmitted or stored, achieving a compression ratio of 16:5. A reconstructed image is formed by replacing each block by the ECAM-chip-generated codeword.

However, there are time when the output pattern of the ECAM chip is not one of the 32 codewords, in which case an all-white block is generated instead. Figures 16.12 and 16.13 illustrate two original binary images and their ECAM-chip-reconstructed images. It is obvious that the reconstructed binary image is not as good as the original, yet this is the price paid for reduced bandwidth. In addition, any real application would optimize the codewords for less distortion.

Working on the above task, the ECAM chip performs one associative recall operation on a 4×4 block in less than 3 μsec (this includes communication time between the ECAM chip and the host computer). This projects to about 49 msec for a 512×512 binary image, or more than 20 images per second—fast enough for real-time applications. If one simulated the ECAM on a serial digital computer, it would take 3072 simple instructions (multiply or add instructions) plus other complex operations for the computer to perform one associative recall operation. Therefore, in terms of associative recall operations, the ECAM chip runs faster than a 1024-MIPS serial digital computer.

16.8. SUMMARY

In this chapter we have proposed a model for a group of neural-associative memories called the recurrent correlation associative memories (RCAM). These RCAMs are asymptotically stable as long as their weighting functions are continuous and monotone nondecreasing. In particular, a new high-capacity RCAM called the exponential correlation associative memory (ECAM) was presented. The asymptotic storage capacity of the ECAM scales *exponentially* with the length of memory patterns. It was also found that under certain conditions the asymptotic storage capacity of the ECAM meets the ultimate upper bound for the capacity of associative memories. Moreover, the asymptotic storage capacity of the ECAM whose exponentiation neurons have fixed dynamic range is found to be proportional to that dynamic range.

Simulation results confirming the theoretical findings about the attraction radius and the storage capacity of the ECAM were also presented. A VLSI chip based on the ECAM model was fabricated and tested. The ECAM chip was shown to perform almost as well as computer simulation. The speed of the chip was measured by employing it to do vector quantization on binary images. It was found that the ECAM chip

(a)   (b)

Fig. 16.12. Comparison of (a) the original girl image and (b) the reconstructed image after vector quantization by the ECAM chip.

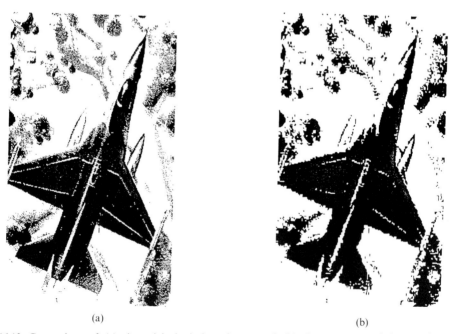

(a)   (b)

Fig. 16.13. Comparison of (a) the original airplane image and (b) the reconstructed image after vector quantization by the ECAM chip.

can process binary images in real time, i.e., about twenty 512×512 images every second. We believe that the RCAM model and the ECAM chip in particular provide a fast and efficient way for solving many associative-recall problems, such as vector quantization and pattern recognition.

REFERENCES

Anderson, J. A. (1972). "A Simple Neural Network Generating an Interactive Memory," *Math. Biosci.*, **14**, 197–220.

Bruck, J. and Goodman, J. W. (1987). "A Generalized Convergence Theorem for Neural Networks and Its Applications in Combinatorial Optimization," in *Proc. IEEE Int. Conference on Neural Networks*, San Diego, CA, Vol. III, pp. 649–656.

Chiueh, T. D. and Goodman, R. M. (1988). "High-Capacity Exponential Associative Memory," in *Proc. IEEE Int. Conference on Neural Networks*, San Diego, CA, Vol. I, pp. 153–160.

Chiueh, T. D. and Goodman, R. M. (1990). "VLSI Implementation of a High-Capacity Neural Network Associative Memory," in *Advances in Neural Information Processing Systems 2*, D. S. Touretzky, ed. Morgan Kaufmann, San Mateo, CA, pp. 793–800.

Chiueh, T. D. and Goodman, R. M. (1991). "Recurrent Correlation Associative Memories," *IEEE Trans. Neural Networks*, **2**(2), 275–284.

Chou, P. A. (1988). "The Capacity of the Kanerva Associative Memory Is Exponential," in *Neural Information Processing Systems*, D. Z. Anderson, ed.

American Institute of Physics, New York, pp. 184–191.

Dembo, A. and Zeitouni, O. (1988a). "High Density Associative Memories," in *Neural Information Processing Systems*, D. Z. Anderson, ed. American Institute of Physics, New York, pp. 211–218.

Dembo, A. and Zeitouni, O. (1988b). "General Potential Surfaces and Neural Networks," *Phys. Rev. A*, **37**(6), 2134–2143.

Glasser, L. A. and Dopperpuhl, D. W. (1985). *The Design and Analysis of VLSI Circuits*. Addison-Wesley, Reading, MA.

Gray, R. M. (1984). "Vector Quantization," *IEEE ASSP Magazine*, **1**, 4–29.

Hopfield, J. J. (1982). "Neural Network and Physical Systems with Emergent Collective Computational Abilities," *Proc. Nat. Acad. Sci. U.S.A.*, **79**, 2554–2558.

Hopfield, J. J. (1984). "Neurons with Graded Response Have Collective Computational Properties Like Those of Two-State Neurons," *Proc. Nat. Acad. Sci. U.S.A.*, **81**, 3088–3092.

Kohonen, T. (1972). "Correlation Matrix Memories," *IEEE Trans. Computers*, **C-21**, 353–359.

McEliece, R. J., Posner, E. C., Rodemich, E. R., and Venkatesh, S. S. (1987). "The Capacity of the Hopfield Associative Memory," *IEEE Trans. Information Theory*, **IT-33**, 461–482.

Mead, C. A. (1989). *Analog VLSI and Neural Systems*. Addison-Wesley, Reading, MA.

Psaltis, D. and Park, C. H. (1986). "Nonlinear Discriminant Functions and Associative Memories," in *Neural Networks for Computing*, J. S. Denker, ed. American Institute of Physics, New York, pp. 370–375.

Sayeh, M. R. and Han, J. Y. (1987). "Pattern Recognition Using a Neural Network," in *Proc. SPIE Cambridge Symposium on Optical and Optoelectronic Engineering*, Cambridge, MA.

17

Design of a Bidirectional Associative Memory Chip

KWABENA A. BOAHEN and ANDREAS G. ANDREOU

17.1. INTRODUCTION

Bidirectional associative memories (BAMs) support heteroassociative storage and recall of binary patterns. A four-layer BAM network is shown in Fig. 17.1a. It has two input/output (I/O) layers, A and B, and two hidden layers, G and H. The pathways between layers are weighted projections that connect every unit in one layer to all units in the other. An input pattern at layer A recalls an associated pattern at layer B through the A $\Rightarrow$ G $\Rightarrow$ B pathway. Reciprocally, inputs at layer B elicit patterns at layer A via pathway B $\Rightarrow$ H $\Rightarrow$ A. This network supports truly bidirectional associations, i.e., two-way links between stored pattern pairs. Associative memory architectures and dynamics are reviewed in Chapter 1 and the BAM model is also discussed in Chapter 5 of this volume.

The recurrent pathways between layers A and B produce interesting dynamics in the BAM network. Stable reverberations occur when patterns in these layers reinforce each other, that is,

$$A_j \rightarrow B_j \rightarrow A_j \dots$$

for the jth association (A_j, B_j). Consequently, the network remains in this state even after external inputs are removed. When a new input pattern is applied, the state of the network evolves to the stored pattern that best matches the input. Formally, using column vectors with ± 1 components to represent the patterns, the pattern A, applied at layer A, recalls the association (A_p, B_p) given by

$$A_p^T A = \max_{j=1,\dots,r} A_j^T A \qquad (1)$$

where the max operation is over all stored associations.

Stored associations are encoded using a unary representation at the hidden layers (grandmother cells). A unique unit in each hidden layer is assigned to every association stored; this is essentially a sparse coding of the patterns present at the I/O layers. There is a one-to-one correspondence between weights and stored patterns; the weights of reciprocal connections between unit a_i (b_i) and units g_j and h_j are equal to the ith component of A_j (B_j). In other words, the weight vectors of hidden units g_j and h_j are simply A_j and B_j, respectively. The folded architecture shown in Fig. 17.1b makes it possible to share one stored weight between two reciprocal connections.

The original BAM network proposed by Kosko (1988) had two layers and used a distributed representation. Our four-layer network is mathematically equivalent to Kosko's model (Boahen et al., 1989a, b). From a chip designer's point of view, however, a local representation offers three major advantages over a distributed one.

First, optimal storage efficiency of one information bit per storage is achieved. This is the appropriate measure of hardware efficiency for an implementation based on digital storage because the chip area is proportional to the number of storage cells. For reliable operation (Eq. 1) with random patterns, only up to $n/8$ associations can be stored in two-layer BAM with n neurons in each layer. This translates to an efficiency of $1/4(\log_2 n - 1)$ information bits per storage bit. (Each weight requires $\log_2(n/8) + 1$ bits.)

Second, no weight adjustment circuitry is required; pattern vectors are written directly to the weight storage cells. In a distributed memory, weights are updated by

$$m_{kl} \leftarrow m_{kl} + a_k b_l$$

for each new association. The overhead of this circuitry is especially severe if new associations are programmed only occasionally.

Third, introducing lateral inhibition within the hidden layers prevents recall performance from degrading as the number of stored patterns

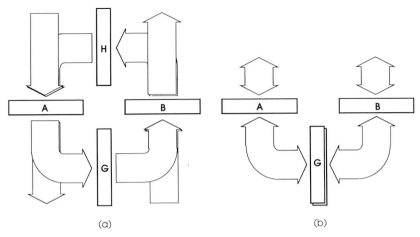

Fig. 17.1. (a) Four-layer bidirectional associative memory. The pathway (arrows) in the foreground allows inputs at layer A to produce outputs at layer B; the one in the background supports recall in the opposite direction. (b) Folded architecture: the hidden layers are merged and reciprocal connections are introduced.

increases. Inhibition enhances the contribution of the best match while reducing the contributions of other patterns. A four-layer BAM with this nonlinear expansion at the hidden layers is equivalent to a higher-order correlation two-layer BAM. However, a distributed higher-order network has very poor storage efficiency. For instance, a second-order network has one-third the storage efficiency of a first-order one, and a third-order network has only 1/15 (Peretto and Niez, 1986).

Poor fault tolerance is an often-cited disadvantage of local representations. Evidently there is a trade-off between storage efficiency and fault tolerance. Distributed representations achieve fault tolerance through redundant storage of information, hence their poor storage efficiency. Concomitantly, they are extremely robust against hard or soft faults in several memory locations (Sivilotti et al., 1985). Local representations can also achieve fault tolerance through redundancy; the user can trade storage efficiency for fault tolerance by storing more than one copy of each association.

Furthermore, robustness against certain failures in the hidden units and their communication lines is achieved by using a reentrant memory array. A hidden unit's operation is verified by performing a recall before committing an association to it. Each hidden unit is assigned an eligibility bit which captures its state when the programming (as opposed to recall) signal is asserted. This bit is used to select the weight storage location of the winning hidden unit for

reentry; the desired new association overwrites the previously recalled one.

The four-layer BAM chip described here uses efficient circuit techniques to realize programmable reciprocal connections and lateral inhibition. A simple current-controlled current conveyor and a novel two-transistor reciprocal junction allow units to interact bidirectionally through a single line. This is accomplished without multiplexing by using *independent* voltage and current signals. A winner-take-all (WTA) circuit similar to the one proposed by Lazzaro et al. (1989) that requires just one communication line is used to implement lateral inhibition. Using these techniques, BAM chips that approach static RAM densities have been realized.

17.2. CIRCUIT TECHNIQUES

A unit's output is represented by a voltage signal; its inputs are represented by current signals. Since currents and voltages may be independently transmitted along the same line, these signal representations allow output and inputs to be communicated using just one line. Voltage output facilitates fan-out, while current input provides summation. Units interact bidirectionally through a reciprocal junction circuit using a simple current conveyor. These circuits were designed for subthreshold operation. The advantages of the subthreshold current-mode approach are outlined in Andreou and Boahen (1989).

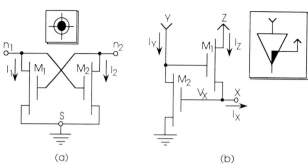

(a)   (b)

Fig. 17.2. (a) Reciprocal junction. Voltage signals applied at node n_1 or n_2 are converted to current signals. The lines that apply voltage signals bring back these currents. The circuit symbol is shown in the insert; the perpendicular lines correspond to nodes n_1 and n_2, (b) Current conveyor. The current, I_X, supplied to node X is conveyed to node Z. The voltage at X, V_X, is determined by the current I_Y. The conveyor's circuit symbol is in the insert; the black triangle depicts current buffering.

17.2.1. Subthreshold Device Model

A simple model for subthreshold conduction in the MOS transistor is given in Mead (1989):

$$I_{ds} = I_0 \left(\frac{W}{L}\right) e^{\kappa V_{gb}/V_T}(e^{-V_{sb}/V_T} - e^{-V_{db}/V_T}) \quad (2)$$

where voltages are with reference to the local substrate or well and the channel length modulation is ignored. The drain current's dependence on the back-gate (well/substrate voltage) becomes explicit when voltages are referred to the source potential:

$$I_{ds} = I_0 \left(\frac{W}{L}\right) e^{[(1-\kappa)V_{bs}]/V_T} e^{\kappa V_{gs}/V_T}$$

$$\times \left(1 - e^{-V_{ds}/V_T} + \frac{V_{ds}}{V_0}\right) \quad (3)$$

I_0 is the zero-bias current for the given device, κ measures the effectiveness of the gate potential in controlling the channel current, and W and L are the channel width and length, respectively. V_0 is the Early voltage which is proportional to L. V_T $(=kT/q)$ is the thermal voltage which is equal to 26 mV at room temperature.[1] Typical parameters for minimum-size device (4 μm × 4 μm) fabricated in a standard digital 2-μm n-well process are: $I_0 = 0.72 \times 10^{-18}$ A, $\kappa = 0.75$, and $V_0 = 15.0$ V. The current changes by a factor of 10 for an 80-mV change in V_{ds} or a 240-mV change in V_{bs}; this relation holds up to about 100 nA.

In the saturation region, $V_{ds} > 4V_T$ ($V_{dsat} \simeq$ 100 mV at room temperature), the device's output conductance and transconductance are given by

$$g_{dsat} = \frac{\partial I_{ds}}{\partial V_{ds}} = \frac{I_{ds}}{V_0} \quad (4)$$

$$g_m = \frac{\partial I_{ds}}{\partial V_{gs}} = \frac{\kappa I_{ds}}{V_T} \quad (5)$$

[1] This is the n-type device equation; the p-type version has voltage signs reversed.

The ratio $A = g_m/g_{dsat} = \kappa V_0/V_T$ measures the gain available from the device. It is independent of current and is 430 for the given parameter values.

17.2.2. Reciprocal Junction

The two-transistor circuit shown in Fig. 17.2a provides bidirectional interaction between units connected to nodes n_1 and n_2. It receives voltage inputs at these nodes and produces current outputs at the same nodes; each transistor acts like a synaptic junction. Interaction is turned on by grounding S, turned off by bringing S to V_{dd}, or modulated in a multiplicative fashion by applying an analog signal to the well.

When S is at V_{dd} one of the devices has a positive gate–source voltage and sources the current:

$$I_{off} = I_0 e^{[\kappa(V_{n_1} - V_{n_2})]/V_T} = I_0 L$$

assuring $V_{n_1} > V_{n_2}$, without loss of generality, and ignoring the body effects. I_{off} is proportional to $L \equiv e^{\kappa V_{n_1}/V_T}/e^{\kappa V_{n_2}/V_T}$, the dynamic range of the current signals. For signals in the range 100 pA to 100 nA, $L = 10^3$, and $I_{off} \approx 1$ fA.

For proper operation in the on state, the devices must be in saturation ($V_{n_1}, V_{n_2} > V_{dsat}$). Then, for a small change in V_{n_1}, the changes in I_1 and I_2 are related by

$$\frac{\Delta I_1}{\Delta I_2} = \frac{g_{dsat_1}}{g_{m_2}} = \frac{1}{A}\frac{I_1}{I_2}$$

This gives $\Delta I_1/I_1 = \frac{1}{430}\Delta I_2/I_2$. Hence, changing V_{n_1} to increase I_2 by 100 percent causes a 0.23 percent change in I_1. So bidirectional communication is possible with less than -50 dB crosstalk.

17.2.3. Current Conveyor

A current-controlled current conveyor is shown in Fig. 17.2b. This circuit has a communication node X, a control node Y, and an output node

Z. The current I_X at the communication node is conveyed to the output node at a potential V_X determined by the control current I_Y. Traditional current conveyors (Smith and Sedra, 1968) use a voltage input to set V_X, i.e. $V_X = V_Y$. The current-controlled conveyor's operation is described by two simple relationships:

$$I_Z = I_X \qquad \mathscr{I}(V_X) = I_Y$$

where the function $\mathscr{I}(V_X)$ converts V_X to a current using a transistor identical to M_2. Thus the communication node can receive one current I_X, and, at the same time, transmit a second current I_Y.

In this circuit, M_1 establishes V_X to make M_2's current equal I_Y. This is accomplished by negative feedback through M_2 which serves as an inverting amplifier. The conductance seen at node X is approximately Ag_{m_1}, i.e., the source conductance of M_1 times M_2's gain A. In addition, M_1 buffers I_X, transferring this current to its high-impedance drain terminal. This negative feedback arrangement is the core of Säckinger's regulated cascode circuit (Säckinger and Guggenbühl, 1990). In Boahen et al., 1989b, the authors proposed its use, in conjunction with the reciprocal junction, for two-way communication.

A current conveyor communicates bidirectionally with one or more conveyors through reciprocal junctions connected to its communication node. I_Y is the outgoing signal; it is replicated at each junction by V_X; and I_Z is the sum of all incoming signals. Changes in I_Z produce changes in V_X and therefore in the copies of I_Y. Small changes are related by

$$\frac{\Delta I_Y}{\Delta I_Z} = \frac{g'_{m_2}}{Ag_{m_1}} = \frac{1}{A}\frac{I_Y}{I_Z}$$

where the copy $I'_Y = I_Y + \Delta I_Y$ and $g'_{m_2}\,(=g_{m_2})$ is the transconductance of the device that mirrors I_Y. Hence, in percentages, the change in I'_Y is 430 times less than in I_Z, just as the previous case.

For large changes in I_Z, M_1's gate–source voltage changes by $(V_T/\kappa)\ln(I_{Z_2}/I_{Z_1})$, and therefore

$$\frac{\Delta I_Y}{I_Y} = \frac{1}{A}\ln\left(\frac{I_{Z_2}}{I_{Z_1}}\right)$$

This means I_Z can change by about a factor of 75 before I'_Y changes by one percent. Characteristics of minimum-sized versions of these circuits, obtained from chips fabricated through foundry services, are shown in Fig. 17.3.

17.2.4. Winner-Takes-All

In this circuit, shown in Fig. 17.4, m current conveyors compete for current supplied to a common line. This current, I_m, is steered to the output of the conveyor with the largest input current; all other outputs are zero. This is an adaptation of Lazzaro and Mead's original circuit (Lazzaro et al., 1989) to provide current outputs.

Input currents are supplied to control nodes, output currents are obtained from output nodes, and the communication nodes are connected together. Each conveyor sees a voltage source at its communication node—not a current source. If $I_Y < \mathscr{I}(V_X)$, M_2 enters the linear region, turning M_1 off (refer to Fig. 17.2b). Otherwise, M_1 adjusts V_X to set $\mathscr{I}(V_X) = I_Y$. Thus the conveyor with the largest input sets the voltage on the common line and conveys I_m.

When the inputs are very similar, the conversion from input to output is *exponential*, with the sum of the outputs normalized to I_m. In this case M_1 remains on and M_2 stays in saturation. The inputs develop voltage signals across M_2's drain conductance, $g_{d\,\mathrm{sat}_2}$; these voltages are converted exponentially to current by M_1. For example, a one-percent input difference produces a voltage difference of 0.15 V, so the corresponding outputs differ by a factor of 75.

If two conveyors, with inputs $I_Y \pm \frac{1}{2}\Delta I_Y$, are competing for I_m, their outputs are $I_Z \pm \frac{1}{2}\Delta I_Z$, where $I_Z = \frac{1}{2}I_m$ and the differential gain for small signals is

$$\frac{\Delta I_Z}{\Delta I_Y} = \frac{g_{m_1}}{g_{d\,\mathrm{sat}_2}} = A\frac{I_Z}{I_Y}$$

Hence, the small-signal differential gain is 430 for normalized inputs and outputs.

17.3. CIRCUIT DYNAMICS

Dynamic behavior of the current conveyor is determined by corner frequencies associated with its communication and control nodes. The communication node's capacitance arises from the interconnects and reciprocal junctions; capacitance is added at the control node to tailor the conveyor's response. A small-signal conveyor model is introduced and used to find optimal choices for the control node capacitance. It is also used to analyse a two-layer network of current conveyors and a WTA network. A simple small-signal transistor model consisting of a transconductance g_m, controlled by v_{gs}, in parallel with a conductance $g_{d\mathrm{sat}}$ is used here; it is

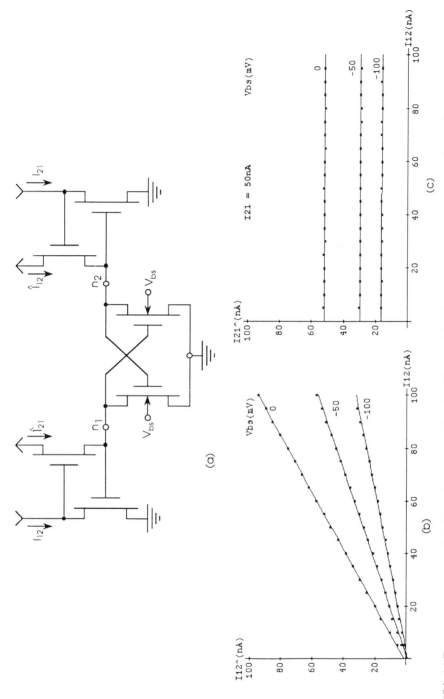

Fig. 17.3. (a) Two conveyors communicate through a reciprocal junction. They send signals I_{12} and I_{21} and receive signals $\hat{I}_{21}$ and $\hat{I}_{12}$. The weight is determined by the well voltage, V_{bs}. Experimental data were obtained by stepping I_{12} from 5 nA to 100 nA, with I_{21} held at 50 nA, for V_{bs} values of 0, −50 mV, and −100 mV. (b) $\hat{I}_{12}$ varied linearly with I_{12}; the slopes are 0.93, 0.57, and 0.33. (c) $\hat{I}_{21}$ remained constant; the slopes are less than −0.004. $\hat{I}_{21}/I_{21}$ equals 1.04, 0.59, and 0.33, showing that the weighting is symmetric.

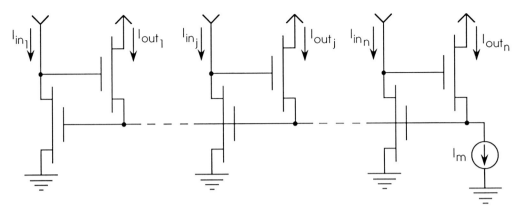

Fig. 17.4. Winner-takes-all (WTA) circuit. m current signals are applied to m current conveyors. The conveyor that has the largest input conveys the current I_m supplied to a common line; the other conveyors have zero output.

assumed that the communication and control node capacitances are much larger than the intrinsic device capacitances.

17.3.1. Small-Signal Conveyor Model

The signals i_X, i_Y, and v_Z are viewed as the conveyor's inputs; they produce output signals v_X, v_Y, and i_Z (see Fig. 17.2b). Its small-signal behavior is described by[2]

$$\begin{bmatrix} v_X \\ v_Y \\ i_Z \end{bmatrix} = \begin{bmatrix} Z_X & Z_{XY} & A_{XZ} \\ Z_{YX} & Z_Y & A_{YZ} \\ A_{ZX} & A_{ZY} & Y_Z \end{bmatrix} \begin{bmatrix} i_X \\ i_Y \\ v_Z \end{bmatrix}$$

The outputs of interest are v_X, which communicates i_Y, and i_Z which should equal i_X. Ignoring the dependence on v_Z, we have

$$\begin{bmatrix} v_X \\ i_Z \end{bmatrix} = \begin{bmatrix} Z_X & Z_{XY} \\ A_{ZX} & A_{ZY} \end{bmatrix} \begin{bmatrix} i_X \\ i_Y \end{bmatrix}$$

The transimpedance Z_{XY} converts the outgoing signal i_Y to a voltage, while the incoming signal i_X appears at the supply node with a gain of A_{ZX}. Ideally, $Z_{XY} = 1/g_{m_2}$ and $A_{ZX} = 1$. The impedance Z_X seen at the communication node and the gain A_{ZY} from the control node to the supply node produce interference between incoming and outgoing signals; they should equal zero. In practice, all the parameters are finite and frequency dependent. They are characterized by corner frequencies $\omega_X = g_{m_1}/C_X$ and $\omega_Y = g_{m_2}/C_Y$,

[2] In our notation, Z_Q or $1/Y_Q$ is the impedance seen at node Q, Z_{QR} is a transimpedance at node Q controlled by node R, and A_{QR} is a voltage or current gain from R to Q. By convention, positive current flows into the circuit.

associated with nodes X and Y. The gain provided by the voltage follower M_1 has dropped to one-half at ω_X; the inverting amplifier M_2 has unity gain at ω_Y.

In the s-domain, the conveyor's small-signal parameters are given by

$$\left. \begin{array}{l} Z_X(s) \approx \left(\dfrac{1}{Ag_{m_1}}\right)\left(\dfrac{1 + A\rho s}{1 + \rho s + \rho s^2}\right) \\[12pt] Z_{XY}(s) \approx \left(\dfrac{1}{g_{m_2}}\right)\left(\dfrac{1}{1 + \rho s + \rho s^2}\right) \end{array} \right\} \quad (6)$$

$$\left. \begin{array}{l} A_{ZX}(s) \approx -\dfrac{1 + \rho s}{1 + \rho s + \rho s^2} \\[12pt] A_{ZY}(s) \approx \left(\dfrac{g_{m_1}}{g_{m_2}}\right)\left(\dfrac{s}{1 + \rho s + \rho s^2}\right) \end{array} \right\} \quad (7)$$

where the complex frequency variable s is in units of ω_X and $\omega_Y = (1/\rho)\omega_X$; the approximations $A \gg 1$ and $\rho A \gg 1$ were used. Poles occur at $s = \frac{1}{2}[-1 \pm \sqrt{(1 - 4/\rho)}]$; if $\rho < 4$ they are complex. Optimal conveyor behavior occurs when $\rho = 2$; Bode plots of the transfer functions are shown in Fig. 17.5.

17.3.2. Conveyor Responses

The impedance seen at the communication node, $Z_X(j\omega)$, equals $1/Ag_{m_1}$, at DC as shown in the previous section. Above $\omega_X/\rho A$, or ω_Y/A, its magnitude increases because M_2's gain starts decreasing. When ω_X is exceeded, C_X causes the impedance to decrease. It reaches a maximum magnitude of $1/g_{m_1}$ at $\hat{\omega}_X = \omega_X/\sqrt{\rho}$. At the

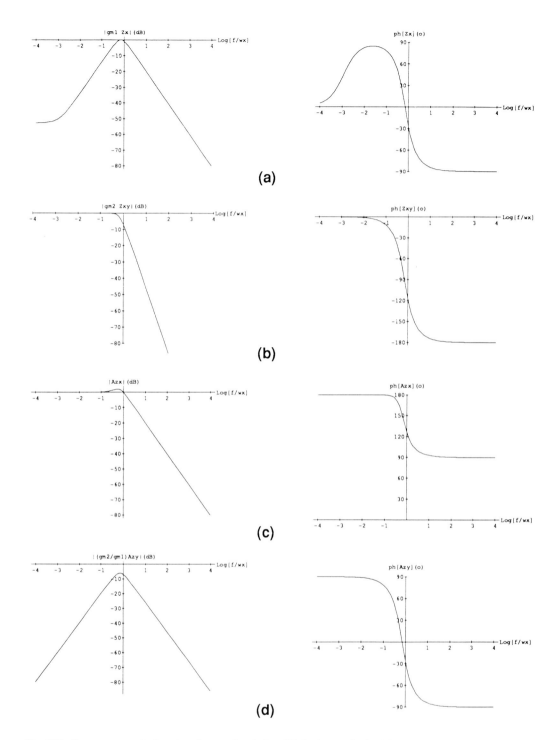

Fig. 17.5. Conveyor transfer functions for $\rho = 2$ and $A = 430$; both magnitude (left) and phase (right) are shown. The frequency scale is in units of ω_X. (a) Z_X, the communication node impedance. (b) Z_{XY}, the transimpedance from control node to communication node. (c) A_{ZX}, the current gain from communication node to supply node. (d) A_{ZY}, the current gain between control and communication nodes.

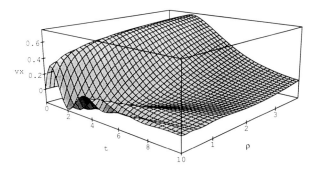

Fig. 17.6. Response of voltage at node X to a current step applied there at $t = 0$. $v_X(t)$ is plotted for values of ρ from 0.2 to 3.9; amplitude is in units of i_u/g_{m_1}, where i_u is the height of the step input, and time is in units of $1/\omega_X = C_X/g_{m_1}$. The settling time is large for both small and large values of ρ; it is minimized round 2.

maximum, $Z_X(j\omega)$'s phase is zero and its magnitude is A times its DC value, producing a 53-dB peak (see Fig. 17.5a).

Ringing occurs in the voltage at node X when $Z_X(s)$ has complex poles. A unit current step generates the voltage signal

$$v_X(t) = \frac{1}{Ag_{m_1}} + \frac{e^{-\frac{1}{2}\omega_X t}}{g_{m_1}\sqrt{(4/\rho - 1)}}$$
$$\times \{2 \sin \omega_0 t - [\sqrt{(4/\rho - 1)} \cos \omega_0 t + \sin \omega_0 t]/A\}$$

$$\approx \frac{1}{Ag_{m_1}} + \frac{2e^{-\frac{1}{2}\omega_X t} \sin \omega_0 t}{g_{m_1}\sqrt{(4/\rho - 1)}} \qquad (8)$$

where $\omega_0 = (\omega_X/2)\sqrt{(4/\rho - 1)}$. The responses for various values of ρ are shown in Fig. 17.6. The settling time is minimized around $\rho \approx 2$.

The transimpedance from the control node to the communication node, $Z_{XY}(j\omega)$, equals $1/g_{m_2}$ at low frequencies. At high frequencies, C_Y shunts i_y and the follower's gain decreases, so there is a 40-dB per decade roll-off. The break-frequency equals $\hat{\omega}_X$. A peak appears in the response if $\rho < 2$; when $\rho = 2$ the response is maximally flat and settling time is minimized (see Fig. 17.5b).

The current gain between the communication and output nodes, $A_{ZX}(j\omega)$, is unity for frequencies up to ω_X; but the response peaks just before the break frequency. When $\rho = 2$, the peak is 2 dB and occurs at $0.55\omega_X$. Above ω_X, C_X shunts i_X, and so the gain decreases at 20 dB per decade (see Fig. 17.5c). Note that $(1 + A_{ZX})i_X$ gives the current in C_X; equating the voltage across C_X to $Z_X i_X$, yields the relationship

$$A_{ZX} = Z_X g_{m_1} s - 1$$

Since s is the ratio between ω and ω_X, $g_{m_1}s$ gives the admittance of C_X at ω.

The current gain from the control node to the output node, $A_{ZY}(j\omega)$, is zero at DC. For AC signals, the transimpedance Z_{XY} develops voltage

signals at the communication node which produce currents in C_X; these currents appear at the supply node. In fact

$$A_{ZY} = Z_{XY}g_{m_1}s$$

C_X introduces a zero at DC, so the gain initially increases at 20 dB per decade, reaches $g_{m_1}/\rho g_{m_2}$ $(= C_X/C_Y)$ at $\hat{\omega}_X$, and then rolls off 20 dB per decade (see Fig. 17.5d).

To summarize, the current conveyor provides a bidirectional communication channel as illustrated in Fig. 17.7. Its bandwidth is $\omega_X/\sqrt{\rho}$ for outgoing signals (I_Y to I'_Y), and ω_X for incoming ones (I_X to I'_X). When $\rho = 2$, the outgoing channel is maximally flat and settling time is minimized for both channels. The interference between these channels, i.e., the fraction of the other signal that is added, can be as high as g'_{m_2}/g_{m_1} $(= I'_Y/I_X)$ for the outgoing one and C_X/C_Y for the incoming one; this happens at the frequency $\omega_X/\sqrt{\rho}$. These reflections introduce undesired feedback paths

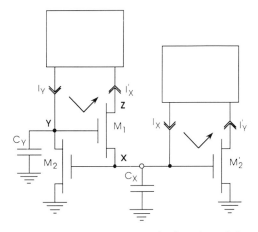

Fig. 17.7. Bidirectional communication channel between two current-mode circuits (boxes). The bent arrows depict channel interference, or reflections, caused by capacitances at the conveyor's communication and control nodes.

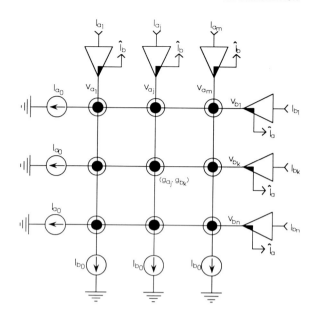

Fig. 17.8. Two-layer network. There are m current conveyors in the first layer and n in the second; they are fully connected by an $n \times m$ array of reciprocal junctions. Interference between incoming and outgoing signals at the conveyors' communication nodes produces signal loops. It is shown that the fixed bias currents I_{a_0} and I_{b_0} are necessary to obtain a loop gain less than unity.

around circuits that interface bidirectionally with the conveyor.

17.3.3. Network's Loop Gain

A fully connected two-layer network of current conveyors and reciprocal junctions is shown in Fig. 17.8. Conveyors in layer A send out currents $I_{a_1}, \ldots, I_{a_m}$; those in layer B transmit $I_{b_1}, \ldots, I_{b_n}$. Thus, all conveyors in layer A receive the same current

$$\hat{I}_b = I_{b_0} + \sum_{k=1}^{m} I_{b_j}$$

Similarly, all conveyors in layer B receive $\hat{I}_a$, the sum of I_{a_j} plus I_{a_0}; I_{a_0} and I_{b_0} are fixed bias currents. The reciprocal junction connecting conveyors a_j and b_k has two small-signal transconductances controlled by a_j and b_k, respectively: $g_{a_j} = \kappa I_{a_j}/V_T$ and $g_{b_k} = \kappa I_{b_k}/V_T$.

Conveyor a_j presents a small-signal impedance Z_{a_j} to the network. At resonance, $Z_{a_j} \rightarrow 1/G_a$, where $G_a = \kappa \hat{I}_b/V_T$ is the transconductance of its buffering device. In this case, all layer A conveyors present the same impedance to the network; they also see the same network admittance. This symmetry dictates that they develop the same signals v_a.

The signals developed at layer B will be

$$v_b = - \sum_{j=1}^{m} \frac{g_{a_j}}{G_b} v_a$$

assuming resonance occurs there too. These signals feed back to layer A, producing

$$v_a' = - \sum_{k=1}^{n} \frac{g_{b_k}}{G_a} v_b$$

Therefore, the loop gain $A_L \equiv v_a'/v_a$ is

$$A_L = \frac{1}{G_a G_b} \sum_{j=1}^{m} g_{a_j} \sum_{k=1}^{n} g_{b_k}$$

$$= \frac{\sum_{j=1}^{m} I_{a_j}}{I_{a_0} + \sum_{j=1}^{m} I_{a_j}} \frac{\sum_{k=1}^{n} I_{b_k}}{I_{b_0} + \sum_{k=1}^{n} I_{b_k}} \qquad (9)$$

This is the largest value for A_L, for the worst-case condition where all conveyors resonate at the same frequency and they are fully connected. The fixed bias currents, I_{a_0} and I_{b_0}, ensure that $A_L < 1$.

17.3.4. Winner-Takes-All Response

As a conveyor's input approaches the winning cell's input, the current I_m switches from the winning cell to this cell's output (see Fig. 17.4). When their inputs are the same, say equal to I_Y, their outputs equal $\frac{1}{2}I_m$; therefore, their small-signal parameters are identical. For the inputs $I_Y \pm \frac{1}{2}\Delta I_Y$, the outputs are $\frac{1}{2}I_m \pm \frac{1}{2}\Delta I_Z$, where $\frac{1}{2}\Delta I_Z$ is the current that flows from one conveyor to the other. This current flows across Z_X to cancel the effect of the inputs which act through Z_{XY}.

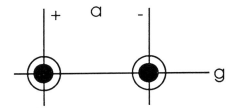

Fig. 17.9. Excitatory or inhibitory reciprocal weight using a pair of junctions which weight signals by 0 (off state) or 1 (on state).

Therefore,

$$\frac{\Delta I_Z}{2} = \frac{Z_{XY}}{Z_X} \frac{\Delta I_Y}{2}$$

Hence, the differential gain is

$$A(s) = \frac{Z_{XY}}{Z_X} = \frac{g_{m_1}}{g_{m_2}} \frac{A}{1 + A_S} \qquad (10)$$

Here, $g_{m_1} = \kappa I_m/2V_T$, $g_{m_2} = \kappa I_Y/V_T$, and s is in units of ω_Y; the response is first-order and independent of ω_X. The gain equals A at DC, for normalized inputs and outputs as shown in the previous section, rolls off 20 dB per decade above ω_Y/A, and becomes unity at ω_Y.

17.4. BAM CIRCUITS

I/O units and hidden units perform nonlinear operations on their inputs. An I/O unit has two states, corresponding to the pattern values ± 1; its state is determined by comparing its excitatory and inhibitory inputs. A hidden unit also has two states; its state is determined by comparing its input with all other units in its layer. These comparisons are performed by WTA cells; a current conveyor supplies the input and communicates the outputs. This current conveyor/WTA combination is named after the cortical pyramidal cell because their functions are similar.

Programmable reciprocal weights are realized with a pair of reciprocal junctions and a dual rail scheme (Sivilotti et al., 1985) as shown in Fig. 17.9. For an excitatory (inhibitory) weight the junction on the excitatory (inhibitory) line is turned on; the other junction is turned off. In the forward direction (a to g), analog inputs are applied as voltages on either the excitatory or inhibitory lines, depending on their sign. If the relevant junction is on, i.e., the input is excitatory

and the weight is excitatory or they are both inhibitory, the input appears as a current on the common line. In the reciprocal direction (g to a), positive analog inputs applied as voltages on the common line appear as current on the excitatory line if the weight is excitatory, otherwise they appear on the inhibitory line. This circuit, together with a flip-flop that stores the junctions' complementary states, is called a BAM cell.

17.4.1. Pyramidal Cell

Cortical pyramidal cells are arranged in layers with projections from one layer to the next. Their apical dendrites receive incoming afferent signals while their axons carry outgoing efferent signals. Pyramidal cell axons have collaterals that branch back and contact basal dendrites of neighboring cells; these lateral interactions are mainly inhibitory. Figure 17.10; shows a cortical pyramidal cell and the pyramidal cell circuit. The circuit has two communication nodes, X_V and X_L: X_V carries afferent and efferent signals, as do apical dendrites and axons of cortical pyramidal cells. X_L mediates lateral inhibitory interactions, mimicking the cortical cell's basal dendrites and recurrent axon collaterals. Inputs extrinsic to the network are applied at node Y_V; the voltage at Y_V is used to monitor the inputs and the WTA output of this cell.

The circuit consists of an n-type current conveyor (M_1, M_2) and a p-type WTA cell (M_3, M_4). The extra transistor (M_5) is needed to stabilize the circuit. When a WTA cell and a current conveyor feed each other, their current-buffering devices, M_1 and M_3, can act as common-source amplifiers. An unstable positive feedback loop results; the devices enter the linear region and X_V and X_L go the the rails. The extra device prevents this by driving node X_V to keep the conveyor's buffer (M_1) in saturation. Although M_5 shunts part of the incoming current, a fixed fraction of the input is passed to the WTA cell.

A pyramidal cell contributes a current I_U to the WTA competition and has a quiescent current of I_U. Therefore, its efferent signal $\mathscr{I}(V_{X_v})$ is 1 (in units of I_U) in the quiescent state but increases to $(m + 1)$ when it is winning, where m is the number of competing cells. Note that the winner's output exceeds the combined output of all the other cells.

Given the capacitances at the communication nodes X_V and X_L, the capacitances at Y_V and Y_L are chosen to optimize the current conveyor's

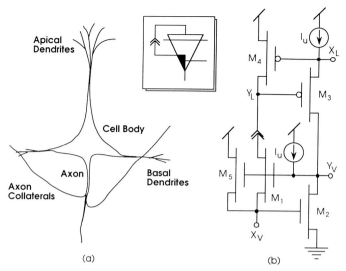

Fig. 17.10. (a) Cortical pyramidal cell. Inputs are received at the apical and basal dendrites; the cell's output is communicated by its axon. (b) Pyramidal cell circuit. An n-type current conveyor communicates inputs and outputs through node X_V and a p-type WTA cell realizes lateral inhibition through node X_L. The circuit symbol includes the current conveyor's control and supply nodes and the WTA's input, as well as the communication nodes.

response and to ensure that the feedback path around the WTA cell has less than unity gain. For the conveyor,

$$\rho_V = \frac{\omega_{X_V}}{\omega_{Y_V}} = \frac{I_{X_V}}{I_{Y_V}} \frac{C_{Y_V}}{C_{X_V}} \qquad (11)$$

This relation is used to determine C_{Y_V} given the ratio I_{X_V}/I_{Y_V} and the desired value for ρ_V.

The loop transmission is given by

$$A_L(j\omega) = A_{ZY_L}(j\omega/\omega_{Y_L}) A_{ZY_V}(j\omega/\omega_{X_V})$$

where A_{ZY_L}, the WTA's differential gain, is given by Eq. (11) and A_{ZY_V}, the conveyor's current gain from Y to Z, is given by Eq. (9). These functions have break-points at $\omega_1 = \omega_{Y_L}/A$ and $\omega_2 = \omega_{X_V}/\sqrt{\rho_V}$, respectively. If $\omega_2 \gg \omega_1$, the loop response increases at 20 dB per decade below ω_1, is flat between ω_1 and ω_2, and rolls off 40 dB per decade above ω_2. Between ω_1 and ω_2, the asymptotic approximation for the gain is

$$A_L = \frac{g_{m_1} g_{m_3}}{g_{m_2} g_{m_4}} \frac{\omega_{Y_L}}{\omega_{X_V}} = c \frac{I_{Z_L}}{I_{Y_V}} \frac{C_{X_V}}{C_{Y_L}}$$

where c is the fraction of I_{X_V} that is passed to the WTA cell and $I_{Y_V} = I_{Z_L} + I_u$. Since $I_{Z_L} = \frac{1}{2}(mI_u)$,

we have

$$A_L = \frac{cm}{m+2} \frac{C_{X_V}}{C_{Y_L}} \qquad (12)$$

Given the desired value for A_L and the number of competing cells, m, this relation is used to find C_{Y_L}.

17.4.2. BAM Cell

The BAM cell has two modes of operation: *recall* and *programming*. In the recall mode, the reciprocal junctions mediate interactions between I/O units and hidden units through the $A\pm$ lines and the H line (see Fig. 17.11a). In the programming mode, one synapse is turned on, and the other is turned off, by setting the flip-flop appropriately; the $A\pm$ lines serve as bit lines and the H line is used as a word line. The BAM cell circuit is identical to a conventional SRAM cell, except for two extra devices, M_1 and M_3, in the reciprocal junctions. These transistors attempt to write zeroes to unselected cells. Nevertheless, SRAM-like operation may be achieved by sizing the pull-ups.

A selected cell is shown in Fig. 17.11b. M_3, M_6, and M_7 are omitted; these devices are off. M_3 remains off, M_6 and M_7 reinforce the new state when they turn on. Assume the n-type devices are the same size and the saturation current of the pull-ups is R_K times that of the n-types

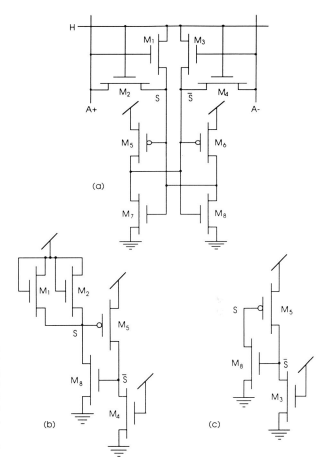

Fig. 17.11. (a) BAM cell circuit. It consists of a pair of reciprocal junctions and a pair of cross-coupled inverters. (b) A selected cell (H high) with $A+$ high and S low. Devices that are off are omitted. The access devices M_1 and M_2 together pull S up to reduce the current in M_5 so that M_4 can bring $\bar{S}$ low. (c) An unselected cell (H low) with both $A+$ and S low. Node S stays at GND, so M_3 cannot overcome M_5.

for equal gate-to-source voltages. Observing that M_1, M_2, and M_5 have the same V_{gs}, it follows that M_5 has $R_K/2$ times the current in M_8. M_4's current, which equals M_8's, will exceed M_5's if $R_K < 2$. An unselected cell is shown in Fig. 17.11c. The extra access device, M_3, attempts to upset the cell. To prevent this, M_5's saturation current must exceed that of M_3; this requires that $R_K > 1$. This analysis also applies if the states of S and $A+$ are reversed—simply interchange the roles of $A+$ and $A-$.

SPICE simulations confirmed that correct operation is possible with ratios that vary by a factor of 2; for 3/3 (width/length) n-type devices, the cell worked correctly for p-type sizes ranging from 7/2 to 13/2. The BAM cell layout is shown in Fig. 17.12. Scalable design rules were used; the separation between p- and n-types is 10λ. The resulting cell area is $38\lambda \times 38\lambda$.

17.5. BAM NETWORK

A BAM network is built using pyramidal cells and BAM cells as shown in Fig. 17.13. The I/O units communicate with external circuitry through their vertical control nodes; unidirectional currents are used as input and the differential voltages serve as outputs. This network finds the hidden unit whose weight vector best matches the input pattern and recalls the association assigned to that unit. The recurrent pathway sustains the new state after external inputs are removed; it also produces hysteresis. Network design involves specifying values for fixed bias currents and pyramidal cell capacitances. SPICE simulations confirmed the network's operation and verified the design procedure; the delays obtained agree with theoretical predictions.

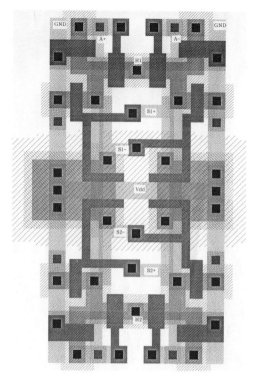

Fig. 17.12. BAM cell layout (two cells are shown). GND, $A+$ and $A-$ run vertically in metal-2; V_{dd} and H run horizontally in metal-1. Contacts to these lines are shared by adjacent cells.

17.5.1. Operation

I/O units supply input to a hidden unit through connecting BAM cells. The winning cell in an I/O unit has an output of $3I_u$; the complementary cell's output is I_u. Hence, if an I/O unit's state

agrees with the BAM cell's, it contributes 3 (in units of I_u), else its contribution is only 1. Accordingly, the hidden unit whose weight vector *best* matches the I/O layer's state gets the largest input. This unit is picked by winner-takes-all competition.

Hidden units feed their output to either the excitatory or the inhibitory input of an I/O unit, depending on the connecting BAM cell. If the winning unit's BAM cell is in the $+1(-1)$ state, it contributes $(r + 1)I_u$ to the excitatory (inhibitory) input, where r is the number of hidden units. In contrast, no more than $(r - 1)I_u$, the total output of the remaining units, is supplied to the complementary input. Accordingly, the I/O layer's state *exactly* matches the winning hidden unit's weight vector.

Analog-valued input patterns are presented by supplying currents to the I/O units. For positive and negative values, unidirectional currents are applied to excitatory and inhibitory pyramidal cells, respectively. Their outputs equal the sum of these currents and the intrinsic ones (from WTA cells). Therefore, the input pattern and the currently recalled one are both projected to the hidden layer. The network changes states only if the former is larger in magnitude.

Specifically, let A_0 be the currently recalled vector, tA_1 be the input vector, pumped up t times, and A_2 be the weight vector that best matches A_1. These vectors have ± 1 components; their units are given by the outputs of the I/O units' WTA cells, i.e., $2I_u \cdot A_1$ forces a transition from A_0 to A_2 if $A_2^T(A_0 + tA_1) > A_2^T(A_0 + tA_1)$ or

$$t(d_{01} - d_{12}) > d_{02}$$

where d_{jk} $(= d_{kj})$ is the Hamming distance

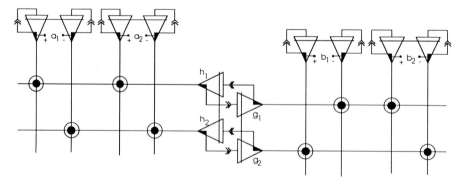

Fig. 17.13. A BAM network with two units in each layer. I/O units have two competing pyramidal cells that compare their excitatory and inhibitory inputs; hidden units have just one and compete with each other for output. A hidden unit's input is buffered by a unit in the other hidden layer. BAM cells connect these units together; only junctions that are "on" are shown. The associations stored are $([+1\ +1]^T, [-1\ +1]^T)$ (first row) and $([-1\ -1]^T, [+1\ -1]^T)$ (second row).

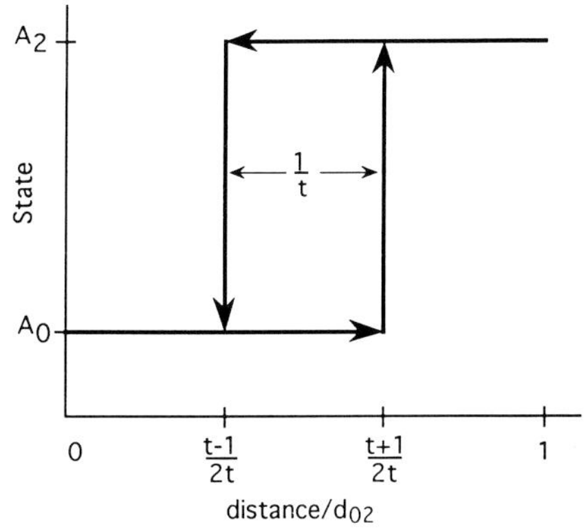

Fig. 17.14. Network recall hysteresis. Two possible states $\mathbf{A}_0$ and $\mathbf{A}_2$ are shown on the vertical axis; d_{02} is the Hamming distance between them. The Hamming distance of the input vector from $\mathbf{A}_0$ is shown on the horizontal axis; its components are t times greater than inputs fed back by the recurrent pathway.

between $\mathbf{A}_j$ and $\mathbf{A}_k$. This condition cannot hold if $d_{12} > d_{01}$ or $t \leq 1$ ($|d_{01} - d_{12}| \leq d_{02} \leq d_{01} + d_{12}$). If $\mathbf{A}_1$ is initially equal to $\mathbf{A}_0$ and, one by one, the bits that differ between $\mathbf{A}_1$ and $\mathbf{A}_2$ are flipped, a state transition occurs when

$$\frac{d_{12}}{d_{01}} = \frac{t-1}{t+1}$$

For instance, when $t = 3$, the input's distance from the target must be half that to the current vector. In the absence of feedback, however, these distances need only be equal. Effectively, the recurrent pathway introduces hysteresis over a region $(1/t)d_{02}$, as shown in Fig. 17.14. It also biases recall in favor of a vector similar to the current one. This behavior is useful in applications where the network interfaces directly with analog environmental sensors.

17.5.2. Design

A network with n units per I/O layer and r units per hidden layer is connected by two $n \times r$ BAM cell arrays. Each BAM cell supplies 0, I_u, or $(r + 1)I_u$ to an I/O unit and I_u or $3I_u$ to a hidden unit; it also adds capacitances C_A and C_H to their communication nodes. Therefore, an I/O unit receives 0 to $2rI_u$ and has rC_A capacitance at its vertical communication node; hidden units get between nI_u and $3nI_u$ and have nC_H. Knowing the input ranges and the communication node capacitances, fixed bias currents and control node capacitances may be specified.

Table 17.1. Control node capacitances for pyramidal cells. n and r are the number of units in the I/O and hidden layers; c is the fraction of the input passed to the WTA cell; C_A and C_H are the capacitances per BAM cell on the A and H lines.

Unit	I/O	Hidden
C_{Y_V}	$3C_A/2$	rC_H
C_{Y_L}	crC_A	$2cnC_H$

Choosing a fixed bias of rI_u for the hidden layers and zero for the I/O layers gives a network loop gain of two-thirds (from Eq. 9). The first choice limits the variation of the I/O units' inputs to a factor of 3—from rI_u to $3rI_u$. The hidden units' inputs vary over a similar range: from nI_u to $3nI_u$. The second choice was made to avoid adding current sources at the hidden layer; these sources would have to be well matched to avoid producing errors in picking the maximum.

The chosen control node capacitances are given in Table 17.1. For the I/O units, the ratio I_{X_V}/I_{Y_V} varies from $\frac{2}{3}r$ to $2r$ and $C_{X_V} = rC_A$. So, with $C_{Y_V} = \frac{3}{2}C_A$, ρ_V varies between 1 and 3; see Eq. (11). Choosing $C_{Y_L} = crC_A$ gives a loop gain of one-half since $m = 2$ for the I/O units; see Eq. (12). For the hidden units, I_{X_V}/I_{Y_V} ranges from $2n/(r + 1)$ to $2n$, assuming a winning unit receives at least $2nI_u$ and a losing one gets at most $2nI_u$. Given that $C_{X_V} = nC_H$, choosing $C_{Y_V} = rC_H$ means ρ_V varies from 2 to $2r$, assuming $r \gg 1$; see Eq. (11). Choosing $C_{Y_L} = 2cnC_H$ gives a loop gain of less than one-half; see Eq. (12).

Unfortunately, except for C_{Y_V} in the I/O units, the control node capacitances must be scaled up as the network's size increases. MOS capacitors are used on the chip; the control node potential is large enough to strongly invert the channel. In 2-μm CMOS technology with 400-Å gate oxide, the MOS capacitor has about 0.9 fF/μm^2, $C_A \approx$ 20 fF, and $C_H \approx 30$ fF. Therefore, each BAM cell adds 55 μm^2 to the area of the capacitors; this represents a 4 percent overhead.

17.5.3. Speed

Delays arise from the conveyors' incoming and outgoing channels, and the WTA cells. The former may be modeled as first-order low-pass with corner frequencies at ω_X and $\omega_X/\sqrt{\rho_V}$, respectively. Their rise times are given by $t_r = 2.2/\omega_c$, where ω_c is the corner frequency. So the delay due to the conveyor is

$$t_{CC} = 2.2(1 + \sqrt{\rho_V})\frac{C_{X_V}}{g_{m_1}} = 2.2(1 + \sqrt{\rho_V})\frac{V_T}{\kappa v}$$

where $v = I_{X_V}/C_{X_V}$ is the slew rate; it is between I_u/C_A and $3I_u/C_A$ for the I/O units (replace C_A with C_H for the hidden units) and is independent of the network's size. To obtain the longest delay, let $v \equiv I_u/C_A$. For $I_u = 40$ nA and $C_A = 20$ fF, $v = 2V/\mu S$, and with $\rho_V = 2$, the delay is only 92 nsec! Such speed is possible because of the small voltage signals, typically about 50 mV.

For the WTA, the current available to slew the input voltage is the difference between the input and the present maximum. This signal, ΔI, must produce a two-diode-drop voltage swing, ΔV. For the I/O units the capacitance is $C_{Y_L} = crC_A$ and, typically, $\Delta I = crI_u$, so the slew rate is v. Since ΔV is about 2 V, the delay, $t_{WTA} = \Delta V/v$, is 1 μsec. The same relationship holds for the hidden units with $v = I_u/2C_H$. WTA circuits that require only 200 mV swings are being developed to reduce the delay to 100 nsec.

Summing delays due to the conveyor and the WTA gives 1.1 μsec and 2.1 μsec for I/O and hidden units, respectively. Therefore, signals take 6.4 μsec to propagate around the network. The external inputs are removed after this period and the outputs may be read. The recall time is dominated by the WTA's delay; faster circuits could reduce it to 1 μsec, with 40 nA unit current.

17.5.4. Simulations

The network in Fig. 17.13 was simulated using SPICE;[3] the results are shown in Fig. 17.15. The SPICE deck was automatically extracted from the layout and included all the parasitic capacitances; all devices were 4 μm × 4 μm. Pyramidal cell capacitances were determined according to the previous section, with $C_A = 37.5$ fF and $C_H = 51.5$ fF; these values include capacitances due to the I/O and hidden units themselves. The unit current was 40 nA and the I/O units received a fixed bias of 80 nA; the supply was 5.0 V.

Initially, g_1 and h_1 are winning and the states of a_1 and b_1 are $+1$ and -1, respectively. At $t = 5$ μsec, 300 nA is applied to a_1's inhibitory pyramidal cell. To see what happens, start at the top left-hand corner of Fig. 17.15 and follow the arrows; refer to Fig. 17.13 for the connectivity.

Moving down the left side, the top graph shows a_1's output voltages; its inhibitory output increases (dotted line). The reciprocal junction

[3] Berkeley SPICE, version 3C1, with the BSIM model (Level 4).

connecting a_1 to a_2 replicates the signal and g_2's input increases at $t = 5.06$ μsec (dotted line in next graph). Now, g_2's input exceeds g_1's and their output voltages change to reflect this (next graph). The reciprocal junctions send g_1's and g_2's outputs to b_1's inhibitory and excitatory inputs, respectively; its inputs change at $t = 7.26$ μsec (next graph).

Moving up the right side, the bottom graph shows b_1's outputs. They are sent to the H units by the same reciprocal junctions; inhibitory and excitatory outputs go to h_1 and h_2, respectively. At $t = 9.25$ μsec, h_2's input increases and h_1's decreases (next graph). Now h_2 starts winning and the H units' outputs change accordingly (next graph). a_1's excitatory and inhibitory inputs come from h_1 and h_2, respectively; they change at $t = 11.55$ μsec (next graph).

Back at the beginning, a_1's outputs respond and are sent to layer G at $t = 13.56$ μsec; it took 8.50 μsec for signals to propagate around the network. g_2's input increases further while g_1's decreases. Therefore, g_2 continues to win after the external input is removed at $t = 15$ μsec. The I/O units have a 2.0-μsec delay, and the hidden units have 2.3 μsec. Theoretically, with v equal to 1.06 V/μsec and 0.78 V/μsec, for I/O and hidden units, respectively, their current conveyor delays are 140 nsec and 195 nsec. For the WTAs, $\Delta V = 1.9$ V, $\Delta I = 40$ nA, with $c = \frac{1}{2}$, and $C_{Y_L} = 41$ fF, which gives a delay of 1.95 μsec. The hidden units have $\Delta I = 80$ nA and $C_{Y_L} = 88$ fF; this gives 2.07 μsec. Therefore, the expected delays are 2.09 μsec and 2.26 μsec; these figures are in agreement with the SPICE results.

Observe that pairs of signals in the same row of Fig. 17.15 are actually communicated through the same node. Therefore, a sudden change in either signal produces a glitch in the other. When the voltage changes, capacitive currents are produced; this explains the positive spike in a_1's inhibitory input at $t = 5$ μsec. When the current changes, the voltage deviates while the conveyor accommodates the new input; this explains the negative spike in h_2's output at $t = 5$ μsec. Notice that this spike is replicated in a_1's input, producing the opposing spike.

The differential outputs at the I/O unit's vertical control nodes are shown in Fig. 17.16; ± 75 mV differential signals appear at the outputs depending on the unit's state. The transitions are caused by changes in the WTA's outputs and the network's inputs, except for those at $t = 5$ μsec and $t = 15$ μsec, which are due to the extrinsic input. Initially, the A layer's state is $[+1 +1]^T$ and the B layer's is $[-1 +1]^T$; recalling the association in the first row. After the external

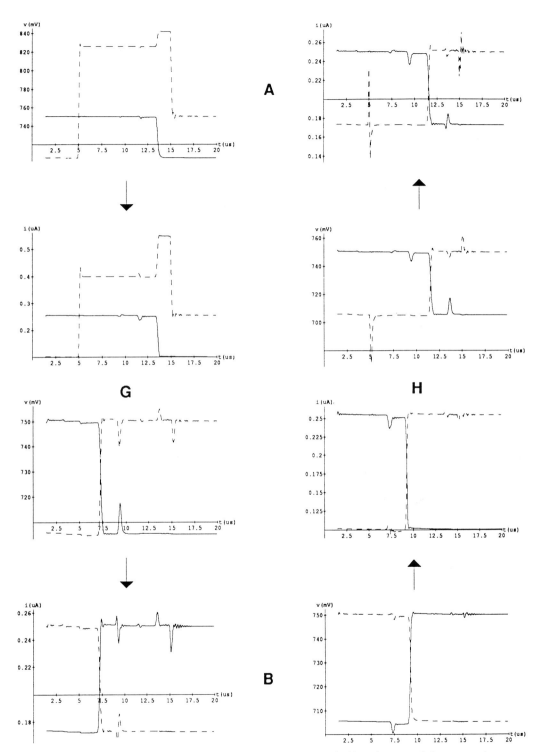

Fig. 17.15. SPICE simulation. Input currents and output voltages of I/O units a_1 and b_1 are at the top and the bottom, respectively; their inhibitory signals are in dotted lines. The hidden units are in the middle; g_2's and h_2's signals are in dotted lines.

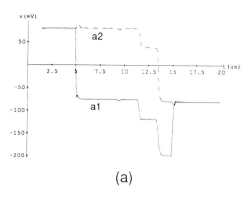
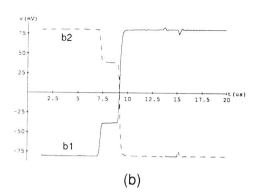

(a)   (b)

Fig. 17.16. SPICE simulation (continued). Differential voltages measured at the vertical control nodes of the I/O units. (a) A units, (b) B units.

input is removed they are $[-1-1]^T$ and $[+1-1]^T$, respectively; recalling the association stored in the second row.

17.6. SUMMARY

A design for a bidirectional associative memory chip has been developed; this second-generation design evolved from our first attempt (Boahen et al., 1989a) and from the experience of others (Sivilotti et al., 1985; Graf and de Vegvan, 1987; Verleysen et al., 1989). A complete test chip including latches for reentrant programming and custom digital I/O pads that interface directly with the I/O units is now in fabrication. This chip has 11 224 transistors and 40 pads; the die area is 2.2 mm × 2.2 mm in 2-μm technology.

Architectural issues were addressed at two levels. First a network architecture with good storage and recall performance was proposed. Second, an efficient, scalable, and fault-tolerant chip architecture was developed. Large-scale integration of analog circuits was deal with at three levels: signal representations, devices, and circuits. Representations that use interconnects efficiently and minimize the effect of their capacitance were chosen. A region of device operation with minimal power consumption and high transconductance was chosen. Minimalistic circuits were sought to perform communication and computation in a large network.

The effect of device mismatch in large associative memories and the related issue of optimum device geometry and placement need to be addressed. Random fluctuations in the device parameters are bound to average out across the synaptic arrays; however, the units themselves may have to be scaled up to obtain more precision. The effect of systematic variations across the chip also needs to be investigated. Characterization of these process variations (Pavasović, 1990) is invaluable in this respect.

ACKNOWLEDGMENTS

This research was funded by the Independent Research and Development program of the Applied Physics Laboratory; we thank Robert Jenkins for his personal interest in this work. Aleksandra Pavasović kindly provided the device data and Philippe Pouliquen made several useful suggestions.

REFERENCES

Andreou, A. G. and Boahen, A. K. (1989). "Synthetic Neural Circuits using Current-Domain Signal Representations," *Neural Computation*, Vol. 1. MIT Press, Cambridge, MA, pp. 489–501.

Boahen, K. A., Pouliquen, P. O., Andreou, A. G., and Jenkins, R. E. (1989a). "A Heteroassociative Memory using Current-Mode Circuits and Systems," *IEEE Trans. Circuits*, **36**(5), 747–755.

Boachen, K. A., Andreou, A. G., and Pouliquen, P. O. (1989b). "Architectures for Associative Memories using Current-Mode Analog MOS Circuits," *Advanced Research in VLSI: Proc. Dec. Caltech Conference on VLSI*, C. L. Seitz, ed. MIT Press, Cambridge, MA.

Graf, H. P. and de Vegvar, P. (1987). "A CMOS Implementation of a Neural Network Model," *Advanced Research in VLSI: Proc. of the 1987 Stanford Conference*, P. Losleben, ed. MIT Press, Cambridge, MA.

Kosko, B. (1988). "Bidirectional Associative Memories," *IEEE Trans. Systems, Man, and Cybernetics*, **18**, 49–60.

Lazzaro, J., Ryckebusch, S., Mahowald, M. A., and Mead, C. A. (1989). "Winner-Take-All Networks of O(n) Complexity," in *Advances in Neural Information Processing Systems*, Vol. 1, D. S. Touretzky, ed. Morgan Kaufmann, San Mateo, CA, pp. 703–711.

Mead, C. A. (1989). *Analog VLSI and Neural Systems*. Addison-Wesley, Reading, MA.

Pavasović, A. (1990). "Subthreshold Operation of MOSFET Devices in Analog VLSI Circuits," Ph.D. Dissertation, Johns Hopkins University.

Peretto, P. and Niez, J. J. (1986). "Long Term Memory Storage Capacity of Multiconnected Neural Networks," *Biological Cybernetics*, **54**, 53–63.

Säckinger, E. and Guggenbühl, H. (1990). "A High-Swing, High Impedance MOS Cascode Circuit," *IEEE Solid-State Circuits*, **25**(1), 289–298.

Sivilotti, M., Emerling, M., and Mead, C. A. (1985). "A Novel Associative Memory Implemented using Collective Computation," in *Advanced Research in VLSI: Proc. 1985 Chapel Hill Conf.*, H. Fuchs, ed. Computer Science Press, Rockville, Maryland.

Smith, K. C. and Sedra, S. A. (1968). "The Current Conveyor—A New Circuit Building Block," *IEEE Proc.*, **56**, 1368–1369.

Verleysen, M., Sirletti, B., Vandemeulebroecke, A. M., and Jespers, P. G. (1989). "Neural Networks for High-Storage Content-Addressable Memory: VLSI Circuit and Learning Algorithm," *IEEE J. Solid-State Circuits*, **24**(3), 562–569.

18

Optical Implementation of Programmable Associative Memories

FRANCIS T. S. YU

18.1. INTRODUCTION

Computers can solve computational problems thousands of times faster and more accurately than the human brain. However, cognitive tasks such as pattern recognition, language understanding, contextual information retrieval, etc., can be performed more effectively and more efficiently by the human brain. These cognitive tasks are still beyond the reach of modern electronic computers. In this chapter, we describe an optical associative memory which exhibits some simple cognitive/recognition behavior as a step in the direction of realizing more complex optical neural systems.

A fully interconnected associative memory of N neurons has N^2 interconnections. The transfer function of each neuron can be described by a thresholding operation, making the output of the neuron binary, or a nonlinear sigmoidal operation, which gives rise to analog values. As illustrated in Fig. 18.1, the state of the ith neuron in the network can be represented by the following *iterative equation*:

$$u_i = f\left\{ \sum_{j=1}^{N} T_{ij}u_j - \theta_i \right\} \qquad (1)$$

where u_i is the activation potential of the ith neuron, T_{ij} is the ijth entry of the *associative memory matrix* (AMM) connecting the jth neuron and the ith neuron, and f is a nonlinear thresholding function given by

$$f(x) = \begin{cases} 1 & x > 0 \\ 0 & x \le 0 \end{cases} \qquad (2)$$

Thus the operation of a neuron is simply the nonlinear processing of the weighted sum of input signals. The memory output is given by

$$U = f\{TV\} \qquad (3)$$

where U is the output state vector of the memory, f is a componentwise nonlinear thresholding operator, V is the input vector, and T is the AMM.

VLSI technology has been used to implement the above memory operations. For a two-layer network, the number of interconnections is the square of the number of neurons in the network. For example, a fully interconnected 10^4-neuron network requires about 10^8 interconnections, which is beyond the state-of-the-art VLSI technology. On the other hand, optics offers the advantages of parallelism and massive interconnection, and so large-scale optical neural networks have been proposed by Farhat et al. (1985); Athale et al. (1986); Lalanne et al. (1987); Wu et al. (1989); and Yu et al. (1990a, b).

Since light beams propagating in space do not interfere with each other, optical systems provide a large space–bandwidth product. These aspects have prompted us to look into the optical implementation of associative memories. The first optical implementation of an associative neural memory was by Farhat et al. (1985). Since then, a score of optical implementations have been proposed; e.g., the associative memory using liquid crystal light valve (LCLV) by Farhat and Psaltis (1987) and Psaltis et al. (1987); the hybrid optical memory using programmable spatial light modulator (SLM) by Johnson et al. (1987); and the optical ring resonator by Anderson and Erie (1987).

18.2. OPTICAL ASSOCIATIVE MEMORY

Recently, there has been an increasing interest in the implementation of associative memories using optical techniques. An *associative memory* is a process in which the presence of a complete or a partial input should result in the retrieval of a predetermined stored pattern or memory. Here, we look at an electro-optical associative memory

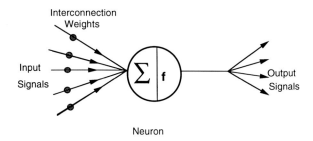

Fig. 18.1. Artificial neuron operation.

using an inner-product model as opposed to an outer-product-based model. One advantage of the inner-product model is that it can be implemented by employing a new class of photonic materials, called *electron trapping* (ET) materials as proposed by Jutamulia et al. (1991).

An optical correlator has been used to perform space-invariant pattern recognition in which a system is equipped with a bank of memories. The match of an input object with a memory pattern is detected by the presence of a strong correlation peak, which is essentially a space-invariant inner product. The correlation peak resembles a point source and can be further used to reconstruct a hologram such that a recalled output can be projected. If the recalled output is identical to the input, the system is categorized as *autoassociative*, whereas, if the recalled output is not the same as the input, the memory is termed *heteroassociative*.

In order to be able to recall different stored memories, the inner product must not be space-invariant. In other words, the location of a noninvariant correlation peak represents the match of the input object with a memory pattern. Therefore, an optical associative memory retrieval process may be realized by performing multiple noninvariant inner products in parallel and then multiplying the inner products with the corresponding memories. Liu et al. (1986) have demonstrated a real-time optical associative retrieval technique. Their technique may be outlined as follows: (1) Compute the inner product of the input pattern vector with each of the stored memories. (2) Multiply each resulting inner product with its corresponding memory. (3) Sum all these products over all memories to produce an activity pattern. (4) The activity pattern is thresholded to produce the final result. If the final result does not belong to one of the stored memories, the thresholded pattern can be fed back as a new input. This process can continue until convergence. Since a correlation peak is proportional to the inner product, this method is equivalent to the holographic method of Paek and Psaltis (1987).

Although the methods mentioned above may approximate a biological neural process, many iterations may be needed to obtain a convergent output. One can avoid using two-dimensional optical thresholding by changing the order of the thresholding operation, thus forming a noniterative association that is akin to the *winner-take-all* algorithm. Even though this latter method may not be closely related to the biological process, it may still be useful in some practical applications. This modification results in the following retrieval equations:

$$o(x, y) = m^k(x, y) \qquad (4)$$

if

$$\iint m^k(x, y)\, i(x, y)\, dx\, dy \geq U \qquad (5)$$

where $i(x, y)$ and $o(x, y)$ are the input and output patterns, respectively. $m^k(x, y)$ is the kth stored memory, and U is an arbitrary constant. In fact, this algorithm is similar to that of the symbolic substitution by correlation approach of Yu and Jutamulia (1987). It is akin to the method of controlled nonlinearity in the correlation domain proposed by Athale et al. (1986) as well as the holographic associative memory of Owechko et al. (1987).

In this WTA algorithm, there are two operations required to retrieve a stored memory $o(x, y)$: (1) Compute the inner product of the input pattern with each memory, and (2) if the inner product equals or exceeds a threshold value, then the output pattern should be the corresponding memory. Furthermore, by employing Parseval's theorem, the inner product of the input pattern with a stored memory in the space domain is equal to the inner product in the Fourier domain:

$$\iint m^k(x, y)\, i(x, y)\, dx\, dy = \iint M^k(p, q)$$

$$\times I(-p, -q)\, dp\, dq \geq U$$

$$(6)$$

where $M^k(p, q)$ and $I(p, q)$ are the Fourier spectra of $m^k(x, y)$ and $i(x, y)$, respectively, and U is an arbitrary constant.

Since $m^k(x, y)$ and $i(x, y)$ are positive real numbers, and their Fourier spectra are generally complex, the realization of Eq. (6) is nontrivial. However, its association can be given as follows:

$$\iint |M^k(p, q)|^2 |I(p, q)|^2 \, dp \, dq \geq U' \qquad (7)$$

where U' is a new threshold value.

Notice that the left-hand side of Eq. (7) can be used for object identification and counting as demonstrated by Jutamulia et al. (1982) as follows:

$$\iint |M^k(p, q)|^2 |I(p, q)|^2 \, dp \, dp =$$

$$|\mathscr{F}\,[|M^k|^2 \, |I|^2]|_{x=0, \, y=0} \qquad (8)$$

where $\mathscr{F}$ represents the Fourier transform operation. Alternately, Eq. (8) can be written as follows:

$$|\mathscr{F}\,[(M^k I^*)(M^k I^*)^*]|_{x=0, \, y=0}$$

$$= |[\mathscr{F}\,(M^k I^*)] \otimes [\mathscr{F}\,(M^k I^*)]|_{x=0, \, y=0}$$

$$= |[m^k \otimes i] \otimes [m^k \otimes i]|_{x=0, \, y=0} \qquad (9)$$

where $*$ and $\otimes$ denotes the complex conjugate and the correlation operation, respectively. In this equation, we note that the autocorrelation peak value is in fact the cross-correlation of $m^k(x, y)$ and $i(x, y)$. The result of Eq. (9) is proportional to the number of peaks from the first correlation $(m^k \otimes i)$. This result can be used for object identification and counting applications, as pointed out by McAulay et al. (1990). Notice that this result can be used for counting specific objects, whereas Eq. (7) can be used for pattern matching (i.e., counting a specific object either zero or one).

We shall mention that there are two basic matching schemes, namely, XNOR and AND. In Boolean algebra, it is denoted as

$$A \text{ XNOR } B = (A \text{ AND } B) \text{ OR } (\bar{A} \text{ and } \bar{B}) \qquad (10)$$

where the bar denotes the complement operation. We note that this equation will result in different inner-product calculations such that the AND scheme performs only half of the matching performed in the XNOR scheme. However, the essence of the information is not reduced, since $(\bar{A} \text{ AND } \bar{B})$ and $(A \text{ AND } B)$ are not totally independent.

In fact, the two matching schemes based on the XNOR and AND operations should be applied in different circumstances. For symbolic computation, if the XNOR matching scheme is applied, a digit 0 is just as much information as a 1. However, for a 2-D associative memory dealing with imaging and object recognition, the AND (multiplicative) matching scheme is preferred. In the XNOR scheme, the match between 0 of the background in a memory scene with 0 of the background in an input scene will contribute positively to the inner product. This contribution is generally not wanted for associative imaging memories because we want to obtain a high ratio between auto-inner-product and cross-inner-product.

In a real-world situation, the input pattern can come from a uniform distribution. For example, a totally bright input consists of only 1s. In this case, the space domain XNOR scheme will always detect half matches between input and all stored memories at the threshold level provided that a memory consists of an equal number of 1s and 0s. On the other hand, the AND scheme will always detect full matches for all memories. The Fourier domain approach can avoid these problems by simply blocking the intense DC component.

18.3. OPTICAL IMPLEMENTATION OF ASSOCIATIVE MEMORIES

Equation (1) can be applied to one-dimensional memory patterns. For two-dimensional patterns, N^2 interconnections are required, and the corresponding memory dynamics are given by

$$v_{lk}(n + 1) = f \left[\sum_{i=1}^{N} \sum_{j=1}^{N} T_{lkij} v_{ij}(n) \right] \qquad (11)$$

where v_{lk} represents the state of the lkth neuron in an $N \times N$ space and $[T_{lkij}]$ is a four-dimensional associative memory matrix (AMM). Since the AMM can be partitioned into an array of two-dimensional submatrices $[T_{11ij}]$, $[T_{12ij}]$, $\ldots [T_{NNij}]$, a four-dimensional associative memory matrix can be displayed in an $N^2 \times N^2$ two-dimensional format shown in Fig. 18.2. This two-dimensional representation has drawn attention to optical implementation of AMMs. Recently, an optical architecture was proposed by Farhat and Psaltis (1987) in which they used a basic matrix–vector optical processor. This system utilizes a linear array of LEDs, which we used as an input pattern composer, and is connected to a synaptic mask by the lenslet array shown in Fig. 18.3. To provide the AMM with adaptive learning capabilities, a fine-resolution programmable spatial light modulator (SLM) with a large number of distinguishable gray levels is needed. However, currently available SLMs

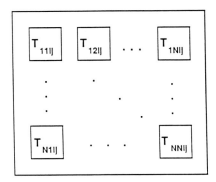

Fig. 18.2. Partition of a 4-D weight matrix T_{lkij} into an array of submatrices.

have a relatively small space–bandwidth product and have very limited gray levels which makes it difficult to implement high accuracy synaptic weights. Furthermore, the interconnection part has to be realized either electronically or optically using an array of light-integrating elements, which would either slow down the operation speed or pose severe optical alignment problems. To alleviate these problems, we describe a hybrid-optical architecture using *inexpensive* pocket-sized liquid-crystal televisions originally proposed by Yu et al. (1990a).

A major feature of this architecture, in contrast to the matrix–vector processor of Farhat and Psaltis, is that the positions of the input feature composer and the synaptic mask have been interchanged. By using this arrangement, the summation of the interconnections between the input vector and the memory submatrices can be performed optically.

Liquid crystal televisions (LCTVs) have been used as viable spatial light modulators (SLMs) in various optical signal processing and computing architectures. Based on the current advances of LCTV devices, the imaging quality has been improved close to that of commercially available high-resolution video monitors. For instance, the contrast ratio of a recent LCTV with built-in thin-film transistors can be higher than 30:1. It can have a dynamic range of approximately 16 gray levels and can be continuously adjusted. In the optical architecture that we discuss in this chapter, a Hitachi C5-LC1 LCTV is used to display the AMM and a Seiko LVD-202 LCTV is used to display the input patterns. The resolution of these LCTV's is 256×420 pixels and 200×330 pixels, respectively.

An 8×8 fully interconnected LCTV neural network is shown in Fig. 18.4, in which an 80-watt xenon-arc lamp is used as the incoherent source. The lenslet array consists of 8×8 lenses, where each lens images each of the AMM submatrices onto LCTV2 to provide the interconnections between the AMM and the input pattern. Thus the AMM submatrices are superimposed (multiplied) onto the input pattern to establish the interconnection part of Eq. (11); i.e.,

$$\sum_{i=1}^{N} \sum_{j=1}^{N} T_{lkij} v_{ij}(n) \qquad (12)$$

After LCTV2, the transmitted light field is collected by an imaging lens, which is focused at the lenslet array and is imaged onto a charge-coupled-device (CCD) array detector. The detected signals are sent to an electronic thresholding circuit, and the final results are fed back to LCTV2 for the next iteration. Note that the data

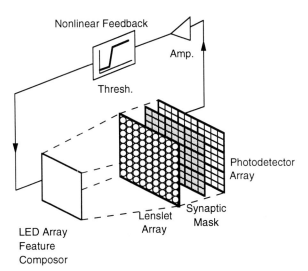

Nonlinear Feedback

Amp.

Thresh.

Photodetector Array

Synaptic Mask

Lenslet Array

LED Array Feature Composor

Fig. 18.3. Two-dimensional optical associative memory neural net.

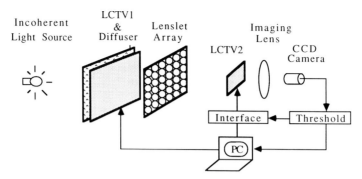

Fig. 18.4. An LCTV optical associative memory neural net.

flow in the optical system is primarily controlled by a microcomputer. For instance, the AMM and the input pattern can be written onto LCTV1 and LCTV2 and the computer can also make decisions based upon the output results.

18.4. HOPFIELD ASSOCIATIVE MEMORY

An associative memory belongs to a class of fundamental neural networks for which a complete or a partial pattern of a known input recalls a stored associated pattern. Recollection in an associative memory is performed by matrix multiplication of the AMM and the input vector. The reader is referred to Chapter 1 for details on the recording and retrieval in single layer static and dynamic associative memories.

Hopfield (1982) proposed a simple dynamic associative memory model where an association can be retrieved from a distorted input after several nonlinear iterations. The Hopfield memory models the retrieval process as an iterative thresholded matrix–vector multiplication and is given by

$$V_i \rightarrow 1 \quad \text{if } \sum_{j=1}^{N} T_{ij}V_j \begin{cases} \geq 0 \\ < 0 \end{cases} \quad (13)$$
$$V_i \rightarrow 0$$

where V_i and V_j are the binary output and the binary input, respectively. This iterative process is repeated by using the preceding output as the new input, until convergence. As seen in Chapter 1, the Hopfield model employs the following memory storage/recording recipe:

$$T_{ij} = \begin{cases} \sum_{m=1}^{N} (2V_i^m - 1)(2V_j^m - 1) & \text{for } i \neq j \\ 0 & \text{for } i = j \end{cases}$$

$$(14)$$

where V_i^m and V_j^m are ith and jth elements of the mth unipolar binary memory vectors.

Some features of the Hopfield model are: (1) The AMM is synthesized based on an outer-product operation through the supervised learning rule expressed in Eq. (14). (2) The retrieval dynamics involves a nonlinear thresholding function. (3) The desired output is computed from a distorted or an incomplete input through retrieval iterations. (4) The AMM does not necessarily change during the retrieval phase. The most important feature of the model is that it provides correct recognitions and associations for distorted or partial input signals.

We have experimentally studied the performance of the Hopfield model using the LCTV neural network of Fig. 18.4. We stored the letters "A," "B," "W," and "X" as memories based on the outer-product operation of Eq. (14). Each of the letters occupies an 8×8 pixel array as shown in Fig. 18.5a. The positive and the negative parts of the AMM are shown in Figs. 18.5b and 18.5c, respectively. As depicted in Fig. 18.5d, a partial image of "A" is fed to the input LCTV2 of the optical associative memory. By sequentially displaying the positive and the negative parts of the AMM on LCTV1, the output signal arrays can be picked up by the CCD detector and then sent to the microcomputer for subtraction and thresholding. A partially recovered pattern is obtained at the output end, as shown in the middle of Fig. 18.5d. Since the output pattern is not completely recovered, this result is fed back to the input LCTV2 for the next iteration. The last picture in Figure 18.5d shows the final output of this process. Mention should be made that the positive and the negative parts of the AMM could have been added with bias, which leads to a single-step operation.

(a)

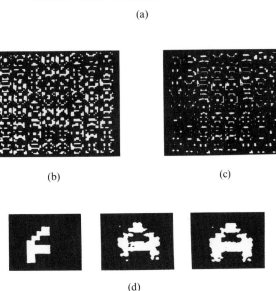

(b)

(c)

(d)

Fig. 18.5. Experimental results of the Hopfield model: (a) training set; (b), (c) positive and negative AMMs; (d) reconstructed result.

18.5. ORTHOGONAL PROJECTION ASSOCIATIVE MEMORY

The error correcting capability of the Hopfield model is effective with the assumption that the stored vectors are mutually independent. The correction ability decreases rapidly as the number of stored patterns increases. According to McEliece et al. (1987), the number of stored vectors M in the Hopfield model should be sufficiently smaller than the number of neurons N in the network, that is

$$M < \frac{N}{4 \ln N} \quad (15)$$

However, in practice, the stored vectors are generally not fully independent. This produces ambiguous output results.

We note that orthogonalization techniques have been used in associative memory for digital image processing. We now illustrate the use of the orthogonal projection (OP) algorithm for improving the error-correcting capability of the optical associative memory of Fig. 18.4.

Let us consider a set of M N-dimensional vectors $\mathbf{V}^{(m)}$ which will be used to construct an AMM. The basic concept of the OP algorithm is to project each vector $\mathbf{v}^{(m_0)}$ of the vector set $\{\mathbf{V}^{(m)}\}$ onto the orthogonal subspace spanned by the independent vectors $\mathbf{V}^{*(m)}$, $m = 1, 2, \ldots, m_0 - 1$. The orthogonal projection vector can be described by the Gram–Schmidt orthogonalization procedure (Golub and Van Loan, 1983) as

$$\mathbf{V}^{*(m_0)} = \mathbf{V}^{(m_0)} - \sum_{m=1}^{m_0 - 1} \frac{(\mathbf{V}^{(m)}, \mathbf{V}^{*(m)})}{\|\mathbf{V}^{*(m)}\|} \mathbf{V}^{*(m)} \quad (16)$$

where $(\mathbf{V}^{(m)}, \mathbf{V}^{*(m)})$ denotes the inner product, and $\|\mathbf{V}^{*(m)}\|$ is the norm of $\mathbf{V}^{*(m)}$.

Hence, for a given matrix $\mathbf{T}^{(m-1)}$, the recursive computation of the memory matrix $\mathbf{T}^{(m)}$ can be expressed as

$$\mathbf{T}^{(m)} = \begin{cases} \mathbf{T}^{(m-1)} + [\mathbf{U}^{(m)} - \mathbf{T}^{(m-1)}\mathbf{V}^{(m)}] \dfrac{\mathbf{V}^{*(m)\mathrm{T}}}{\|\mathbf{V}^{*(m)}\|} \\ \qquad\qquad\qquad\quad \text{for } \|\mathbf{V}^{*(m)}\| \neq 0, \quad (17) \\ \mathbf{T}^{(m-1)} \qquad\qquad\qquad\quad \text{otherwise,} \end{cases}$$

where $\mathbf{U}^{(m)}$ stands for the desired output vector, and the initial memory matrix $\mathbf{T}^{(0)}$ can be either zero or an identity matrix.

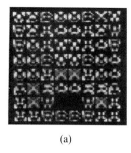

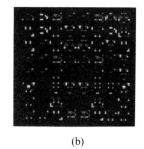

(a)   (b)

(c)   (d)

Fig. 18.6. Orthogonal projection model: (a), (b) positive and negative AMMs using OP algorithm; (c) reconstruction using OP algorithm; (d) reconstruction using Hopfield model.

Using the OP algorithm described above, *computer simulations* of the proposed optical AMM were conducted and the result is shown in Fig. 18.6. Note that the same set of English letters shown in Fig. 18.5a is used as the reference (stored) set. By applying Eqs. (16) and (17), the constructed positive and negative parts of the AMM are as shown in Figs. 18.6a and 18.6b, respectively. The reconstructions of a partial pattern "A" by using the OP and Hopfield models are shown in Figs. 18.6c and 18.6d, respectively. The successive patterns in these figures represent the output results obtained from successive iterations. From these results we see that the OP algorithm is more robust and that it has a higher convergence speed compared to the Hopfield model. In this example, the OP algorithm requires only two iterations to obtain the correct result, whereas the Hopfield model converges to a local minimum, which gives rise to an erroneous result as shown in Fig. 18.6d.

A numerical analysis of the robustness of an 8×8-neuron one-layer neural network is evaluated. Twenty-six 8×8-pixel images of the English letters are used as the reference patterns. The average Hamming distance among the reference patterns is about 26 pixels, and the minimum distance is about 4 pixels. Let us assume that the input patterns are corrupted in additive random noise, where the input signal-to-noise ratios (SNRs) are chosen to be about 5 dB (i.e., 50 percent noise), 7 dB (i.e., 33 percent noise), and 10 dB (i.e., 10 percent noise), respectively. Figure 18.7 shows the output pixel errors against the number of stored patterns for various values of input SNRs. From this figure, we see that the Hopfield model becomes unstable for a storage capacity beyond five letters, whereas using the OP algorithm, the associative memory can retrieve all the letters with a 50 percent input noise level. We also note that the error-correction capability decreases substantially as input noise and the number of stored patterns increase. Nevertheless, the OP algorithm provides better error-correction capability than the Hopfield model.

18.6. MULTILEVEL RECOGNITION ALGORITHM

For the case where the Hamming distances between the stored vectors are very small, the above associative memories generate poor results. As an example, consider the four letters T, I, O, and G stored in the AMM as shown in Fig. 18.8a.

Although the partial input image of G (as shown in Fig. 18.8b) contains a distinguishing feature, the Hopfield model fails to retrieve the pattern G and converges to a spurious memory

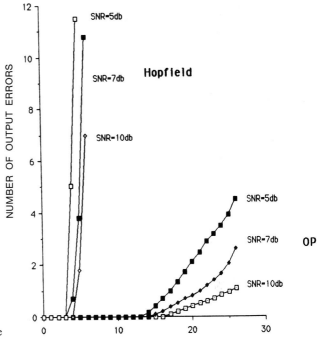

Fig. 18.7. Performance comparison of the Hopfield and the OP AMMs. Total pixel number = 64.

NUMBER OF STORED PATTERNS

(local minimum). We also note that, in some cases, the output will not converge to a correct result even when the input is one of the reference patterns.

From the preceding example we see that the smaller the Hamming distances among the stored patterns, the smaller the associative memory's error-correction capability. However, it is also known that the less information stored in the memory, the more effective the memory is in correcting input error. Using the advantages of the programmability of the proposed optical AMM described in the preceding sections, a multilevel recognition (MR) algorithm can be developed. This algorithm adopts the *tree search strategy* that increases the error correction ability by reducing the number of vectors stored in each memory matrix as proposed by Hecht–Nielsen (1986). The MR algorithm first classifies the reference patterns into subgroups and then develops a tree search strategy based on the similarity (i.e., Hamming distance) of the reference patterns. A smaller number of reference patterns can be stored in the AMM built for each subgroup. The MR algorithm then changes the AMM as a function of the Hamming distances between the intermediate result and the patterns in different subgroups. In this manner, the storage capacity is not limited by the size of the associative memory. However, the trade-off is that the processing speed is slowed due to the need to rewrite the AMMs.

As an example, we consider once more the

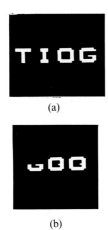

(a)

(b)

Fig. 18.8. (a) Four English letters stored in the AMM. (b) Output obtained by Hopfield AMM.

(a)

(b)   (c)

Fig. 18.9. (a) Three subgroups of letters used in the MR algorithm. (b) Reconstruction of letter G by T_{TO} and T_{OG}, respectively.

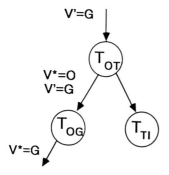

Fig. 18.10. Flow chart of the MR algorithm.

four English letters shown in Fig. 18.8. According to the similarity of the patterns, the above four letters can be classified into two groups, namely, {O, G} and {T, I}. Suppose T and O are arbitrarily selected from these two subgroups to form a root group {T, O}, as shown in Fig. 18.9a. Instead of constructing an AMM T_{TIOG}, three submemory matrices, T_{TO}, T_{OG}, and T_{TI}, are composed and stored in the microcomputer. In the first demonstration, the AMM T_{OT} is displayed on LCTV1 and an input vector V', which represents the partial image of G, is presented on input LCTV2 of the optical associative memory. After converging to a stable state, the output V^* is compared, based on the Hamming distance, to the stored patterns O and T. If the Hamming distance between V^* and pattern O is shorter than that between V^* and pattern T, the AMM T_{OG} will be used to replace the matrix T_{OT}. Subsequently, the input vector V^* is fed into the network for a new round of iteration, such that the letter G is retrieved as shown in Fig. 18.9c.

A flow chart diagram to illustrate this tree search operation is shown in Fig. 18.10.

18.7. INTERPATTERN ASSOCIATIVE MEMORY

Neural networks have been shown to be very effective in pattern-recognition applications. For the recognition of a given set of reference patterns, there are two basic approaches to constructing an AMM. The first is *intrapattern association*, which emphasizes the association of elements within each reference pattern. For example, the Hopfield model outer products of the reference patterns are added to form the AMM. This type of approach may lead to poor recognition performance if the reference patterns are not mutually independent. The second scheme is *interpattern association* (IPA) in which the AMM is constructed by emphasizing the association among the reference patterns. Suppose that the references are similar patterns (e.g., human faces, fingerprints, handwritten characters). Here, the special features of the patterns become very important for pattern recognition. Therefore, it is advantageous to consider the relationships between the special and common features among the reference patterns for the construction of the AMM as proposed by Lu et al. (1990).

For example, consider a set of three overlapping patterns A, B, and C as presented in the Venn diagram shown in Fig. 18.11. These patterns can be divided into seven subspaces in which I, II, and III are the special subspaces of patterns A, B, and C, respectively. IV, V, and VI are the common subspaces of A and B, B and C, C and A, respectively. VII is the common subspace

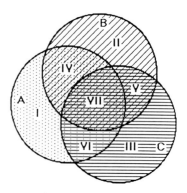

Fig. 18.11. Common and special subspaces of three reference patterns.

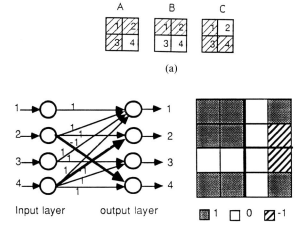

Fig. 18.12. Example of an IPA memory: (a) three reference patterns; (b) one-layer neural net; (c) AMM.

Input layer   output layer

(b)

(c)

of A, B, and C. The rest of the pattern space can be defined as an empty set $\emptyset$. These subspaces can be expressed by the following logic functions:

$$I = A \wedge \overline{(B \vee C)}$$

$$II = B \wedge \overline{(A \vee C)} \qquad V = (B \wedge C) \wedge \overline{A}$$

$$III = C \wedge \overline{(A \vee B)} \qquad VI = (C \wedge A) \wedge \overline{B}$$

$$IV = (A \wedge B) \wedge \overline{C} \qquad VII = (A \wedge B \wedge C) \wedge \overline{\emptyset}$$

$$(18)$$

where $\wedge$, $\vee$, and $\overline{}$ stand for the logic AND, OR, and NOT operations, respectively.

Let us now start building the interconnections between the input and the output neurons. By using this set of logic functions we can determine excitatory, inhibitory, and null interconnections. For instance if an input neuron in the VII subspace is in an "on" state, this neuron can only excite the output neurons within subspace VII and has no connection with the output neurons in other subspaces. On the other hand, if an input neuron in subspace V is "on," it will excite the output neurons in subspaces V and VII but inhibit the output neurons in subspace I. Furthermore, if an input neuron is "on" in subspace I, it will excite all the output neurons in pattern A (i.e., subspaces I, IV, VI, and VII) but inhibit the output neurons in $(B \vee C) \wedge \overline{A}$ (i.e., subspaces II, III, and V).

By using the *principle of induction*, the above logic function rule can be extended to M reference patterns, as given by

$$X = P \wedge \overline{Q} \qquad (19)$$

where

$$P = p_1 \wedge p_2 \wedge \cdots \wedge p_n$$

$$Q = q_1 \vee q_2 \vee \cdots q_m$$

where $p_1, p_2, \ldots, p_n$ and $q_1, q_2, \ldots, q_m$ are reference patterns, and $n + m = M$ is the total number of reference patterns. The input neurons in subspace X must excite (i.e., have positive connections with) all the output neurons in P subspaces and, inhibit (i.e., have negative connections with) all output neurons in $Q \wedge \overline{P}'$ subspaces, where P' is defined by

$$P' = p_1 \vee p_2 \vee \cdots \vee p_n \qquad (20)$$

and has no connection with the output neurons in the remaining subspaces. For simplicity, if the connection strengths (i.e., interconnection weights) are assumed to be 1 for positive connections, -1 for negative connections, and 0 for no connection, then the IPA neural network can be constructed in a simple *three-state* structure.

To illustrate further the construction of the IPA model, let A, B, and C represent three 2×2 array patterns shown in Fig. 18.12a. The pixel-pattern relationship is given in Table 18.1. It is apparent that pixel 1 represents the common feature of patterns A, B, and C; pixel 2 is the common feature of A and B; pixel 3 is the common feature of A and C; and pixel 4 represents the special feature of C.

By applying the logic operations of Eq. (18), a tristate interconnection neural network can be constructed, as illustrated in Fig. 18.12b. This is a one-layer neural network of four input and four output neurons, where each neuron represents

Table 18.1. Pixel–Pattern Relationship of Three Reference Patterns.

Pattern	Pixel			
	1	2	3	4
A	1	1	1	0
B	1	1	0	0
C	1	0	1	1

one pixel in the reference patterns. For example, the first input neuron corresponds to pixel 1 of the input pattern and excites only the first output neuron. The second input neuron (the corresponding pixels belong to patterns **A** and **B**) excites both the first and second output neurons, while it inhibits the fourth output neuron, which belongs to the special subspace of pattern **C**.

We can now partition the 4-D AMM into a 2-D submatrix array, as illustrated in Fig. 18.12c. The AMM can be divided into four blocks, where each block corresponds to one output neuron. The four elements in one block represent the four neurons at the input end. For example, since all four elements in the upper-left block have a value of 1, it indicates that any one of the four input neurons can excite the first output neuron. As another example, in the upper-right block the first and third elements are 0, the second element has a value of 1, and the fourth element is -1. Thus the interconnections can be determined such that the first and third input neurons have no connection to the second output neuron, the second input neuron excites the second output neuron, and the fourth input neuron inhibits the second output neuron.

It is the differences rather than the similarities among patterns that are used for pattern recognition. Similar to some other neural network learning algorithms, the Hopfield model constructs the AMM by correlating the elements within each pattern and ignoring the relationships among the patterns. For example, the AMM of the Hopfield model for three reference patterns **A**, **B**, and **C** can be written by an outer-product representation, such as

$$\mathbf{T} = \mathbf{A}\mathbf{A}^T + \mathbf{B}\mathbf{B}^T + \mathbf{C}\mathbf{C}^T \qquad (21)$$

where the superscript T represents the transpose of the pattern vectors.

If input pattern **A** is applied to the neural

system, the output pattern vector would be

$$\mathbf{V} = \mathbf{T}\mathbf{A}$$

$$= \mathbf{A}(\mathbf{A}^T\mathbf{A}) + \mathbf{B}(\mathbf{B}^T\mathbf{A}) + \mathbf{C}(\mathbf{C}^T\mathbf{A}) \qquad (22)$$

in which $\mathbf{A}^T\mathbf{A}$ represents the autocorrelation of pattern **A**, while $\mathbf{B}^T\mathbf{A}$ and $\mathbf{C}^T\mathbf{A}$ are the cross-correlations between **A** and **B**, and **A** and **C**, respectively. If the differences (e.g., Hamming distances) among patterns **A**, **B**, and **C** are sufficiently large, then the autocorrelation among **A**, **B**, and **C** will be much larger than their cross-correlations, i.e.,

$$\mathbf{A}^T\mathbf{A} \gg \mathbf{B}^T\mathbf{A} \qquad \mathbf{A}^T\mathbf{A} \gg \mathbf{C}^T\mathbf{A} \qquad (23)$$

Thus, from Eqs. (22) and (23) pattern **A** has a larger weighting factor than patterns **B** and C. Here, the two rightmost terms in Eq. (22) are considered as noise disturbances. It is therefore apparent that by choosing a suitable threshold value, pattern **A** can be reconstructed at the output end of the network.

On the other hand, if patterns **A**, **B**, and **C** are very similar, the inequalities of Eq. (23) will no longer hold and the Hopfield model fails because the threshold value is ill-defined.

Computer simulations using an 8×8-neuron memory are performed for the Hopfield and other AMMs models. The reference patterns considered are the 26 capital English letters lined up in sequence based on their similarities. Figure 18.13 shows the error rates as a function of the number of stored reference patterns for various input signal-to-noise (SNR) ratios. We note that the Hopfield model fails at about 4 patterns, whereas the IPA model is quite stable at about 12 stored letters, for 7 dB SNR. As for the noiseless input, the IPA model can perform even better and produces correct results for all 26 stored letters, whereas the Hopfield model starts making significant errors when the number of stored patterns increases beyond 4 patterns.

It is interesting to show the experimental results obtained from the preceding LCTV optical network. We used the letters B, P, and R as the training set for constructing the AMM, as shown in Fig. 18.14a. The positive and the negative parts of the AMM for the IPA model are shown in Figs. 18.14b and 18.14c, and for the Hopfield AMMs are shown in Figs. 18.14d and 18.14e, respectively. Comparing these two AMMs, it can be seen that the IPA AMM has fewer interconnections and requires fewer gray levels. The IPA AMM requires only three gray levels, whereas the Hopfield AMM needs $2M + 1$ gray levels. This is significant for an SLM

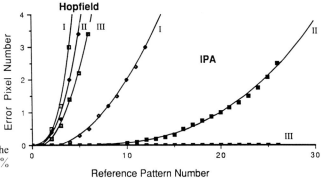

Fig. 18.13. Comparison of the IPA and the Hopfield models: (I) 10% noise level; (II) 5% noise level; (III) no noise.

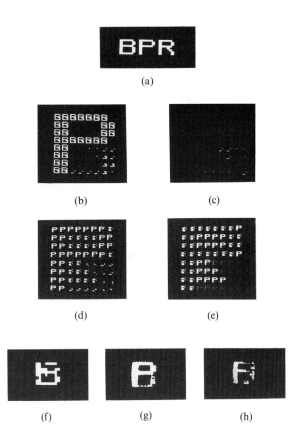

Fig. 18.14. Experiments of an optical associative memory: (a) three similar reference patterns; (b), (c) positive and negative AMMs of the IPA model; (d), (e) positive and negative AMMs of the Hopfield model; (f) input pattern, SNR = 7 dB; (g) pattern reconstruction using IPA model; (h) result using Hopfield model.

implementation. An input pattern that is embedded in 30% random noise (SNR = 7 dB) is shown in Fig. 18.14f. The output results are shown in Figs. 18.14g and 18.14h for the IPA and Hopfield models, respectively. Once again, we see that the IPA model performs better even for stored patterns within the storage capacity of the Hopfield model.

18.8. HETEROASSOCIATIVE MEMORY

We now discuss a heteroassociative (HA) memory using IPA. The concept of the IPA model is to determine whether the pixels in the pattern space belong to special or common subspaces and then decide the excitatory and inhibitory interconnections based on a simple

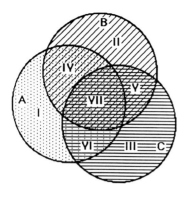

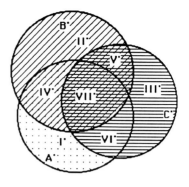

Input pattern space S1   Output pattern space S2

Fig. 18.15. Heteroassociation model. S_1 and S_2 are the input and the output pattern spaces, respectively.

logical relationship. Let us now apply the IPA algorithm to the heteroassociation model, as described by Yu et al. (1990b). For example, the overlapping input–output training sets shown in Fig. 18.15 can be divided into subspaces, where I, II, and III are the special subspaces of input patterns **A**, **B**, and **C**, respectively, while IV, V and VI are the common subspaces for **A** and **B**, **B** and **C**, **C** and **A**, and VII is the common subspace for **A**, **B** and **C**. Similarly, the output pattern space has the same structure. These subspaces of the input–output spaces can be determined by the following logical functions:

$$\text{I} = A \wedge \overline{(B \vee C)} \qquad \text{I}' = A' \wedge \overline{(B' \vee C')}$$

$$\text{II} = B \wedge \overline{(A \vee C)} \qquad \text{II}' = B' \wedge \overline{(A' \vee C')}$$

$$\text{III} = C \wedge \overline{(A \vee B)} \quad \text{III}' = C' \wedge \overline{(A' \vee B')}$$

$$\text{IV} = (A \wedge B) \wedge \bar{C} \quad \text{IV}' = (A' \wedge B') \wedge \bar{C}'$$

$$\text{V} = (B \wedge C) \wedge \bar{A} \quad \text{V}' = (B' \wedge C') \wedge \bar{A}'$$

$$\text{VI} = (C \wedge A) \wedge \bar{B} \quad \text{VI}' = (C' \wedge A') \wedge \bar{B}'$$

$$\text{VII} = (A \wedge B \wedge C) \wedge \bar{\varnothing}$$

$$\text{VII}' = (A' \wedge B' \wedge C') \wedge \bar{\varnothing}$$

$$(24)$$

Based on these simple logical operations, a heteroassociation memory can be constructed.

Let us consider the input-to-output neuron relationship as a mapping process for every pixel in the input space to the output space. For example, when a neuron (e.g., representing a pixel) in the input subspace VII is "on," it implies that this neuron will excite all the neurons within

the output space VII' for which it has no connection with the neurons in other output subspaces in S_2. However, when a neuron in the input subspace V is "on," it will excite the neurons in the output subspaces V' and VII', but inhibit the neurons in I'. Similarly, logical operations can also be applied to neurons in subspaces IV and VI.

For the case when a neuron is "on" in input subspace I, it implies that pattern A appears at the input end, and this neuron can excite all the neurons in output pattern A' (i.e., subspaces I', IV', VI' and VII'), and inhibit the neurons in subspaces $(B' \vee C') \wedge \bar{A}'$ (i.e., subspaces II', III' and V'). Similarly, the neurons in subspaces II or III would excite output patterns B' or C', and inhibit I', III' and VI', or I', II' and IV', respectively. Thus we see that a hetero-association memory matrix (HAMM) can be constructed by simple logical rules.

Clearly, by applying the *induction principle*, heteroassociative logical operations can be extended for M reference patterns, as was done in the preceding section.

To illustrate the construction of the HAMM, let us assume that $(\mathbf{A}, \mathbf{B}, \mathbf{C})$ and $(\mathbf{A}', \mathbf{B}', \mathbf{C}')$ are the input–output training patterns, shown in Figs. 18.16a and 18.16b. The corresponding input and output pattern pixel relationships are given in Tables 18.2 and 18.3, respectively. As can be seen from the input pattern space, pixel 1 (upper-left) is the common feature of **A**, **B**, and **C**, pixel 2 is the common feature of **A** and **B**, pixel 3 is the common feature of **A** and **C**, while pixel 4 is the special feature of **C**. Likewise, from the output pattern space, pixel 3 is the special feature of **B**', and so on.

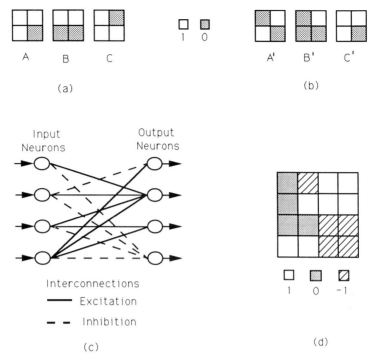

Fig. 18.16. Construction of a HAMM using IPA Model; (a) input–output training sets; (b) a tristate neural network; (c) HAMM.

Table 18.2. Input Pixel–Pattern Relationship.

	Pixel			
Pattern	1	2	3	4
A	1	1	1	0
B	1	1	0	0
C	1	0	1	1

Table 18.3. Output Pixel–Pattern Relationship.

	Pixel			
Pattern	1	2	3	4
A′	0	1	1	0
B′	0	1	0	0
C′	1	1	1	0

A tristate-interconnection neural network can therefore be constructed as shown in Fig. 18.16c. Note that the second output neuron representing the common feature of **A′**, **B′**, and **C′** has positive interconnections to all input neurons. The fourth output neuron representing a common feature of **A′**, **B′**, and **C′** is subjected to inhibition from all input neurons, and these represent the common features of **A**, **B**, and **C** in the input pattern space. The corresponding HAMM is constructed as shown in Fig. 18.16d.

Next, we consider the implementation of the heteroassociation model in the preceding LCTV optical neural net for character translation. A set of input–output training patterns is shown in Fig. 18.17a, where the upper and lower rows are the corresponding Arabic and Chinese numerals, respectively. The positive and negative parts of the HAMMs for translating Arabic numerals to Chinese numerals are shown in Figs. 18.17b and 18.17c, respectively. Figures 18.17d and 18.17e show a partial Arabic numeral and its retrieved Chinese numeral obtained after only one iteration. Thus we see that the

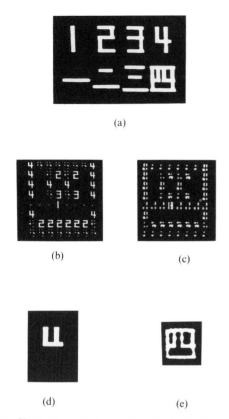

(a)

(b)   (c)

(d)   (e)

Fig. 18.17. Character translations: (a) Arabic–Chinese numeral training sets; (b) positive (left sides) and negative (right sides) parts of the HAMM; (c) partial input Arabic numeral; (d) the translated Chinese numeral.

heteroassociative memory can indeed perform pattern translations.

18.9. SPACE–TIME SHARING ASSOCIATIVE MEMORY

For a fully interconnected neural network, every neuron has to be interconnected to all other neurons; for instance, 1000 neurons would require one million interconnections. Thus a high-resolution SLM is required for realizing such massive interconnections. However, the resolution of currently available SLMs is rather limited, which poses an obstacle in developing a practical optical memory for large-scale operation. In this section, we discuss a space–time sharing technique as described by Yu et al. (1991) to alleviate this constraint.

For an $N \times N$ neuron network, the retrieval equation is given by

$$V_{lk}(n+1) = f\left[\sum_{i=1}^{N}\sum_{j=1}^{N} T_{lkij}U_{ij}(n)\right] \quad (25)$$

where n stands for the nth iteration, f represents a nonlinear operator, V_{lk} and U_{ij} represent the state of lkth and ijth neurons respectively, and T_{lkij} is the connection strength from the lkth to the ijth neuron. $[T_{lkij}]$ is the AMM synaptic matrix, which can be partitioned into an array of $N \times N$ submatrices, as described above.

We use the LCTV optical associative memory shown in Fig. 18.4 for our discussion. The AMM and the input pattern are displayed onto the LCTV1 and LCTV2, respectively. Each lens in the lenslet array images a specific AMM submatrix onto the input LCTV2 to establish the proper interconnections. We limit the resolution of the LCTV1 to an $R \times R$ array of pixels and limit the lenslet array to an $L \times L$ array of neurons. Also, we assume that the size of the AMM (i.e., $N \times N$) is larger than the resolution of the LCTV1, i.e., $N^2 > R$. Since the AMM cannot be represented entirely by LCTV1, we consider two cases:

Case I. For $N^2 > R$ and $LN < R$, we let $D = \text{int}(N/L)$, where $\text{int}(\cdot)$ is an integral value. We assume that the AMM is partitioned into $D \times D$ sub-AMMs and each sub-AMM consists of $L \times L$ submatrices of $N \times N$ size, as shown in Fig. 18.18. In this case, we see that in order to realize Eq. (25), $D \times D$ sequential operations of the sub-AMMs are required. Thus, it is apparent that a smaller associative memory can handle a larger space–bandwidth product (SBP) input pattern through sequential operation of the sub-AMM.

Case II. For $N^2 > R$ and $LN > R$, we let $D = \text{int}(N/L)$ and $d = \text{int}(LN/N)$. In this case, the submatrices within the sub-AMMs are further divided into $d \times d$ submatrices, and each submatrix has size $(N/d) \times (N/d)$, as shown in Fig. 18.19. Thus the retrieval equation can be written in the following form:

$$V_{lk}(n+1) = f\left[\sum_{p=0}^{d-1}\sum_{q=0}^{d-1}\right.$$
$$\left. \times\left(\sum_{i=pd+1}^{(p+1)(N/d)}\sum_{j=qd+1}^{(q+1)(N/d)} T_{lkij}U_{ij}(n)\right)\right] \quad (26)$$

If the input pattern is also partitioned into $d \times d$ submatrices and each submatrix is $(N/d) \times (N/d)$

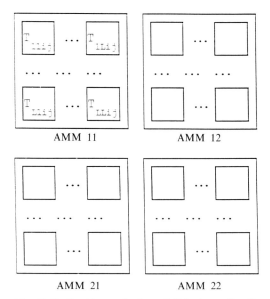

Fig. 18.18. Partition of the AMM into $D \times D$ sub-AMMs, for $D = 2$.

in size, then by sequentially displaying each of the AMM submatrices with respect to the input submatrices onto the LCTV1 and LCTV2, respectively, a very large SBP pattern can be processed using a smaller association memory. It is noted that the price we pay for achieving a larger bandwidth operation in this case is prolonging the processing time by $D^2 \times d^2$.

Generally speaking, the processing time increases as the square function of the space–bandwidth product of the input pattern, and is given by

$$t_2 = \left(\frac{N_2 \times N_2}{N_1 \times N_1}\right)^2 t_1 = \left(\frac{N_2}{N_1}\right)^4 t_1 \quad (N_2 > N_1 \geq L)$$

(27)

where t_2 and t_1 are the processing times for the input patterns with $N_2 \times N_2$ and $N_1 \times N_1$ resolution elements, respectively. The relationship described in the preceding equation is plotted in Fig. 18.20. For instance, if the resolution elements of the input pattern is increased four times in each dimension, i.e. $N_2 = 4N_1$, the processing time will be $4^4 = 256$ times longer.

For experimental demonstration, we show that patterns with 12×12 resolution elements can be processed using a 6×6-neuron network. A 240×480-element color LCTV (LCTV1) is used to display the AMM. Since each color pixel is composed of red, green, and blue elements, the resolution is actually reduced to 240×160 pixels. In the experiment, however, we have used 2×2 pixels for each interconnection weight and so the resolution of the LCTV1 is essentially reduced to 120×80 elements.

By taking $N = 12$, $L = 6$, $R = 80$, it follows that $N^2 = 144 > R$ and $LN = 72 < R$, with $D = \text{int}(N/L) = 2$. The AMM can be divided into a 2×2 sub-AMM array which can be sequentially displayed on the LCTV1. Assuming that an input pattern is displayed on LCTV2, the

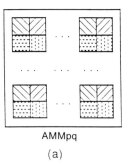

AMMpq

(a)

Fig. 18.19. Partition of a sub-AMM into $d \times d$ submatrices, for $d = 2$: (a) The pqth sub-AMM; (b) $d \times d$ smaller submatrices.

(b)

T_2/T_1

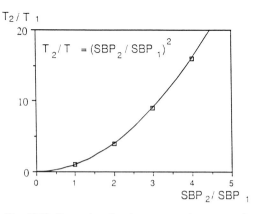

Fig. 18.20. Processing time increases as the square of the SBP of the input pattern, for $SBP_2 = N_2 \times N_2$ and $SBP_1 = N_1 \times N_1$.

signals collected by the CCD camera can be thresholded and then composed to produce an output pattern, that has a SBP four times larger than that of the optical associative memory.

One of the experimental results is shown in Fig. 18.21. The training set, which includes four cartoon patterns, is shown in Fig. 18.21a. Here, each pattern is limited by a 12×12-pixel matrix. Figure 18.21b shows a partial image of the second cartoon figure with 12×5 pixels blocked out as the input pattern. The sub-AMMs are then displayed sequentially onto LCTV1. Different

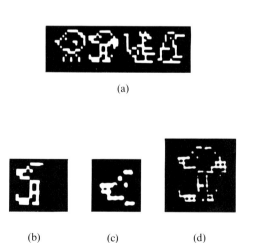

Fig. 18.21. Processing of a 12×12-element by a 6×6-neural net: (a) four reference patterns stored in the AMM; (b) partial input patterns; (c) one of the four 6×6 sub-output arrays; (d) composed output pattern.

parts of the output pattern are obtained and one of the output parts is shown in Fig. 18.21c. The final composed output pattern using this technique is given in Fig. 18.21d.

To further demonstrate the larger-scale operation, a 24×24 neuron AMM is used. Since $N = 24$, $L = 6$, $R = 80$, $N^2 = 576 > R$, and $LN = 94 > R$, we take $D = \text{int}(D/L) = 4$ and $d = \text{int}(LN/R) = 2$. We assume that the AMM is partitioned into 4×4 sub-AMMs and each sub-AMM is divided into 2×2 smaller sub-matrices. In this case the input pattern is also divided into 2×2 submatrices and each sub-matrix is 12×12 in size. It is apparent that by sequentially displaying the submatrices of the AMM and the input submatrices onto LCTV1 and LCTV2, a 24×24 output pattern can be obtained.

Figure 18.22a shows a training set consisting of four 24×24-pixel cartoon patterns. A partial image of the third cartoon figure shown in Fig. 18.22b is used as the input pattern and is divided into 2×2 matrices of 12×12 size during retrieval processing. Four of the sixteen 6×6 output parts are shown in Fig. 18.22c. By going through all the partitions of the AMM, an output pattern is composed as shown in Fig. 18.22d. Notice that the whole process takes $D^2 \times d^2 = 64$ operations using the LCTV optical neural network.

18.10. SUMMARY

We have presented an LCTV-based optical neural network architecture in which several models of associative memories have been implemented. These are the Hopfield, the orthogonal projection, the multilevel algorithm, the interpattern-association, and the space–time sharing models. We have shown that the IPA and orthogonal projection models provide better performance because of their increased storage capacities. Although the space–time sharing model is capable of processing a wide space–bandwidth input, it negatively affects the processing speed of the associative memory.

By exploiting the massive connectivity and parallel processing capabilities of optics and utilizing the strengths of both electronics and photonics, we have implemented a hybrid optical associative memory. We hope that this chapter will provide some basic design concepts and implementations for future work in optical associative memories.

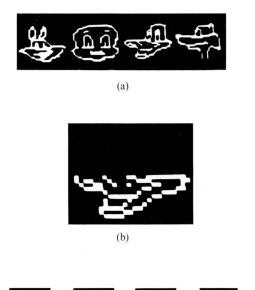

(a)

(b)

(c)

(d)

Fig. 18.22. Simulated results; processing of a 24 × 24-element pattern by a 6 × 6-network: (a) reference patterns stored in the AMM; (b) partial input pattern; (c) four of the sixteen 6 × 6 sub-output arrays; (d) composed output pattern.

REFERENCES

Anderson, D. Z. and Erie, M. C. (1987). "Resonator Memories and Optical Novelty Filter," *Opt. Eng.*, **26**, 434–444.

Athale, R. A., Szu, H. H., and Friedlander, C. B. (1986). "Optical Implementation of Associative Memory with Controlled Nonlinearity in the Correlation Domain," *Opt. Lett.*, **11**, 482–484.

Farhat, N. H. and Psaltis, D. (1987). "Optical Implementation of Associative Memory Based on Models of Neural Networks," *Optical Signal Processing*, J. L. Horner, ed. Academic Press, New York, pp. 129–162.

Farhat, N. H., Psaltis, D., Prata, A., and Paek, E. (1985). "Optical Implementation of the Hopfield Model," *Appl. Opt.*, **24**, 1469–1475.

Golub, G. H. and Van Loan, C. F. (1983). *Matrix Computation*. Johns Hopkins Press, Baltimore.

Hecht-Nielsen, R. (1986). "Performance Limits of Optical, Electro-Optical, and Electronic Neurocomputers," *Proc. SPIE*, **634**, 277–306.

Hopfield, J. J. (1982). "Neural Network and Physical System with Emergent Collective Computational Abilities," *Proc. Nat. Acad. Sci. U.S.A.*, **79**, 2554–2558.

Johnson, K. M., Handschy, M. A. and Pagano-Stauffer, L. A. (1987). "Optical Computing and Image Processing with Ferroelectric Liquid Crystals," *Opt. Eng.*, **26**, 385–391.

Jutamulia, S., Fujii, H., and Asakura, T. (1982). "Double Correlation Technique for Pattern Recognition and Counting," *Opt. Commun.*, **43**, 7–11.

Jutamulia, S., Storti, G. M., Lindmayer, J., and Seiderman, W. (1991). "Use of Electron Trapping Materials in Optical Signal Processing. 2: Two-Dimensional Associative Memory," *Appl. Opt.*, **30**, 2879–2884.

Lalanne, P., Taboury, J., Saget, J. C., and Chavel, P. (1987). "An Extension of the Hopfield Model Suitable for Optical Implementation," *Proc. SPIE*, **813**, 27–30.

Liu, H. K., Kung, S. Y., and Davis, J. A. (1986). "Real-Time Optical Associative Retrieval Technique," *Opt. Eng.*, **25**, 853–856.

Lu, T., Xu, X., Wu, S., and Yu, F. T.S. (1990). "A Neural Network Model Using Inter-Pattern Association (IPA)," *Appl. Opt.*, **29**, 284–288.

McAulay, A. D., Wang, J., and Ma, C. (1990). "Optical Heterassociative Memory Using Spatial Light Rebroadcasters," *Appl. Opt.*, **29**, 2067–2073.

McEliece, R. J., Posner, E. C., Rodemich, E. R., and Venkatesh, S. S. (1987). "The Capacity of the Hopfield Associative Memory," *IEEE Trans. Information Theory*, **IT-33**, 461–462.

Owechko, Y., Dunning, G. J., Marom, E., and Soffer, B. H. (1987). "Holographic Associative Memory with Nonlinearities in the Correlation Domain," *Appl. Opt.*, **26**, 1900–1910.

Paek, E. G. and Psaltis, D. (1987). "Optical Associative Memory Using Fourier Transform Holograms," *Opt. Eng.*, **26**, 428–433.

Psaltis, D., Yu, J., Gu, X. G., and Lee, H. (1987). "Optical Neural Nets Implemented with Volume Holograms," *Optical Computing, Tech. Digest Series*, **11**, 129–137.

Wu, S., Lu, T., Xu, X., and Yu, F. T. S. (1989). "An Adaptive Optical Neural Network Using a High Resolution Video Monitor," *Micro. Opt. Tech. Lett.*, **2**, 252–257.

Yu, F. T. S. and Jutamulia, S. (1987). "Implementation of Symbolic Substitution Logic Using Optical Associative Memories," *Appl. Opt.*, **26**, 2293–2294.

Yu, F. T. S., Lu, T., Yang, X., and Gregory, D. A. (1990a). "Optical Neural Network with Pocket-Sized Liquid-Crystal Televisions," *Opt. Lett.*, **15**, 863–865.

Yu, F. T. S., Lu, T., and Yang, X. (1990b). "Optical Implementation of Hetero-Association Neural Network with Inter-pattern Association Model," *Int. J. Opt. Comput.*, **1**, 129–140.

Yu, F. T. S., Yang, X., and Lu, T. (1991). "Space-Time Sharing Optical Neural Network," *Opt. Lett.*, **16**, 247–249.

19

Optical Learning Neurochips for Pattern Recognition, Classification, and Association

KAZUO KYUMA, JUN OHTA and YOSHIKAZU NITTA

19.1. INTRODUCTION

Neural networks contain large numbers of nonlinear processing elements called neurons, which are massively interconnected by synapses (Rumelhart et al., 1986). In the human brain, it is estimated that the number of neurons is of the order of 10^{10} to 10^{11}, each making 10^3 to 10^4 synaptic connections with neighboring neurons. One of the most interesting features of neural networks is their learning ability. The learning is performed by adjusting the connection weights (synaptic weights) between neurons, in order to obtain the desired input/output relations for pattern classifications, associations, predictions, and so on.

Such neural network computations are very time consuming if implemented on conventional sequential computers. There is therefore a strong need for the development of neural network hardware for real-time applications (Abu-Mostafa et al., 1987). Among several approaches for hardware implementation, opto-electronic neural networks (Abu-Mostafa et al., 1987; Farhat, 1987; Kyuma et al., 1990) employing advanced semiconductor technologies are quite attractive because of their innate parallelism, massive interconnectivity, and large-scale integration capabilities.

The fundamental computation in neural networks is a matrix–vector multiplication. By employing an optical synaptic network that performs the matrix–vector multiplication in full parallelism, Psaltis and Farhat demonstrated an optical associative memory based on the discrete Hopfield model for the first time in 1985 (Psaltis and Farhat, 1985). Subsequently, a great many optical neural networks have been reported for many applications using a variety of optical devices (Yariv et al., 1986; Soffer et al., 1986; Anderson, 1986; Paek and Psaltis, 1987; Farhat, 1987; Wagner and Psaltis, 1987; Kyuma et al., 1988; Ohta et al., 1989a; Yamamura et al., 1990).

However, most of the optical neural networks reported so far suffer from one or more of the following problems:

1. The optical systems are assembled on an optical table using discrete devices; thus they are bulky and have unstable performance due to mechanical vibration. The development of optical neurochips is indispensable if small and stable neural systems are desired.
2. The neural network models that have been developed through computer simulations by theorists are not necessarily suitable for optical implementation due to the difficulty of implementing high-accuracy synaptic weights.
3. Spatial light modulators (SLMs) do not perform efficiently as adjustable synaptic elements. High-accuracy, nonvolatile, analog, and small-size SLMs, suitable for integration with other devices, are needed to implement on-chip learning.

Until recently, much of the effort has been toward the integration of optical devices in the form of neurochips, employing GaAs/AlGaAs technologies (Ohta et al., 1989b, 1990; Nitta et al., 1989), silicon technologies (Rietman et al., 1989; Agranat and Yariv, 1987), and ferro-electric liquid-crystal technologies (Johnson and Moddel, 1989; Jared and Johnson, 1991). Our own first-generation optical neurochips consisted of three layers: a light-emitting diode (LED) array, a static (fixed) interconnection mask, and a photodetector (PD) array (Ohta et al., 1989b; Nitta et al., 1989; Ohta et al., 1990). These array devices were integrated on a GaAs substrate. Several kinds of neurochips have been developed, including an optical associative neurochip with 32 neurons (Ohta et al., 1989b), and an optical neurochip with 90 neurons that recognizes 26 characters of the English alphabet (Ohta et al., 1990). Neural network models suitable for optical implementation, such as the quantized learning model, have been developed (Takahashi et al.,

1990, 1991; Oita et al., 1990; Balzer et al., 1991). However, the first-generation optical neurochips did not have any learning capability because the synaptic weights had to be set during the fabrication process. This major limitation is mainly due to a lack of suitable SLMs.

In this chapter, we describe second-generation optical neurochips that have the ability to learn and acquire knowledge from the external "world" in real time (Kyuma et al., 1991; Ohta et al., 1991; Nitta et al., 1992a). A major breakthrough has been the development of a variable-sensitivity photodetector (VSPD) that has the combined functions of analog SLM and a photodiode (Nitta et al., 1991). In Section 19.2 we start with an introduction to optical neural networks, including basic architectures for optical neurons and synaptic networks, and a general description of optical neurochips. In Section 19.3 the principle of operation and the fundamental characteristics of the VSPD are described. A recently observed phenomenon of the function of VSPD as analog memory (Nitta et al., 1992b) is also discussed. Section 19.4 describes two kinds of optical neurochips which have been developed recently. The first type (type 1) has on-chip learning ability. The second type (type-2) has an internal analog memory in which the synaptic weights are stored, as well as an on-chip learning ability. In Section 19.5, the optical implementations of a back-propagation (BP) model and a mode competition model are described, using neurochips of type 1. The application of an optical competition network to recognition of Japanese stamps is also discussed (Banzhaf et al., to be published). Section 19.6 deals with the details of the optical neurochip with internal memory, as well as the optical implementation of the BP model using this neurochip. The focus is particularly placed on the principle of operation for the analog memory function. In Section 19.7, some discussion of the integration density (number of neurons and synapses per chip) is presented. Section 19.8 is a chapter summary.

19.2. OPTICAL NEURAL NETWORK

19.2.1. Optical Neuron

Figure 19.1a shows a simplified model of the neuron. The functional role of the neuron is threefold: (1) summation of its inputs, which are weighted by passing through synapses; (2) nonlinear transformation of this summation; and (3) transmission of its output to other neurons. Figure 19.1b shows the optoelectronic neuron

model. The input optical signals are transmitted through a spatial light modulator (SLM) that corresponds to the synapses, and are summed by a photodetector (PD). The sum of the inputs is processed by an electrical nonlinear element such as a thresholding comparator. The neuron's output is converted back to an optical signal by a light-emitting diode (LED), and propagates in free space towards other neurons.

19.2.2. Optical Synaptic Network

A synaptic network that interconnects many neurons with analog weights is mathematically equivalent to a matrix–vector multiplier. Figure 19.2 shows a schematic diagram of an optical matrix–vector multiplier which consists of an LED array, a two-dimensional SLM, and a PD array. As shown in this figure, the light from the jth LED is spread out horizontally by a cylindrical lens (not shown in Fig. 19.2 for simplicity) to illuminate the corresponding jth row of the SLM. The light passing through the ith column of the SLM is focused vertically onto the ith element of the PD array by another cylindrical lens (not shown in this figure either). The output photocurrent from the ith PD, u_i, is,

$$u_i = \sum \eta W_{ij} v_j \tag{1}$$

Where the scalars v_j, u_i, and matrix W_{ij} represent the output intensity of jth LED, the photocurrent of the ith PD, and the optical transmittance of (i, j)th pixel of the SLM. η is PD sensitivity parameter.

When we use this multiplier for neural network implementation, the vector $\mathbf{v}$ corresponds to the vector of neuron states. The matrix $\mathbf{W}$ and vector $\mathbf{v}$ correspond to the synaptic interconnection weight matrix and the weighted sum vector, respectively.

Thus, the matrix–vector multiplication is accomplished in full parallelism. In comparison to the electronic approach, the large fan-out capability and the nonexistence of electromagnetic interference in massively parallel optical systems are advantages. Also, the freedom of connectivity in 3-D space is another attractive feature of optical systems, which is rather difficult to realize using the present electonic (bulk and/or integrated) technologies.

19.2.3. Optical Neurochip

An optical neurochip, which is a 3-D optical integrated circuit of the matrix–vector multiplier described above, is schematically illustrated in Fig. 19.3a. The optical neurochip essentially

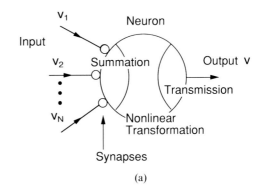

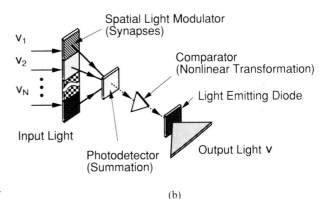

Fig. 19.1. (a) Simplified model of a neuron.
(b) Opto-electronic neuron model.

(b)

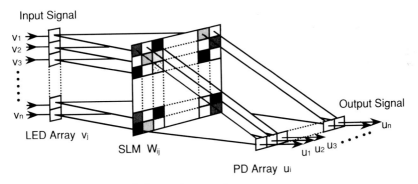

Fig. 19.2. Schematic diagram of an optical synaptic network.

consists of three layers. A linear LED array and a linear PD array are placed in a cross-bar structure with an SLM sandwiched between them that acts as a synaptic interconnection device.

Recently, several types of SLMs have been developed, using liquid-crystal materials, dielectric materials, semiconductor materials, and others. However, one of the problems of presently available SLMs is that they are not suitable for 3-D integration with LED and PD arrays due to large size, restricted characteristics, and/or difficulties in fabrication. For example, a liquid-crystal SLM, which is most conveniently utilized in optical computing systems, has a thickness of about a few millimeters. This large thickness causes serious optical cross-talk, resulting in degradation of performance.

In order to avoid such problems, we propose

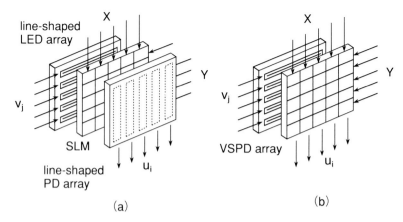

Fig. 19.3. Schematic configurations of a GaAs optical neurochip: (a) three-layer structure using an SLM; (b) two-layer structure using a VSPD array.

a novel type of optical neurochip using a variable-sensitivity photodetector (VSPD) array. The basic structure of the proposed neurochip is illustrated in Fig. 19.3b. The VSPD is a smart detector which has the combined functions of an SLM and a PD. In this case, the variable sensitivity η_{ij} is used as the synaptic weight instead of the optical transmittance W_{ij} of the conventional SLM. The VSPD has many advantages over conventional SLMs, such as lower cross-talk, faster response, analog operation, and monolithic integration capability with LEDs.

19.3. VARIABLE- SENSITIVITY PHOTODETECTOR (VSPD)

19.3.1. Advantages of MSM-VSPD

Several types of VSPDs have been proposed in the literature, including an MSM (metal–semiconductor–metal) type PD (Nitta et al., 1991; MacDonald and Lee, 1991) and an MOS (metal–oxide–semiconductor) type PD called a gate-controlled photodiode (GCPD) (Sun et al., 1989). The principle of operation of the VSPD is that the detection sensitivity η is electrically or optically varied by an external control signal. The MSM-VSPD is quite efficient as an optical synaptic interconnection device on account of the following features:

1. The sensitivity increases monotonically with the applied bias voltage.
2. The direction of the photocurrent is reversed by changing the polarity of the applied voltage. This unique property permits the implementation of both excitatory (positive) and inhibi-

tory (negative) synapses in one VSPD element.
3. Because of its planar structure, an MSM-VSPD is suitable for monolithic planar integration and vertical (3-D) integration with other devices such as MOSFETs (metal–oxide-semiconductor field-effect transistor) and LEDs.
4. The MSM-VSPD has a simple structure which makes it suitable for large-scale integration.
5. The MSM-VSPD works as an analog memory (Nitta et al., 1992b) as well as a variable-sensitivity photodiode. This feature is useful in implementing adaptive neural nets where storage capabilities are required.

Variable, analog, and bipolar synaptic interconnections are possible with a single VSPD element, whereas conventional optical systems using the matrix–vector multiplier of Fig. 19.2 require two optical modulation elements corresponding to the excitatory and inhibitory synapses. In the following, the basic characteristics of the MSM-VSPD are described.

19.3.2. Variable Sensitivity

A schematic diagram of the VSPD is shown in Fig. 19.4. It consists of a photodiode having two Schottky contacts on a GaAs substrate (Sze et al., 1971; Nakajima et al., 1990). Under light illumination, photocarriers are generated in the depletion layer between the Schottky contacts. On application of an electric field across the interdigital electrodes, these photocarriers drift to the electrodes and contribute to the photocurrent. The principle of operation of the variable sensitivity is that the photocurrent is modulated

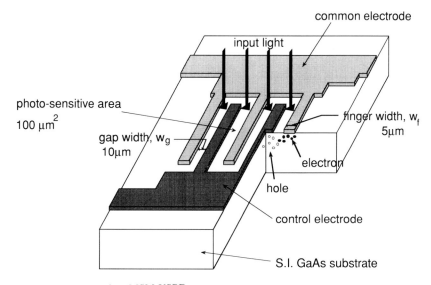

Fig. 19.4. Schematic diagram of an MSM-VSPD.

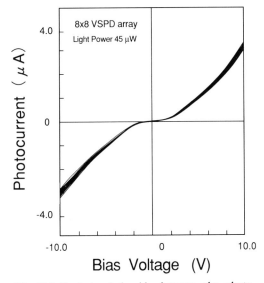

Fig. 19.5. Typical relationship between the photo-current and the bias voltage.

by the applied transverse electric field. The relations between the photocurrent and applied electrode voltage (such as those related to sensitivity and linearity) depend on the gap width w_g, electrode finger width w_f, and properties of the GaAs surface. For a detailed analysis the reader is referred to Ohta et al., 1991.

The VSPDs are fabricated by evaporating aluminum interdigital Schottky contacts on a Cr–O doped semi-insulating GaAs substrate. The thickness of the A1 film is about 300 nm. The whole size of the photosensitive area is $100 \times 100 \ \mu m^2$. The gap width w_g and the finger width w_f of the electrodes are, respectively, 10 μm and 5 μm. These parameters are optimized so that high sensitivity and wide dynamic range are obtained.

Figure 19.5 shows the typical experimental relationship between the photocurrent and the applied bias voltage V_b under constant illumination power of 45 μW for several VSPD elements. A halogen lamp is used as the illuminating source. The photocurrent, which is proportional to the detection sensitivity η, has a close to linear relation with the bias voltage. A sensitivity η of 0.3 A/W was obtained with $V_b = 10$ V. The uniformity of the sensitivity is 5 percent at $V_b = 10$ V, for an 8×8 VSPD.

The relation between the photocurrent and the bias voltage is symmetric about zero. That is, the direction of the photocurrent is reversed by changing the polarity of the voltage. This is because the MSM-VSPD has a symmetric structure about the photosensitive semiconductor area. As described later, this feature is very useful for implementing optical neural networks, because one VSPD element works as a synaptic device with both polarities.

The photocurrent increases linearly with the bias voltage V_b for $|V_b| > 1$ V. This voltage is called the onset voltage. The low-sensitivity region near zero bias voltage is undesirable. To

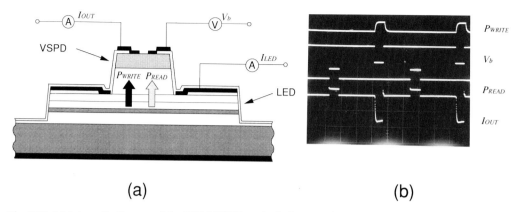

Fig. 19.6. (a) Schematic diagram of the MSM-VSPD vertically integrated with the LED array. (b) A photograph showing the "write" and "readout" operations of the analog memory.

reduce the magnitude of the onset voltage, w_g needs to be increased, which in turn decreases the sensitivity η.

The response time is less than 0.1 μs, which corresponds to 640 MCPS (million connections per second) for a 64-synapse chip (optical learning chip, type 1) and 10.24 GCPS for 1024-synapse chip (optical learning chip with internal memory, type 2), operating at 10 MHz.

The dark current is less than 1 nA at $V_b = 5$ V. This low dark current is caused by the Schottky barrier of the Al–GaAs contact. The breakdown voltage is higher than 15 V. These measured results indicate that good Schottky barriers are formed at the Al–GaAs interfaces.

19.3.3. Memory Functions

As described, when the bias voltage is changed, the photocurrent is modulated. The principle of operation of the memory function is that information, that is, the light power and/or the applied voltage, is stored in the VSPD through effects on its sensitivity when a light signal and the bias voltage are applied simultaneously.

Figure 19.6a shows a schematic configuration of a VSPD that is vertically integrated with an LED. The measured characteristics of the memory function is also shown in Fig. 19.6b. The LED provides two kinds of light pulses to the VSPD: a write pulse P_{WRITE} and a readout pulse P_{READ}. The first trace is a write pulse used to store information in the VSPD by changing its sensitivity. The second trace is a bias voltage V_b to the electrodes that is synchronized with P_{WRITE}. In the write phase, the photocurrent I_{OUT} from the VSPD is proportional to the product of the optical power of P_{WRITE} and V_b, as shown in the

bottom trace. In the readout phase, the readout pulse P_{READ} is injected into the VSPD from the same LED as in the write phase, but without bias voltage, $V_b = 0$. The readout photocurrent I_{OUT}, which is shown in the bottom trace, corresponds to the stored information $V_b \times P_{WRITE}$. It should be noted that the direction of the current flow in the readout phase is reversed compared to that of the write phase. This result means that the internal electric field, which is built in the undoped GaAs absorption layer, has the opposite direction to the applied bias electric field. Both positive and negative analog values can be stored by changing the polarity of the applied voltage in the write phase.

Figure 19.7 shows the photocurrent I_{OUT} in the readout phase as a function of the voltage V_b in the write phase. The input optical power P_{WRITE} of the write pulse was fixed to 1 μW with an injection current of 3 mA. I_{OUT} was measured by varying V_b from 0 to 5 V, as a parameter of the readout power P_{READ} from 0 to 10 μW. It is found that I_{OUT} is nearly proportional to V_b below 3 V, but that it is saturated over the range of 4 V. It is also found that I_{OUT} is proportional to P_{READ}, which indicates that the readout pulse does not affect the internal electric field.

Figure 19.8 shows the VSPD storage characteristics. After a bias voltage of 5 V and a write pulse of 10 μW are applied, a readout pulse P_{READ} of 10 μW is injected into the VSPD with a frequency of 1 kHz. The decay time τ, which is defined as the time at which I_{OUT} is attenuated to $1/e$ of its initial value, is about 20 minutes.

The physical mechanism of the analog memory is not completely understood at present and is under study. However, according to our preliminary experiments, the mechanism is based on

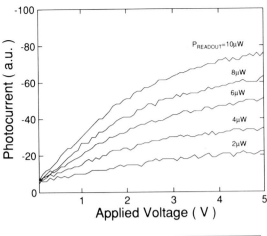

Fig. 19.7. Photocurrent in the readout phase as a function of the bias voltage in the write phase.

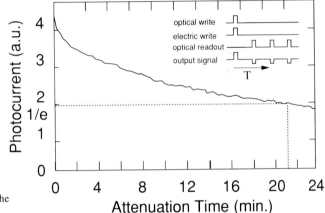

Fig. 19.8. Storage characteristics of the memory as a function of time.

the optoelectronic ionization/relaxation properties of impurities at the Al/GaAs interface.

19.4. OPTICAL NEUROCHIP FABRICATION

Two types of optical neurochips were fabricated. One type was designed as an optical learning chip (type 1), where external Si memories were used to store the synaptic weights during the learning phase. The other type was designed as an optical learning chip with analog memory (type 2), where the chip itself has the internal memories for storing synaptic weights.

19.4.1. Optical Learning Chip (Type 1)

As mentioned above, the optical neurochips are composed of a linear LED array and a 2-D VSPD array. An optical learning neurochip with eight neurons was fabricated. A schematic diagram of the neurochip is shown in Fig. 19.9. The epitaxial wafer of the LED array was grown by molecular beam epitaxy. The LED array consisted of eight lines with 500-μm spacing. Eight LED elements with 30×30 μm^2 aperture were located on each line with a spacing of 500 μm. The eight LED lines were connected to the corresponding electric pads for communication with the external electronics. The total area of the LED array was 4×4 mm^2. The standard deviation of the optical emission intensity from the mean value was within 3 percent.

The VSPD element has the same structure as described in Section 19.3. The VSPDs are located on a matrix with a spacing of 500 μm in both directions. On one side the interdigital electrodes of the VSPD elements were individually connected to the 64 wire bonding pads for application of independent external bias voltages. On

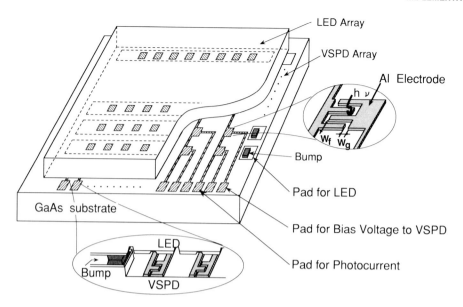

Fig. 19.9. Schematic configuration of an optical learning neurochip (type 1).

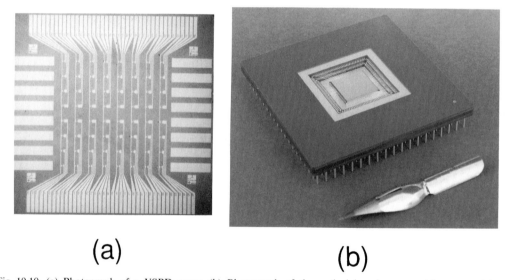

Fig. 19.10. (a) Photograph of a VSPD array. (b) Photograph of the optical learning neurochip mounted in the LSI package. The chip size is 6 × 6 mm².

the other side the electrodes were mutually connected at each row in order to sum the photocurrent from the VSPD elements. The total area was 4 × 4 mm². Figure 19.10a shows a photograph of the VSPD array.

The LED chip and the VSPD chip were attached to each other in a hybrid integrated structure by the flip-chip bonding technique. The number of electric pads was 80, that is to say, 8 for driving the LED array, 8 for collecting the photocurrent, and 64 for addressing the VSPD elements. Figure 19.10b shows the integrated neurochip mounted in a conventional LSI package.

VSPD Array

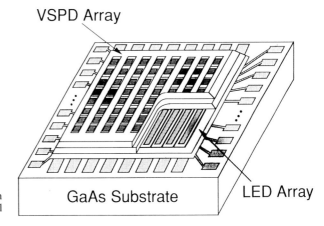

Fig. 19.11. Schematic configuration of an optical learning neurochip with internal analog memory (type 2).

GaAs Substrate

LED Array

19.4.2. Optical Learning Chip with Internal Memory

A neurochip with internal analog memory that implements 32 neurons was fabricated by using monolithic integration techniques developed recently (Nitta et al., 1992a). The structure of the neurochip is illustrated in Fig. 19.11. The epitaxial wafer was grown on an n-type GaAs substrate by molecular beam epitaxy, first growing the layers for the LED array, then successively growing the layers for the VSPD array. A semi-insulating layer was introduced between the LED layers and the VSPD layers for electrical isolation.

After the crystal growth, the wafer was processed into arrays as shown in Fig. 19.11 by conventional photolithographic and etching techniques. The LED array has almost the same structure as that mentioned above, except that the number of lines is 32, the aperture of the LED is $60 \times 60 \ \mu m^2$, and the spacing is 160 μm. The 32 LED lines were connected to the corresponding electrical pads.

The VSPD element has a gap width $w_g = 5 \ \mu m$ and a finger width $w_f = 5 \ \mu m$ (see Fig. 19.4). The electrical wiring is slightly different from that of the type 1 chip. That is, on one side the interdigital electrodes are mutually connected at each column to provide the common bias voltage, instead of individual addressing. The details of the addressing method will be described in Section 19.6. The total number of electrical pads is 96: 32 for driving the LED array, 32 for applying the common bias voltage to the VSPD lines, and 32 for collecting the output photocurrent. The total chip size is $6.3 \times 6.3 \ mm^2$. Figure 19.12 shows a photograph of the VSPD array.

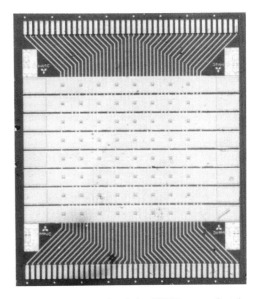

Fig. 19.12. Photograph of the VSPD array for the optical learning neurochip with analog memory. The chip size is $6 \times 6 \ mm^2$.

19.5. NEURAL NETWORKS USING OPTICAL NEUROCHIPS

19.5.1. Implementation of Multilayer Perceptron Net

Learning experiments for a three-layer BP network were performed using the type-1 optical neurochip. The experimental setup and the network structure are shown in Fig. 19.13. The number of neurons in the input, hidden, and output layers are $N^{(1)} = 8$, $N^{(2)} = 8$, $N^{(2)} = 3$,

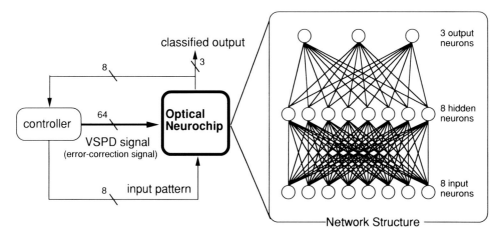

Fig. 19.13. Experiments using the type 1 optical neurochip. The experimental setup and the neural network structure (BP model) are schematically illustrated.

respectively, where the 1, 2, and 3 signify the input, hidden, and output layer, respectively. In order to implement a three-layer network with one chip having only eight neurons and 64 synapses, the time division multiplexing (TDM) technique (Oita et al., 1990) is employed, where the processing is divided into two stages. The first stage implements the processing between the input-to-hidden layers, and the second stage implements the processing between the hidden-to-output layers. The synaptic weights and the neuron states are updated during the learning phase by changing the applied voltage to the VSPD electrodes from − 10 to 10 V and the injection current from 0 to 20 mA, respectively.

A digital computer is used in this experiment to present the training patterns, to perform nonlinear thresholding, to calculate and generate error signals, and to display results. The sigmoid function is used as the nonlinear transfer function. A/D and D/A converters with 8-bit resolution are used as the input/output interface with the computer. Each voltage and output photocurrent signal was divided into 128 levels for positive values and similarly for negative values, while the injection current was divided into 256 levels.

The optical neurochip is used as a BP-based pattern classifier in which the training patterns are classified into three categories. A set of 4–16 random patterns with 8-bit binary codes is used as the training signals. In the learning phase, one of the training signals is presented to the input layer. Next, the feed-forward computation is carried out, while the synaptic weights, which are stored in the computer, are addressed on the optical neurochip. If the output of the classifica-

tion is not correct, the synaptic weights are modified according to the BP algorithm (which is implemented on the computer). Then the next training signal is presented. This process is repeated until the error at the output for all the training signals approaches zero.

The learning curve obtained in the experiments is shown in Fig. 19.14, together with the computer simulation results. The recognition rate, which is averaged over 100 randomly selected patterns for each training signal, is plotted as a function of the number of learning cycles. The learning rate ρ and the momentum coefficient α were set to 0.4 and 0.7, respectively. It is shown that all the patterns are correctly classified after 100 pre-sentations, and the experimental results agree very well with the computer simulations.

19.5.2. Implementation of Winner-Takes-All Associative Memory

In the winner-takes-all network, each prototype (memory) pattern is assigned to one of the neurons. The principle of operation of this network is that, in response to an input pattern, the neuron with the closest weight vector to the input vector survives and dominates through competition with other neurons. The index of the winning neuron is then used to construct the recollected pattern/memory (Pankove et al., 1990).

The winner-takes-all network is depicted in Fig. 19.15. Three layers with the feed-forward synaptic connections, A_{ik}, B_{kj}, and the feedback synaptic connections in the hidden competitive layer constitute the network. A set of the

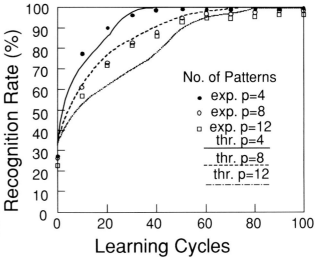

Fig. 19.14. Experimental and computer-simulated learning curves. The recognition rate is plotted as a function of the number of learning cycles.

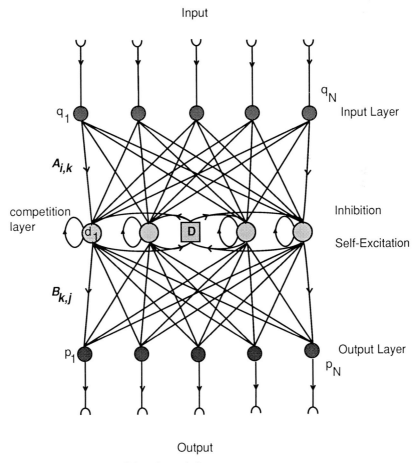

Fig. 19.15. A three-layer winner-takes-all-based associative memory.

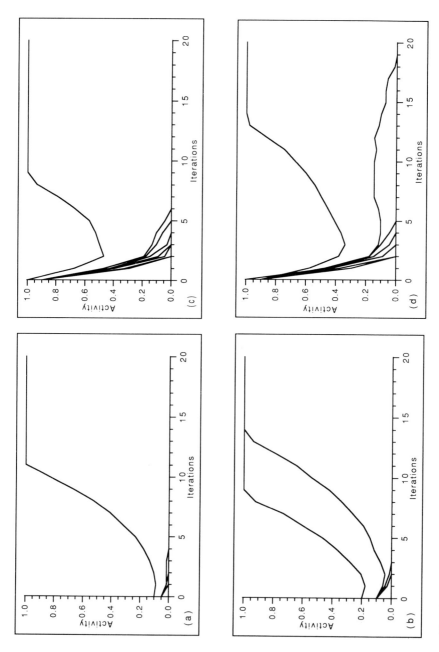

Fig. 19.16. Computer simulation results of the mode competitive process. Weak competition: (a) correct winner; (b) two winners (not desirable). Medium competition: (c) correct winner. Strong competition: (d) wrong winner due to noise.

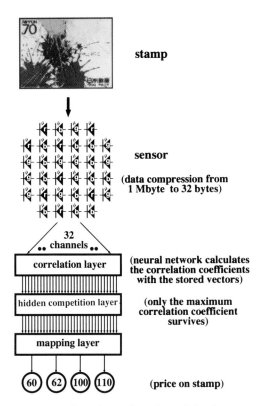

stamp

sensor

(data compression from
1 Mbyte to 32 bytes)

32 channels

correlation layer

(neural network calculates
the correlation coefficients
with the stored vectors)

hidden competition layer

(only the maximum
correlation coefficient
survives)

mapping layer

(60) (62) (100) (110)   (price on stamp)

Fig. 19.17. Schematic configuration of the Japanese stamp recognition system using the competition-based associative memory.

prototype patterns is stored in the synaptic matrix A_{ik} in advance. The matrix A_{ik} is determined so that the correlations between all the stored patterns and the input pattern are obtained at the hidden competitive layer. The feedback connections inhibit each neuron with certain competition strength that depends on the overall activity of all the competing neurons and self-excite all neurons as illustrated in Fig. 19.15. As a result, after the presented pattern is mapped to the activities of the neurons in the hidden competitive layer by A_{ik}, subsequent competition between the neurons decides which neuron is most dominant. The role of the connection B_{kj} is to perform an association between the hidden neuron activity and stored associative patterns/memories.

Figure 19.16 shows typical computer simulation results of the competitive process under various parameter settings and initial neuron states. If the competition is too weak, more than one neuron survives. On the other hand, if the competition is too strong, all the neurons are

suppressed and an undesirable neuron finally emerges as the winner due to the instability of the system. By setting the competition strength in the proper range, the network behaves correctly for all initial conditions.

We have applied the optical learning chip described to the recognition of Japanese postage stamps (Banzhaf et al., to be published). Figure 19.17 shows an overview of the entire system, which is constructed of a preprocessor of the input patterns and the associative memory network described above.

The processing is done as follows. First, a test stamp is presented to the system. A special hardware sensor consisting of a 2-D array of color-sensitive photodetectors performs robust data compression from 1 Mbyte to 32 bytes by extracting both color and intensity information from the stamps. Second, the resulting data-compressed signal is fed into the neural network, which recognizes the presented stamp and classifies it into one of four price classes. It should be noted that, in Fig. 19.16, the number of neurons in the input layer is assumed to be 32. According to our computer simulations, 32 stamps were successfully recognized.

The experiments were made using eight samples of stamps. The number of neurons in the input, competitive, and output layers were, respectively, 8, 8, and 4. In order to implement the network, the TDM technique was employed. The computation of the weighted sums, that is, the computation between the input and competitive layers, was performed in the first stage of processing. Here, the compressed information of the 8-D input vector, which is obtained through a simulated sensor, is used as the input signal. The computation of the competitive dynamics is performed as the second stage of processing in the optical neurochip. Figure 19.18 shows the typical computation results in the optical hardware. We made two experiments where two arbitrary and mutually exclusive sets of stamps are used as prototype patterns. In both cases, a recognition rate of 100% is achieved.

19.6. NEURAL NETWORKS USING OPTICAL LEARNING CHIP WITH INTERNAL MEMORY (TYPE 2)

19.6.1. Matrix Addressing Scheme

In the type 1 optical neurochip, the electrodes of all the VSPD elements must be individually connected with the external computer in order to change the sensitivity of the VSPD elements.

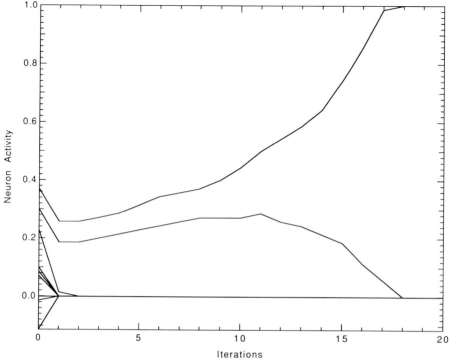

Fig. 19.18. Typical competition process obtained by the experiments employing the optical neurochip.

This requirement limits the integration density of the array because the possible number of electrical pins is limited to a few hundred by present LSI packaging technologies.

One method to solve this problem and to realize large-scale integration is the use of a matrix addressing scheme, where one FET is used as a switching device to access the corresponding VSPD element. We have recently reported this kind of active matrix addressing operation in which the GaAs VSPD array is monolithically integrated with the InGaAs/GaAs FET array (Koshiba et al., 1992). Here, an external digital memory was utilized to store the synaptic weights. However, the use of the external memory is disadvantageous in some applications because communication between the neurochip and the external memory makes the learning time slow, and makes the total system large and costly.

In this section, we propose a novel type of matrix addressing scheme using the optical memory, which enables the operation of the 2-D VSPD array without external memory during the learning phase. The structure of this neurochip has already been shown in Fig. 19.11. The principle of operation for matrix addressing is illustrated in Fig. 19.19. In the write phase

(corresponding to the learning phase), the current is injected into the LED elements in the first row (all LEDs at this row turn on and emit light towards the VSPD elements). At the same time, the write voltages, which correspond to the signals (synaptic weights) to be stored, are applied to the VSPD elements of every column in parallel. In this manner, the write signals are memorized in the VSPD elements of the first row as the sensitivity η_{ij} ($j = 1, 2, \ldots, n$). The same procedure is sequentially repeated for the second row through the nth row in one frame scan. As a result, the sensitivities of the 2-D VSPD array are memorized in the form of matrix η_{ij} ($i, j = 1, 2, \ldots, n$).

In the readout (retrieval) phase, the readout signals v_j (states of neurons) are produced by the LED elements in parallel, and the photocurrent u_i (product of neuron states and synaptic weights), which is proportional to the matrix-vector product,

$$u_i = \sum_{j=1}^{n} \eta_{ij} v_j \qquad (2)$$

is obtained from the VSPD array.

Experimental results for the 32-neuron

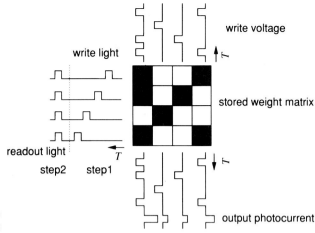

Fig. 19.19. Principle of operation for the matrix addressing scheme using analog memory.

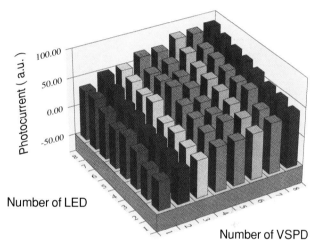

Fig. 19.20. Typical experimental results, showing that the sensitivity of the individual VSPD elements is varied by the matrix addressing scheme.

(32 × 32-synapse chip), shows that every VSPD elements in the 2-D array is individually accessed and that the signals are stored by using the matrix addressing scheme. A part of the chip, that is, 8 × 8 elements among 32 × 32, is utilized as shown in Fig. 19.20. In the experiment, when the sensitivity of a certain VSPD element is stored, the write voltage is increased by 0.25 V for the neighboring elements along both the row and the column. This figure indicates that the matrix addressing is successfully achieved with analog values.

19.6.2. Implementation of Multilayer Perceptron Net

The BP-based net is again employed to demonstrate the optical neurochip with internal memory. The experimental setup and the structure of the network are shown in Fig. 19.21, which is similar to Fig. 19.13. The points of difference from Fig. 19.13 are summarized as follows. First, in order to implement a three-layered network, the chip is spatially divided into two regions: input-to-hidden (1, 2) and hidden-to-output (2, 3) regions. The number of neurons in these layers are $N^{(1)} = 8$, $N^{(2)} = 8$, and $N^{(3)} = 3$, respectively. Then the total numbers of the LED and VSPD elements for implementing the network are, respectively, 16 and 88. Second, the nonlinear processing is done in parallel by 11 $(N^{(2)} + N^{(3)})$ external electronic circuits. These circuits are custom-designed HIC (hybrid integrated circuits) which allow us to have programmable activation slopes. Third, during the learning process, the synaptic weights, which are updated by applying

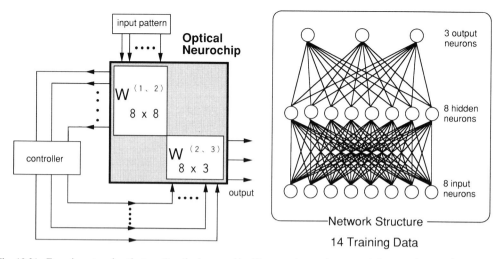

Fig. 19.21. Experiments using the type 2 optical neurochip. The experimental setup and the neural network structure (BP model) are schematically illustrated.

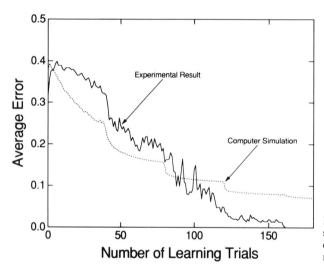

Fig. 19.22. Experimental and computer-simulated learning curves. The mean-squares error (MSE) is plotted as a function of the number of learning cycles.

voltages from −5 V to 5 V to the VSPD electrodes and by keeping the injection current into the LED elements at 3 mA, are memorized in the VSPD array. The same A/D and D/A converters as in Section 19.5 are used as the input/output interface with the computer. The main role of the computer is to present the training signals, to calculate the error signals, and to display the results.

We have used the optical neurochip for pattern classification, where 14 training patterns with 8 bits each are classified into three categories. In the learning phase, if the output signal is not correct, the synaptic weights are modified according to the BP algorithm, and voltages cor-

responding to the error signals are applied to the VSPD electrodes to update the stored weights. The remainder of the process is similar to that reported in Section 19.5.1.

Experimental results are shown in Fig. 19.22. The mean-squares error (MSE), $\delta_k^2 = (t_k - o_k)^2$, where t_k is a desired output (training signal) and o_k is an obtained output, is used to evaluate learning performance. The goal of the BP learning algorithm is to minimize the MSE. In the learning phase, the initial states of the weights are randomly set.

The experimental MSE, which is averaged over many trials for all the training signals, is plotted as a function of the number of learning cycles

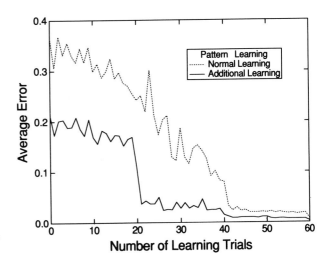

Fig. 19.23. Experimental results for the supplementary learning scheme.

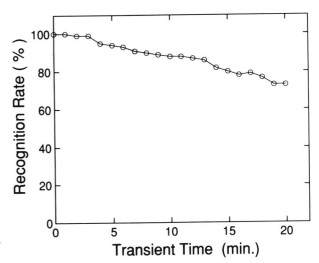

Fig. 19.24. Experimental holding characteristics of the internal analog memory.

in Fig. 19.22. The learning rate ρ and the acceleration coefficient α were set to 0.4 and 0.7, respectively. It is shown that the input patterns are correctly classified into the three categories after about 100 presentations.

19.6.3. Supplementary Learning

Since this neurochip has an internal memory with 20 minutes store time, it has a unique ability of supplementary learning. That is, any supplementary knowledge that is different from the knowledge stored already can be acquired through learning. This ability is similar to that of the human brain in the sense that the human being does not learn many things simultaneously, but learns them gradually during the course of his

daily life. Figure 19.23 shows the experimental results, where 4 training patterns among 8 are stored through the initial learning phase and the remaining 4 patterns are stored successively through the second phase. The solid line shows the learning curve for the second stage. For comparison, the experimental results for the case where all the 8 training signals are stored in a single learning phase are also shown as a dotted line.

After the learning is over, the stored memory is gradually relaxed and the MSE increases. Figure 19.24 shows the recognition rate as a function of time after the learning of 10 training patterns is completed. By storing the acquired weights in an off-chip memory after learning and by refreshing the weights in the internal memory,

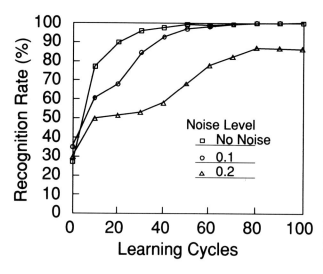

Fig. 19.25. Experimental effect of noise on learning performance. The recognition rate is plotted as a function of the number of learning cycles.

the recognition rate can be easily maintained at close to 100%.

19.7. INTEGRATION DENSITY

In this section we discuss the limits on integration density of neurons. There are three limiting factors: signal-to-noise ratio (SNR), optical cross-talk, and power dissipation. It is found from computer simulations that the first two factors are dominant. With increase in the integration density, the size of the devices becomes smaller. This means that the incident optical power at each VSPD element decreases, and so the SNR degrades. Also, on increasing the integration density, optical cross-talk between neighboring VSPD elements increases.

Let us start with the effects of the SNR on the learning performance and the integration density. Figure 19.25 shows the experimental results on learning performance when noise is intentionally introduced into the system. Noise with a Gaussian distribution is generated by the computer and added to the output from the optical neuro-chip. The noise level is defined as the standard deviation of the Gaussian distribution normalized by the maximum output value. As shown in Fig. 19.25, the degradation in learning perform-ance is small when the noise level is less than 0.1 which corresponds to SNR = 20 dB. Assuming that the dominant noise is due to shot noise in the VSPDs, the maximum integration density is estimated to exceed 2000 neurons/cm^2.

Figure 19.26 shows computer simulation results for optical crosstalk as a function of integration density (neurons/cm^2), where the

parameter is the air-gap between the LED array and the VSPD array. The optical cross-talk is defined as the ratio of the stray light at the VSPD element that leaks from the neighboring LED elements to the signal light that arrives from the LED just below the VSPD. The maximum integration density depends on the neural net-work models utilized. In the case of the feedback Hopfield model (see Chapter 1 for additional details on this model), the cross-talk for obtaining adequate performance without any serious degra-dation is about − 10 dB, according to computer simulations. Therefore, the maximum integration density is roughly 1000 neurons/cm^2 when the gap is 2 μm. On the other hand, for the feed-forward BP-based model, the integration density is larger than 2000 neurons/cm^2 because the tolerated cross-talk is − 5 dB. This is due to the corrective adaptive learning effects, where noise is treated as additional error during the learning phase.

19.8. SUMMARY

We have described optical neurochips and their application in neural network implementations. These optical neurochips make use of optics, which includes parallelism and global high-density interconnectivity. Two kinds of optical neurochips have been fabricated utilized 3-D optical integration technologies. The first (type 1) neurochip has learning capability through the variable-sensitivity photodetectors (VSPD) that are used to implement synaptic elements. Using a neurochip with 8 neurons (8 × 8 synapses), the back-propagation-based multilayer perceptron

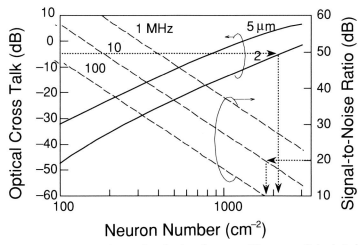

Fig. 19.26. Theoretical estimation of the integration density of neurons. The cross-talk is plotted as a function of the number of neurons per unit area (cm^2).

net and the winner-takes-all associative memory were successfully implemented.

The second type (type 2) neurochip has analog memory as well as the capability to learn, so the knowledge acquired through learning is stored in the chip's internal memory. This memory behavior is based on an optical phenomenon due to impurities in the Al/GaAs interface that we have discovered recently. A method for addressing the 2-D VSPD array and storing the synaptic weights is given. Using the fabricated neurochip with 32 neurons (32 × 32 synapses), we have succeeded in demonstrating pattern classification based on the BP model. We also give a theoretical limit on integration density of our neurochips, which is about 2000 neurons/cm^2.

Some features of our neurochips are their potential abilities in high-density integration and ultrafast processing. A learning and retrieval processing speed faster than 600 MCPS has been obtained. By increasing the integration density, a processing speed of 1 TCPS may be achieved.

We focused our attention on optical neurochips consisting of LED and VSPD arrays for neural network implementation. Other possible applications of VSPD arrays include artificial retina devices with edge detection and/or motion detection for real images (Lange et al., 1992) and direct image association (Zhang et al., 1992).

ACKNOWLEDGMENTS

We thank Drs. S. Tai and W. Banzhaf, and Messers. E. Lange and T. Toyoda for their cooperation in this work.

REFERENCES

Abu-Mostafa, Y. S. and Psaltis, D. (1987). "Optical Neural Computers," *Scientific American*, **255**(3), 88–95.

Agranat, A. and Yariv, A. (1987). "Semiparallel Microelectronic Implementation of Neural Network Models Using CCD Technology," *Electron. Lett.*, **23**, 580–581.

Anderson, D. Z. (1986). "Coherent Optical Eigenstate Memory," *Opt. Lett.*, **11**, 56–58.

Balzer, W., Takahashi, M., Ohta, J., and Kyuma, K. (1991). "Weight Quantization in Boltzmann Machines," *Neural Networks*, **4**, 405–409.

Banzhaf, W., Lange, E., Oita, M., Ohta, J., Kyuma, K., and Nakayama, T. (in press). "Optical Implementation of a Competitive Network," in *Fast Intelligent Systems*, B. Soucek, ed. Wiley Series in 6th Generation Computing Technologies. Wiley, New York, in press.

Farhat, N. H. (1987). "Optoelectronic Analogs of Self-programming Neural Nets: Architectures and Methodologies for Implementing Fast Stochastic Learning by Simulated Annealing," *Appl. Opt.*, **26**, 5093–5103.

Jared, D. A. and Johnson, K. M. (1991). "Optically Addressed Thresholding Very-Large-Scale-Integration/Liquid-Crystal Spatial Light Modulators," *Opt. Lett.*, **16**, 967–969.

Johnson, K. M. and Moddel, G. (1989). "Motivations for Using Ferroelectric Liquid Crystal Spatial Light Modulators in Neurocomputing," *Appl. Opt.*, **28**, 4888–4899.

Koshiba, Y., Ohta, J., Tai, S., Toyoda, T., and Kyuma, K. (1992). "Active Matrix Addressing Operation of a Dynamic Optical Neurochip with Monolithically Integrated InGaAs/GaAs MSM-FETs," *OEC'92, Optoelectronics Conference, Technical Digest*, **1601–4**, pp. 154–155. Makuhari, Japan, July 1992.

Kyuma, K., Ohta, J., Kojima, K., and Nakayama, T. (1988). "Optical Neural Networks, Systems and Devices Technologies," *Optical Computing 88, Proc. SPIE*, **963**, 475–484.

Kyuma, K., Mitsunaga, K., and Ohta, J. (1990). "Optical Implementation of Neural Networks," *Optical Interconnections and Networks, Proc. SPIE*, **1281**, 124–135.

Kyuma, K., Nitta, Y., Ohta, J., Tai, S., and Takahashi, M. (1991). "The First Demonstration of an Optical Learning Chip," in *Optical Computing 1991*, Washington D.C., Tech. Dig. Series, Vol. 6. Optical Society of America, pp. 291–294.

Lange, E., Funatsu, E., Hara, K., and Kyuma, K. (1992). "Optical Neurochips for Direct Image Processing," *Proc. Spring Meeting of IEICE, D83*, Tokyo.

MacDonald, R. I. and Lee, S. S. (1991). "Photodetector Sensitivity Control for Weight Setting in Opto-electronic Neural Networks," *Appl. Opt.*, **30**, 176–179.

Nakajima, K., Iida, T., Sugimoto, K., Kan, H., and Mizushima, Y. (1990). "Properties and Design Theory of Ultrafast GaAs Metal–Semiconductor–Metal Photodetector with Symmetrical Schottky Contacts," *IEEE Trans. Electron. Dev.*, **37**, 31–35.

Nitta, Y., Ohta, J., and Kyuma, K. (1989). "GaAs/AlGaAs Optical Interconnection Chip for Neural Networks," *Japan J. Appl. Phys.*, **28**, 2101–2103.

Nitta, Y., Ohta, J., Tai, S., and Kyuma, K. (1991). "Variable Sensitivity Photodetector Using Metal–Semiconductor–Metal Structure for Optical Neural Networks," *Opt. Lett.*, **16**, 611–613.

Nitta, Y., Ohta, J., Tai, S., and Kyuma, K. (1992a). "Optical Neurochip with Learning Capability," *IEEE Photon. Technol. Lett.*, **4**, 247–249.

Nitta, Y., Ohta, J., Tai, S., and Kyuma, K. (1992b). "Proposal for an Optical Neurochip with Internal Analogue Memory and its Fundamental Characteristics," *Japan J. Appl. Phys.*, **31**(2), L1182–L1184.

Ohta, J., Tai, S., Oita, M., Kuroda, K., Kyuma, K., and Hamanaka, K. (1989a). "Optical Implementation of an Associative Neural Network Model with a Stochastic Process," *Appl. Opt.*, **28**, 2426–2428.

Ohta, J., Takahashi, M., Nitta, Y., Tai, S., Mitsunaga, K., and Kyuma, K. (1989b). "GaAs/AlGaAs Optical Synaptic Interconnection Device for Neural Network," *Opt. Lett.*, **14**, 844–846.

Ohta, J., Kojima, K., Nitta, Y., Tai, S., and Kyuma, K. (1990). "Optical Neurochip Based on a Three-Layered Feedforward Model," *Opt. Lett.*, **15**, 1362–1364.

Ohta, J., Nitta, Y., Tai, S., Takahashi, M., and Kyuma, K. (1991). "Variable Sensitivity Photodetector for Optical Neural Networks," *IEEE J. Lightwave Technol.*, **9**, 1747–1754.

Oita, M., Takahashi, M., Tai, S., and Kyuma, K. (1990). "Character Recognition Using Dynamic Opto-electronic Neural Network with Unipolar Binary Weights," *Opt. Lett.*, **15**, 1227–1229.

Paek, E. G. and Psaltis, D. (1987). "Optical Associative Memory Using Fourier Transform," *Opt. Eng.*, **26**, 428.

Pankove, J., Radehaus, C., and Wagner, K. (1990). "Winner-Take-All Neural Net with Memory," *Electron. Lett.*, **26**, 349–350.

Psaltis, D. and Farhat, N. (1985). "Optical Information Processing Based on an Associative-memory Model of Neural Nets with Thresholding and Feedback," *Opt. Lett.*, **10**, 98–100.

Rietman, E. A., Frye, R. C., Wong, C. C., and Kornfeld, C. D. (1989). "Amorphous Silicon Photoconductive Arrays for Artificial Neural Networks," *Appl. Opt.*, **28**, 3474–3478.

Rumelhart, D. E., McClelland, J. L., and the PDP Research Group (1986). *Parallel Distributed Processing*, Vols. I & II. MIT Press, Cambridge, MA.

Soffer, B., Dunning, G., Owechko, Y., and Marom, E. (1986). "Associative Holographic Memory with Feedback Using Phase-conjugate Mirrors," *Opt. Lett.*, **11**, 118–120.

Sun, C. C., Wieder, H. H., and Chang, W. S. C. (1989). "A New Semiconductor Device—The Gate-controlled Photodiode, Device Concept and Experimental Results," *IEEE J. Quantum Electron.*, **25**, 896–903.

Sze, S. M., Coleman, D. J. Jr., and Loya, A. (1971). "Current Transport in Metal–Semiconductor–Metal (MSM) Structures," *Solid-State Electron.*, **14**, 1209–1218.

Takahashi, M., Kojima, K., Oita, M., Tai, S., and Kyuma, K. (1990). "Proposal of a Quantized Learning Rule and its Application to Optical Neural Network for 26-Character Alphabet Recognition," in *Conference Record Int. Topical Meeting on Optical Computing*, Kobe, Japan, 10114, pp. 321–322.

Takahashi, M., Oita, M., Tai, S., Kojima, K., and Kyuma, K. (1991). "A Quantized Back Propagation Learning Rule and Its Application to Optical Neural Networks," *Opt. Comput. Process.*, **1**, 175–182.

Wagner, K. and Psaltis, D. (1987). "Multilayer Optical Learning Networks," *Appl. Opt.*, **26**, 5061–5076.

Yamamura, A., Kobayashi, S., Neifeld, M. K., and Psaltis, D. (1990). "An Optoelectronic Multi-layer Network," in *Conference Record Int. Topical Meeting on Optoelectronic Multi-layer Network*," in *Conference Record Int. Topical Meeting on Optical Computing*, Kobe, Japan, 104A, pp. 152–153.

Yariv, A., Kwong, S.-K., and Kyuma, K. (1986). "Demonstration of an All-optical Associative Holographic Memory," *Appl. Phys. Lett.*, **48**, 1114–1116.

Zhang, W., Ishii, T., Takashashi, M., and Kyuma, K. (1992). "$N \times N$ to N Pattern Associative Memory Using Optoelectronic Neurochips," *Opt. Lett.*, **17**, 673–675.

Index